TECHNICAL CALCULUS

PRENTICE-HALL SERIES IN TECHNICAL MATHEMATICS

Frank L. Juszli, editor

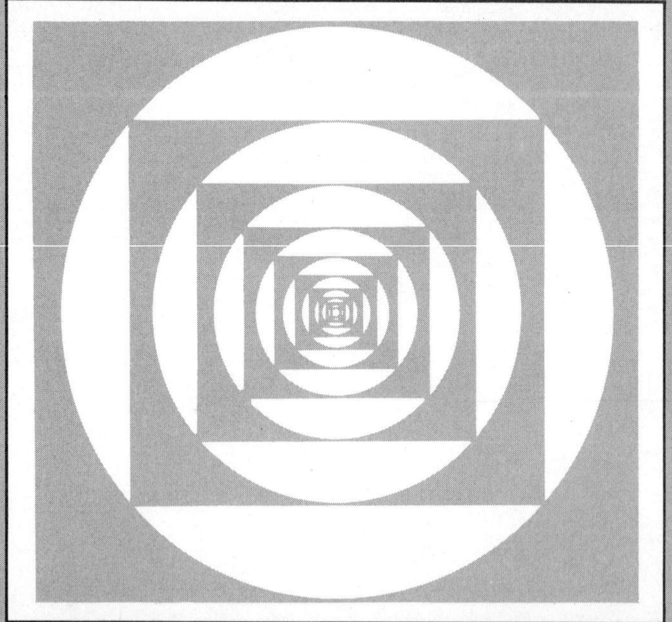

TECHNICAL CALCULUS

Paul Calter

Professor of Mathematics
Vermont Technical College

PRENTICE HALL, Englewood Cliffs, New Jersey 07632

Library of Congress Cataloging-in-Publication Data

Calter, Paul.
Technical calculus.

(Prentice-Hall series in technical mathematics)
Includes indexes.
1. Calculus. I. Title. II. Series.
QA303.C1793 1988 515 87–7246
ISBN 0–13–898149–3

Editorial/production supervision: Mary Carnis
Cover design: Ben Santora
Manufacturing buyer: Lorraine Fumoso

Printed in the United States of America

10 9 8 7 6 5 4 3 2 1

ISBN 0-13-898149-3 025

Prentice-Hall International (UK) Limited, *London*
Prentice-Hall of Australia Pty. Limited, *Sydney*
Prentice-Hall Canada Inc., *Toronto*
Prentice-Hall Hispanoamericana, S.A., *Mexico*
Prentice-Hall of India Private Limited, *New Delhi*
Prentice-Hall of Japan, Inc., *Tokyo*
Simon & Schuster Asia Pte. Ltd., *Singapore*
Editora Prentice-Hall do Brasil, Ltda., *Rio de Janeiro*

Contents

Preface xi

1

The Straight Line 1

1–1 Length of a Line Segment *2*
1–2 Slope and Angle of Inclination *7*
1–3 Equation of a Straight Line *17*
Chapter Test *29*

2

The Conic Sections 32

2–1 Circle *34*
2–2 Parabola *41*
2–3 Ellipse *54*
2–4 Hyperbola *67*
Chapter Test *79*

3

Derivatives of Algebraic Functions 81

3–1 Limits *82*
3–2 The Derivative *91*
3–3 Rules for Derivatives *101*
3–4 Derivative of a Function Raised to a Power *108*
3–5 Derivatives of Products and Quotients *113*
3–6 Derivatives of Implicit Relations *119*
3–7 Higher-Order Derivatives *123*
Chapter Test *124*

4

Graphical Applications of the Derivative 125

4–1 Tangents and Normals *126*
4–2 Maximum and Minimum Points *131*
4–3 Inflection Points *138*
4–4 Approximate Solution of Equations by Newton's Method *140*
4–5 Curve Sketching *144*
Chapter Test *150*

5

More Applications of the Derivative *152*

5–1 Rate of Change *154*
5–2 Motion of a Point *158*
5–3 Related Rates *166*
5–4 Applied Maximum–Minimum Problems *175*
5–5 Differentials *186*
Chapter Test *191*

Derivatives of Trigonometric, Logarithmic, and Exponential Functions 194

6–1 Derivatives of the Sine and Cosine Functions *195*
6–2 Derivative of the Tangent, Cotangent, Secant, and Cosecant *201*
6–3 Derivatives of the Inverse Trigonometric Functions *206*
6–4 Derivatives of the Logarithmic Functions *209*
6–5 Derivative of the Exponential Function *216*
Chapter Test *221*

7 Integration 223

7–1 The Indefinite Integral *224*
7–2 Integral of a Power Function *232*
7–3 Constant of Integration *235*
7–4 Use of a Table of Integrals *238*
7–5 Applications to Motion *246*
7–6 Application to Electric Circuits *252*
Chapter Test *255*

8 The Definite Integral 257

8–1 The Definite Integral *258*
8–2 Area under a Curve *262*
8–3 Finding Areas by Means of the Definite Integral *270*
Chapter Test *283*

9 Applications of the Definite Integral 285

9–1 Volumes of Integration *286*
9–2 Length of Arc *297*
9–3 Area of Surface of Revolution *301*
9–4 Average and Root-Mean-Square Values *304*
Chapter Test *307*

10 Centroids and Moments 309

10–1 Centroids *310*
10–2 Fluid Pressure *322*
10–3 Work *326*
10–4 Moment of Inertia *331*
Chapter Test *339*

11 Methods of Integration 340

11–1 Integration by Parts *341*
11–2 Integrating Rational Fractions *346*
11–3 Integrating by Means of Algebraic Substitution *354*
11–4 Integrating by Trigonometric Substitution *357*

11–5 Improper Integrals *362*
11–6 Approximate Integration *365*
Chapter Test *373*

12 First-Order Differential Equations 374

12–1 Differential Equations *375*
12–2 Variables Separable *379*
12–3 Exact Differential Equations *383*
12–4 Homogeneous First-Order Differential Equations *386*
12–5 First-Order Linear Differential Equations *388*
12–6 Geometric Applications of Differential Equations *394*
12–7 Exponential Growth and Decay *399*
12–8 Series *RL* and *RC* Circuits *403*
Chapter Test *408*

13 Second-Order Differential Equations 410

13–1 Variables Separable *411*
13–2 Second-Order Equations with Constant Coefficients and Right-Hand Side Zero *413*
13–3 Second-Order Equations with Right Side Not Zero *421*
13–4 Mechanical Vibrations *427*
13–5 *RLC* Circuits *437*
Chapter Test *448*

14 The Laplace Transform 450

14–1 The Laplace Transform of a Function *451*
14–2 Inverse Transforms *458*
14–3 Solving Differential Equations by the Laplace Transform *461*
14–4 Electrical Applications *464*
Chapter Test *472*

15 Numerical Solution of Differential Equations 474

15–1 First-Order Differential Equations *475*
15–2 Second-Order Equations *482*
Chapter Test *486*

Infinite Series 487

16–1 Sequences and Series *489*
16–2 Maclaurin's Series *496*
16–3 Taylor's Series *502*
16–4 Accuracy of Computation *507*
16–5 Operations with Power Series *509*
Chapter Test *513*

17 Fourier Series 515

17–1 Writing a Fourier Series *516*
17–2 Waveform Symmetries *522*
17–3 Waveforms with Period of $2L$ *530*
17–4 A Numerical Method *532*
Chapter Test *538*

Appendices 539

A Summary of Facts and Formulas 539

B Conversion Factors 577

C Table of Integrals 582

D Summary of BASIC 586

E Answers to Selected Problems 593

Index to Applications 621

General Index 623

To Amy and Carl

Preface

This book is intended mainly for students at two-year Technical Colleges and Community Colleges. It contains enough material for a one-semester calculus course, as well as for a second semester (usually called Calculus II or Advanced calculus). A one-semester calculus course will typically cover the topics in chapters one through nine, while the second semester, if offered, will choose topics from the remaining eight chapters.

Much of this book is taken from my *Technical Mathematics with Calculus,* with extensive revision and addition. However, there is much brand new material, including Differential equations, both first- and second-order, the Laplace Transform, Numerical Solution of Differential Equations, Infinite Series, and Fourier Series.

A mathematics book is never easy reading, so much care has gone into making the material as clear as possible. We follow an intuitive rather than a rigorous approach, and give information in small segments. Marginal notes, numerous illustrations, and careful page layout are designed to make the material as interesting and easy to follow as possible.

Exercises: Practice is essential for learning mathematics, as with other skills, so we include a large number of exercises. Those given after each section are graded by difficulty and grouped by type, to allow practice on a particular area. However the problems in the Chapter Tests are scrambled as to type and difficulty.

Answers to all odd-numbered problems are given in the Answer Key, Appendix E, while complete solutions to every problem are contained in the Solutions Manual, available to instructors.

Formulas: Each important formula, both mathematical and technical, is boxed and numbered in the text. At the risk of having this text called a ''cookbook,'' we also list these formulas in Appendix A as the Summary of Facts and Formulas. This listing can function as a ''handbook,'' for a calculus course and for other courses as well, and provides a common thread between chapters. We hope it will also help a student to see interconnections that might otherwise be overlooked.

Common Errors: An instructor quickly learns the pitfalls and traps that ''get'' students year after year. Many of these are boxed in the text and are labelled ''Common Error.''

Applications: We include discussion in the text of many technical applications, such as the use of calculus for analyzing motion problems or electric circuits. They are included for classes that wish to cover those topics, as well as for motivation, to show that mathematics has real-world uses. Space does not permit a full discussion of each application. It is assumed that students have sufficient background before attempting the more difficult calculus problems in such areas, and that they will get technical help not offered in this text.

The Index to Applications should help in finding specific applications.

Computer: There are certain calculations, such as numerical integration or numerical solution of differential equations, that are best done by computer. For these we give computer methods in the text. Elsewhere, computer problems are suggested at the end of exercises, as enrichment activities. As with technical applications, space does not permit the teaching of programming here, and we assume that a student who attempts one of these has suitable computer background. We do, however, provide a Summary of the BASIC Language in Appendix D. Complete programs are given in the answer key and solutions manual. A diskette containing all the programs is available to instructors, who may copy it for their students.

Supplements: The Solutions Manual contains worked out solutions to each problem in the text, and listings of all computer programs. A Computer Diskette, prepared by Michael Calter, contains each program from the text, ready-to-run.

Acknowledgements: For reviewing the manuscript and making valuable suggestions I want to thank Franklin Blou, Essex County College; Martin Horowitz, Queensbourough Community College; Glenn Jacobs, Greenville Technical College; and Frank Scalzo, Queensborough Community College. Reviewers of the earlier edition of this material include my colleagues at Vermont Technical College: Byron Angell, Walt Granter, John Knox, and Don Nevin, as well as David Bashaw, New Hampshire Technical Institute; Elizabeth Bliss, Trident Technical College; Ellen Kowalczyk, Madison Area Technical College; Donald Reichman, Mercer County Community College; Ursula Rodin, Nashville State Technical Institute; Edward W. Seabloom, Lane Community College; and Robert Seaver, Lorain Community College. I am further grateful to Michael Calter for producing the computer diskette.

Special thanks go to Frank Juszli, editor of the Prentice-Hall series in Technical Mathematics, for his review of the entire manuscript, past and present, and for his valuable suggestions. The tedious job of solving the many exercise problems was done by Brett Benner, Ken Berkey, Keith Crowe, Jim Davis, Nancy Davis, Kelly Dennehy, Mel Emerson, Ellie Germain, Robert Morel, Jeffery Sloan, and Ray Wells. Finally, it was all put together by Mary Carnis, Production Editor at Prentice-Hall. Thank you all.

Paul Calter
Randolph, Vermont

The Straight Line

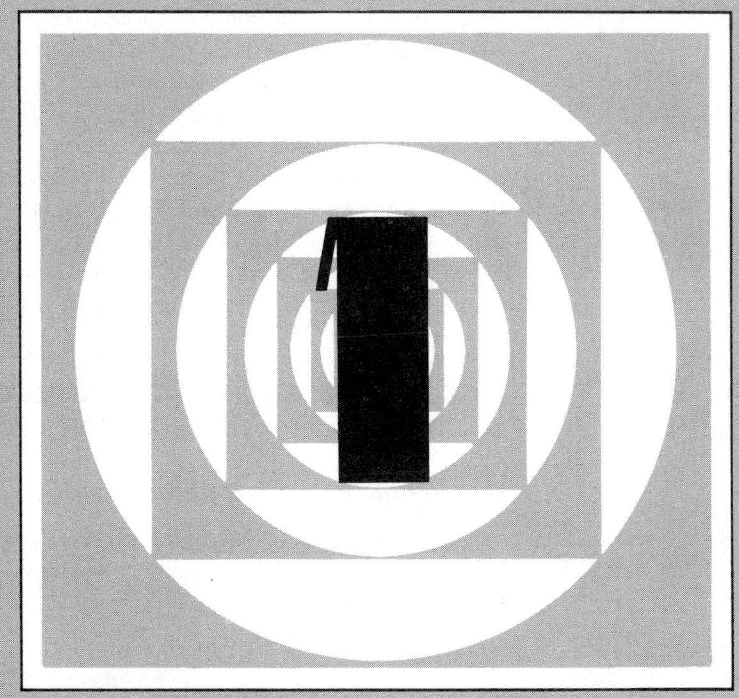

This chapter begins our study of *analytic geometry,* which is continued in Chapter 2 on the conic sections. Here we place geometric figures, such as points, lines, circles, and so on, on coordinate axes, where they may be studied using the methods of algebra.

We start by computing *lengths of line segments* and then study the idea of *slope* of a line and of a curve, an idea that will be crucial to our understanding of the derivative in Chapter 3. We then learn to write the equation of a straight line, just as we will learn to write the equations of circles, parabolas, ellipses, and hyperbolas in the following chapter. Why? Because these equations enable us to understand and to use these geometric figures in ways that are not possible without the equations.

1-1. LENGTH OF A LINE SEGMENT

Line Segments on or Parallel to a Coordinate Axis

When we speak about the length of a line segment, or about the distance between two points, we usually mean the *magnitude* of that length or distance.

To find the magnitude of the length of a line segment lying on or parallel to the x (or y) axis, simply subtract the abscissa (or ordinate) of either endpoint from the abscissa (or ordinate) of the other endpoint, and take the absolute value of this result.

EXAMPLE 1: The magnitudes of the lengths of the lines in Fig. 1-1 are

(a) $AB = 5 - 2 = 3$
(b) $PQ = 5 - (-2) = 7$
(c) $RS = -1 - (-5) = 4$

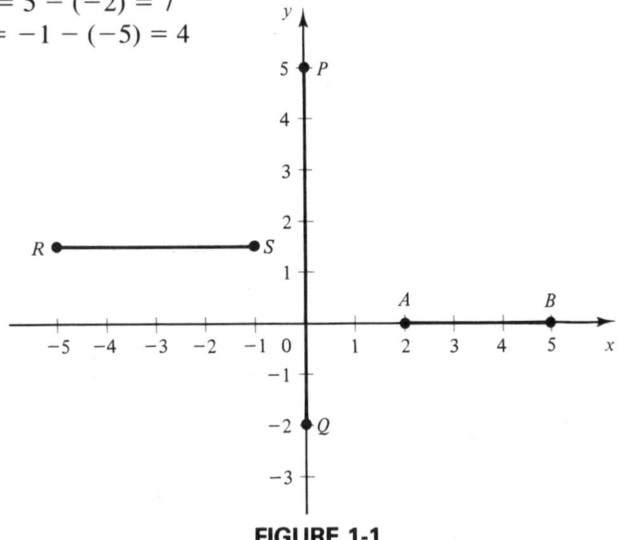

FIGURE 1-1

Directed Distance

Note that directed distance applies only to a line parallel to a coordinate axis.

Sometimes when speaking about the length of a line segment or the distance between two points, it is necessary to specify *direction* as well as the magnitude. When we specify the directed distance AB, for example, we mean the distance *from A to B*. This is sometimes written $\overline{AB}$.

EXAMPLE 2: For the two points $P(5, 2)$, and $Q(1, 2)$, the directed distance PQ is $PQ = 1 - 5 = -4$ and the directed distance QP is $QP = 5 - 1 = 4$.

Increments

Let us say that a particle is moving along a curve (Fig. 1-2) from point P to point Q. In doing so, its abscissa changes from x_1 to x_2. We call this change an *increment* in x, and label it Δx (read *delta x*). Similarly,

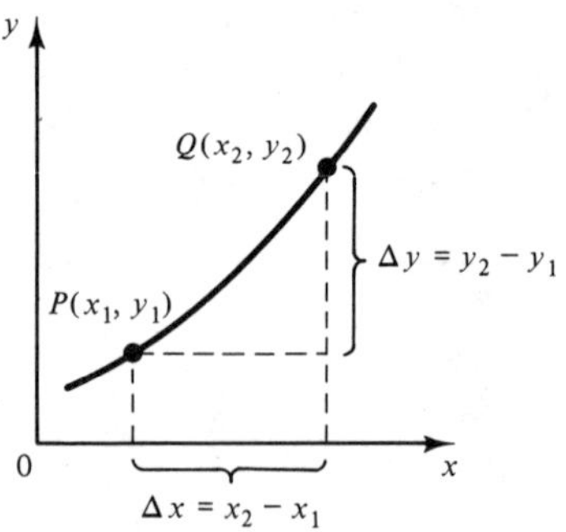

FIGURE 1-2. Increments

the ordinate changes from y_1 to y_2 and is labeled Δy. The increments are found simply by subtracting the coordinates at P from those at Q.

Increments	$\Delta x = x_2 - x_1$ and $\Delta y = y_2 - y_1$	**301**

EXAMPLE 3: A particle moves from $P_1(2, 5)$ to $P_2(7, 3)$. The increments in its coordinates are

$$\Delta x = x_2 - x_1 = 7 - 2 = 5$$

and

$$\Delta y = y_2 - y_1 = 3 - 5 = -2$$

Distance Formula

We now find the length of a line inclined at any angle, such as PQ in Fig. 1-3. We first draw a horizontal line through P and drop a perpendicular from Q, forming a right triangle PQR. The sides of the triangle are d, Δx, and Δy. By the Pythagorean theorem,

$$d^2 = (\Delta x)^2 + (\Delta y)^2 = (x_2 - x_1)^2 + (y_2 - y_1)^2$$

FIGURE 1-3. Length of a line segment.

Since we want the magnitude of the distance, we take only the positive root and get

Distance Formula	$d = \sqrt{(\Delta x)^2 + (\Delta y)^2} = \sqrt{(x_2 - x_1)^2 + (y_2 - y_1)^2}$	**247**

EXAMPLE 4: Find the length of the line segment with the endpoints (3, −5) and (−1, 6).

Solution: Let us give the first point (3, −5) the subscripts 1, and the other point the subscripts 2 (it does not matter which we label 1). So

$$x_1 = 3, \qquad y_1 = -5, \qquad x_2 = -1, \qquad y_2 = 6$$

Substituting into the distance formula, we obtain

$$d = \sqrt{(-1 - 3)^2 + [6 - (-5)]^2} = \sqrt{(-4)^2 + (11)^2} = 11.7 \quad \text{(rounded)}$$

Common Error	Do not take the square root of each term separately. $d \neq \sqrt{(x_2 - x_1)^2} + \sqrt{(y_2 - y_1)^2}$

EXERCISE 1

Length of a Line Segment

Find the distance between the points. Keep three significant digits.

1. (5, 0) and (2, 0)

2. (0, 3) and (0, −5)

3. (−2, 0) and (7, 0)

4. (−4, 0) and (−6, 0)

5. (0, −2.74) and (0, 3.86)

6. (55.34, 0) and (25.38, 0)

7. (7, 2) and (3, 2)

8. (−1, 6) and (−1, 3)

9. (−5, −5) and (−5, −6)

10. (9, −2) and (−9, −2)

11. (5.59, 3.25) and (8.93, 3.25)

12. (−2.06, −5.83) and (−2.06, −8.34)

13. (3, 9) and (−3, 5)

14. (−6, −7) and (−9, −2)

15. (−2.9, 5.3) and (5.8, −3.7)

16. (8.38, −3.95) and (2.25, −4.99)

17. (−47, 34) and (55, −48)

18. (−1.1, 4.2) and (4.2, −1.1)

Find the directed distance AB.

19. $A(3, 0)$; $B(5, 0)$

20. $A(3.95, -2.07)$; $B(-3.95, -2.07)$

21. $B(8, -2)$; $A(-8, -5)$

22. $A(-9, -2)$; $B(17, -2)$

23. $B(-6, -6)$; $A(-6, -7)$

24. $B(11.5, 3.68)$; $A(11.5, -5.38)$

25. Find the length of girder AB in Fig. 1-4.

26. Find the distance between the holes in Fig. 1-5.

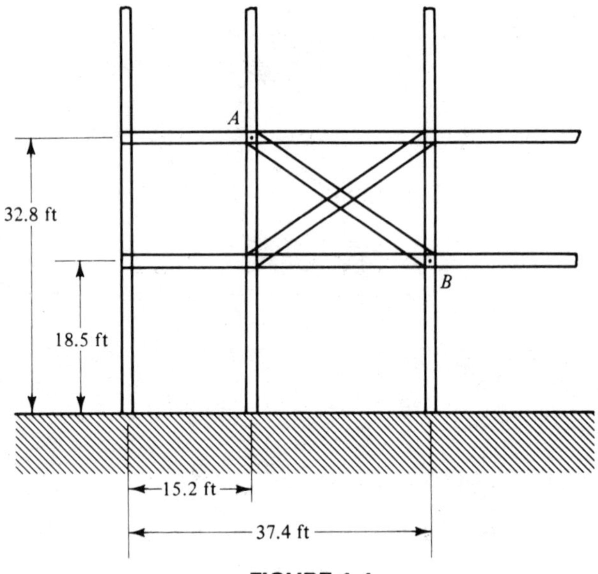

FIGURE 1-4

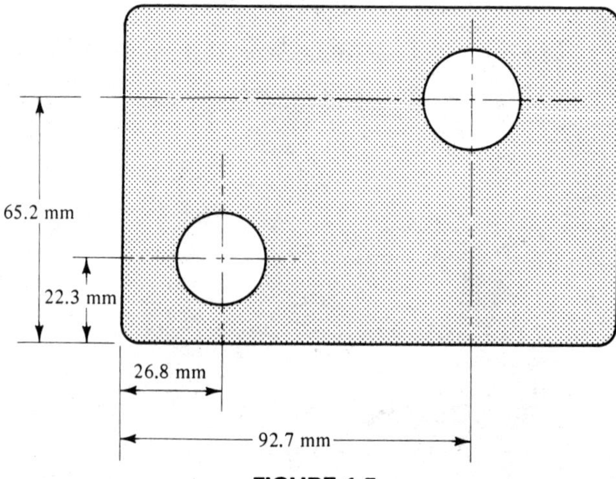

FIGURE 1-5

27. A triangle has vertices at (3, 5), (−2, 4), and (4, −3). Find the length of each side. Then compute the area using Hero's formula (Eq. 138). Work to three significant digits.

Computer

We don't intend to teach computer in this text, but we include an occasional problem for those who know simple programming and have access to a terminal.

28. Write a program that will accept as input the coordinates of the vertices of a triangle, and then compute and print the length of each side and the area of the triangle.

1-2. SLOPE AND ANGLE OF INCLINATION

Definition

Referring back to Fig. 1-3, we call the horizontal distance between the points P and Q the *run* from P to Q, which is equal to Δx for this line. The vertical distance Δy between the same point is called the *rise*. The *slope* of the line PQ is the ratio of the rise to the run. It is usually given the symbol m.

Slope	$m = \dfrac{\text{rise}}{\text{run}} = \dfrac{\Delta y}{\Delta x} = \dfrac{y_2 - y_1}{x_2 - x_1}$	**248**

The slope is equal to the rise divided by the run.

EXAMPLE 1: Find the slope of the line connecting the points (−3, 5) and (4, −6) (Fig. 1-6).

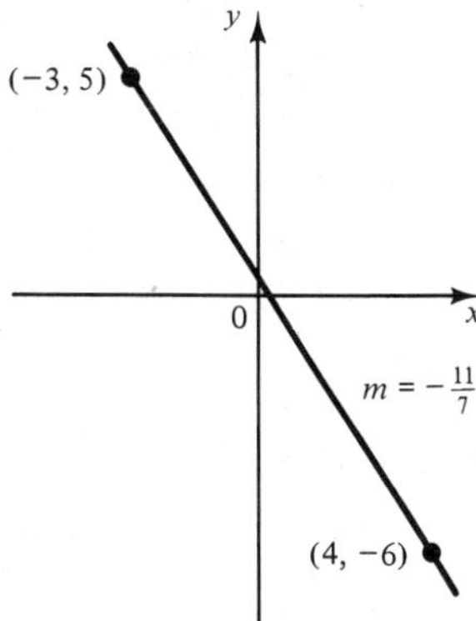

FIGURE 1-6

Solution: As when using the distance formula, it does not matter which is called point 1 and which is point 2. Let us choose

$$x_1 = -3, \qquad y_1 = 5, \qquad x_2 = 4, \qquad y_2 = -6$$

Then by Eq. 248,

$$\text{slope } m = \frac{-6 - 5}{4 - (-3)} = \frac{-11}{7} = -\frac{11}{7}$$

EXAMPLE 2: Find the slope of the line in Fig. 1-7.

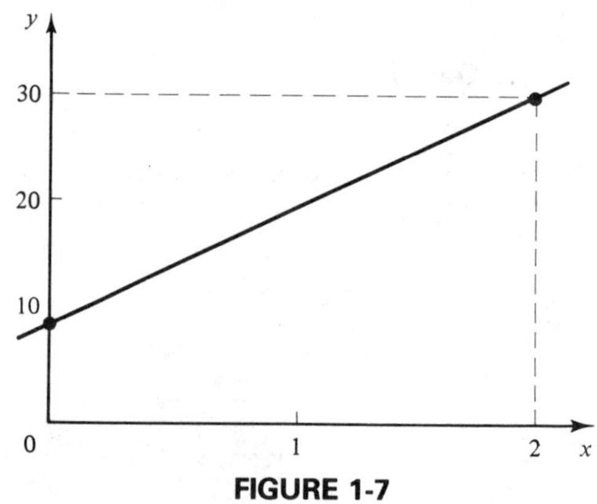

FIGURE 1-7

Solution: The rise is 30 − 10 = 20 and the run is 2 − 0 = 0, so

$$m = \frac{20}{2} = 10$$

(Do not be thrown off by the different scales on each axis).

Common Error	Be careful not to mix up the subscripts. $$m \neq \frac{y_2 - y_1}{x_1 - x_2}$$

Horizontal and Vertical Lines

For any two points on a horizontal line, the values of y_1 and y_2 in Eq. 248 are equal, making the slope equal to zero. For a vertical line, the values of x_1 and x_2 are equal, giving division by zero. Hence the slope is undefined for a vertical line. The slopes of various lines are shown in Fig. 1-8.

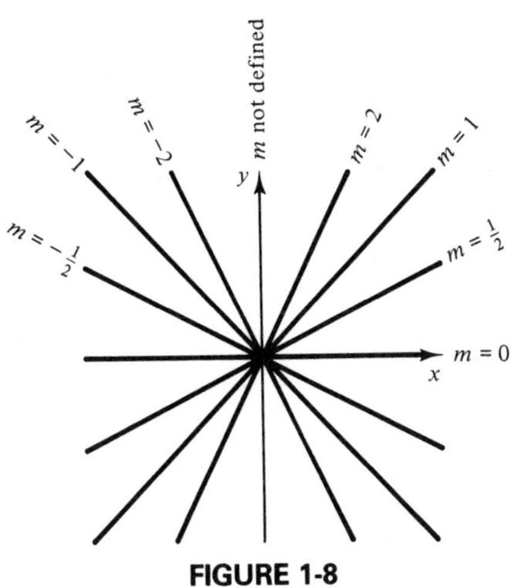

FIGURE 1-8

Angle of Inclination

The smallest positive angle that a line makes with the positive x direction is called the *angle of inclination* θ of the line (Fig. 1-9). A line parallel to the x axis has an angle of inclination of zero. Thus the angle of inclination can have values from 0° up to 180°. From Fig. 1-3,

$$\tan\theta = \frac{\text{opposite side}}{\text{adjacent side}} = \frac{y_2 - y_1}{x_2 - x_1} = \frac{\text{rise}}{\text{run}} = \frac{\Delta y}{\Delta x}$$

But this is the slope m of the line, so

$m = \tan\theta$ $0° \leqq \theta < 180°$	**249**

The slope of a line is equal to the tangent of the angle of inclination θ *(except when* $\theta = 90°$*).*

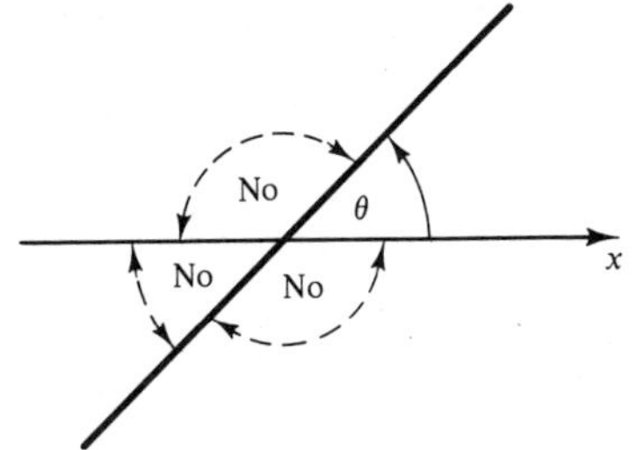

FIGURE 1-9. Angle of inclination, θ.

EXAMPLE 3: The slope of a line having an angle of inclination of 50° (Fig. 1-10) is, by Eq. 249,

$$m = \tan 50° = 1.192 \quad \text{(rounded)}$$

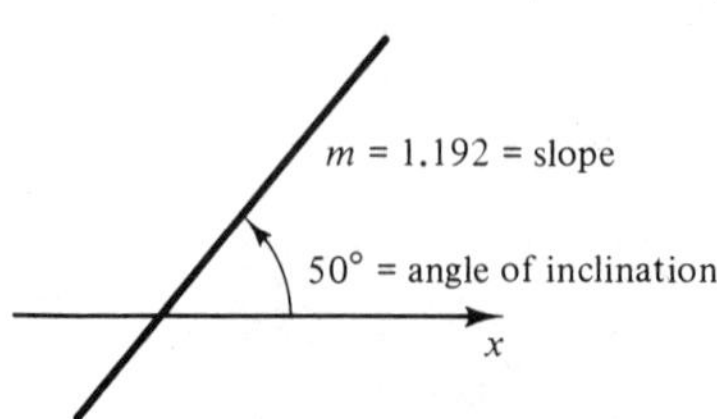

FIGURE 1-10

EXAMPLE 4: Find the angle of inclination of a line (Fig. 1-11) having a slope of 3.

Solution: By Eq. 249, $\tan \theta = 3$, so

$$\theta = \arctan 3 = 71.6° \quad \text{(rounded)}$$

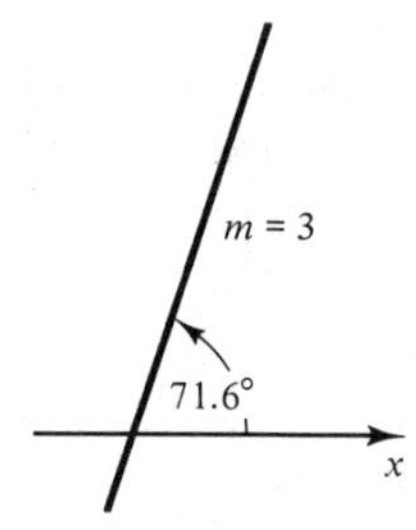

FIGURE 1-11

When the slope is negative, our calculator will give us a negative angle, which we then use to obtain a positive angle of inclination less than 180°.

EXAMPLE 5: For a line having a slope of -2 (Fig. 1-12), arctan $(-2) = -63.4°$ (rounded), so

$$\theta = 180° - 63.4° = 116.6°$$

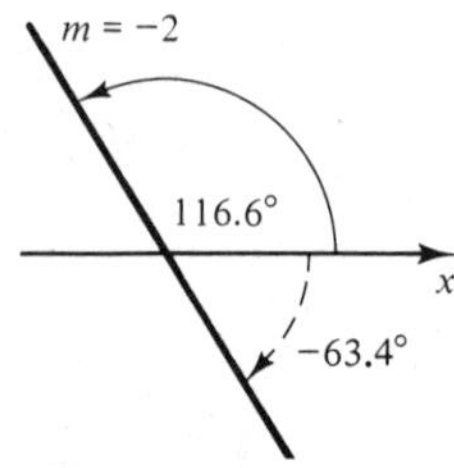

FIGURE 1-12

EXAMPLE 6: Find the angle of inclination of the line passing through (−2.47, 1.74) and (3.63, −4.26) (Fig. 1-13).

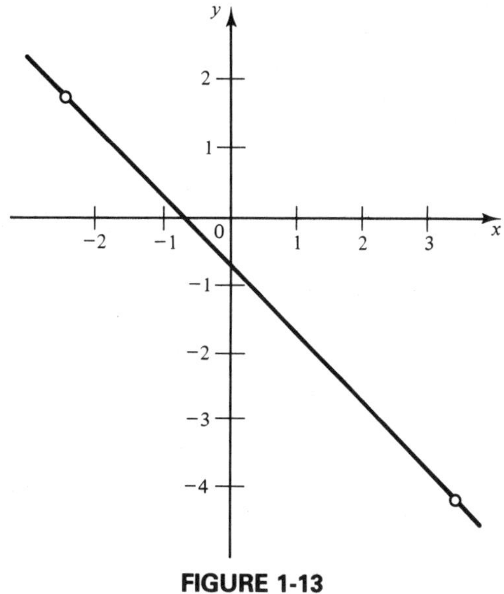

FIGURE 1-13

Solution: The slope, from Eq. 248, is

$$m = \frac{-4.26 - 1.74}{3.63 - (-2.47)} = -0.984$$

from which $\theta = 135.5°$.

When *different scales* are used for the x and y axes, the angle of inclination *will appear distorted.*

EXAMPLE 7: The angle of inclination of the line in Fig. 1-7 is $\theta = \arctan 10 = 84.3°$. Notice that the angle in the graph appears much smaller than 84.3°, due to the different scales on each axis.

Slopes of Parallel and Perpendicular Lines

Parallel lines have, of course, *equal* slopes.

Slopes of Parallel Lines	$m_1 = m_2$	**258**

Parallel lines have equal slopes.

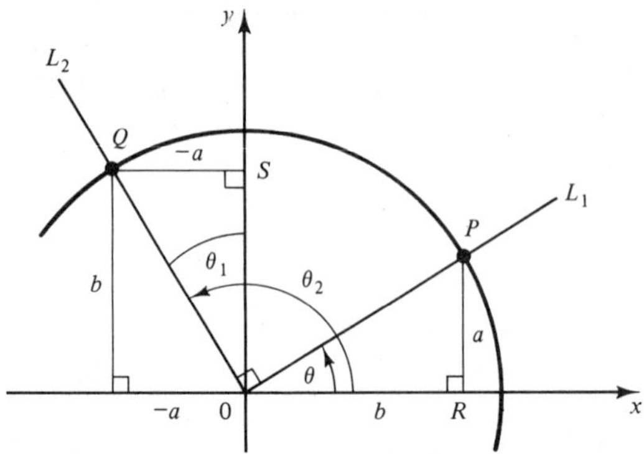

FIGURE 1-14. Slopes of perpendicular lines.

Now consider line L_1, with slope m_1 and angle of inclination θ_1 (Fig. 1-14) and a *perpendicular* line L_2 with slope m_2 and angle of inclination $\theta_2 = \theta_1 + 90°$. Note that $\angle QOS = \theta_2 - 90° = \theta_1 + 90° - 90° = \theta_1$, so right triangles OPR and OQS are congruent. Thus the magnitudes of their corresponding sides a and b are equal. The slope m_1 of L_1 is a/b, and the slope m_2 of L_2 is $-b/a$, so

Slopes of Two Perpendicular Lines	$m_1 = -\dfrac{1}{m_2}$	**259**

The slope of a line is the negative reciprocal of the slope of a perpendicular to that line.

EXAMPLE 8: Any line perpendicular to a line whose slope is 5 has a slope of $-\frac{1}{5}$.

Common Error	The minus sign in Eq. 259 is often forgotten. $$m_1 \neq \frac{1}{m_2}$$

EXAMPLE 9: Is angle ACB (Fig. 1-15) a right angle?

Solution: The slope of AC is

$$\frac{8 - 3}{5 - (-15)} = \frac{5}{20} = \frac{1}{4}$$

Thus to be perpendicular BC must have a slope of -4. Its actual slope is

$$\frac{-10 - 8}{10 - 5} = \frac{-18}{5} = -3.6$$

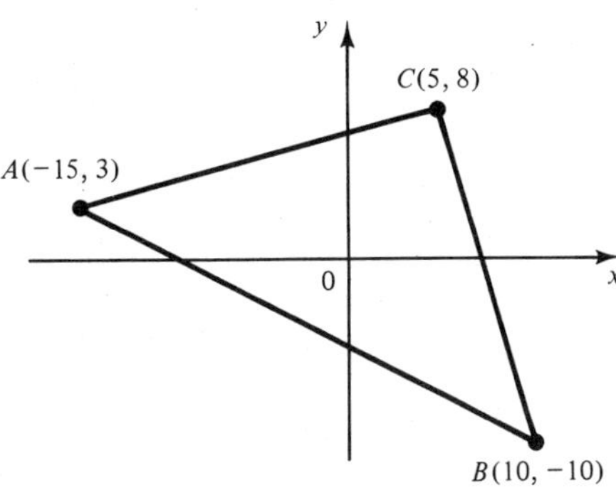

FIGURE 1-15

So while the lines *appear* to be perpendicular, we have shown that they are not.

Angle of Intersection between Two Lines

Figure 1-16 shows two lines L_1 and L_2 intersecting at an angle ϕ, measured counterclockwise from line 1 to line 2. We want a formula for ϕ in terms of the slopes m_1 and m_2 of the lines. By Eq. 142, θ_2 equals the sum of the angle of inclination θ_1 and the angle of intersection ϕ. So $\phi = \theta_2 - \theta_1$. Taking the tangent of both sides gives $\tan \phi = \tan (\theta_2 - \theta_1)$, or

$$\tan \phi = \frac{\tan \theta_2 - \tan \theta_1}{1 + \tan \theta_1 \tan \theta_2}$$

by the trigonometric identity for the tangent of the difference of two angles (Eq. 169). But the tangent of an angle of inclination is the slope, so

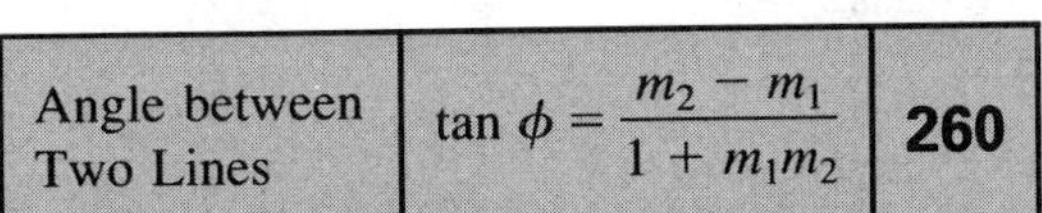

Angle between Two Lines	$\tan \phi = \dfrac{m_2 - m_1}{1 + m_1 m_2}$	**260**

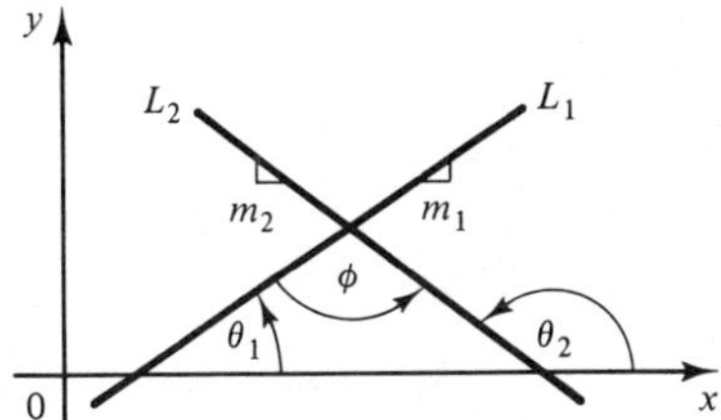

FIGURE 1-16. Angle of intersection between two lines.

EXAMPLE 10: The tangent of the angle of intersection between line 1 having a slope of 4, and line 2 having a slope of -1, is

$$\tan \phi = \frac{-1 - 4}{1 + 4(-1)} = \frac{-5}{-3} = \frac{5}{3}$$

from which $\phi = \arctan \frac{5}{3} = 59.0°$ (rounded). This angle is measured counterclockwise from line 1 to line 2, as shown in Fig. 1-17.

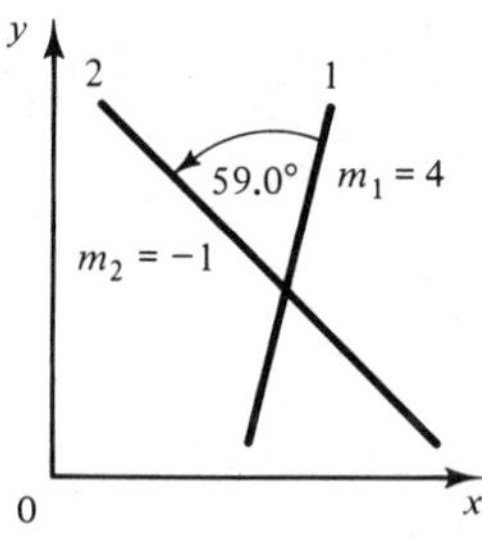

FIGURE 1-17

Tangents and Normals to Curves

We know what the tangent to a circle is, and probably have an intuitive idea of what the *tangent to a curve* is. We'll speak a lot about tangents, so let us try to get a clear idea of what one is.

To determine the tangent line at P (Fig. 1-18), we start by selecting a second point Q, on the curve. As shown, Q can be on either side of P. The line PQ is called a *secant line*. We then let Q approach P (from either side). If the secant lines PQ approach a single line T as Q approaches P, then line T is called the tangent line at P.

In Chapter 3 you will see that the slope gives the *rate of change* of the function, an extremely useful quantity to find. You will see that it is found by taking the *derivative* of the function.

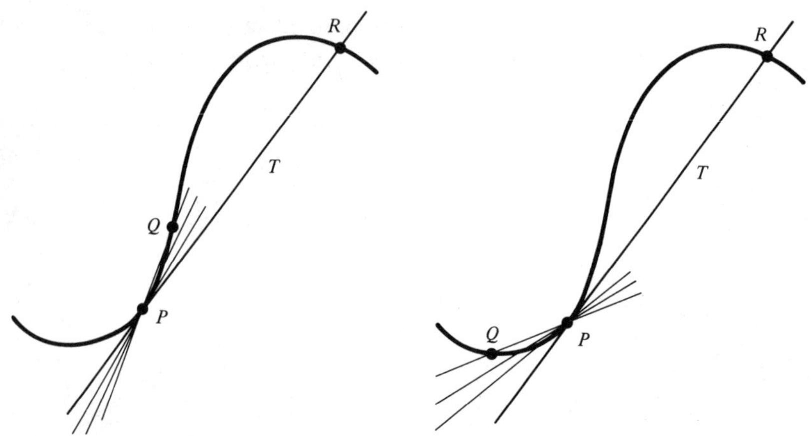

FIGURE 1-18. A tangent line T at point P, here defined as the limiting position of a secant line PQ, as Q approaches P.

Note that the tangent line can intersect the curve at more than one point, such as point R. However, at points "near" P, the tangent line intersects the curve just once while a secant line intersects the curve twice. The slope of a *curve* at some point P is defined as the *slope of the tangent* to the curve drawn through that point.

EXAMPLE 11: Plot the curve $y = x^2$ for integer values of x from 0 to 4. By eye, draw the tangent to the curve at $x = 2$ and find the approximate value of the slope.

Solution: We make a table of point pairs.

x	0	1	2	3	4
y	0	1	4	9	16

The plot and the tangent line are shown in Fig. 1-19. Using the scales on the x and y axes, we measure a rise of 12 units for the tangent line in a run of 3 units. The slope of the tangent line is then

$$m_t \cong \frac{12}{3} = 4$$

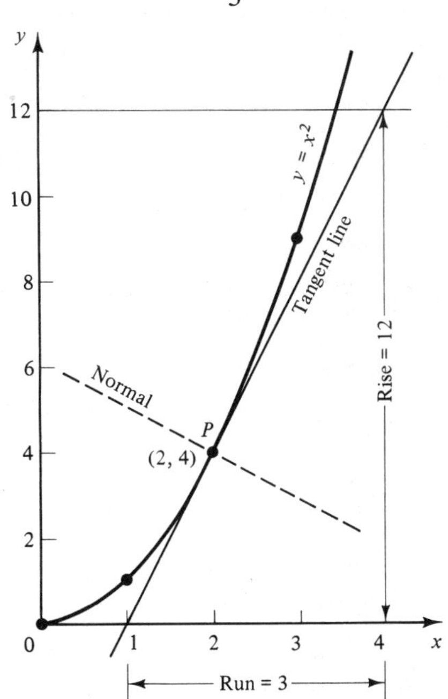

FIGURE 1-19. Slope of a curve.

The *normal* to a curve at a given point is the line perpendicular to the tangent at that point, as in Fig. 1-19. Its slope is the negative reciprocal of the slope of the tangent. For Example 11, $m_n = -\frac{1}{4}$.

EXERCISE 2

Slope and Angle of Inclination

Find the slope of the line passing through the given points.

Work to three significant digits here.

1. (8, 3) and (−2, 4)

2. (−3, 7) and (5, 1)

3. (−2, a) and (a, 3)

4. (5.2, 8.3) and (−2.8, −1.4)

5. (284, 184) and (338, 402)

6. (−2.14, −7.36) and (−1.37, −2.08)

Find the slope of the line having the given angle of inclination.

7. 38.2°

8. 77.9°

9. 1.83 rad

10. 58°14′

11. 156.3°

12. −47.2°

Find the angle of inclination, in decimal degrees, of a line having the given slope.

13. $m = 3$

14. $m = 1.84$

15. $m = -4$

16. $m = -2.75$

17. $m = 0$

18. $m = -15$

Find the slope of a line perpendicular to a line having the given slope.

19. $m = 5$

20. $m = 2$

21. $m = 4.8$

22. $m = -1.85$

23. $m = -2.85$

24. $m = -5.372$

Find the slope of a line perpendicular to a line having the given angle of inclination.

25. 58.2°

26. 1.84 rad

27. 136°44′

Find the angle of inclination, in decimal degrees, of a line passing through the given points.

28. (5, 2) and (−3, 4)

29. (−2.5, −3.1) and (5.8, 4.2)

30. (6, 3) and (−1, 5)

31. (x, 3) and ($x + 5$, 8)

Angle between Two Lines

32. Find the angle of intersection between line 1 having a slope of 1 and line 2 having a slope of 6.

33. Find the angle of intersection between line 1 having a slope of 3 and line 2 having a slope of −2.

34. Find the angle of intersection between line 1 having an angle of inclination of 35° and line 2 having an angle of inclination of 160°.

35. Find the angle of intersection between line 1 having an angle of inclination of 22° and line 2 having an angle of inclination of 86°.

Tangents and Normals to Curves

Plot the functions from $x = 0$ to $x = 4$. Graphically find the slope of the curve at $x = 2$.

36. $y = x^3$

37. $y = -x^2 + 3$

38. $y = \dfrac{x^2}{4}$

Applications

39. The distance between two stakes on a slope is taped at 2055 ft, and the angle of the slope with the horizontal is 12.3°. Find the horizontal distance between the stakes.

40. What is the angle of inclination with the horizontal of a roadbed that rises 15 ft in each 250 ft, measured horizontally?

41. How far apart must two stakes on a 7° slope be placed so that the horizontal distance between them is 1250 m?

In some fields, such as highway work, the word *grade* is used instead of slope. It is usually expressed as a percent: *percent grade* = *100* × *slope*. Thus, a 5% grade rises 5 units for every 100 units of run.

42. On a 5% road grade, at what angle is the road inclined to the horizontal? How far does one rise in traveling upward 500 ft, measured along the road?

43. A straight tunnel under a river is 755 ft long and descends 12 ft in this distance. What angle does the tunnel make with the horizontal?

44. A straight driveway slopes downward from a house to a road and is 28 m in length. If the angle of inclination from the road to the house is 3.6°, find the height of the house above the road.

45. An escalator is built so as to rise 2 m for each 3 m of horizontal travel. Find its angle of inclination.

46. A straight highway makes an angle of 4.5° with the horizontal. How much does the highway rise in a distance of 2500 ft, measured along the road?

1-3. EQUATION OF A STRAIGHT LINE

We want to be able to write the equation of a straight line, given any two pieces of information about the line. We may be given, say, the slope of the line and a point through which it passes. Instead, we may know two points through which the line passes, or some other pieces of data.

The equation of a straight line, like any equation, can be written in several different forms. We'll find it convenient to develop one form for each of the ways in which the data about the line may be presented. Thus there is a point-slope form for when we know a point on the line and the slope, a two-point form, and so on. We start with the slope-intercept form.

Slope-Intercept Form

Suppose that $P(x, y)$ is any point on a straight line. We seek an equation that links x and y in a functional relationship, so that, for any x, a value of y can be found. We can get such an equation by applying the definition of slope to our point P, and some *known* point on the line. For a line that intersects the y axis b units from the origin (Fig. 1–20) the rise is $y - b$ and the run is $x - 0$, so by Eq. 248,

$$m = \frac{y - b}{x - 0}$$

Simplifying, $mx = y - b$, or

Slope-Intercept Form	$y = mx + b$	**253**

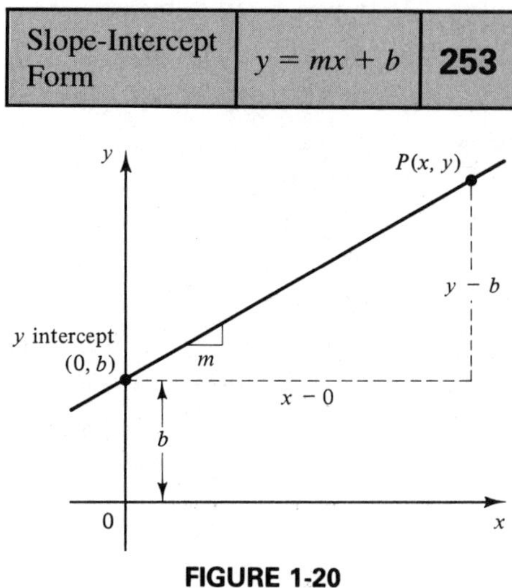

FIGURE 1-20

This is called the slope-intercept form, because the slope m and y intercept b are easily identified in the equation. For example, in the equation $y = 2x + 1$,

$$y = 2x + 1$$

slope ↑ (2)   ↑ y intercept (+ 1)

EXAMPLE 1: Find the slope and y intercept of the line $5x - 2y + 3 = 0$ (Fig. 1-21).

Solution: We first put the equation into slope-intercept form by solving for y,

$$2y = 5x + 3$$

$$y = \frac{5}{2}x + \frac{3}{2}$$

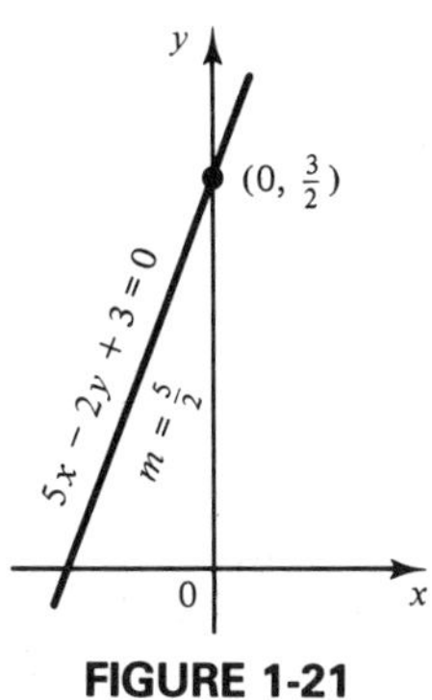

FIGURE 1-21

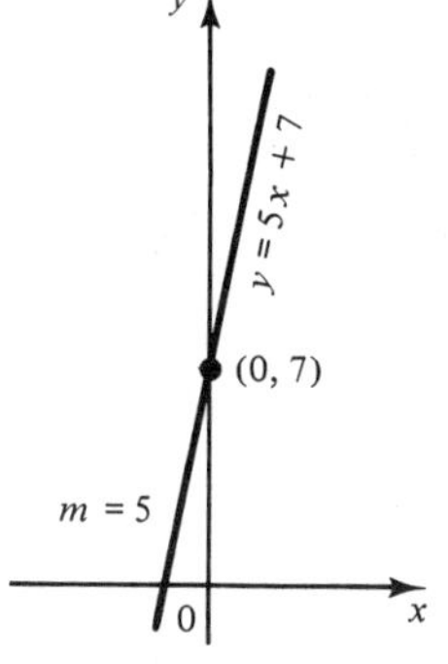

FIGURE 1-22

So

$$m = \frac{5}{2} \quad \text{and} \quad b = \frac{3}{2}$$

EXAMPLE 2: Write the equation of the line having a slope of 5 and a y intercept of 7 (Fig. 1-22).

Solution: Substituting $m = 5$ and $b = 7$ into Eq. 253, we obtain

$$y = 5x + 7$$

General Form

The slope-intercept form is the equation of a straight line in *explicit* form, $y = f(x)$. We can also write the equation of a straight line in *implicit* form, $f(x, y) = 0$ simply by transposing all terms to one side of the equal sign and simplifying. We usually write the x term first, then the y term, and finally the constant term. This form is referred to as the *general form* of the equation of a straight line.

General Form	$Ax + By + C = 0$	**250**

where A, B, and C are constants.

You will see as we go along that there are several forms for the equation of a straight line, and we use the one that is most convenient in a particular problem. But to make it easy to compare answers, we usually rearrange the equation into general form.

EXAMPLE 3: Change the equation $y = \frac{6}{7}x - 5$ from slope intercept form to general form.

Solution: Subtracting y from both sides,

$$0 = \frac{6}{7}x - 5 - y$$

Multiplying by 7 and rearranging,

$$6x - 7y - 35 = 0$$

Point-Slope Form

Let us write an equation for a line of slope m, but which passes through a given point (x_1, y_1) which is *not* in general on either axis, as in Fig. 1-23. Again using the definition of slope (Eq. 248),

Point-Slope Form	$m = \dfrac{y - y_1}{x - x_1}$	**255**

This form of the equation of a straight line is most useful when we know the slope of a line and one point through which the line passes.

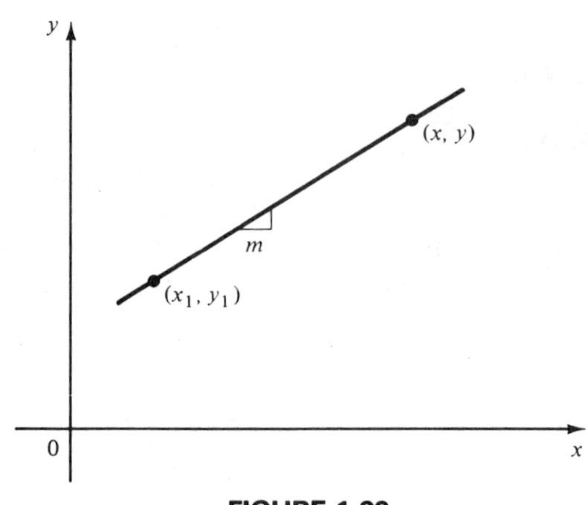

FIGURE 1-23

EXAMPLE 4: Write the equation in general form of the line having a slope of 2 and passing through the point (1, −3) (Fig. 1-24).

Solution: Substituting $m = 2$, $x_1 = 1$, $y_1 = -3$ into Eq. 255,

$$2 = \frac{y - (-3)}{x - 1}$$

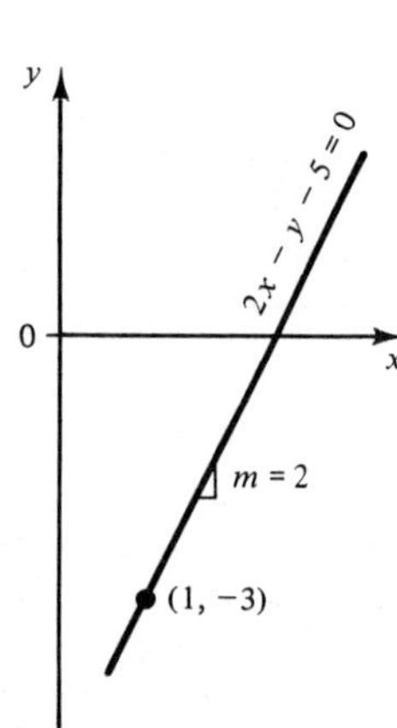

FIGURE 1-24

Multiplying by $x - 1$,

$$2x - 2 = y + 3$$

or, in general form,

$$2x - y - 5 = 0$$

EXAMPLE 5: Write the equation in general form of the line passing through the point (3, 2) and perpendicular to the line $y = 3x - 7$ (Fig. 1-25).

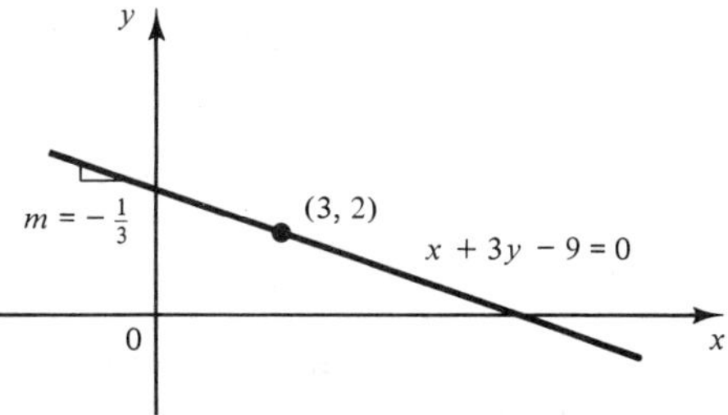

FIGURE 1-25

Solution: The slope of the given line is 3, so the slope of our perpendicular line is $-\frac{1}{3}$. Using the point-slope form,

$$-\frac{1}{3} = \frac{y-2}{x-3}$$

Going to the general form, $x - 3 = -3y + 6$, or

$$x + 3y - 9 = 0$$

EXAMPLE 6: Write the equation in general form of the line passing through the point (3.15, 5.88) and having an angle of inclination of 27.8° (Fig. 1-26).

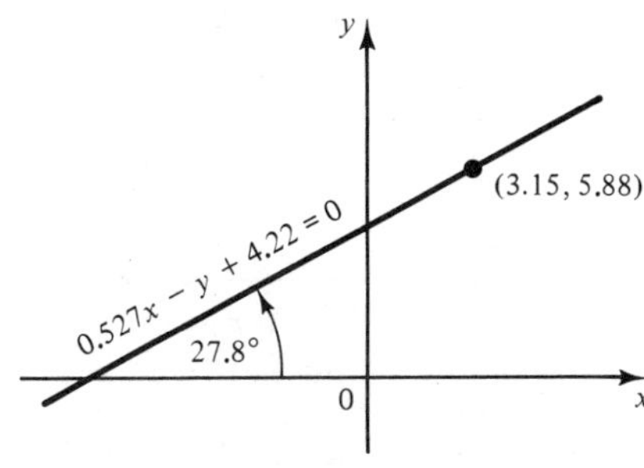

FIGURE 1-26

Solution: From Eq. 249, $m = \tan 27.8° = 0.527$. From Eq. 255,

$$0.527 = \frac{y - 5.88}{x - 3.15}$$

Changing to general form, $0.527x - 1.66 = y - 5.88$, or

$$0.527x - y + 4.22 = 0$$

Two-Point Form

If two points on a line are known, the equation of the line is easily written using the two-point form, which we now derive. If we call the points P_1 and P_2 in Fig. 1-27 the slope of the line is

$$m = \frac{y_2 - y_1}{x_2 - x_1}$$

The slope of the line segment connecting P_1 with any other point P on the same line is

$$m = \frac{y - y_1}{x - x_1}$$

Since these slopes must be equal, we get

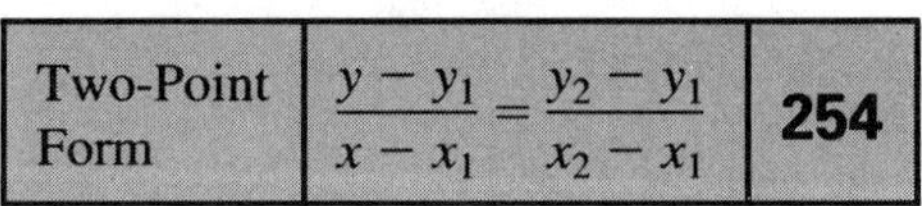

Two-Point Form	$\dfrac{y - y_1}{x - x_1} = \dfrac{y_2 - y_1}{x_2 - x_1}$	**254**

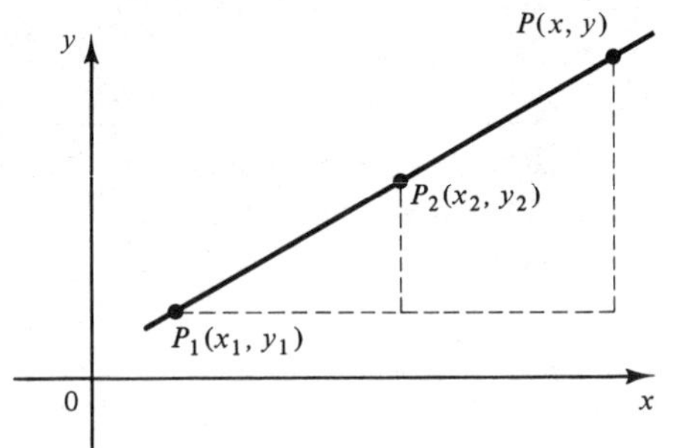

FIGURE 1-27

EXAMPLE 7: Write the equation in general form of the line passing through the points $(1, -3)$ and $(-2, 5)$ (Fig. 1-28).

Solution: Calling the first given point P_1 and the second P_2 and substituting into Eq. 254,

$$\frac{y-(-3)}{x-1}=\frac{5-(-3)}{-2-1}=\frac{8}{-3}$$

$$\frac{y+3}{x-1}=-\frac{8}{3}$$

Putting the equation into general form, $3y + 9 = -8x + 8$, or

$$8x + 3y + 1 = 0$$

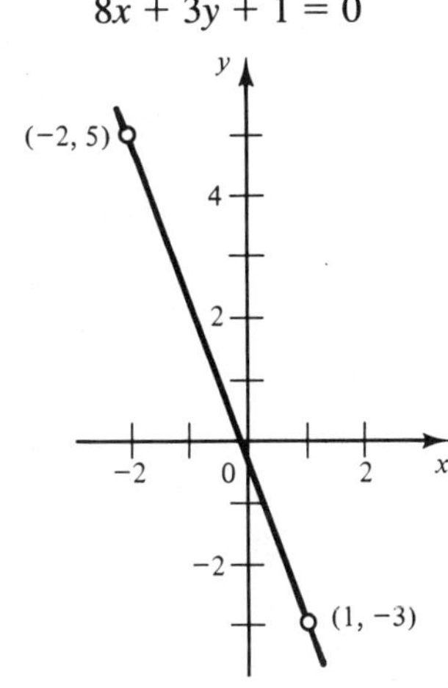

FIGURE 1-28

Lines Parallel to the Coordinate Axes

Line 1 in Fig. 1-29 is parallel to the x axis. Its slope is 0 and it cuts the y axis at $(0, b)$. From the point-slope form, $y - y_1 = m(x - x_1)$, we get $y - b = 0(x - 0)$, or

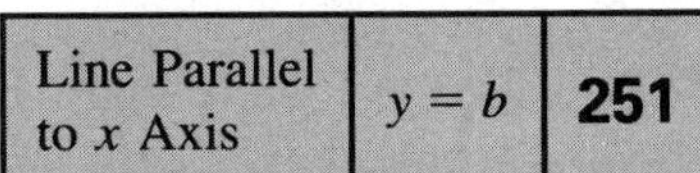

Line Parallel to x Axis	$y = b$	251

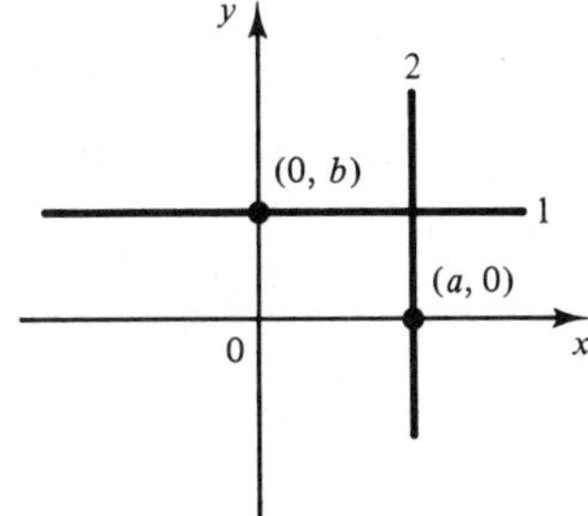

FIGURE 1-29

Line 2 has an undefined slope, but we get its equation by noting that $x = a$ at every point on the line, *regardless of the value of y*. Thus

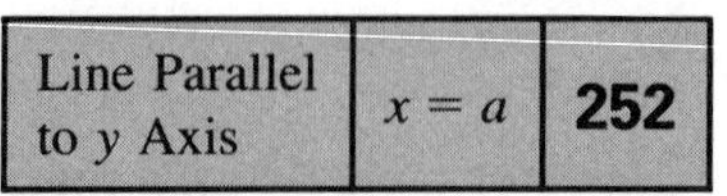

Line Parallel to y Axis	$x = a$	252

EXAMPLE 8
(a) A line that passes through the point (5, −2) and is parallel to the x axis has the equation $y = -2$.
(b) A line that passes through the point (5, −2) and is parallel to the y axis has the equation $x = 5$.

EXERCISE 3

Equation of a Straight Line

Find the slope and y intercept for each equation.

1. $y = 3x - 5$

2. $y = 7x + 2$

3. $y = -\frac{1}{2}x - \frac{1}{4}$

4. $y = -x + 4$

5. $y = 6$

6. $3x - 2y = 4$

7. $x + 2y - 3 = 0$

8. $3x - \frac{1}{2}y - 5 = 0$

9. $2(x + 3) - 1 = 3(y - 2)$

10. $\frac{2x}{3} - \frac{3y}{5} = 4$

Write the equation, in general form, of each line.

11. slope = 4; y intercept = −3

12. slope = −1; y intercept = 2

13. slope = 2.25; y intercept = −1.48

14. slope = r; y intercept = p

15. passes through points (3, 5) and (−1, 2)

16. passes through points (4.24, −1.25) and (3.85, 4.27)

17. slope = −4, passes through (−2, 5)

18. slope = −2, passes through (−2, −3)

19. x intercept = 5, y intercept = −3

20. x intercept = −2, y intercept = 6

21. y intercept = 3, parallel to $y = 5x - 2$

22. y intercept = −2.3, parallel to $2x - 3y + 1 = 0$

23. y intercept = −5, perpendicular to $y = 3x - 4$

24. y intercept = 2, perpendicular to $4x - 3y = 7$

25. passes through (−2, 5), parallel to $y = 5x - 1$

26. passes through (4, −1), parallel to $4x - y = -3$

27. passes through (−4, 2), perpendicular to $y = 5x - 3$

28. passes through (6, 1), perpendicular to $6y - 2x = 3$

29. y intercept = 4, angle of inclination = 48°

30. y intercept = −3.52, angle of inclination = 154°44′

31. passes through (4, −1), angle of inclination = 22.8°

32. passes through (−2.24, 5.17), angle of inclination = 68°14′

33. passes through (5, 2) and is parallel to the x axis.

34. passes through (−3, 6) and is parallel to the y axis.

35. line A in Fig. 1-30

36. line B in Fig. 1-30

37. line C in Fig. 1-30

38. line D in Fig. 1-30

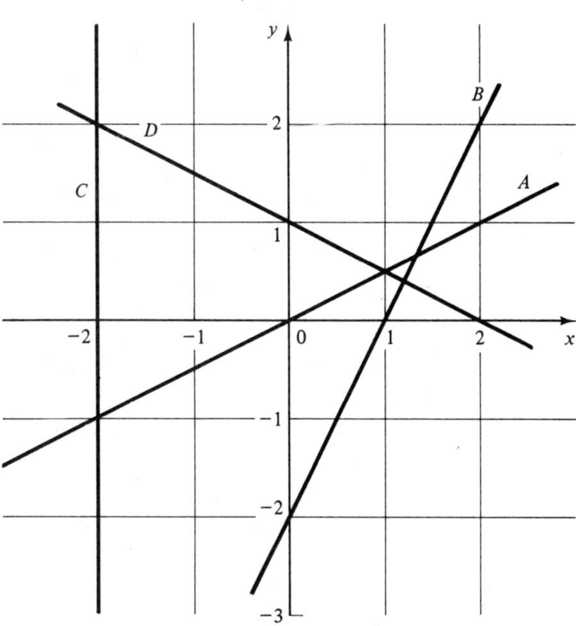

FIGURE 1-30

Resistance Change with Temperature

39. The resistance of metals is a linear function of the temperature (Fig. 1-31) for certain ranges of temperature. The slope of the line is $R_1\alpha$, where α is the temperature coefficient of resistance at temperature t_1. If R is the resistance of any temperature t, write an equation for R as a function of t.

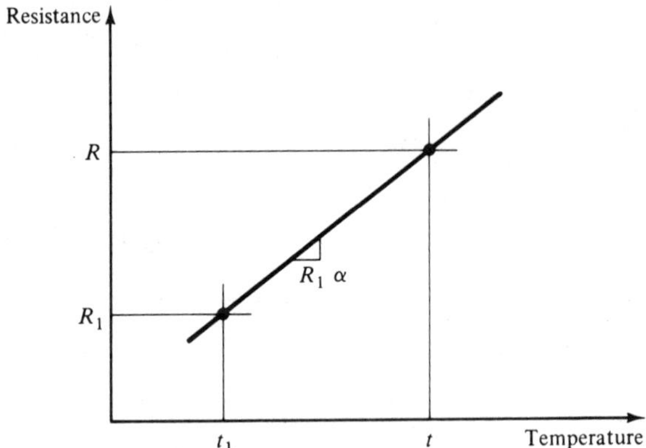

FIGURE 1-31. Resistance change with temperature.

40. Using a value for α of $\alpha = 1/234.5t_1$ (for copper), find the resistance of a copper conductor at 75.0°C if its resistance at 20.0°C is 148.4 Ω.

41. If the resistance of the copper conductor in Problem 40 is 1255 Ω at 20.0°C, at what temperature will the resistance be 1265 Ω?

Spring Constant

42. A spring whose length is L_0 with no force applied (Fig. 1-32) stretches an amount x with an applied force of F, where $F = kx$. The constant k is called the spring constant. Write an equation, in slope-intercept form, for F in terms of k, L, and L_0.

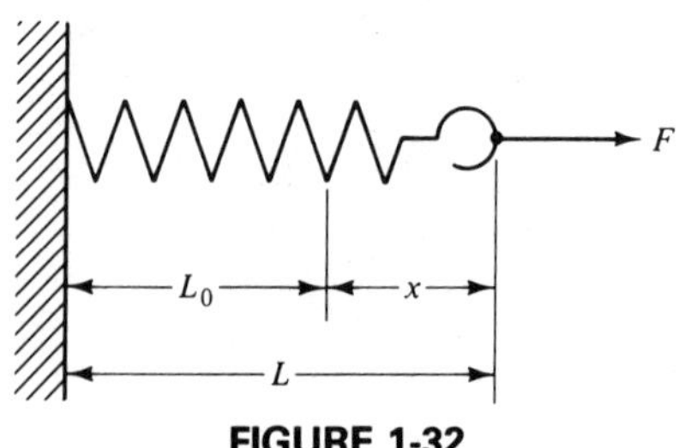

FIGURE 1-32

43. What force would be needed to stretch a spring ($k = 14.5$ lb/in.) from an unstretched length of 8.50 in. to a length of 12.50 in.?

Velocity of Uniformly Accelerated Body

44. When a body moves with constant acceleration a (such as in free fall), its velocity v at any time t is given by $v = v_0 + at$, where v_0 is the initial velocity. Note that this is the equation of a straight line. If a body has a constant acceleration of 2.15 m/s^2 and has a velocity of 21.8 m/s at 5.25 s, find (a) the initial velocity and (b) the velocity at 25.0 s.

Thermal Expansion

45. When a bar is heated, its length will increase from an initial length L_0 at temperature t_0 to a new length L at temperature t. The plot of L versus t is a straight line (Fig. 1-33) with a slope of $L_0\alpha$, where α is the coefficient of thermal expansion. Derive the equation $L = L_0(1 + \alpha\ \Delta t)$, where Δt is the change in temperature, $t - t_0$.

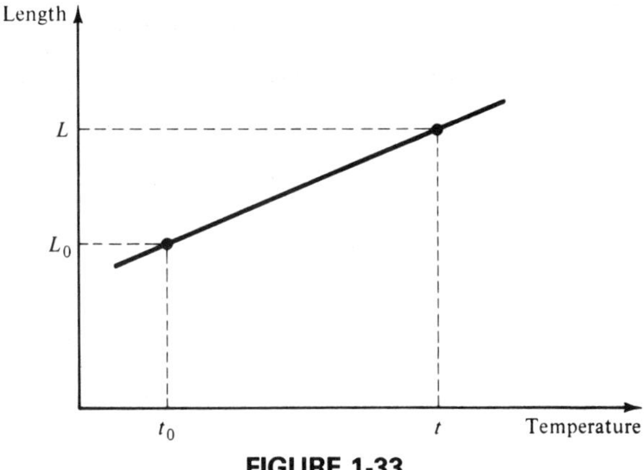

FIGURE 1-33

46. A steel pipe is 21.50 m long at 0°C. Find its length at 75.0°C if α for steel is 12.0×10^{-6} per Celsius degree.

Fluid Pressure

47. The pressure at a point located at a depth x ft from the surface of the liquid varies directly as the depth. If the pressure at the surface is 20.6 lb/in.^{2} and increases by 0.432 lb/in.^{2} for every foot of depth, write an equation for P as a function of the depth x (in feet). At what depth will the pressure be 30.0 lb/in.^{2}?

48. A straight pipe slopes downward from a reservoir to a water turbine (Fig. 1-34). The pressure head at any point in the pipe, expressed in feet, is equal to the vertical distance between the point and the surface of the reservoir. If the

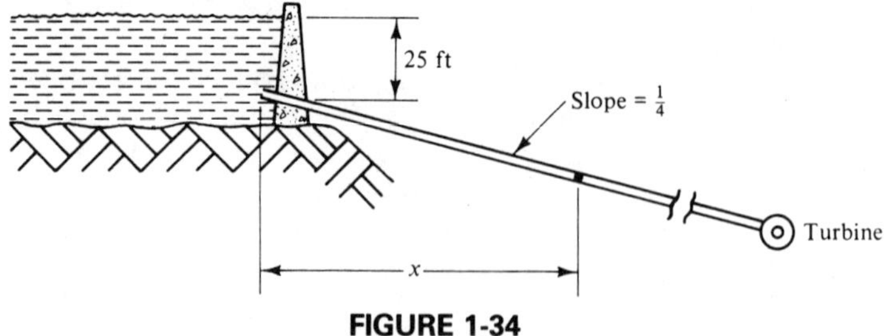

FIGURE 1-34

reservoir surface is 25 ft above the upper end of the pipe, write an expression for the head H as a function of the horizontal distance x. At what distance x will the head be 35 ft?

Temperature Gradient

49. Figure 1-35 shows a uniform wall whose inside face is at temperature t_i and whose outside face is at t_0. The temperatures within the wall plot as a straight line connecting t_i and t_o. Write the equation $t = f(x)$ of that line, taking $x = 0$ at the inside face, if $t_i = 25°C$ and $t_o = -5°C$. At what x will the temperature be 0°C? What is the slope of the line?

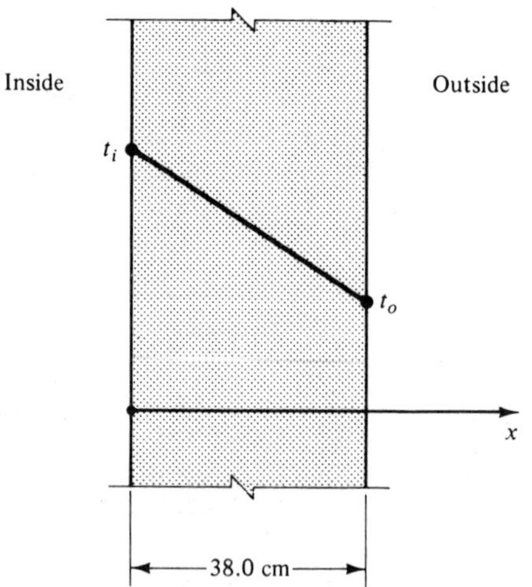

FIGURE 1-35. Temperature drop in a wall. The slope of the line (in °C/cm) is called the *temperature gradient.* The amount of heat flowing through the wall is proportional to the temperature gradient.

Hooke's Law

50. The increase in length of a wire in tension is directly proportional to the applied load P. Write an equation for the length L of a wire that has an initial length of 3.00 m and which stretches 1.00 mm for each 12.5 N. Find the length of the wire with a load of 750 N.

Computer

51. Write a program for the *false position method,* which is used to find the approximate roots to an equation $f(x) = 0$. We seek the value R at which the graph of $y = f(x)$ crosses the x axis (Fig. 1-36). We must make two initial guesses, x_1 and x_2, which lie on either side of R. Using these, we find the ordinates y_1 and y_2, from which we get the slope m of the line P_1P_2 from

$$m = \frac{y_2 - y_1}{x_2 - x_1} = \frac{y_2 - 0}{x_2 - x_3}$$

From this we compute the value x_3 at which a straight line from $P_1(x_1, y_2)$ to $P_2(x_2, y_2)$ would cross the x axis.

$$x_3 = x_2 - y_2\left(\frac{x_2 - x_1}{y_2 - y_1}\right)$$

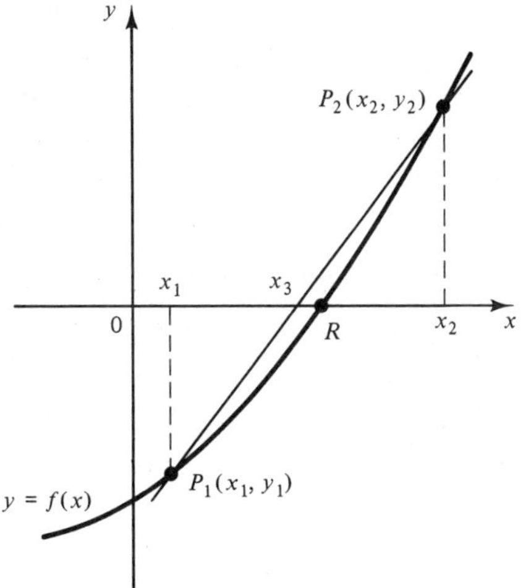

FIGURE 1-36. False position method.

This value, x_3, becomes our second approximation, replacing either x_1 or x_2, depending on which side of the root x_3 lies, and we repeat the computation until we get as close to R as we wish.

CHAPTER TEST

1. Find the distance between the points (3, 0) and (7,0).
2. Find the distance between the points (4, −4) and (1, −7).
3. Find the slope of the line perpendicular to a line that has an angle of inclination of 34.8°.
4. Find the angle of inclination in degrees of a line with a slope of −3.

5. Find the angle of inclination of a line perpendicular to a line having a slope of 1.55.
6. Find the angle of inclination of a line passing through $(-3, 5)$ and $(5, -6)$.
7. Find the slope of a line perpendicular to a line having a slope of $a/2b$.
8. Write the equation of a line having a slope of -2 and a y intercept of 5.
9. Find the slope and y intercept of the line $2y - 5 = 3(x - 4)$.
10. Write the equation of a line having a slope of $2p$ and a y intercept of $p - 3q$.
11. Write the equation of the line passing through $(-5, -1)$ and $(-2, 6)$.
12. Write the equation of the line passing through $(-r, s)$ and $(2r, -s)$.
13. Write the equation of the line having a slope of 5 and passing through the point $(-4, 7)$.
14. Write the equation of the line having a slope of $3c$ and passing through the point $(2c, c - 1)$.
15. Write the equation of the line having an x intercept of -3 and a y intercept of 7.
16. Find the acute angle between two lines, if one line has a slope of 1.5, and the other has a slope of 3.4.
17. Write the equation of the line that passes through $(2, 5)$ and is parallel to the x axis.
18. Find the angle of intersection between line 1 having a slope of 2 and line 2 having a slope of 7.
19. Find the directed distance AB between the points $A(-2, 0)$ and $B(-5, 0)$.
20. Find the angle of intersection between line 1 having an angle of inclination of 18° and line 2 having an angle of inclination of 75°.
21. Write the equation of the line that passes through $(-3, 6)$ and is parallel to the y axis.
22. Find the increments in the coordinates of a particle that moves along a curve from $(3, 4)$ to $(5, 5)$.
23. Find the area of a triangle with vertices at $(6, 4)$, $(5, -2)$, and $(-3, -4)$.

In Problems 24–33, a tangent T, of slope m, and a normal N are drawn to a curve at the point $P(x_1, y_1)$ (Fig. 1-37). Show the following.

24. The equation of the tangent is $y - y_1 = m(x - x_1)$.
25. The equation of the normal is $x - x_1 + m(y - y_1) = 0$.

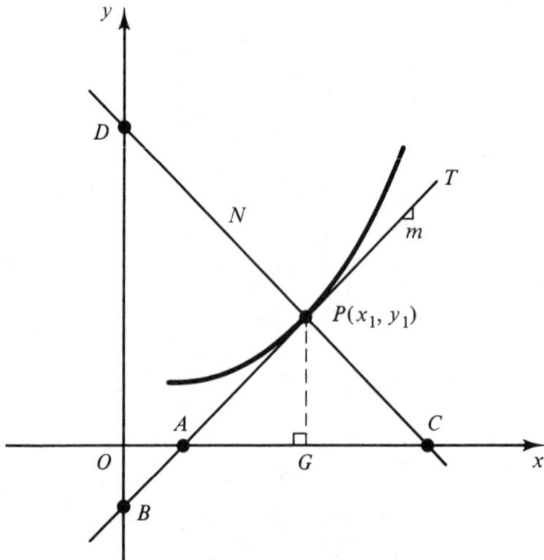

FIGURE 1-37. Tangent and normal to a curve.

26. The x intercept OA of the tangent is $x_1 - y_1/m$.

27. The y intercept OB of the tangent is $y_1 - mx_1$.

28. The length of the tangent from P to the x axis is

$$PA = \frac{y_1}{m}\sqrt{1 + m^2}$$

29. The length of the tangent from P to the y axis is

$$PB = x_1\sqrt{1 + m^2}$$

30. The x intercept OC of the normal is $x_1 + my_1$.

31. The y intercept OD of the normal is $y_1 + x_1/m$.

32. The length of the normal from P to the x axis is

$$PC = y_1\sqrt{1 + m^2}$$

33. The length of the normal from P to the y axis is

$$PD = \frac{x_1}{m}\sqrt{1 + m^2}$$

The Conic Sections

The extremely useful curves we study in this chapter are called *conic sections,* or just *conics,* because each can be obtained by passing a plane through a right circular cone, as in Fig. 2-1.

When the plane is perpendicular to the cone's axis, it intercepts a *circle*. When the plane is tilted a bit, but not so much as to be parallel to an element (a line on the cone that passes through the vertex) of the cone, we get an *ellipse*. When the plane is parallel to an element of the cone, we get a *parabola;* and when the plane is parallel to the cone's

You might want to try making these shapes with modeling clay or damp sand. Also, what position of the plane would give a straight line? A point?

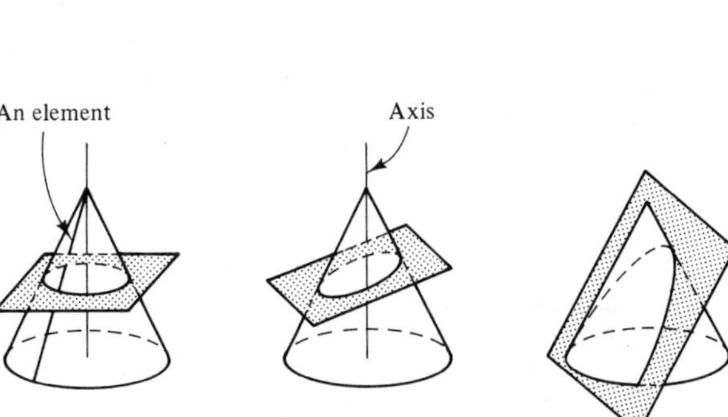

FIGURE 2.1. Conic sections.

axis, we get the two-branched *hyperbola*. In this section we will write equations for each of these curves, and use the equations to solve some interesting problems.

2-1. CIRCLE

Definition	A *circle* is a plane curve all points of which are at a fixed distance (the *radius*) from a fixed point (the *center*).	267

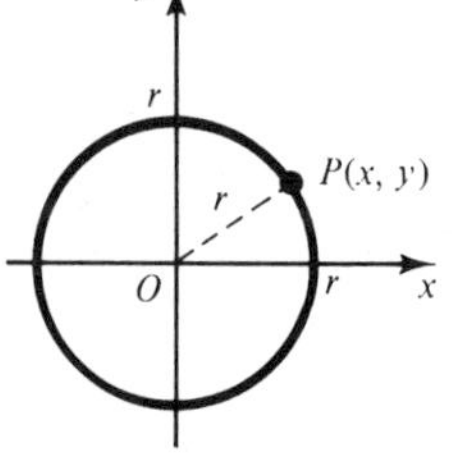

FIGURE 2-2. Circle with center at origin.

Standard Equation of a Circle

Let us write the equation of a circle of radius r whose center is at the origin (Fig. 2-2). Let x and y be the coordinates of any point P on the circle. Then by the definition of a circle, the distance OP must be constant and equal to r. But by the distance formula (Eq. 247),

$$OP = \sqrt{x^2 + y^2} = r$$

Squaring, we get

Note that both x^2 and y^2 have the same coefficient. Otherwise, it is not a circle.

Standard Equation, Circle of Radius r: Center at Origin	$x^2 + y^2 = r^2$	268

EXAMPLE 1: The equation of a circle of radius 3, whose center is at the origin is

$$x^2 + y^2 = 3^2 = 9$$

Circle with Center Not at Origin

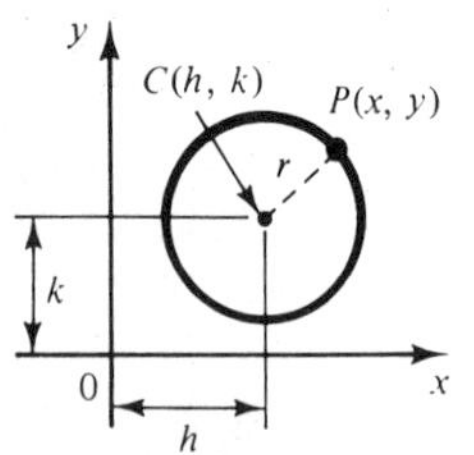

FIGURE 2-3. Circle with center at (h, k).

Figure 2-3 shows a circle whose center has the coordinates (h, k). We can think of the difference between this circle and the one in Fig. 2-2 as having its center moved or *translated* h units to the right and k units upward. Our derivation is similar to the preceding one.

$$CP = r = \sqrt{(x - h)^2 + (y - k)^2}$$

Squaring, we get

Standard Equation, Circle of Radius r Center at (h, k)	$(x - h)^2 + (y - k)^2 = r^2$	269

EXAMPLE 2: Write the equation of a circle of radius 5 whose center is at (3, −2).

Solution: (We substitute into Eq. 269 with $r = 5$, $h = 3$, and $k = -2$:

$$(x-3)^2 + [y-(-2)]^2 = 5^2$$
$$(x-3)^2 + (y+2)^2 = 25$$

EXAMPLE 3: Find the radius and the coordinates of the center of the circle $(x+5)^2 + (y-3)^2 = 16$.

Solution: We see that $r^2 = 16$, so the radius is $r = 4$. Also, since

$$x - h = x + 5$$
$$h = -5$$

and since

$$y - k = y - 3$$
$$k = 3$$

so the center is at (−5, 3).

Common Error	It is easy to get the signs of h and k wrong. In the preceding example, do *not* take $h = +5$ and $k = -3$

Translation of Axes

We see that Eq. 269 for a circle with its center at (h, k) is almost identical with Eq. 268 for a circle with center at the origin, except that *x has been replaced by* $(x - h)$ *and y has been replaced by* $(y - k)$. This same substitution will, of course, work for curves other than the circle, and we will use it to translate or *shift* the axes for the other conic sections.

Translation of Axes	To translate or shift the axes of a curve to the left by a distance h and downward by a distance k, replace x by $(x - h)$ and y by $(y - k)$ in the equation of the curve.	**262**

The General Second-Degree Equation

A second-degree equation in x and y which has all possible terms would have an x^2 term, a y^2 term, an xy term, and all terms of lesser degree as well. They are usually written in the following order:

There are six constants in this equation, but only five are independent. We can divide through by any constant and thus eliminate it.

General Second-Degree Equation	$Ax^2 + Bxy + Cy^2 + Dx + Ey + F = 0$	**261**

where A, B, C, D, E, and F are constants.

General Form of the Equation of a Circle

If we expand Eq. 269 we get

$$(x - h)^2 + (y - k)^2 = r^2$$

$$x^2 - 2hx + h^2 + y^2 - 2ky + k^2 = r^2$$

Rearranging gives $x^2 + y^2 - 2hx - 2ky + (h^2 + k^2 - r^2) = 0$, or

General Equation of a Circle	$x^2 + y^2 + Dx + Ey + F = 0$	**270**

where D, E, and F are constants. Comparing this with the general second-degree equation, we see that the general second-degree equation

$$Ax^2 + Bxy + Cy^2 + Dx + Ey + F = 0$$

represents a circle if $B = 0$, and $A = C$.

EXAMPLE 4: Write the equation of Example 3 in general form.

Solution: The equation, in standard form, was

$$(x + 5)^2 + (y - 3)^2 = 16$$

Expanding, we obtain

$$x^2 + 10x + 25 + y^2 - 6y + 9 = 16$$

$$x^2 + y^2 + 10x - 6y + 25 + 9 - 16 = 0$$

or

$$x^2 + y^2 + 10x - 6y + 18 = 0.$$

Changing from General to Standard Form

When we want to go from general form to standard form, we must *complete the square,* both for x and for y.

EXAMPLE 5: Write the equation $2x^2 + 2y^2 - 18x + 16y + 60 = 0$ in standard form. Find the radius and center, and plot the curve.

Solution: We first divide by 2:

$$x^2 + y^2 - 9x + 8y + 30 = 0$$

and separate the x and y terms:

$$(x^2 - 9x \quad) + (y^2 + 8y \quad) = -30$$

Completing the square on the left and compensating on the right gives

$$\left(x^2 - 9x + \frac{81}{4}\right) + (y^2 + 8y + 16) = -30 + \frac{81}{4} + 16$$

$$\left(x - \frac{9}{2}\right)^2 + (y + 4)^2 = \left(\frac{5}{2}\right)^2$$

So

$$r = \frac{5}{2}, \qquad h = \frac{9}{2}, \qquad k = -4$$

The circle is shown in Fig. 2-4.

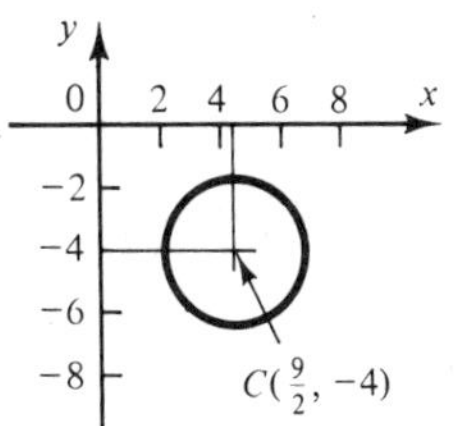

FIGURE 2-4

Equation of a Circle Found from Three Conditions

Since the equation of a circle either in standard form or in general form has three arbitrary constants, we must have three independent pieces of information in order to write the equation. The given information is substituted into whichever of the two forms of the circle equation is most appropriate, and the equations are solved simultaneously.

EXAMPLE 6: Write the equation of the circle passing through the points (4, 6), (−2, −2), and (−4, 2).

Solution: Since each of these points is on the circle, each must satisfy the general equation of the circle (Eq. 270). So we substitute the coordinates of each in turn into Eq. (270) and get three equations in three unknowns.

At (4, 6), $16 + 36 + 4D + 6E + F = 0$

At (−2, −2), $4 + 4 - 2D - 2E + F = 0$

and at (−4, 2), $16 + 4 - 4D + 2E + F = 0$

Collecting terms,

$$4D + 6E + F = -52 \quad (1)$$

$$-2D - 2E + F = -8 \quad (2)$$

$$-4D + 2E + F = -20 \quad (3)$$

Subtracting (2) from (1) gives

$$6D + 8E = -44 \quad (4)$$

and subtracting (3) from (2) gives

$$2D - 4E = 12 \quad (5)$$

Now multiplying (5) by two and adding to (4), we get

$$10D = -20$$

$$D = -2$$

Substituting back gives $E = -4$ and $F = -20$, so our equation is

$$x^2 + y^2 - 2x - 4y - 20 = 0$$

If you now complete the square, you will find that this is a circle whose center is at (1, 2) with a radius of 5.

EXERCISE 1

Equation of a Circle

Write the equation of each circle in standard form.

1. center at (0,0), radius = 7

2. center at (0, 0), radius = 4.82

3. center at (2, 3), radius = 5

4. center at (5, 0), diameter = 10

5. center at (5, −3), radius = 4

6. center at (0, −2), radius = 11

Find the center and radius of each circle.

7. $x^2 + y^2 = 49$

8. $x^2 + y^2 = 64.8$

9. $(x - 2)^2 + (y + 4)^2 = 16$

10. $(x + 5)^2 + (y - 2)^2 = 49$

11. $(y + 5)^2 + (x - 3)^2 = 36$

12. $(x - 2.22)^2 + (y + 7.16)^2 = 5.93$

13. $x^2 + y^2 - 8x = 0$

14. $x^2 + y^2 - 2x - 4y = 0$

15. $x^2 + y^2 - 10x + 12y + 25 = 0$

Write the equation of each circle in general form.

16. passes through $(0, 0)$, $(0, -1)$, and $(-6, -1)$

17. passes through $(1, 3)$, $(1, 2)$, and $(2, 5)$

18. center at $(1, 1)$, passes through $(4, -3)$

19. whose diameter joins the points $(-2, 5)$ and $(6, -1)$

20. passes through origin, x intercept $= 4$, and y intercept $= 6$

Tangent to a Circle

Write the equation of the tangent to each circle at the given point. (*Hint:* The slope of the tangent is the negative reciprocal of the slope of the radius to the given point.)

21. $x^2 + y^2 = 25$ at $(4, 3)$

22. $(x - 5)^2 + (y - 6)^2 = 100$ at $(11, 14)$

23. $x^2 + y^2 + 10y = 0$ at $(-4, -2)$

Intercepts

Find the x and y intercepts for each circle. (*Hint:* Set x and y, in turn, equal to zero to find the intercepts.)

24. $x^2 + y^2 - 6x + 4y + 4 = 0$

25. $x^2 + y^2 - 5x - 7y + 6 = 0$

Intersecting Circles

Find the point(s) of intersection. (*Hint:* Solve each pair of equations simultaneously.)

26. $x^2 + y^2 - 10x = 0$ and $x^2 + y^2 + 2x - 6y = 0$

27. $x^2 + y^2 - 3y - 4 = 0$ and $x^2 + y^2 + 2x - 5y - 2 = 0$

28. $x^2 + y^2 + 2x - 6y + 2 = 0$ and $x^2 + y^2 - 4x + 2 = 0$

Applications

29. Prove that any angle inscribed in a semicircle is a right angle.

30. Write the equation of the circle in Fig. 2-5, taking the axes as shown. Use your equation to find A and B.

Even though you may be able to solve some of these with only the Pythagorean theorem, we suggest that you use analytic geometry for the practice.

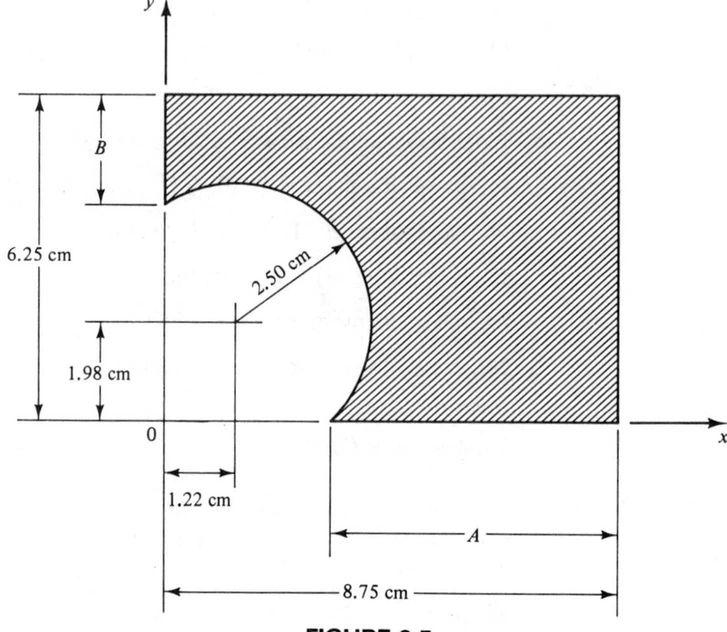

FIGURE 2-5

31. Write the equations for each of the circular arches in Fig. 2-6, taking the axes as shown. Solve simultaneously to get the point of intersection P, and compute the height h of the column.

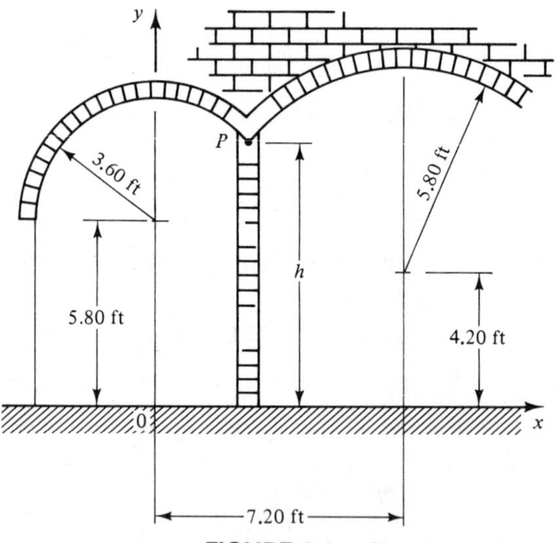

FIGURE 2-6. Circular arches.

32. Write the equation of the centerline of the circular street shown in Fig. 2-7 taking the origin at the intersection 0. Use your equation to find the distance y.

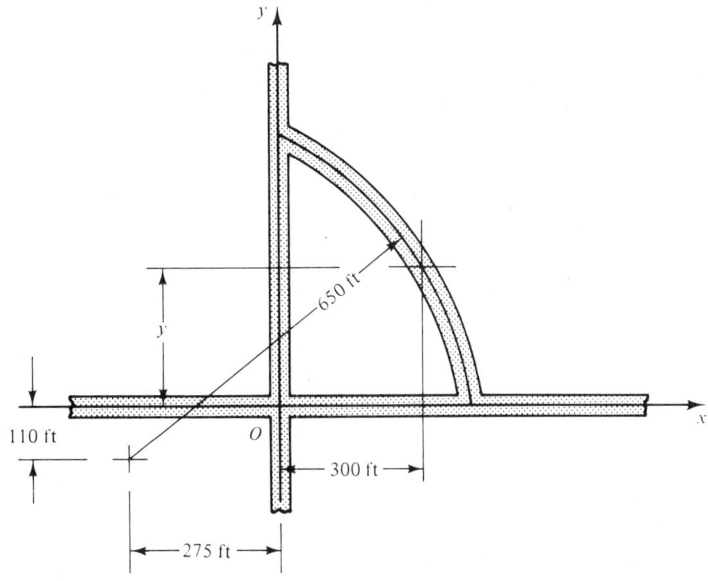

FIGURE 2-7. Circular street.

33. Write the equation of one side of the Gothic arch shown in Fig. 2-8. Use the equation to find the width w of the arch at a height of 3.0 ft.

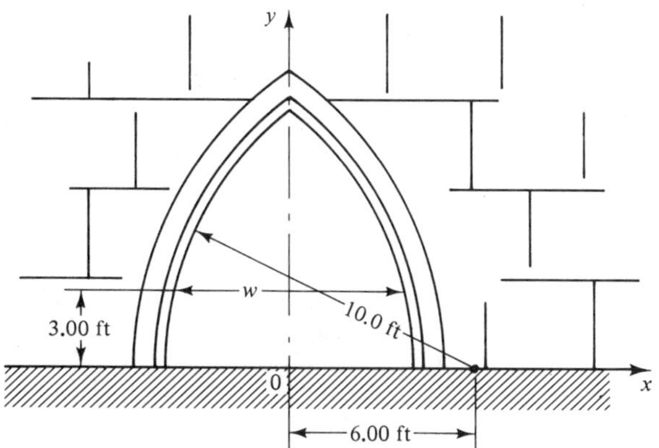

FIGURE 2-8. Gothic arch.

2-2 PARABOLA

Definition	A *parabola* is the set of points in a plane each of which is equidistant from a fixed point, the *focus*, and a fixed line, the *directrix*.	271

Such a set of points, connected with a smooth curve, is plotted in Fig. 2-9, and shows the typical shape of the parabola. The parabola has an

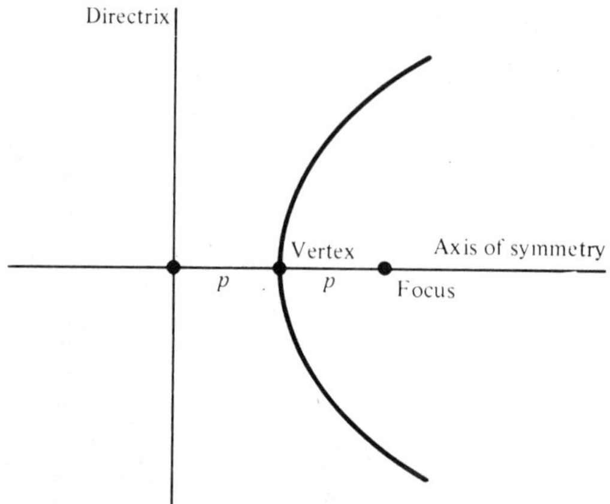

FIGURE 2-9. Parabola.

axis of symmetry which it intersects at the *vertex*. The distance p from directrix to vertex is equal to the directed distance from the vertex to the focus.

Standard Equation of the Parabola

Let us place the parabola on coordinate axes with the vertex at the origin (Fig. 2-10) and with the axis of symmetry along the x axis. Choose any

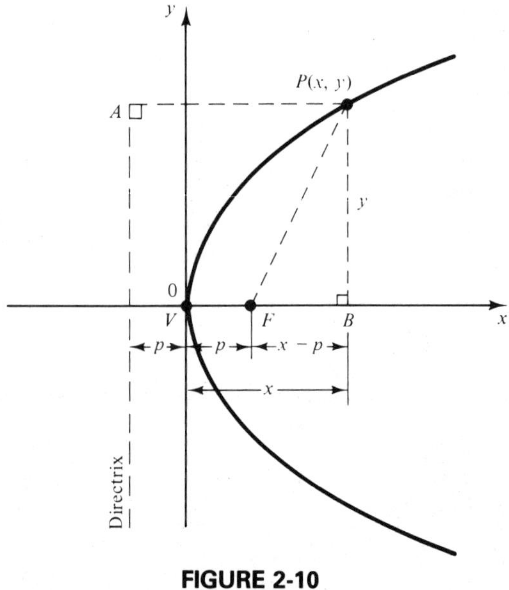

FIGURE 2-10

point P on the parabola. Then by the definition of a parabola, $FP = AP$. But in right triangle FBP,

$$FP = \sqrt{(x - p)^2 + y^2}$$

and also,

$$AP = p + x$$

But since $FP = AP$,

$$\sqrt{(x - p)^2 + y^2} = p + x$$

Squaring both sides yields

$$(x - p)^2 + y^2 = p^2 + 2px + x^2$$
$$x^2 - 2px + p^2 + y^2 = p^2 + 2px + x^2$$

Collecting terms, we get the standard equation of a parabola:

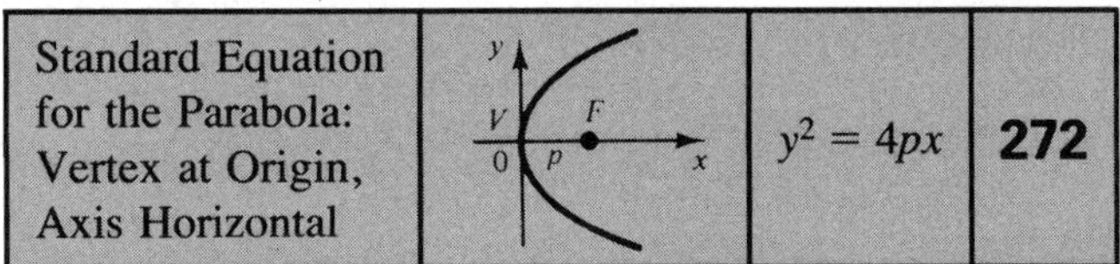

We have defined p as the directed distant VF from the vertex V to the focus F. Thus is p is positive, the focus must lie to the right of the vertex, and hence the parabola opens to the right. Conversely, if p is negative, the parabola opens to the left.

EXAMPLE 1: Find the coordinates of the focus of the parabola $2y^2 + 7x = 0$.

Solution: Subtracting $7x$ from both sides and dividing by 2,

$$y^2 = -3.5x$$

Thus $4p = -3.5$ and $p = -0.875$. Since p is negative, the parabola opens to the left. The focus is thus 0.875 unit to the left of the vertex (Fig. 2-11) and so has the coordinates $(-0.875, 0)$.

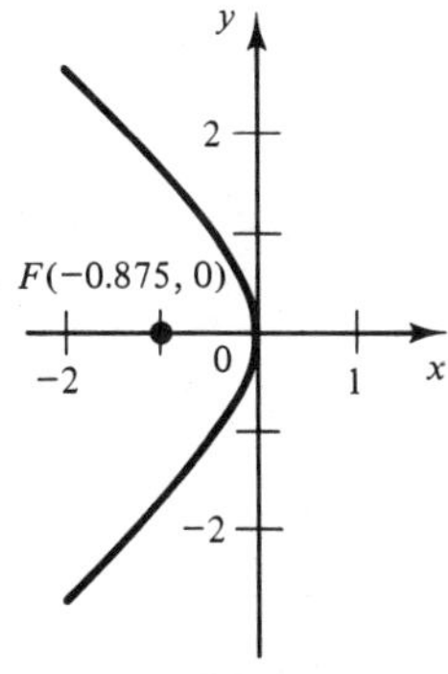

FIGURE 2-11

EXAMPLE 2: Find the coordinates of the focus and write the equation of a parabola whose vertex is at the origin, that has a horizontal axis of symmetry, and that passes through the point $(5, -6)$.

Solution: Our given point must satisfy the equation $y^2 = 4px$. Substituting 5 for x and -6 for y gives

$$(-6)^2 = 4p(5)$$

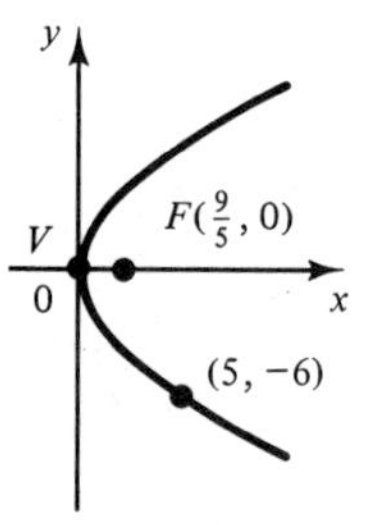

FIGURE 2-12

So $4p = \frac{36}{5}$ and $p = \frac{9}{5}$. The equation of the parabola is then $y^2 = 36x/5$, or

$$5y^2 = 36x$$

Since the axis is horizontal and p is positive, the focus must be on the x axis and is a distance p ($\frac{9}{5}$) to the right of the origin. The coordinates of F are thus ($\frac{9}{5}$, 0) (Fig. 2-12).

Parabola with Vertical Axis

The standard equation for a parabola having a vertical axis is obtained by switching the positions of x and y in Eq. 272,

Note that only one variable is squared in any parabola equation. This gives us the best way to recognize such equations.

Standard Equation for the Parabola: Vertex at Origin, Axis Vertical	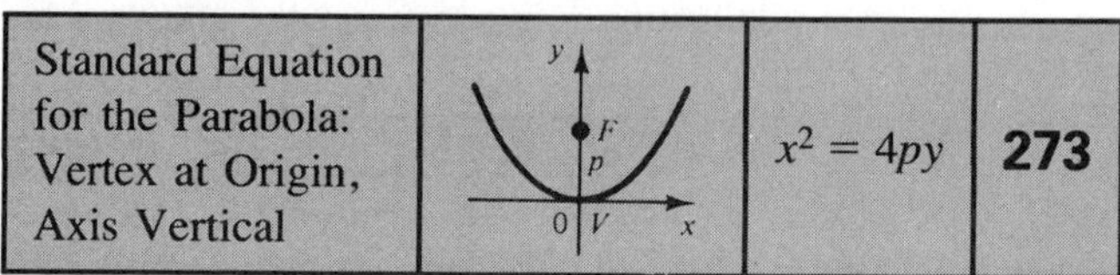	$x^2 = 4py$	**273**

When p is positive, the parabola opens upward; when p is negative, it opens downward.

EXAMPLE 3: A parabola has its vertex at the origin and passes through the points (3, 2) and (−3, 2). Write its equation and find the focus.

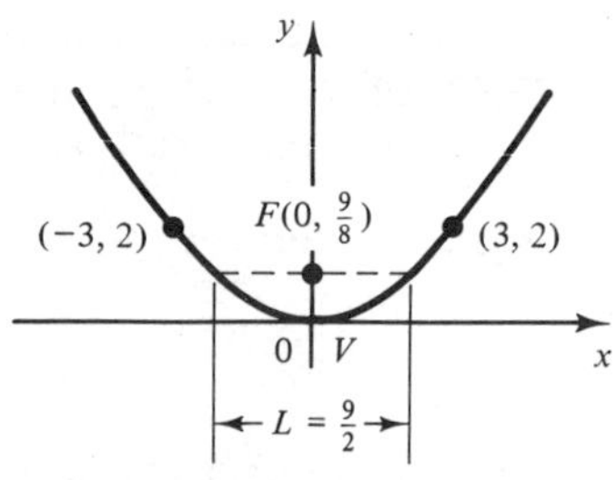

FIGURE 2-13

Solution: A sketch (Fig. 2-13) shows that the axis must be vertical, so our equation is of the form $x^2 = 4py$. Substituting 3 for x and 2 for y gives

$$3^2 = 4p(2)$$

from which $4p = \frac{9}{2}$ and $p = \frac{9}{8}$. Our equation is then $x^2 = 9y/2$ or $2x^2 = 9y$. The focus is on the y axis at a distance $p = \frac{9}{8}$ from the origin, so its coordinates are $(0, \frac{9}{8})$.

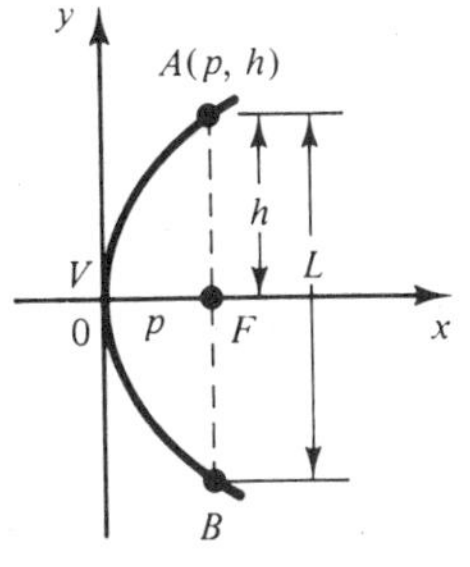

FIGURE 2-14

Length of the Latus Rectum

The *latus rectum* of a parabola is a line through the focus which is perpendicular to the axis of symmetry, such as line AB in Fig. 2-14. We will find the length L of the latus rectum by substituting the coordinates (p, h) of point A into Eq. 272:

$$h^2 = 4p(p) = 4p^2$$

$$h = \pm 2p$$

The length of the latus rectum is also called the *focal width*. It is useful for making a quick sketch of the parabola.

The length L of the latus rectum is twice h, so

Parabola; Length of Latus Rectum	$L = \lvert 4p \rvert$	**277**

The latus rectum has a length four times the distance from vertex to focus.

EXAMPLE 4: The focal width for the parabola in Fig. 2-13 is $\frac{9}{2}$, or 4.5 units.

Shift of Axes

As with the circle, when the vertex of the parabola is not at the origin but at (h, k), our equations will be similar to Eqs. 272 and 273 except that x is replaced with $x - h$ and y is replaced with $y - k$.

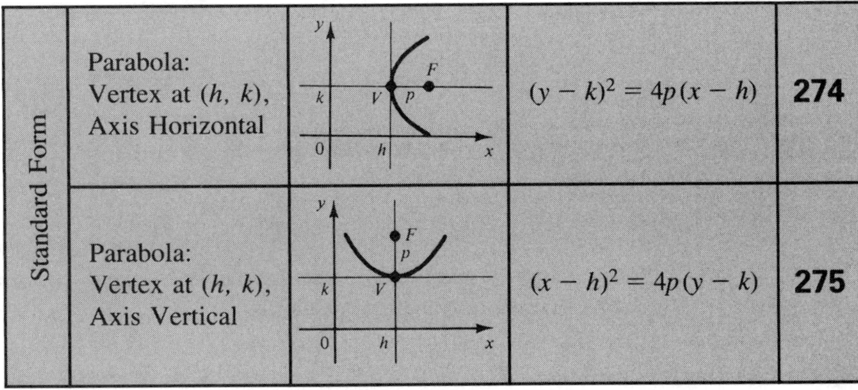

Standard Form				
	Parabola: Vertex at (h, k), Axis Horizontal	y, F, k, V, p, 0, h, x	$(y - k)^2 = 4p(x - h)$	**274**
	Parabola: Vertex at (h, k), Axis Vertical	y, F, p, k, V, 0, h, x	$(x - h)^2 = 4p(y - k)$	**275**

EXAMPLE 5: Find the vertex, focus, focal width, and equation of the axis for the parabola $(y - 3)^2 = 8(x + 2)$.

Solution The given equation is of the same form as Eq. 274, so the axis is horizontal. Also, $h = -2$, $k = 3$, and $4p = 8$. So the vertex is at $V(-2, 3)$ (Fig. 2-15). Since $p = \frac{8}{4} = 2$, the focus is 2 units to the right of the vertex, at $F(0, 3)$. The focal width is $4p$, so $L = 8$. The axis is horizontal and 3 units from the x axis, so its equation is $y = 3$.

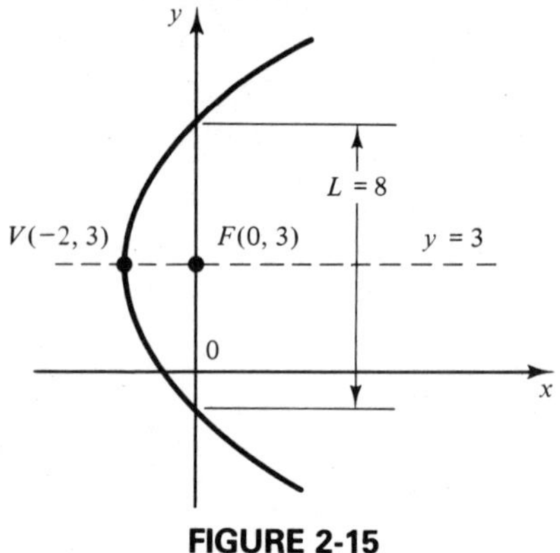

FIGURE 2-15

EXAMPLE 6: Write the equation of a parabola that opens upward, with vertex at $(-1, 2)$, and which passes through the point $(1, 3)$ (Fig. 2-16). Find the focus and the focal width.

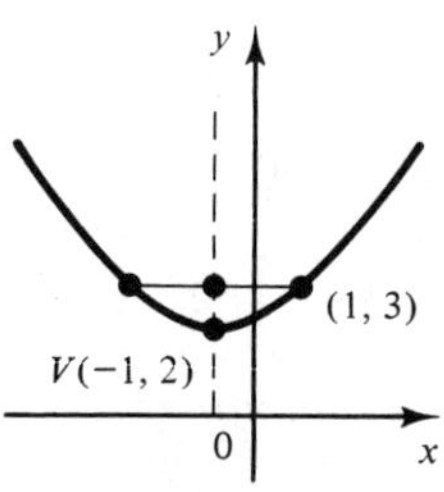

FIGURE 2-16

Solution: We substitute $h = -1$ and $k = 2$ into Eq. 275,

$$(x + 1)^2 = 4p(y - 2)$$

Now, since $(1, 3)$ is on the parabola, these coordinates must satisfy our equation. Substituting yields

$$(1 + 1)^2 = 4p(3 - 2)$$

Solving for p, we obtain $2^2 = 4p$, or, $p = 1$. So the equation is

$$(x + 1)^2 = 4(y - 2)$$

The focus is p units above the vertex, at $(-1, 3)$. The length of the latus rectum is, by Eq. 277,

$$L = |4p| = 4(1) = 4 \text{ units}$$

General Equation of the Parabola

We get the general equation of the parabola by expanding the standard equation (Eq. 274)

$$(y - k)^2 = 4p(x - h)$$

$$y^2 - 2ky + k^2 = 4px - 4ph$$

or

$$y^2 - 4px - 2ky + (k^2 + 4ph) = 0$$

which is of the form

Compare this with the general second-degree equation.

General Equation of a Parabola with Horizontal Axis	$Cy^2 + Dx + Ey + F = 0$	**276**

We see that the equation of a parabola having a horizontal axis of symmetry has a y^2 term but no x^2 term. Conversely, the equation for a parabola with vertical axis has an x^2 term but not a y^2 term. The parabola is the only conic for which this is true.

If the coefficient B of the xy term in the general second-degree equation were not zero, it would indicate that the axis of symmetry was *rotated* by some amount, and is no longer parallel to a coordinate axis. The presence of an xy term indicates rotation of the ellipse and hyperbola as well.

Completing the Square

As with the circle, we go from general to standard form by completing the square.

EXAMPLE 7: Find the vertex, focus, and latus rectum for the parabola

$$x^2 + 6x + 8y + 1 = 0$$

Solution: Separating the x and y terms, we have

$$x^2 + 6x = -8y - 1$$

Completing the square, we obtain

$$x^2 + 6x + 9 = -8y - 1 + 9$$

Factoring yields

$$(x + 3)^2 = -8(y - 1)$$

which is the form of Eq. 275, with $h = -3$, $k = 1$, and $p = -2$.

The vertex is $(-3, 1)$. Since the parabola opens downward (Fig. 2-17), the focus is 2 units below the vertex, at $(-3, -1)$. The length of the latus rectum is $4p$ or 8 units.

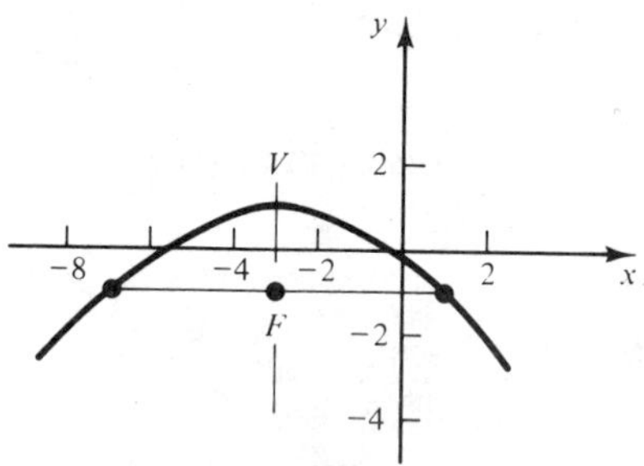

FIGURE 2-17

EXERCISE 2

Parabola with Vertex at Origin

Find the focus and focal width of each parabola.

1. $y^2 = 8x$

2. $x^2 = 16y$

3. $7x^2 + 12y = 0$

4. $3y^2 + 5x = 0$

Write the equation of each parabola.

5. focus at $(0, -2)$

6. passes through $(25, 20)$, axis horizontal

7. passes through $(3, 2)$ and $(3, -2)$

8. passes through $(3, 4)$ and $(-3, 4)$

Parabola with Vertex Not at Origin

Find the vertex, focus, the focal width, and the equation of the axis for each parabola.

9. $(y - 5)^2 = 12(x - 3)$

10. $(x + 2)^2 = 16(y - 6)$

11. $3x + 2y^2 + 4y - 4 = 0$

12. $y^2 + 8y + 4 - 6x = 0$

13. $y - 3x + x^2 + 1 = 0$

14. $x^2 + 4x - y - 6 = 0$

Write the equation, in general form, for each parabola.

15. vertex at (1, 2), $L = 8$, axis is $y = 2$, opens to the right

16. axis is $y = 3$, passes through (6, −1) and (3, 1)

17. passes through (−3, 3), (−6, 5), and (−11, 7), axis horizontal

18. vertex (0, 2), axis is $x = 0$, passes through (−4, −2)

19. axis is $y = -1$, passes through (−4, −2) and (2, 1)

Construction of a Parabola

20. Use the definition of the parabola as the set of points equidistant from a fixed point and a fixed line, to construct a parabola. Let the distance between the focus and the directrix be 2.0 in. (Fig. 2-18). Then draw a line L parallel to the directrix at some arbitrary distance, say 3.0 in. Then with that same (3.0 in.) distance as radius and F as center, use a compass to draw arcs intersecting L at P_1 and P_2. Each of these points is now at the same distance (3.0 in.) from F and from the directrix, and are hence points on the parabola. Repeat the construction with distances other than 3.0 in. to get more points on the parabola.

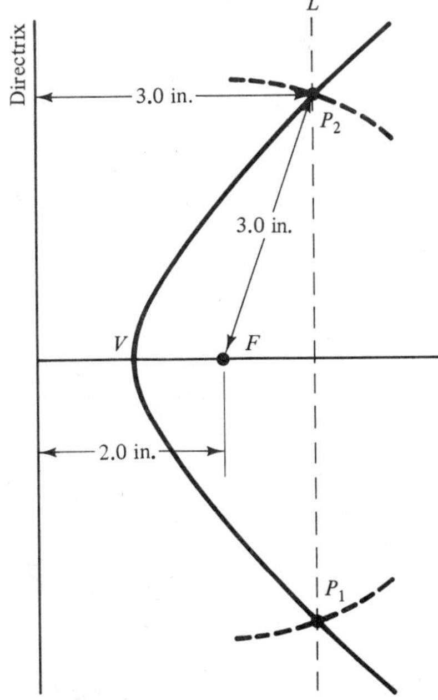

FIGURE 2-18. Construction of a parabola.

Trajectories

21. A ball thrown into the air (Fig. 2-19), will, neglecting air resistance, follow a parabolic path. Write the equation of the path, taking axes as shown. Use your equation to find the height of the ball when it is at a horizontal distance of 95 ft from O.

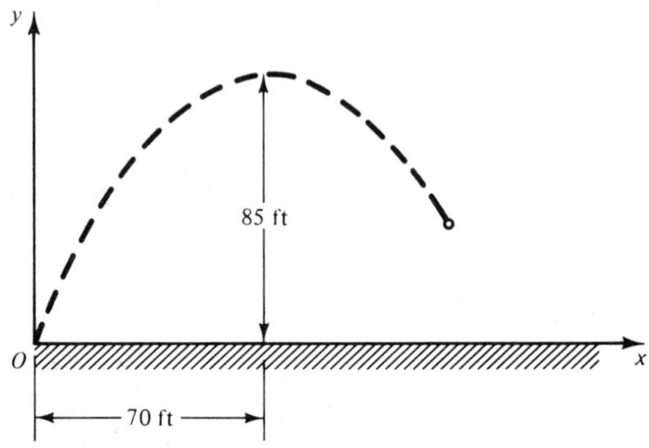

FIGURE 2-19. Ball thrown into the air.

22. Some comets follow a parabolic orbit with the sun at the focal point (Fig. 2-20). Taking axes as shown, write the equation of the path if the distance p is 75 million kilometers.

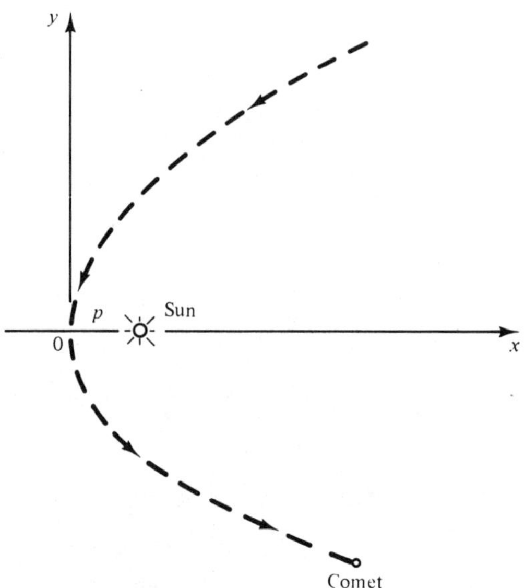

FIGURE 2-20. Path of a comet.

23. An object dropped from a moving aircraft (Fig. 2-21) will follow a parabolic path if air resistance is negligible. A weather instrument released at a height of 3520 m is observed to strike the water at a distance of 2150 m from the point of release. Write the equation of the path, taking axes as shown. Find the height of the instrument when x is 1000 m.

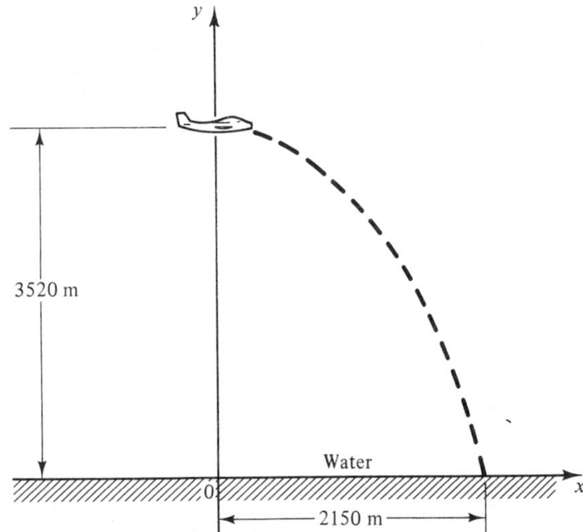

FIGURE 2-21. Object dropped from an aircraft.

Parabolic Arch

24. A 10-ft-high truck passes under a parabolic arch (Fig. 2-22). Find the maximum distance x that the side of the truck can be from the center of the road.

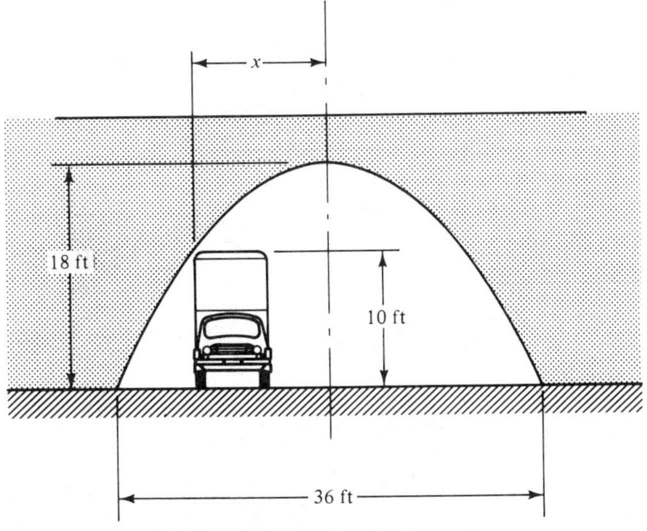

FIGURE 2-22. Parabolic arch.

A cable will hang in the shape of a parabola if its weight per horizontal foot is constant.

25. Assuming the bridge cable AB (Fig. 2-23) to be a parabola, write its equation, taking axes as shown.

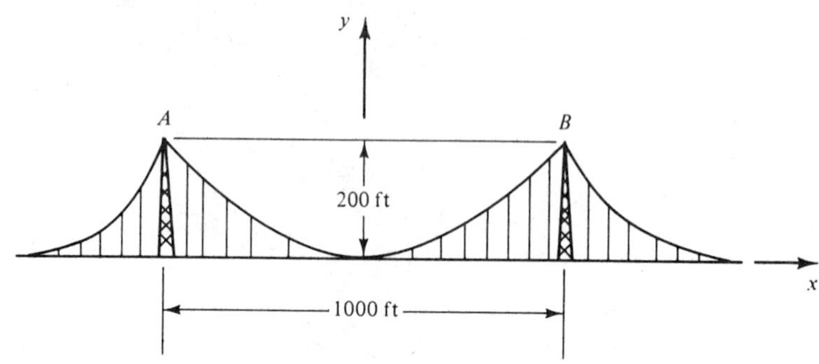

FIGURE 2-23. Parabolic bridge cable.

26. A parabolic arch supports a roadway (Fig. 2-24). Write the equation of the arch, taking axes as shown. Use your equation to find the vertical distance from the roadway to the arch at a horizontal distance of 50 m from the center.

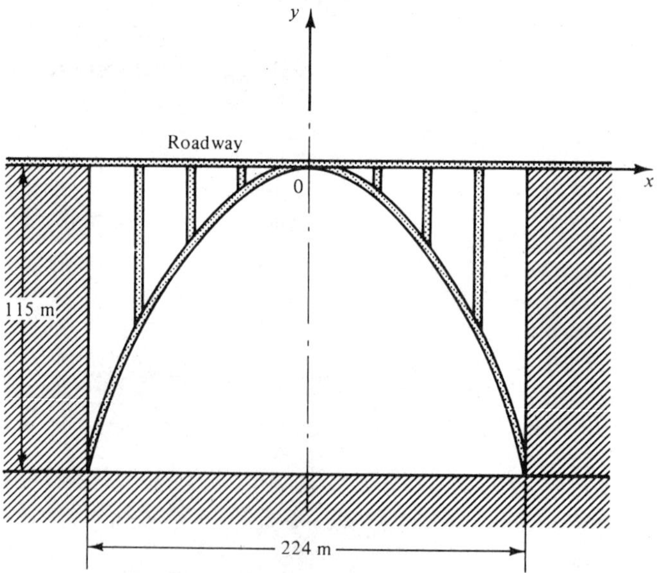

FIGURE 2-24. Parabolic arch.

Parabolic Reflector

27. A certain solar collector consists of a long panel of polished steel bent into a parabolic shape (Fig. 2-25), which focuses sunlight onto a pipe P that runs the length of the collector. At what distance x should the pipe be placed?

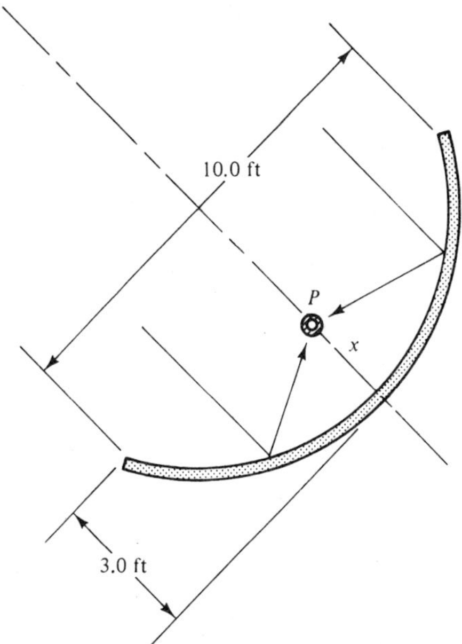

FIGURE 2-25. Parabolic solar collector.

28. A parabolic collector for receiving television signals from a satellite is shown in Fig. 2-26. The receiver R is at the focus, 1.0 m from the vertex. Find the depth d of the collector.

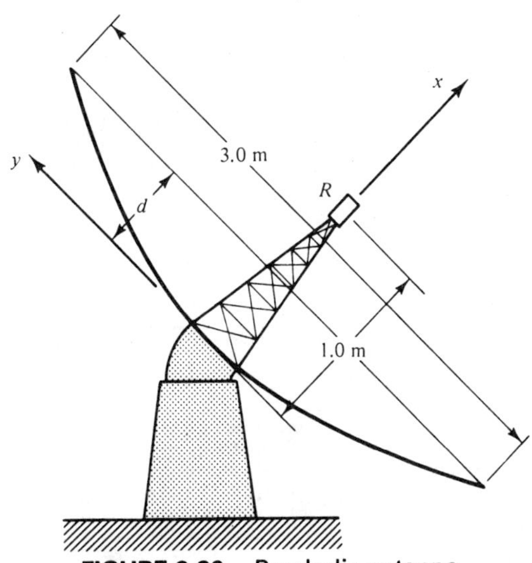

FIGURE 2-26. Parabolic antenna.

Vertical Highway Curves

29. A parabolic curve is to be used at a dip in a highway. The road dips 32.0 m in a horizontal distance of 125 m, and then rises to its previous height in another 125 m. Write the equation of the curve of the roadway, taking the origin at the bottom of the dip and the y axis vertical.

30. Write the equation of the vertical highway curve in Fig. 2-27 taking axes as shown

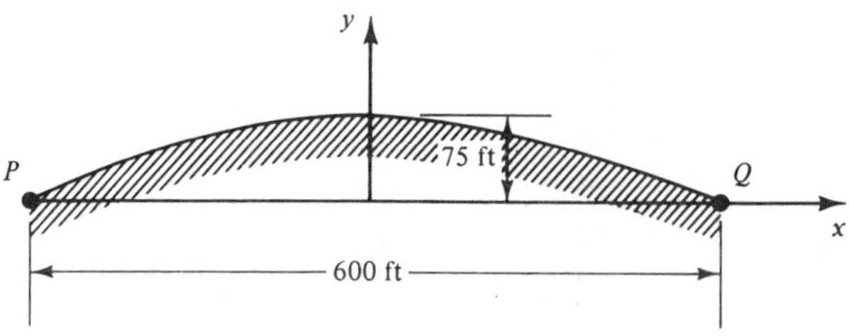

FIGURE 2-27. Road over a hill.

Beams

31. A simply supported beam with a concentrated load at its midspan (Fig. 2-28), will deflect in the shape of a parabola. If the deflection at the midspan is 1.0 in., write the equation of the parabola (called the elastic curve), taking axes as shown.

32. Using the equation found in Problem 31, find the deflection of the beam in Fig. 2-28 at a distance of 10.0 ft from the left end.

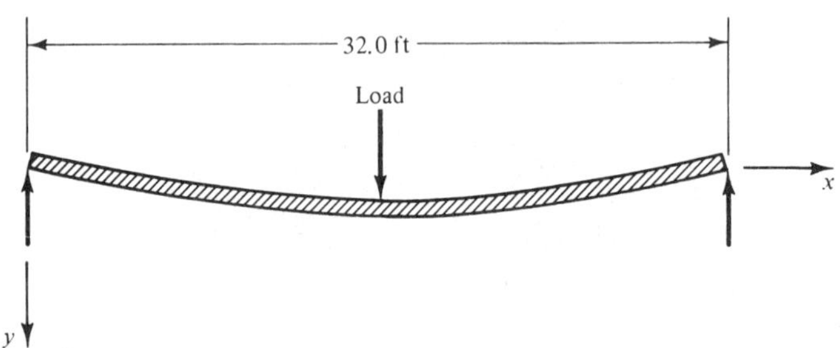

FIGURE 2-28. Deflection of a beam.

2-3. ELLIPSE

Definition	An *ellipse* is a plane curve which is the set of all points such that the sum of the distances from each point on the ellipse to two fixed points (called the *foci*) is constant.	279

Such a set of points, connected with a smooth curve, is plotted in Fig. 2-29, and shows the typical shape of the ellipse. An ellipse has two axes of symmetry, the *major axis* and the *minor axis* which intersect at the *center* of the ellipse. A *vertex* is a point where the ellipse crosses the major axis.

It is often convenient to speak of *half* the lengths of the major and minor axes, and these are called the *semimajor* and *semiminor* axes, whose lengths we label a and b, respectively.

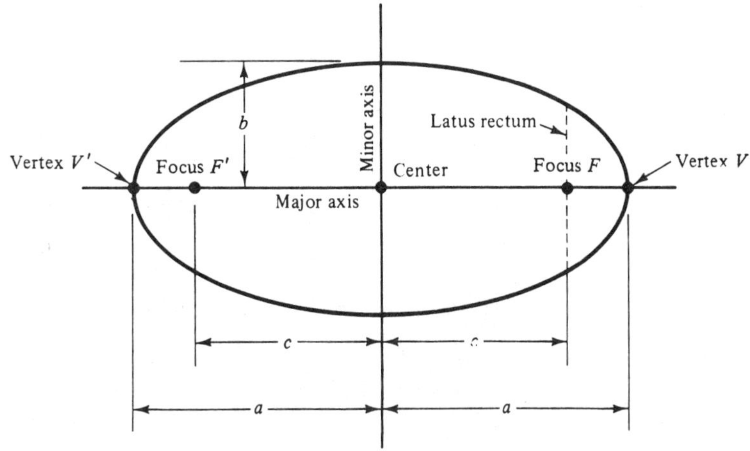

FIGURE 2-29. Ellipse.

Distance to Focus

Before deriving an equation for the ellipse let us first write an expression for the distance c from the center to a focus, in terms of the semimajor axis a and the semiminor axis b.

From our definition of an ellipse, if P is any point on the ellipse, then

$$PF + PF' = k \tag{1}$$

where k is a constant. If P is taken at a vertex V, then (1) becomes

$$VF + VF' = k = 2a \tag{2}$$

Substituting back into (1),

$$PF + PF' = 2a \tag{3}$$

Figure 2-30 shows our point P moved to the intersection of the ellipse and the minor axis. Here PF and PF' are equal. But since their sum is $2a$, PF and PF' must each equal a. By the Pythagorean theorem, $c^2 + b^2 = a^2$, or,

$$c^2 = a^2 - b^2$$

So

Ellipse: Distance from Center to Focus	$c = \sqrt{a^2 - b^2}$	**285**

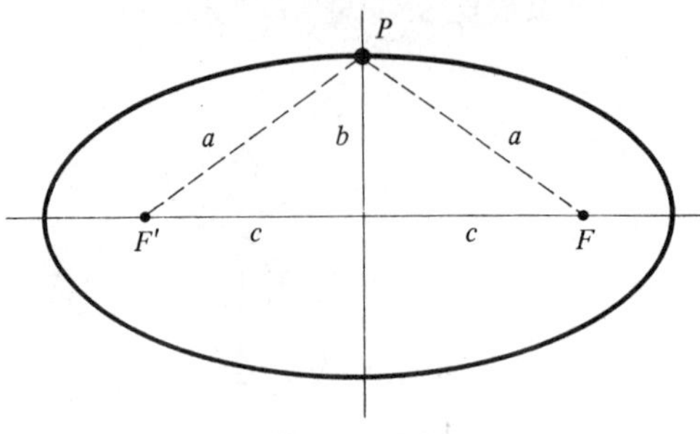

FIGURE 2-30

Ellipse with Center at the Origin

Let us place an ellipse on coordinate axes with its center at the origin and major axis along the x axis, as in Fig. 2-31. If $P(x, y)$ is any point on the ellipse, then by the definition of the ellipse,

$$PF + PF' = 2a$$

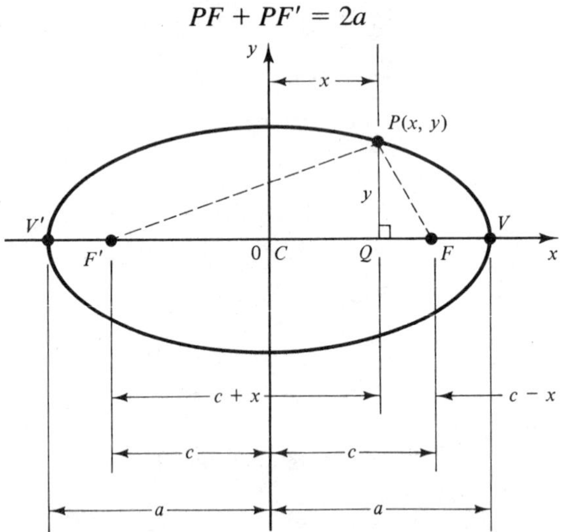

FIGURE 2-31. Ellipse with center at origin.

A line from a focus to a point on the ellipse is called a *focal radius*.

To get PF and PF' in terms of x and y, we first drop a perpendicular from P to the x axis. Then in triangle PQF,

$$PF = \sqrt{(c - x)^2 + y^2}$$

and in triangle PQF',

$$PF' = \sqrt{(c + x)^2 + y^2}$$

Substituting yields

$$PF + PF' = \sqrt{(c - x)^2 + y^2} + \sqrt{(c + x)^2 + y^2} = 2a$$

Rearranging, we obtain

$$\sqrt{(c + x)^2 + y^2} = 2a - \sqrt{(c - x)^2 + y^2}$$

Squaring, and expanding the binomials, we have

$$x^2 + 2cx + c^2 + y^2 = 4a^2 - 4a\sqrt{(c - x)^2 + y^2} + c^2 - 2cx + x^2 + y^2$$

Collecting terms, we get

$$cx = a^2 - a\sqrt{(c - x)^2 + y^2}$$

Dividing by a and rearranging

$$a - \frac{cx}{a} = \sqrt{(c - x)^2 + y^2}$$

Squaring both sides again yields

$$a^2 - 2cx + \frac{c^2x^2}{a^2} = c^2 - 2cx + x^2 + y^2$$

Collecting terms, we get

$$a^2 + \frac{c^2x^2}{a^2} = c^2 + x^2 + y^2$$

But $c = \sqrt{a^2 - b^2}$. Substituting, we get

$$a^2 + \frac{x^2}{a^2}(a^2 - b^2) = a^2 - b^2 + x^2 + y^2$$

or

$$a^2 + x^2 - \frac{b^2x^2}{a^2} = a^2 - b^2 + x^2 + y^2$$

Collecting terms and rearranging gives us

$$\frac{b^2x^2}{a^2} + y^2 = b^2$$

Finally, dividing through by b^2, we get the standard form of the equation of an ellipse:

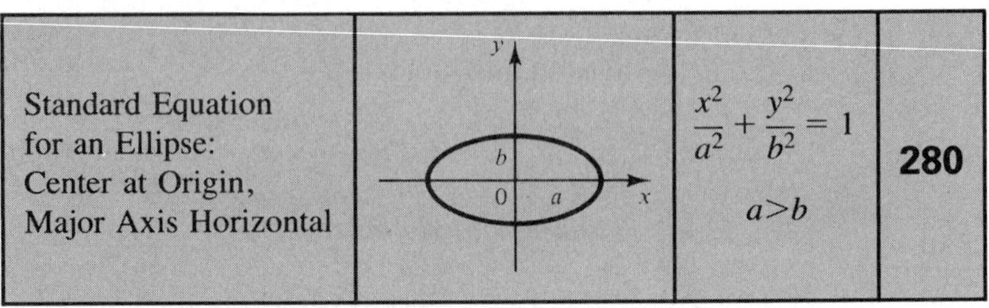

where a is half the length of the major axis, and b is half the length of the minor axis.

EXAMPLE 1: Find the vertices and foci for the ellipse $16x^2 + 36y^2 = 576$.

Solution: To be in standard form, our equation must have 1 (unity) on the right side. Dividing by 576 and simplifying,

$$\frac{x^2}{36} + \frac{y^2}{16} = 1$$

from which $a = 6$ and $b = 4$. The vertices are then $V(6, 0)$ and $V'(-6, 0)$ (Fig. 2-32). The distance c from the center to a focus is

$$c = \sqrt{6^2 - 4^2} = \sqrt{20} \cong 4.47$$

so the foci are $F(4.47, 0)$ and $F'(-4.47, 0)$.

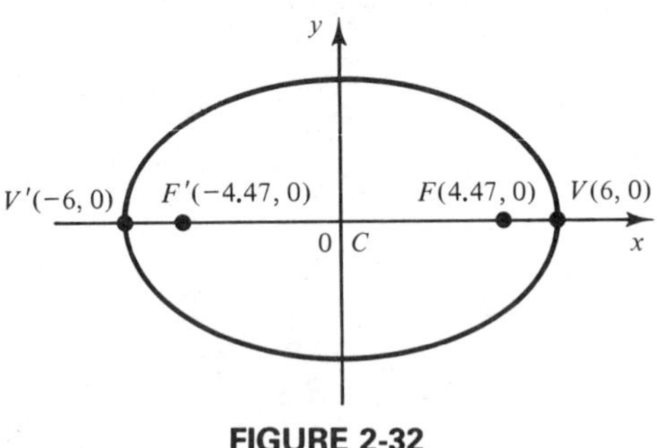

FIGURE 2-32

Major Axis Vertical

When the major axis is vertical rather than horizontal, the only effect on the standard equation is to interchange the positions of x and y:

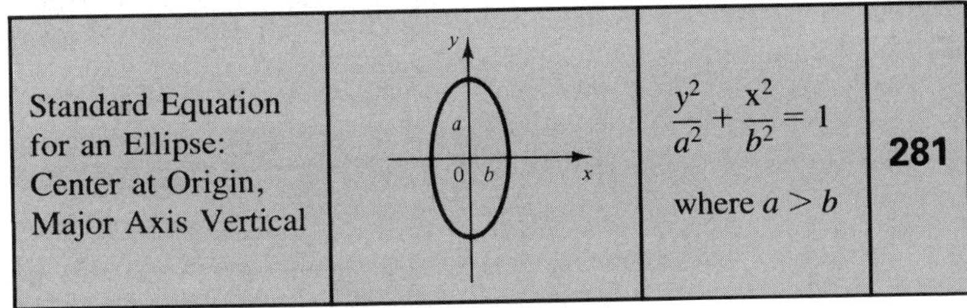

Standard Equation for an Ellipse: Center at Origin, Major Axis Vertical		$\dfrac{y^2}{a^2} + \dfrac{x^2}{b^2} = 1$ where $a > b$	**281**

Notice that the quantities a, b, and c are dimensions of the ellipse itself, and do not change meaning as the ellipse is turned or shifted. Therefore, the equation for the distance from center to focus still holds for an ellipse in any position.

EXAMPLE 2: Find the lengths of the major and minor axes and the distance to the focus for the ellipse

$$\frac{x^2}{16} + \frac{y^2}{49} = 1$$

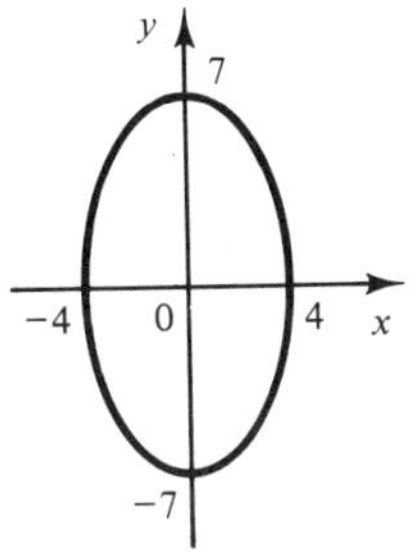

Solution: How can we tell which denominator is a^2 and which is b^2? It is easy: a is always greater than b. So $a = 7$ and $b = 4$. Thus the major and minor axes are 14 units and 8 units long (Fig. 2-33). From Eq. 285,

$$c = \sqrt{a^2 - b^2} = \sqrt{49 - 16} = \sqrt{33}$$

FIGURE 2-33

EXAMPLE 3: Write the equation of an ellipse with center at the origin, whose major axis is 12 units on the y axis, and whose minor axis is 10 units (Fig. 2-34).

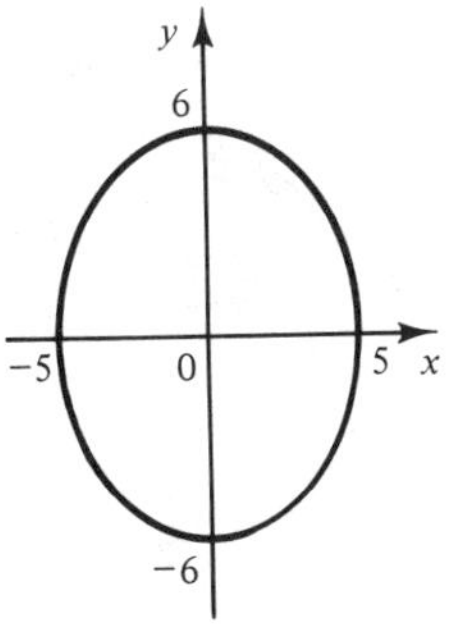

Solution: Substituting into Eq. 281, with $a = 6$ and $b = 5$, we obtain

$$\frac{y^2}{36} + \frac{x^2}{25} = 1$$

FIGURE 2-34

EXAMPLE 4: An ellipse whose center is at the origin and whose major axis is on the x axis has a minor axis of 10 units and passes through the point (6, 4) (Fig. 2-35). Write the equation of the ellipse in standard form.

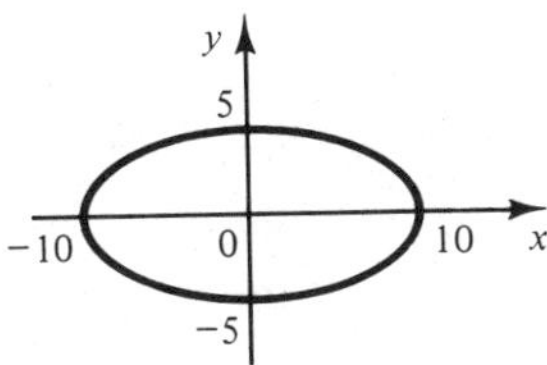

FIGURE 2-35

Solution: Our equation will have the form of Eq. 280. Substituting, with $b = 5$, we have

$$\frac{x^2}{a^2} + \frac{y^2}{25} = 1$$

Since the ellipse passes through (6, 4), these coordinates must satisfy our equation. Substituting yields

$$\frac{36}{a^2} + \frac{16}{25} = 1$$

Solving for a^2, we multiply by the LCD, $25a^2$:

$$36(25) + 16a^2 = 25a^2$$
$$9a^2 = 36(25)$$
$$a^2 = 100$$

So our final equation is

$$\frac{x^2}{100} + \frac{y^2}{25} = 1$$

Common Error	Do not confuse a and b in the ellipse equations. The *larger* denominator is always a^2. Also, the variable (x or y) in the same term with a^2 tells the direction of the major axis.

Shift of Axes

Now consider an ellipse whose center is not at the origin, but at (h, k). The equation for such an ellipse will be the same as before, except that x is replaced by $x - h$ and y is replaced by $y - k$.

Standard Form	Ellipse: Center at (h, k), Major Axis Horizontal	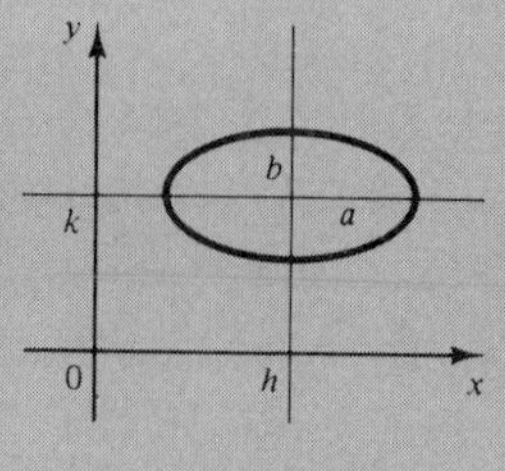	$\frac{(x-h)^2}{a^2} + \frac{(y-k)^2}{b^2} = 1$ $a > b$	**282**

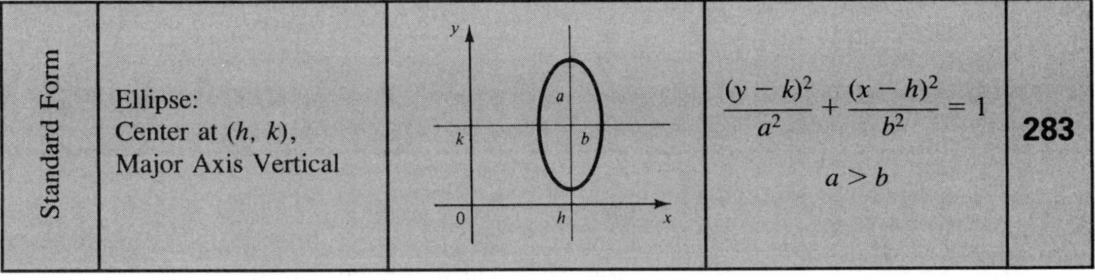

Standard Form	Ellipse: Center at (h, k), Major Axis Vertical		$\dfrac{(y-k)^2}{a^2} + \dfrac{(x-h)^2}{b^2} = 1$ $a > b$	283

EXAMPLE 5: Find the center, vertices, and foci for the ellipse

$$\frac{(x-5)^2}{9} + \frac{(y+3)^2}{16} = 1$$

Solution: From the given equation, $h = 5$ and $k = -3$, so the center is $C(5, -3)$ (Fig. 2-36). Also, $a = 4$ and $b = 3$, and the major axis is vertical because a is with the y term. Thus the vertices are $V(5, 1)$ and $V'(5, -7)$. The distance c to the foci is

$$c = \sqrt{4^2 - 3^2} = \sqrt{7} \cong 2.66$$

so the foci are $F(5, -0.34)$ and $F'(5, -5.66)$.

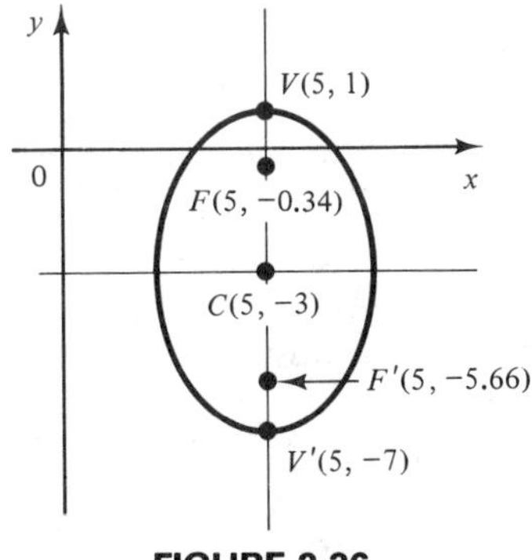

FIGURE 2-36

EXAMPLE 6: Write the equation in standard form of an ellipse with a vertical major axis 10 units long, center at (3, 5), and whose distance between focal points is 8 units (Fig. 2-37).

Yes, Eq. 285 still applies here.

Solution: From the information given, $h = 3$, $k = 5$, $a = 5$, and $c = 4$. By Eq. 285,

$$b = \sqrt{a^2 - c^2} = \sqrt{25 - 16} = 3 \text{ units}$$

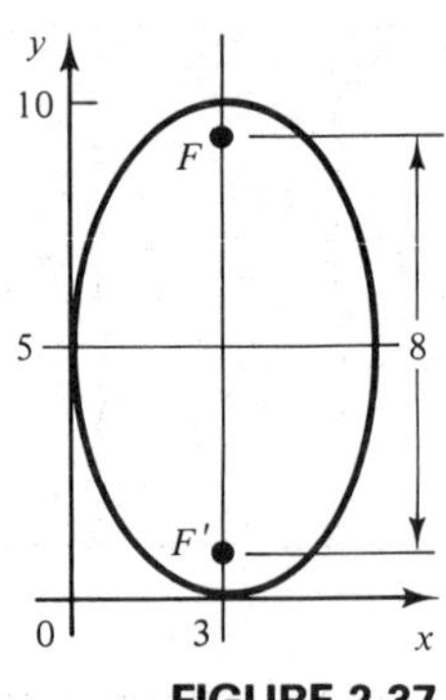

FIGURE 2-37

Substituting into Eq. 283, we get

$$\frac{(y-5)^2}{25}+\frac{(x-3)^2}{9}=1$$

General Equation of the Ellipse

As we did with the circle, we now expand the standard equation for the ellipse to get the general equation. Starting with Eq. 282,

$$\frac{(x-h)^2}{a^2}+\frac{(y-k)^2}{b^2}=1$$

we multiply through by a^2b^2, and expand the binomials:

$$b^2(x^2-2hx+h^2)+a^2(y^2-2ky+k^2)=a^2b^2$$

$$b^2x^2-2b^2hx+b^2h^2+a^2y^2-2a^2ky+a^2k^2-a^2b^2=0$$

or

$$b^2x^2+a^2y^2-2b^2hx-2a^2ky+(b^2h^2+a^2k^2-a^2b^2)=0$$

which is of the form:

General Form of the Equation of an Ellipse	$Ax^2+Cy^2+Dx+Ey+F=0$	**284**

From this we see that the general equation of second degree represents an ellipse with axes parallel to the coordinate axes if $B = 0$, and if A and C are different but have the same sign.

Completing the Square

As before, we go from general to standard form by completing the square.

EXAMPLE 7: Find the center, foci, vertices, and major and minor axes for the ellipse $9x^2 + 25y^2 + 18x - 50y - 191 = 0$.

Solution: Grouping the x terms and the y terms gives us

$$(9x^2+18x)+(25y^2-50y)=191$$

Factoring yields

$$9(x^2+2x\qquad)+25(y^2-2y\qquad)=191$$

Completing the square, we obtain

$$9(x^2 + 2x + 1) + 25(y^2 - 2y + 1) = 191 + 9 + 25$$

Factoring gives us

$$9(x + 1)^2 + 25(y - 1)^2 = 225$$

Finally, dividing by 225, we have

$$\frac{(x + 1)^2}{25} + \frac{(y - 1)^2}{9} = 1$$

We see that $h = -1$ and $k = 1$, so the center is at $(-1, 1)$. Also, $a = 5$, so the major axis is 10 units and is horizontal, and $b = 3$, so the minor axis is 6 units. A vertex is located 5 units to the right of the center, at (4, 1), and 5 units to the left, at (−6, 1). From Eq. 285,

$$c = \sqrt{a^2 - b^2} = \sqrt{25 - 9} = 4$$

So the foci are at (3, 1) and (−5, 1). This ellipse is shown in Fig. 2-38.

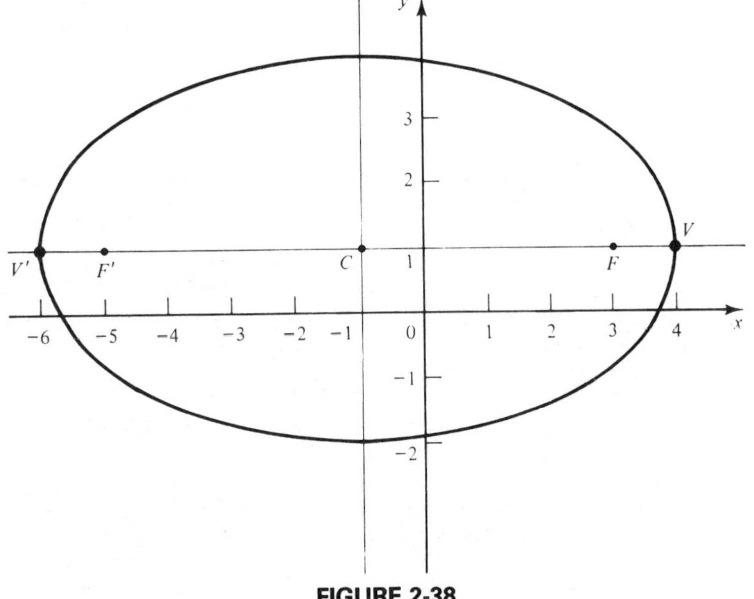

FIGURE 2-38

Common Error	In Example 7 we needed 1 to complete the square on x, but added 9 to the right side, because the expression containing the 1 was multiplied by a factor of 9. Similarly for y, we needed 1 on the left but added 25 to the right. It is very easy to forget to multiply by those factors.

Focal Width

The focal width L, or length of the latus rectum, is the width of the ellipse through the focus (Fig. 2-39). The sum of the focal radii PF and PF' from a point P at one end of the latus rectum must equal $2a$, so $PF' = 2a - L/2$, or, squaring both sides,

$$(PF')^2 = 4a^2 - 2aL + \frac{L^2}{4} \tag{1}$$

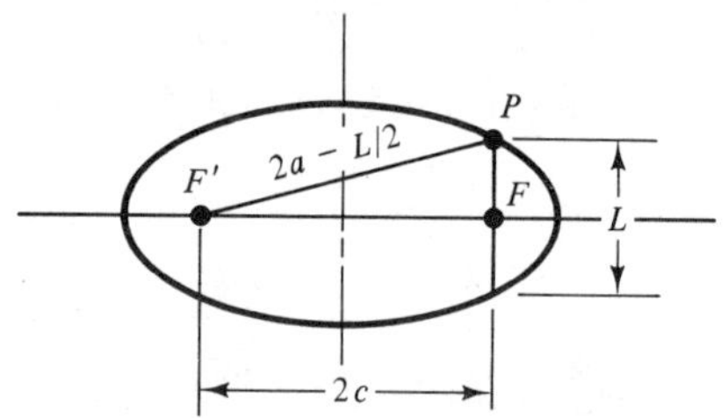

FIGURE 2-39. Derivation of focal width.

But in right triangle PFF',

$$(PF')^2 = \left(\frac{L}{2}\right)^2 + (2c)^2 = \frac{L^2}{4} + 4c^2 = \frac{L^2}{4} + 4a^2 - 4b^2 \tag{2}$$

since $c^2 = a^2 - b^2$. Equating (1) and (2) and collecting terms gives $2aL = 4b^2$, or

As with the parabola, the main use for L is for quick sketching of the ellipse.

Focal Width of Ellipse	$L = \dfrac{2b^2}{a}$	**286**

The focal width of an ellipse is twice the square of the semiminor axis divided by the semimajor axis.

EXAMPLE 8: The focal width of an ellipse that is 25 m long and 10 m wide is

$$L = \frac{2(5^2)}{12.5} = \frac{50}{12.5} = 4 \text{ m}$$

EXERCISE 3

Ellipse with Center at Origin

Find the coordinates of the vertices and foci for each ellipse.

1. $\frac{x_2}{25} + \frac{y_2}{16} = 1$ **2.** $x^2 + 2y^2 = 2$

3. $3x^2 + 4y^2 = 12$ **4.** $64x^2 + 15y^2 = 960$

5. $4x^2 + 3y^2 = 48$ **6.** $8x^2 + 25y^2 = 200$

Write the equation of each ellipse.

7. vertices at $(\pm 5, 0)$, foci at $(\pm 4, 0)$

8. vertical major axis 8 units long, a focus at $(0, 2)$

9. horizontal major axis 12 units long, passes through $(3, \sqrt{3})$

10. passes through $(4, 6)$ and $(2, 3\sqrt{5})$

11. horizontal major axis 26 units long, distance between foci = 24

12. vertical minor axis 10 units long, distance from focus to vertex = 1

13. passes through $(1, 4)$ and $(-6, 1)$

14. distance between foci = 18, sum of axes = 54, horizontal major axis

Ellipse with Center Not at Origin

Find the coordinates of the center, vertices, and foci, and the focal width for each ellipse.

15. $\frac{(x-2)^2}{16} + \frac{(y+2)^2}{9} = 1$ **16.** $\frac{(x+5)^2}{25} + \frac{(y-3)^2}{49} = 1$

17. $5x^2 + 20x + 9y^2 - 54y + 56 = 0$

18. $16x^2 - 128x + 7y^2 + 42y = 129$

19. $7x^2 - 14x + 16y^2 + 32y = 89$

20. $3x^2 - 6x + 4y^2 + 32y + 55 = 0$

21. $25x^2 + 150x + 9y^2 - 36y + 36 = 0$

Write the equation of each ellipse.

22. minor axis = 10, foci at $(13, 2)$ and $(-11, 2)$

23. center at $(0, 3)$, vertical major axis = 12, length of minor axis = 6

24. center at $(2, -1)$, a vertex at $(2, 5)$, length of minor axis = 3

25. center at $(-2, -3)$, a vertex at $(-2, 1)$, a focus halfway between vertex and center

Intersections of Curves

Solve simultaneously to find the points of intersection of each ellipse with the given curve.

26. $3x^2 + 6y^2 = 11$ with $y = x + 1$

27. $2x^2 + 3y^2 = 14$ with $y^2 = 4x$

28. $x^2 + 7y^2 = 16$ with $x^2 + y^2 = 10$

Construction of an Ellipse

29. One way to draw an ellipse is to put two tacks at the focal points (Fig. 2-40) and to adjust a loop of string so that the pencil point passes through the end of the major and minor axes. How far apart must the tacks be placed if the ellipse is to have a major axis of 84 cm and a minor axis of 58 cm?

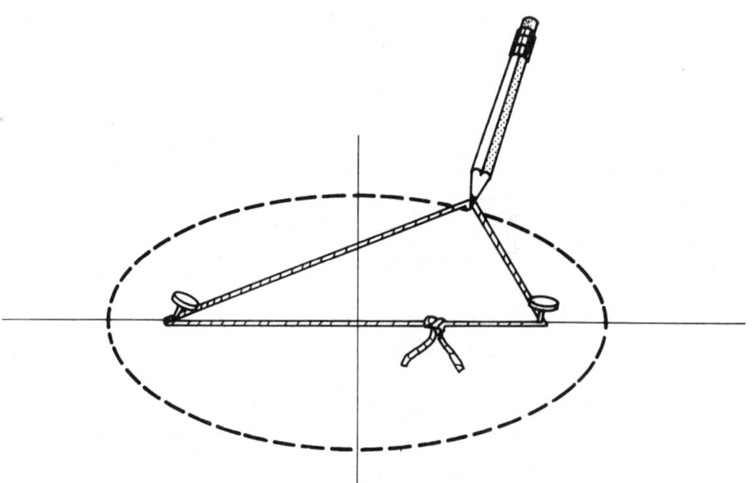

FIGURE 2-40. Construction of an ellipse.

Applications

30. A certain bridge arch is in the shape of half an ellipse 120 ft wide and 30.0 ft high. At what horizontal distance from the center of the arch is the height equal to 15.0 ft?

31. A curved mirror in the shape of an ellipse will reflect all rays of light coming from one focus onto the other focus (Fig. 2-41). A certain spot heater is to be made with a heating element at A and the part to be heated at B, contained within an ellipsoid (a solid obtained by rotating an ellipse about one axis). Find the width x of the chamber if its length is 25 cm and the distance from A to B is 15 cm.

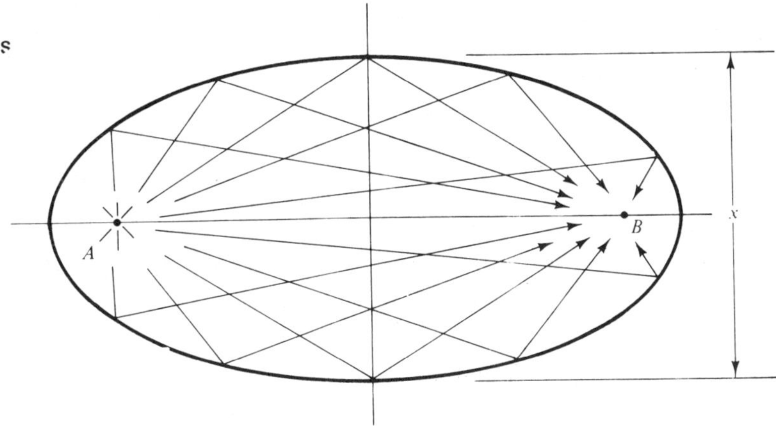

FIGURE 2-41. Focusing property of the ellipse.

An *astronomical unit*, AU, is the distance between the earth and the sun, about 92.6 million miles.

32. The paths of the planets and certain comets are ellipses, with the sun at one focal point. The path of Halley's comet is an ellipse with a major axis of 36.18 AU and a minor axis of 9.12 AU. What is the greatest distance that Halley's comet gets from the sun?

33. An elliptical culvert (Fig. 2-42) is filled with water to a depth of 1.0 ft. Find the width w of the stream.

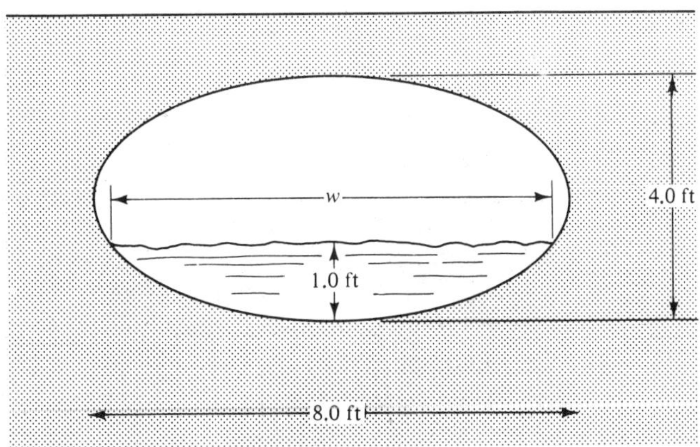

FIGURE 2-42

2-4. HYPERBOLA

Definition	A *hyperbola* (Fig. 2-43) is the set of all points in a plane such that the distances from each point to two fixed points, the *foci*, have a constant difference.	**289**

Figure 2-43 shows the typical shape of the hyperbola. The line passing through the two foci is called the *transverse axis*. The hyperbola crosses the transverse axis at points called the *vertices*. The *center* of a hyperbola is on the transverse axis, midway between the vertices. The *conjugate axis* passes through the center and is perpendicular to the transverse axis.

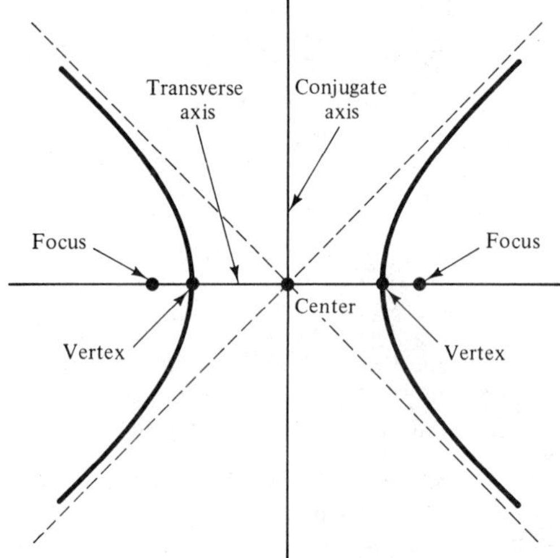

FIGURE 2-43. Hyperbola.

Standard Equation of the Hyperbola

We place the hyperbola on coordinate axes, with its center at the origin and its transverse axis on the x axis (Fig. 2-44). Let a be half the transverse axis, and let c be half the distance between foci. Now take any point P on the hyperbola and draw the focal radii PF and PF'. Then by the definition of the hyperbola, $|PF' - PF| = \text{constant}$. The constant can be found by moving P to the vertex V, where

$$\begin{aligned}|PF' - PF| &= VF' - VF \\ &= 2a + V'F' - VF \\ &= 2a\end{aligned}$$

since $V'F'$ and VF are equal. So $|PF' - PF| = 2a$. But in right triangle $PF'D$,

$$PF' = \sqrt{(x + c)^2 + y^2}$$

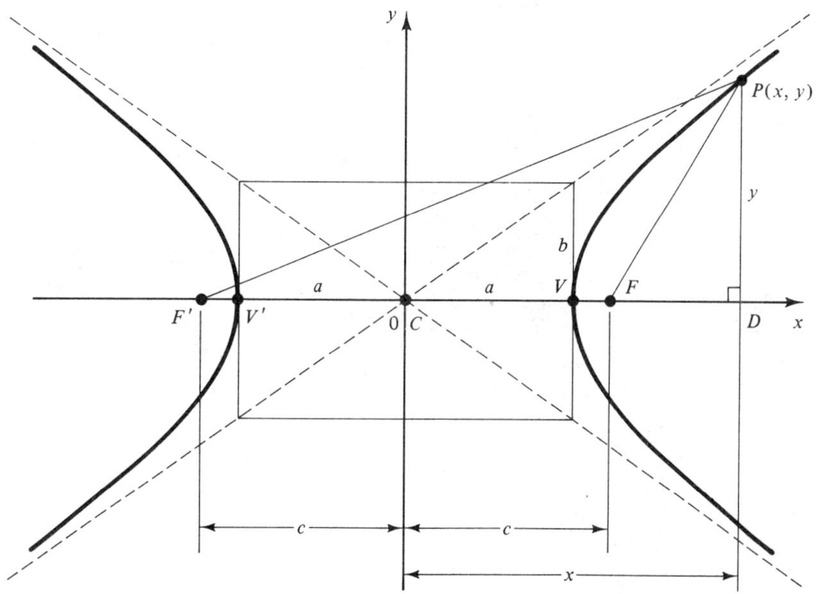

FIGURE 2-44. Hyperbola with center at origin.

and in right triangle *PFD*,

$$PF = \sqrt{(x - c)^2 + y^2}$$

so

$$\sqrt{(x + c)^2 + y^2} - \sqrt{(x - c)^2 + y^2} = 2a$$

We'll soon give a geometric meaning to the quantity *b*.

Notice that this equation is almost identical to the one we had when deriving the equation for the ellipse in Sec. 2-3, and the derivation is almost identical as well. However, to eliminate c from our equation, we define a new quantity b, such that $b^2 = c^2 - a^2$, or

Hyperbola: Distance from Center to Focus	$c = \sqrt{a^2 + b^2}$	**295**

Try to derive these equations. Follow the same steps we used for the ellipse.

The remainder of the derivation is left as an exercise. The resulting equations are very similar to those for the ellipse.

Standard Form	Hyperbola: Center at Origin	Transverse Axis Horizontal		$\frac{x^2}{a^2} - \frac{y^2}{b^2} = 1$	290
		Transverse Axis Vertical		$\frac{y^2}{a^2} - \frac{x^2}{b^2} = 1$	291

Asymptotes

Solving Eq. 290 for y gives $y^2 = b^2(x^2/a^2 - 1)$, or

$$y = \pm b\sqrt{\frac{x^2}{a^2} - 1}$$

As x gets large, the 1 under the radical sign becomes insignificant in comparison with x^2/a^2, and the equation for y becomes

$$y = \pm \frac{b}{a}x$$

This is the equation of a straight line having a slope of b/a. Thus as x increases, the branches of the hyperbola more closely approach straight lines of slopes $\pm b/a$.

If the distance from a point P on a curve to some line L approaches zero as the distance from P to the origin increases without bound, then L is called an *asymptote*. Thus the hyperbola has two asymptotes, of slopes $\pm b/a$.

For a hyperbola whose transverse axis is *vertical,* the slopes of the asymptotes can be shown to be $\pm a/b$.

Slope of the Asymptotes	Transverse Axis Horizontal	slope $= \pm\frac{b}{a}$	296
	Transverse Axis Vertical	slope $= \pm\frac{a}{b}$	297

Looking back at Fig. 2-44, we can now give meaning to the quantity b. If an asymptote has a slope b/a, it must have a rise of b in a run equal to a. Thus b is the distance, perpendicular to the transverse axis, from vertex to asymptote.

Common Error	The asymptotes are not (usually) perpendicular. The slope of one **is not** the negative reciprocal of the other.

Graphing the Hyperbola

A good way to start a sketch of the hyperbola is to draw a rectangle whose dimension along the transverse axis is $2a$ and whose dimension along the conjugate axis is $2b$. The asymptotes are then drawn along the diagonals of this rectangle. Half the diagonal of the rectangle has a length $\sqrt{a^2 + b^2}$, which is equal to c, the distance to the foci. Thus an arc of radius c will cut the transverse axes at the focal points.

As with the parabola and ellipse, a perpendicular through a focus connecting two points on the hyperbola is called a *latus rectum*. Its length, called the focal width, is $2b^2/a$, the same as for the ellipse.

EXAMPLE 1: Find the vertices, foci, semiaxes, slope of the asymptotes, and focal width, and graph the hyperbola $64x^2 - 49y^2 = 3136$.

Solution: We go to standard form by dividing by 3136,

$$\frac{x^2}{49} - \frac{y^2}{64} = 1$$

This is the form of Eq. 290, so that the transverse axis is horizontal, with $a = 7$ and $b = 8$. We draw a rectangle of width 14 and height 16 (shown shaded in Fig. 2-45), thus locating the vertices at $(\pm 7, 0)$. Diagonals through the rectangle give us the asymptotes of slopes $\pm\frac{8}{7}$. We locate the foci by swinging an arc of radius c, equal to half the diagonal of the rectangle. Thus

$$c = \sqrt{7^2 + 8^2} = \sqrt{113} \cong 10.6$$

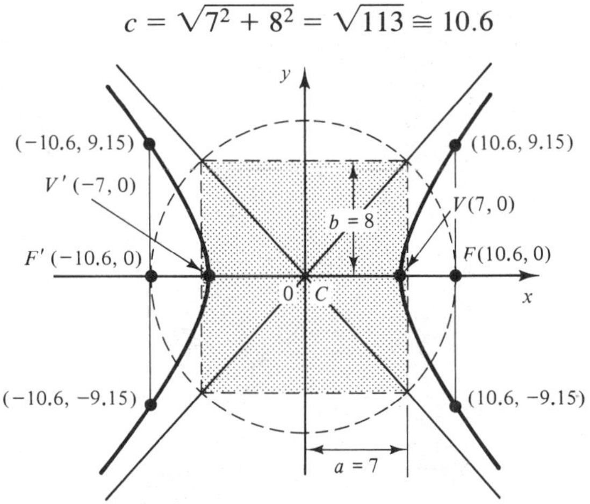

FIGURE 2-45

The foci then are at (± 10.6, 0). We obtain a few more points by computing the focal width,

$$\frac{2b^2}{a} = \frac{2(64)}{7} \cong 18.3$$

This gives us the additional points (10.6, 9.15), (10.6, −9.15), (−10.6, 9.15) and (−10.6, −9.15).

EXAMPLE 2: A hyperbola whose center is at the origin has a focus at (0, −5) and a transverse axis 8 units long (Fig. 2-46). Write the standard equation of the hyperbola and find the slope of the asymptotes.

Solution: The transverse axis is 8, so $a = 4$. A focus is 5 units below the origin, so the transverse axis must be vertical, and $c = 5$. From Eq. 295,

$$b = \sqrt{c^2 - a^2} = \sqrt{25 - 16} = 3$$

Substituting into Eq. 291 gives us

$$\frac{y^2}{16} - \frac{x^2}{9} = 1$$

and from Eq. 297,

$$\text{slope of asymptotes} = \pm\frac{a}{b} = \pm\frac{4}{3}$$

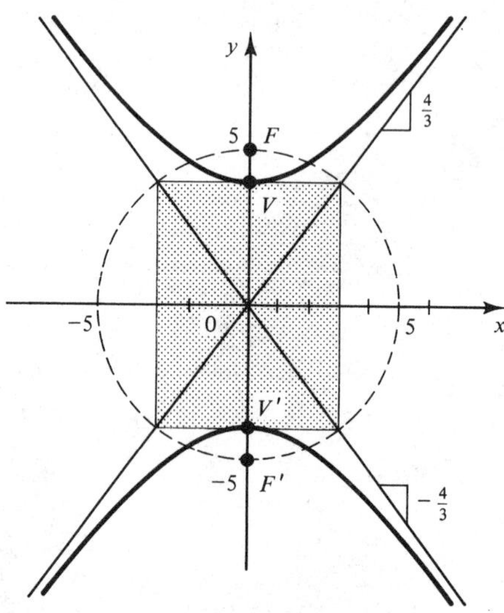

FIGURE 2-46

Common Error	Be sure you know the direction of the transverse axis before computing the slope of the asymptotes. It is $\pm\, b/a$ when the axis is horizontal, but $\pm a/b$ when the axis is vertical.

Shift of Axes

The equations for c, and for the slope of the asymptotes are still valid for these cases.

As with the other conics, we shift the axes to the left by a distance h and downward by a distance k by replacing x by $(x - h)$ and k by $(y - k)$ in Eqs. 290 and 291.

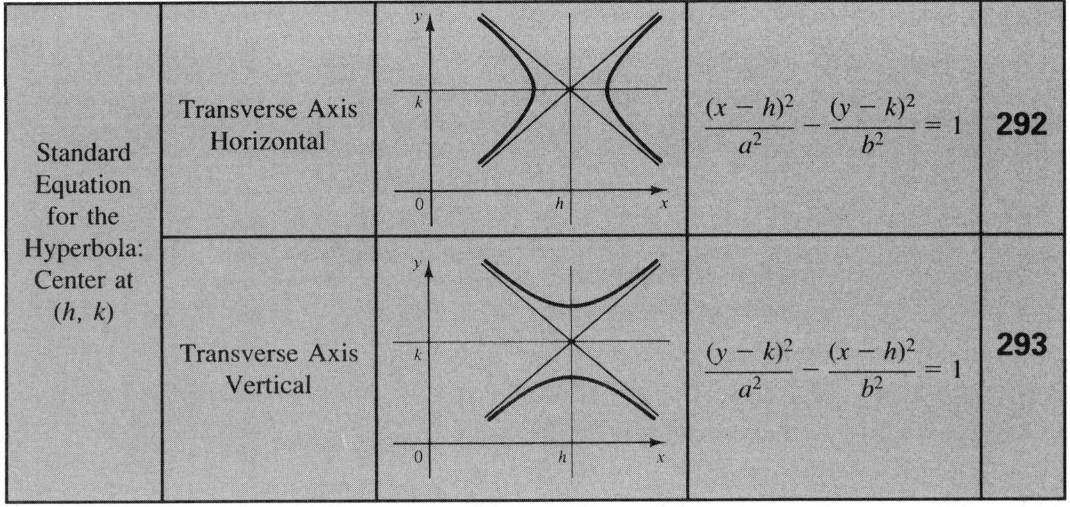

Standard Equation for the Hyperbola: Center at (h, k)	Transverse Axis	Graph	Equation	No.
	Transverse Axis Horizontal		$$\frac{(x-h)^2}{a^2} - \frac{(y-k)^2}{b^2} = 1$$	**292**
	Transverse Axis Vertical		$$\frac{(y-k)^2}{a^2} - \frac{(x-h)^2}{b^2} = 1$$	**293**

Common Error	Do not confuse these equations with those for the ellipse. Here the terms have **opposite signs.** Also, a^2 is always the denominator of the **positive** term, even though it may be smaller than b^2. As with the ellipse, the variable in the same term as a^2 tells the direction of the transverse axis.

EXAMPLE 3: A certain hyperbola has a focus at (1, 0), passes through the origin, has its transverse axis on the x axis, and has a distance of 10 between its focal points. Write its standard equation.

Solution: In Fig. 2-47 we plot the given focus $F(1, 0)$. Since the hyperbola passes through the origin, and the transverse axis also passes through the origin, a vertex must be at the origin as well. Then, since $c = 5$,

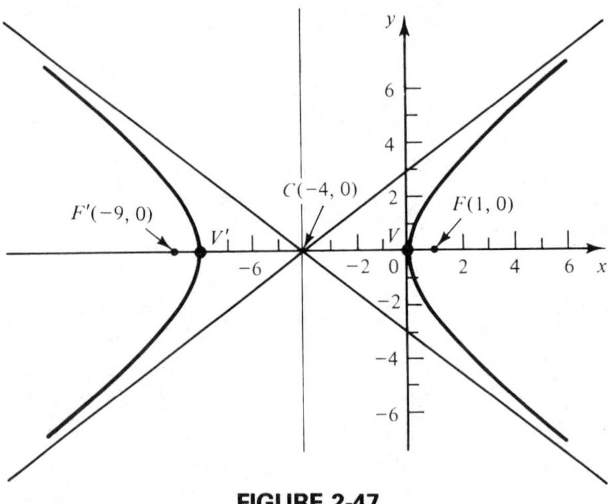

FIGURE 2-47

we go 5 units to the left of F and plot the center, $(-4, 0)$. Thus, $a = 4$ units, from center to vertex. Then by Eq. 295,

$$b = \sqrt{c^2 - a^2} = \sqrt{25 - 16}$$
$$= 3$$

Substituting into Eq. 292 with

$$h = -4, \qquad k = 0, \qquad a = 4, \qquad b = 3$$

we get

$$\frac{(x + 4)^2}{16} - \frac{y^2}{9} = 1$$

as the required equation.

General Equation of a Hyperbola

Its easy to feel overwhelmed by the large number of formulas in this chapter. Study the summary of formulas in Appendix A, where you'll find them side by side and can see where they differ and where they are alike.

Since the standard equations for the ellipse and hyperbola are identical except for the minus sign, it is not surprising that their general equations should be identical except for a minus sign. The general second-degree equation for ellipse or hyperbola is:

General Form of the Equation of a Hyperbola	$Ax^2 + Cy^2 + Dx + Ey + F = 0$	**294**

For the ellipse, A and C have like signs; for the hyperbola, they have opposite signs.

As before, we complete the square to go from general form to standard form.

EXAMPLE 4: Write the equation $x^2 - 4y^2 - 6x + 8y - 11 = 0$ in standard form. Find the center, vertices, foci, and asymptotes.

Solution: Rearranging gives us

$$(x^2 - 6x \qquad) - 4(y^2 - 2y \qquad) = 11$$

Completing the square and factoring,

$$(x^2 - 6x + 9) - 4(y^2 - 2y + 1) = 11 + 9 - 4$$
$$(x - 3)^2 - 4(y - 1)^2 = 16$$

Dividing by 16,

$$\frac{(x-3)^2}{16} - \frac{(y-1)^2}{4} = 1$$

from which $a = 4$, $b = 2$, $h = 3$, and $k = 1$.

From Eq. 295, $c = \sqrt{a^2 + b^2} = \sqrt{16 + 4} \cong 4.47$, and from Eq. 296,

$$\text{slope of asymptotes} = \pm\frac{b}{a} = \pm\frac{1}{2}$$

This hyperbola is plotted in Fig. 2-48.

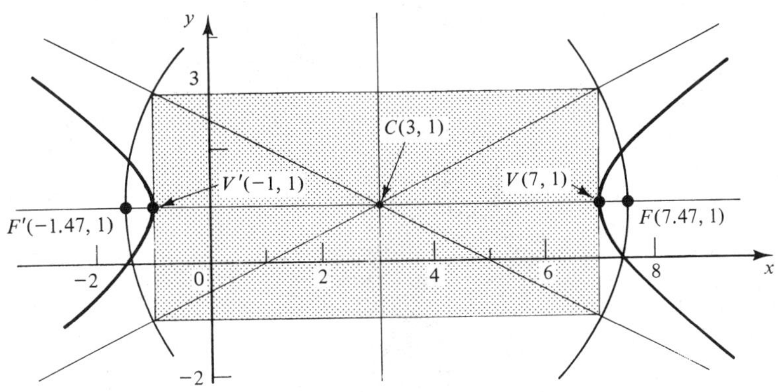

FIGURE 2-48

Hyperbola Whose Asymptotes Are the Coordinate Axes

The graph of an equation of the form

$$xy = k$$

where k is a constant, is a hyperbola similar to those we have studied in this section, but rotated so that the coordinate axes are now the asymptotes, and the transverse and conjugate axes are at 45°.

Hyperbola: Axes Rotated 45°	$xy = k$	**299**

EXAMPLE 5: Plot the equation $xy = -4$.

Solution: We select values for x and evaluate $y = -4/x$:

x	-4	-3	-2	-1	0	1	2	3	4
y	1	$\frac{4}{3}$	2	4	–	-4	-2	$-\frac{4}{3}$	-1

The graph (Fig. 2-49) shows vertices at $(-2, 2)$ and $(2, -2)$. We see from the graph that a and b are equal, and that each is the hypotenuse of a right triangle of side 2. Therefore,

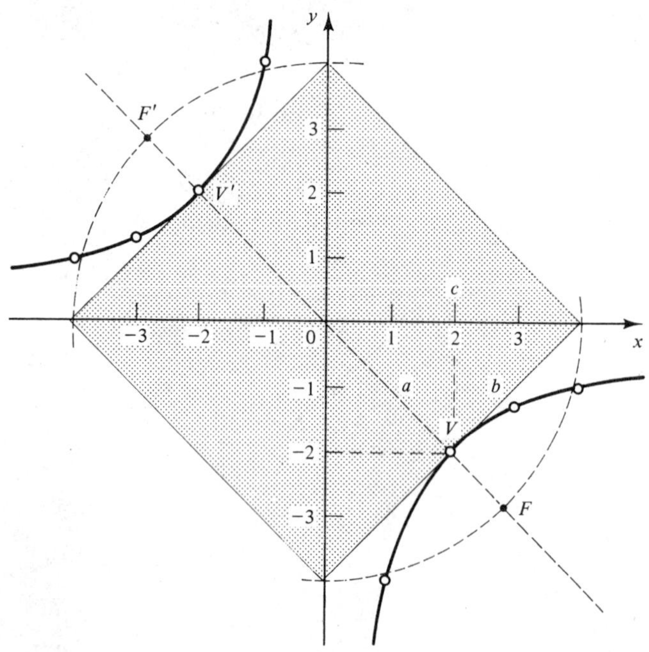

FIGURE 2-49. Hyperbola whose asymptotes are the coordinate axes.

$$a = b = \sqrt{2^2 + 2^2} = \sqrt{8}$$

and from Eq. 295,

$$c = \sqrt{a^2 + b^2} = \sqrt{16} = 4$$

When the constant k in Eq. 299 is negative, the hyperbola lies in the second and fourth quadrants, as in Example 5. When k is positive, the hyperbola lies in the first and third quadrants.

EXERCISE 4

Hyperbola with Center at Origin

Find the vertices, foci, semiaxes, and the slope of the asymptotes for each hyperbola.

1. $\dfrac{x^2}{16} - \dfrac{y^2}{25} = 1$

2. $\dfrac{y^2}{25} - \dfrac{x^2}{49} = 1$

3. $16x^2 - 9y^2 = 144$

4. $9x^2 - 16y^2 = 144$

5. $x^2 - 4y^2 = 16$

6. $3x^2 - y^2 + 3 = 0$

Write the equation of each hyperbola.

7. vertices at $(\pm 5, 0)$, foci at $(\pm 13, 0)$

8. vertices at $(0, \pm 7)$, foci at $(0, \pm 10)$

9. distance between foci = 8, transverse axis = 6 and is horizontal

10. conjugate axis = 4, foci at $(\pm 2.5, 0)$

11. transverse and conjugate axes equal, passes through (3, 5), transverse axis vertical

12. transverse axis = 16 and is horizontal, conjugate axis = 14

13. transverse axis = 10 and is vertical, curve passes through (8, 10)

14. transverse axis = 8 and is horizontal, curve passes through (5, 6)

Hyperbola with Center Not at Origin

Find the center, semiaxes, foci, slope of the asymptotes, and vertices for each hyperbola.

15. $\dfrac{(x-2)^2}{25} - \dfrac{(y+1)^2}{16} = 1$

16. $\dfrac{(y+3)^2}{25} - \dfrac{(x-2)^2}{49} = 1$

17. $16x^2 - 64x - 9y^2 - 54y = 161$

18. $16x^2 + 32x - 9y^2 + 592 = 0$

19. $4y^2 + 8x - 5x^2 - 10y + 19 = 0$ **20.** $y^2 - 6y - 3x^2 - 12x = 12$

Write the equation of each hyperbola in standard form.

21. center at (3, 2), transverse axis = 8 and is vertical, length of the conjugate axis = 4

22. foci at (−1, −1) and (−5, −1), $a = b$

23. foci at (5, −2) and (−3, −2), vertex halfway between center and focus

24. conjugate axis = 8 and is horizontal, center at (−1, −1), length of the transverse axis = 16

Hyperbola Whose Asymptotes Are Coordinate Axes

25. Write the equation of a hyperbola with center at the origin and with asymptotes on the coordinate axes, whose vertex is at (6, 6).

26. Write the equation of a hyperbola with center at the origin and with asymptotes on the coordinate axes, which passes through the point (−9, 2).

Applications

This type of mirror is used in the *Cassegrain* form of reflecting telescope.

27. A hyperbolic mirror (Fig. 2-50) has the property that a ray of light directed at one focus will be reflected so as to pass through the other focus. Write the equation of the mirror shown, taking the axes as indicated.

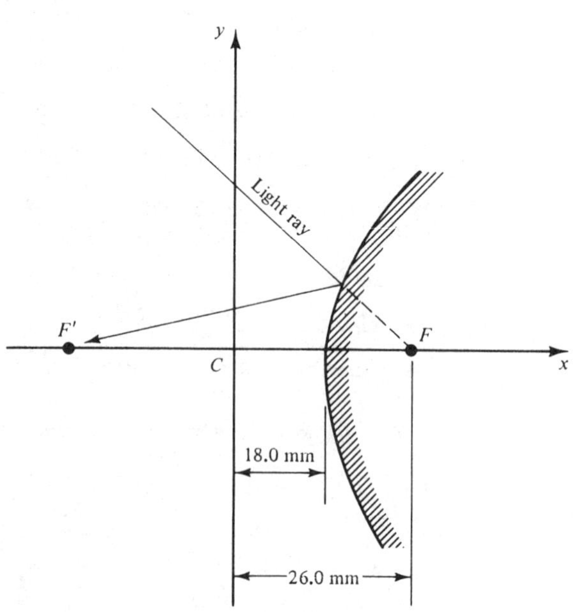

FIGURE 2-50. Hyperbolic mirror.

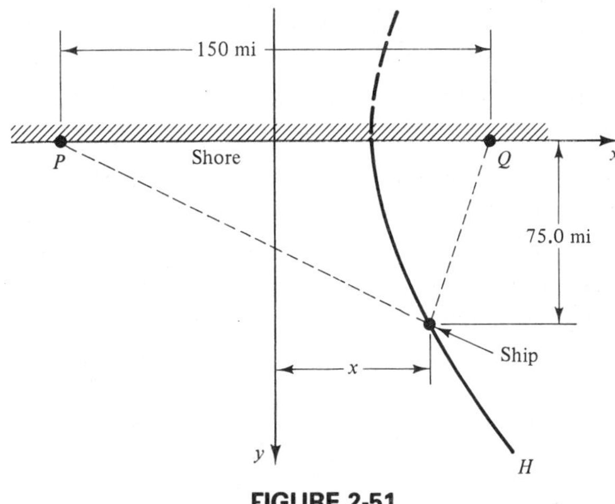

FIGURE 2-51

28. A ship (Fig. 2-51) receives simultaneous radio signals from stations P and Q, on the shore. The signal from station P is found to arrive 375 microseconds (μs) later than that from Q. Assuming that the signals travel at a rate of 0.186 mi/μs, find the distance from the ship to each station. (*Hint:* The ship will be on one branch of a hyperbola H whose foci are at P and Q. Write the equation of this hyperbola, taking axes as shown, and then substitute 75.0 mi for y to obtain the distance x.)

29. Boyle's law states that under certain conditions, the product of the pressure and volume of a gas is constant, or $pv = c$. This equation has the same form as the hyperbola (Eq. 299). If a certain gas has a pressure of 25.0 lb/in.^{2} at a volume of 1000 in.^{3}, write Boyle's law for this gas and make a graph of pressure versus volume.

Computer

30. Write a program that will accept as input the constants A, B, and C in the general second-degree equation (Eq. 261), and then identify and print the type of conic represented.

CHAPTER TEST

Identify the curve represented by each equation. Find, where applicable, the center, vertices, foci, radius, semiaxes, and so on.

1. $x^2 - 2x - 4y^2 + 16y = 19$

2. $x^2 + 6x + 4y = 3$

3. $x^2 + y^2 = 8y$

4. $25x^2 - 200x + 9y^2 - 90y = 275$

5. $16x^2 - 9y^2 = 144$

6. $x^2 + y^2 = 9$

7. Write the equation for an ellipse whose center is at the origin, whose major axis is 20 and is horizontal, and whose minor axis equals the distance between the foci.
8. Write an equation for the circle passing through (0, 0), (8, 0), and (0, −6).
9. Write the equation for a parabola whose vertex is at the origin and whose focus is (−4.25, 0).
10. Write an equation for a hyperbola whose transverse axis is horizontal, with center at (1, 1), passing through (6, 2) and (−3, 1).
11. Write the equation of a circle whose center is (−5, 0) and whose radius is 5.
12. Write the equation of a hyperbola whose center is at the origin, whose transverse axis = 8 and is horizontal, passing through (10, 25).
13. Write the equation of an ellipse whose foci are (2, 1) and (−6, 1), and the sum of the focal radii is 10.
14. Write the equation of a parabola whose axis is the line $y = -7$, whose vertex is 3 units to the right of the y axis, and passing through (4, −5).
15. Find the intercepts of the curve $y^2 + 4x - 6y = 16$.
16. Find the points of intersection of $x^2 + y^2 + 2x + 2y = 2$ and $3x^2 + 3y^2 + 5x + 5y = 10$.
17. A stone bridge arch is in the shape of half an ellipse and is 15.0 m wide and 5.00 m high. Find the height of the arch at a distance of 6.00 m from its center.
18. Write the equation of a hyperbola centered at the origin, where the conjugate axis = 12 and is vertical, and the distance between foci is 13.
19. A parabolic arch is 5.00 m high and 6.00 m wide at the base. Find the width of the arch at a height of 2.00 m above the base.
20. A stone thrown in the air follows a parabolic path, and reaches a maximum height of 56.0 ft in a horizontal distance of 48.0 ft. At what horizontal distances from the launch point will the height be 25.0 ft?

Derivatives of Algebraic Functions

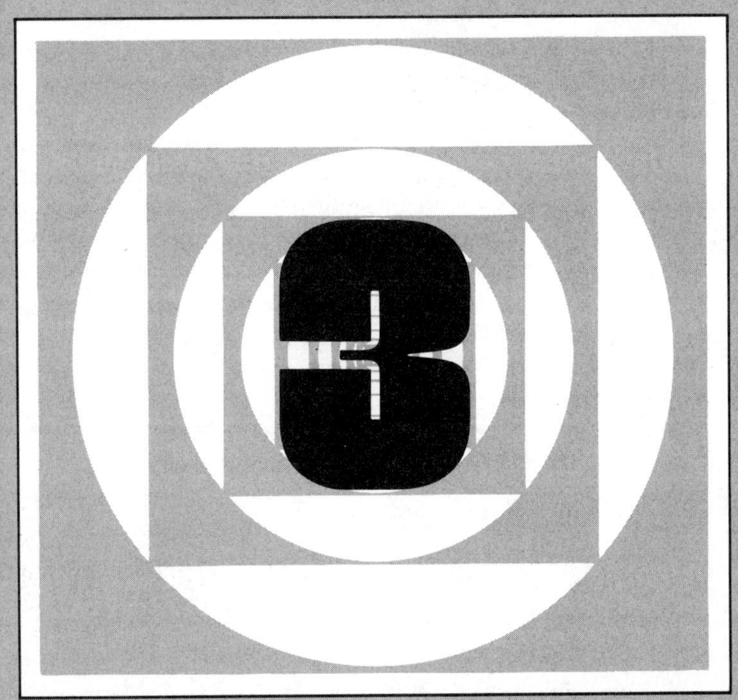

This chapter begins our study of calculus. We start with the limit concept, which is then applied in finding the derivative of a function. We then develop rules with which we can find derivatives quickly and easily.

The derivative is an extremely useful quantity that will enable us to do calculations that were not possible before. But the value of the derivative will not be apparent until the next few chapters, where we give applications, so do not get discouraged here, where we learn the mechanical details of taking the derivative. The payoff will come later.

We'll do derivatives of algebraic functions only, in this chapter, and save the trigonometric, logarithmic, and exponential functions for Chapter 6.

3-1. LIMITS

Limit Notation

Suppose that x and y are related by some function, such as $y = 3x$. Then if we are given a value of x, we can find a corresponding value for y. For example, when x *is* 2, then y *is* 6.

We now want to extend our mathematical language to be able to say what will happen to y *not* when x *is* a certain value, but when x *approaches* a certain value. For example, when x *approaches* 2, then y *approaches* 6. The *notation* we use to say the same thing is

$$\lim_{x \to 2} y = \lim_{x \to 2} (3x) = 6$$

We read this as "the limit, as x approaches 2, of $3x$, is 6."

In general, if the function $f(x)$ approaches some value L as x approaches a, we would indicate that with the notation

Limit Notation	$\lim_{x \to a} f(x) = L$	**300**

Well, why bother with new notation? Why not just say, in the preceding example, that y *is* 6 when x *is* 2? Why is it necessary to creep up on the answer like that?

It is true that limit notation offers no advantage in an example such as the last one. But we really need it when our function *cannot reach* the limit, or is *not even defined* at the limit.

EXAMPLE 1: The function $y = 1/x^2$ is graphed in Fig. 3-1. Even though y never reaches 0, we can still write

$$\lim_{x \to \infty} \frac{1}{x^2} = 0$$

Further, even though our function is not even defined at $x = 0$, because it results in division by zero, we can still write

$$\lim_{x \to 0} \frac{1}{x^2} = \infty$$

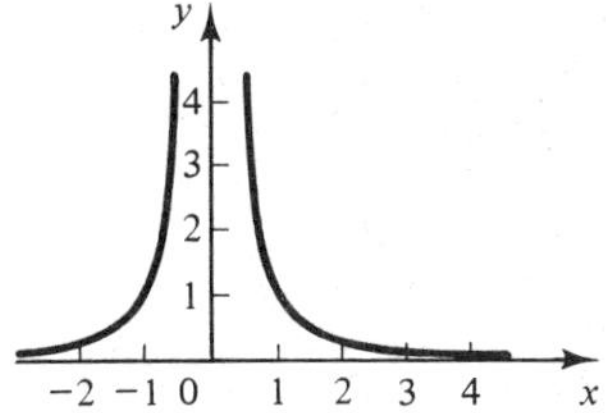

FIGURE 3-1. Graph of $y = 1/x^2$.

EXAMPLE 2: In Sec. 1-2 we defined the tangent line at P as the limiting position of the secant line PQ, as Q approached P. If Δx is the horizontal distance between P and Q (Fig. 3-2), we say that Q approaches P if Δx approaches zero. We can thus say that the slope m_t of the tangent line is the limit of the slope m_s of the secant line as Δx approaches zero. We can now restate this idea in compact form using our new limit notation, as

$$m_t = \lim_{\Delta x \to 0} m_s$$

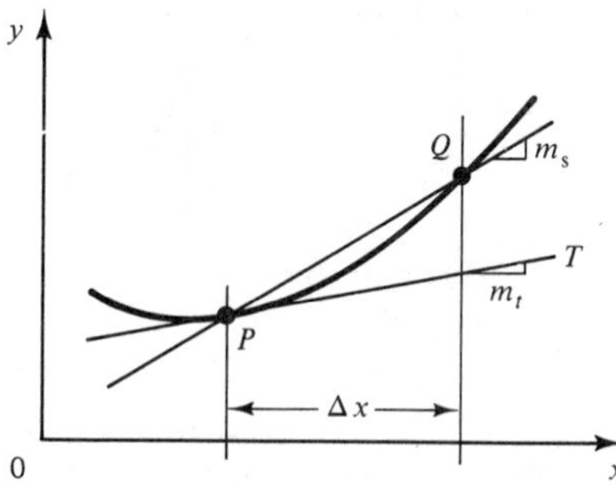

FIGURE 3-2. Tangent to a curve as the limiting position of the secant.

Limits Involving Zero or Infinity

Limits involving zero or infinity can usually be evaluated using the following facts. If C is a nonzero constant, then

(1) $\lim_{x \to 0} Cx = 0$	(4) $\lim_{x \to \infty} Cx = \infty$
(2) $\lim_{x \to 0} \frac{x}{C} = 0$	(5) $\lim_{x \to \infty} \frac{x}{C} = \infty$
(3) $\lim_{x \to 0} \frac{C}{x} = \infty$	(6) $\lim_{x \to \infty} \frac{C}{x} = 0$

EXAMPLE 3

(a) $\lim_{x \to 0} 5x^3 = 0$

(b) $\lim_{x \to 0} \left(7 + \frac{x}{2}\right) = 7$

(c) $\lim_{x \to 0} \left(3 + \frac{25}{x^2}\right) = \infty$

(d) $\lim_{x \to \infty} (3x - 2) = \infty$

(e) $\lim_{x \to \infty} \left(8 + \frac{x}{4}\right) = \infty$

(f) $\lim_{x \to \infty} \left(\frac{5}{x} - 3\right) = -3$

Common Error	Be sure to distinguish between a denominator that is zero and one that is **approaching** zero. In the first case we have division by zero, but in the second case we can get a useful answer.

When we want the limit, as x becomes infinite, of the quotient of two polynomials, such as

$$\lim_{x\to\infty} \frac{4x^3 - 3x^2 + 5}{3x - x^2 - 5x^3}$$

we see that both numerator and denominator become infinite. However, the limit of such an expression can be found if we *divide both numerator and denominator by the highest power of x occurring in either.*

EXAMPLE 4: Evaluate $\lim_{x\to\infty} \frac{4x^3 - 3x^2 + 5}{3x - x^2 - 5x^3}$.

Solution: Dividing both numerator and denominator by x^3 gives us

$$\lim_{x\to\infty} \frac{4x^3 - 3x^2 + 5}{3x - x^2 - 5x^3} = \lim_{x\to\infty} \frac{4 - \frac{3}{x} + \frac{5}{x^3}}{\frac{3}{x^2} - \frac{1}{x} - 5}$$

$$= \frac{4 - 0 + 0}{0 - 0 - 5} = -\frac{4}{5}$$

Limits of the Form 0/0

This entire discussion of limits is to prepare us for the idea of the derivative. There we will have to find the limit of a fraction in which *both the numerator and the denominator approach zero.*

At first glance, when we see a numerator approaching zero, we expect the entire fraction to approach zero. But when we see a denominator approaching zero, we throw up our hands and cry "division by zero." What then do we make of a fraction in which *both* numerator and denominator approach zero?

First, keep in mind that the denominator is not *equal to* zero; it is only *approaching* zero. Second, even though a shrinking numerator would make a fraction approach zero, in this case the denominator is also shrinking. So, in fact, our fraction will not necessarily approach infinity, or zero, but will often approach some useful finite value.

A useful way to investigate limits is by calculator or computer. We simply substitute x closer and closer to the given value, and see if the expression appears to approach a limit, as shown in the next example.

EXAMPLE 5: Evaluate $\lim_{x \to 0} \frac{\sqrt{9 + x} - 3}{3x}$.

Solution: We see that when x *equals* zero, we get

$$f(x) = \frac{\sqrt{9 + 0} - 3}{3(0)} = \frac{3 - 3}{0} = \frac{0}{0}$$

This, of course, is not a *proof* that the limit found is the correct one, or even that a limit exists.

a result that is indeterminate. But what happens when x *approaches* zero? Let us use the calculator to substitute smaller and smaller values of x. Working to five decimal places, we get

x	10	1	0.1	0.01	0.001	0.0001
$f(x)$	0.04530	0.05409	0.05540	0.05554	0.05555	0.05556

So as x approaches zero, $f(x)$ appears to approach $0.0555\overline{5}$ as a limit.

EXAMPLE 6: Find the limit of $(\sin x)/x$ (where x is in radians) as x approaches zero.

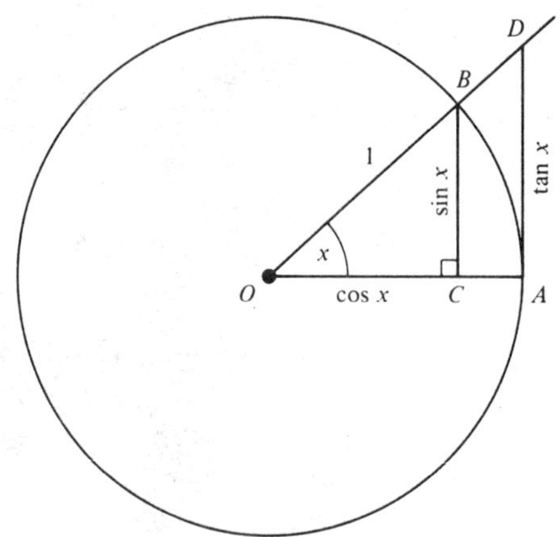

FIGURE 3-3

Solution: This is another case where both numerator and denominator approach zero. We can find the limit of this expression by calculator, substituting smaller and smaller values of x radians and calculating $(\sin x)/x$, (you can try this yourself), but we will use a different approach.

Let x be a small angle, in radians, in a circle of radius 1 (Fig. 3-3), in which

$$\sin x = \frac{BC}{1} = BC, \qquad \cos x = \frac{OC}{1} = OC, \qquad \tan x = \frac{AD}{OA} = AD$$

We see that the area of the sector OAB is greater than the area of triangle OBC but less than the area of triangle OAD. But by Eq. 137,

$$\text{area of triangle } OBC = \frac{1}{2} BC \cdot OC = \frac{1}{2} \sin x \cos x$$

and

$$\text{area of triangle } OAD = \frac{1}{2} \cdot OA \cdot AD = \frac{1}{2} \tan x$$

Also, by Eq. 116,

$$\text{area of sector } OAB = \frac{1}{2} r^2 x = \frac{x}{2}$$

So

$$\frac{1}{2} \sin x \cos x < \frac{x}{2} < \frac{1}{2} \tan x$$

Dividing by $\frac{1}{2} \sin x$ gives us

$$\cos x < \frac{x}{\sin x} < \frac{1}{\cos x}$$

Now letting x approach zero, both $\cos x$ and $1/\cos x$ approach 1. Therefore, $x/\sin x$, which is "sandwiched" between them, also approaches 1.

$$\lim_{x \to 0} \frac{x}{\sin x} = 1$$

We will use this result later when we take the derivative of the sine function.

Taking reciprocals yields

$$\lim_{x \to 0} \frac{\sin x}{x} = 1$$

Sometimes factoring the given expression will help us find a limit.

EXAMPLE 7: Evaluate $\lim_{x \to 3} \dfrac{x^2 - 9}{x - 3}$.

Solution: If we let x equal 3, the denominator becomes zero; hence the function does not exist there. Let us factor the numerator.

$$\lim_{x \to 3} \frac{x^2 - 9}{x - 3} = \lim_{x \to 3} \frac{(x - 3)(x + 3)}{x - 3} = \lim_{x \to 3} \left(\frac{x - 3}{x - 3}\right)(x + 3)$$

Now as x approaches 3, both the numerator and the denominator of the fraction $(x - 3)/(x - 3)$ approach zero. But since the numerator and denominator are equal, the fraction will equal 1 for any nonzero value of x, no matter how small. So

$$\lim_{x\to3}\frac{x^2-9}{x-3} = \lim_{x\to3}(x+3) = 6$$

EXAMPLE 8

$$\lim_{x\to2}\frac{x^2-3x+2}{x-2} = \lim_{x\to2}\frac{(x-2)(x-1)}{x-2}$$

$$= \lim_{x\to2}\left(\frac{x-2}{x-2}\right)(x-1) = \lim_{x\to2}(x-1) = 1$$

Limit Sometimes Depends on Direction of Approach

The limit of $y = 3x$, as x approached 2, was 6. It did not matter whether x approached 2 from below or above (from values less than 2 or from values greater than 2). But sometimes it does matter.

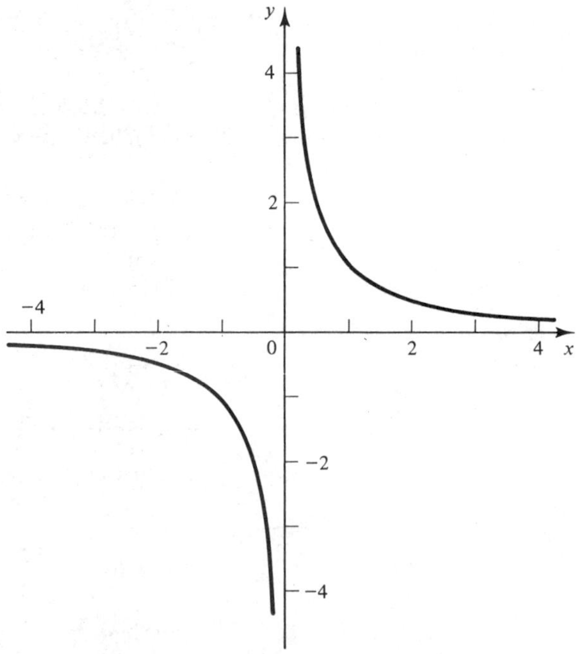

FIGURE 3-4. Graph of $y = 1/x$.

EXAMPLE 9: The function $y = 1/x$ is graphed in Fig. 3-4. When x approaches zero from above, which we write $x \to 0^+$, we get

$$\lim_{x \to 0^+} \frac{1}{x} = +\infty$$

which means that y grows without bound in the positive direction. But when x approaches zero from below, we write

$$\lim_{x \to 0^-} \frac{1}{x} = -\infty$$

which means that y grows without bound in the negative direction.

When the Limit Is an Expression

Our main use for limits will be for finding derivatives in the following sections of this chapter. There we will evaluate limits in which both the numerator and denominator approach zero, and the resulting limit is an *expression* rather than a single number. A limit typical of the sort we will have to evaluate later is given in the following example.

EXAMPLE 10: Evaluate $\displaystyle\lim_{d \to 0} \frac{(x+d)^2 + 5(x+d) - x^2 - 5x}{d}$.

Solution: If we try to set d equal to zero,

$$\frac{(x+0)^2 + 5(x+0) - x^2 - 5x}{0} = \frac{x^2 + 5x - x^2 - 5x}{0} = \frac{0}{0}$$

we get the indeterminate expression 0/0. Instead, let us remove parentheses in the original expression. We get

$$\lim_{d \to 0} \frac{(x+d)^2 + 5(x+d) - x^2 - 5x}{d}$$

$$= \lim_{d \to 0} \frac{x^2 + 2dx + d^2 + 5x + 5d - x^2 - 5x}{d}$$

$$= \lim_{d \to 0} \frac{2dx + d^2 + 5d}{d} = \lim_{d \to 0} (2x + d + 5)$$

after dividing through by d. Now letting d approach zero,

$$\lim_{d \to 0} \frac{(x+d)^2 + 5(x+d) - x^2 - 5x}{d} = 2x + 5$$

EXERCISE 1

Evaluate each limit.

1. $\lim_{x\to 2} (x^2 + 2x - 7)$

2. $\lim_{x\to -1} (x^3 - 3x^2 - 5x - 5)$

3. $\lim_{x\to 2} \dfrac{x^2 - x - 1}{x + 3}$

4. $\lim_{x\to 5} \dfrac{5 + 4x - x^2}{5 - x}$

5. $\lim_{x\to 5} \dfrac{x^2 - 25}{x - 5}$

6. $\lim_{x\to -7} \dfrac{49 - x^2}{x + 7}$

7. $\lim_{x\to 1} \dfrac{x^2 + 2x - 3}{x - 1}$

8. $\lim_{x\to 5} \dfrac{5 + 4x - x^2}{5 - x}$

You may want to evaluate some of these by calculator or computer.

Zero as a Limit

9. $\lim_{x\to 0} (4x^2 - 5x - 8)$

10. $\lim_{x\to 0} \dfrac{3 - 2x}{x + 4}$

11. $\lim_{x\to 0} \dfrac{\sqrt{x} - 4}{\sqrt[3]{x} + 5}$

12. $\lim_{x\to 0} \dfrac{3 + x - x^2}{(x + 3)(5 - x)}$

13. $\lim_{x\to 0} \left(\dfrac{1}{2 + x} - \dfrac{1}{2}\right) \cdot \dfrac{1}{x}$

14. $\lim_{x\to 0} x \cos x$

15. $\lim_{x\to 0} \dfrac{7}{x}$

16. $\lim_{x\to 0} \dfrac{e^x}{x}$

17. $\lim_{x\to 0} \dfrac{\cos x}{x}$

18. $\lim_{x\to 0} \dfrac{\sin x}{\tan x}$

19. $\lim_{x\to 0^+} \dfrac{x + 1}{x}$

20. $\lim_{x\to 0^-} \dfrac{x + 1}{x}$

Infinity as a Limit

21. $\lim_{x\to \infty} \dfrac{2x + 5}{x - 4}$

22. $\lim_{x\to \infty} \dfrac{5x - x^2}{2x^2 - 3x}$

23. $\lim_{x\to \infty} \dfrac{x^2 + x - 3}{5x^2 + 10}$

24. $\lim_{x\to \infty} \dfrac{4 + 2^x}{3 + 2^x}$

25. $\lim_{x\to -\infty} 10^x$

26. $\lim_{x\to \infty} 10^x$

Where the Limit Is an Expression

27. $\lim_{d\to 0} x^2 + 2d + d^2$

28. $\lim_{d\to 0} 2x + d$

29. $\lim_{d\to 0} \frac{(x+d)^2 - x^2}{x^2(x+d)}$

30. $\lim_{d\to 0} \frac{(x+d)^2 - x^2}{d}$

31. $\lim_{d\to 0} \frac{3(x+d) - 3x}{d}$

32. $\lim_{d\to 0} \frac{[2(x+d)+5] - (2x+5)}{d}$

33. $\lim_{d\to 0} 3x + d - \frac{1}{(x+d+2)(x-2)}$

34. $\lim_{d\to 0} \frac{[(x+d)^2+1] - (x^2+1)}{d}$

35. $\lim_{d\to 0} \frac{(x+d)^3 - x^3}{d}$

36. $\lim_{d\to 0} \frac{\sqrt{x+d} - \sqrt{x}}{d}$

37. $\lim_{d\to 0} \frac{(x-d)^2 - 2(x+d) - x^2 + 2x}{d}$

38. $\lim_{d\to 0} \frac{\frac{7}{x+d} - \frac{7}{d}}{d}$

Computer

39. Write a program or use a spreadsheet to compute and print the value of (sin x)/x, where x is in radians. Start with $x = 1$ rad and decrease it each time by a factor of 10, (1, 0.1, 0.01, 0.001, and so forth). Is the value of (sin x)/x approaching a limit?

3-2. THE DERIVATIVE

In this section we use some familiar ideas about motion to introduce the idea of *rate of change,* which then leads us to the definition of the derivative.

Rates of Change

For *uniform motion* (for which the velocity is constant) the graph of displacement versus time gives a straight line (Fig. 3-5a) whose slope

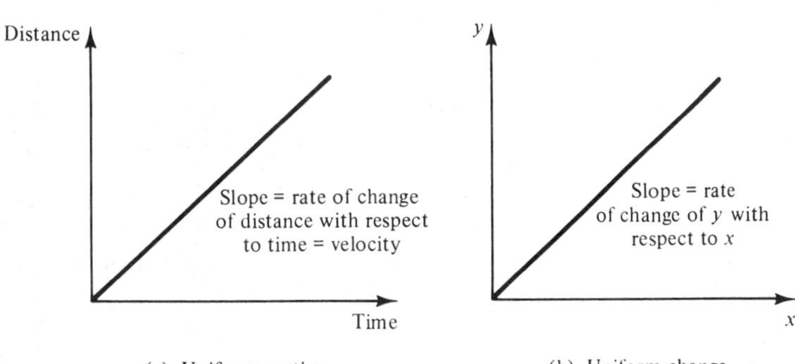

(a) Uniform motion

(b) Uniform change

FIGURE 3-5

is equal to the (constant) velocity. But rates do not have to involve time and distance. To be general, we will use our usual symbols, x for the independent variable and y for the dependent variable (Fig. 3-5b). The slope of this line now gives *the rate of change of* y *with respect to* x.

Average Rate of Change

Now we depart from the uniformly changing quantities studied up to this point, and the departure is a major one. Most quantities in the real world do not change at a constant rate, so we cannot use the simple formula, amount of change = rate × time. The graph of y versus x is no longer a straight line, but is some curve, as in Fig. 3-6 and the slope changes as x changes.

But what can we say about the rate of change when a quantity is changing nonuniformly? We can give the *average rate of change,* over some interval. The average rate of change over some interval is equal to the change in y divided by the change in x, over that interval. For the interval PQ in Fig. 3-7 the average rate is obviously the slope of the line connecting P and Q.

If x were time and y were distance, the average rate of change would be the *average velocity* over that interval.

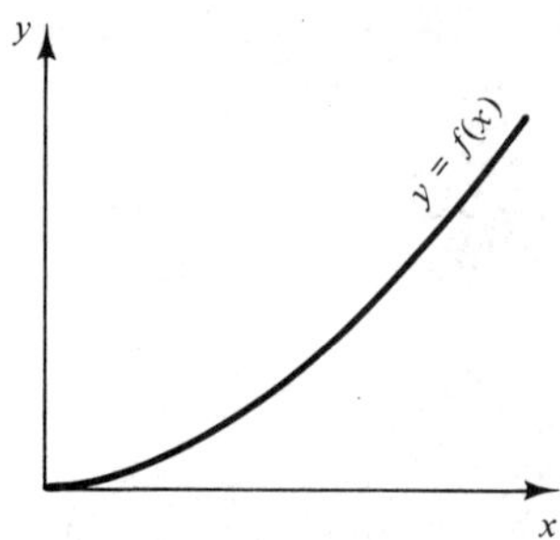

FIGURE 3-6. Nonuniform change.

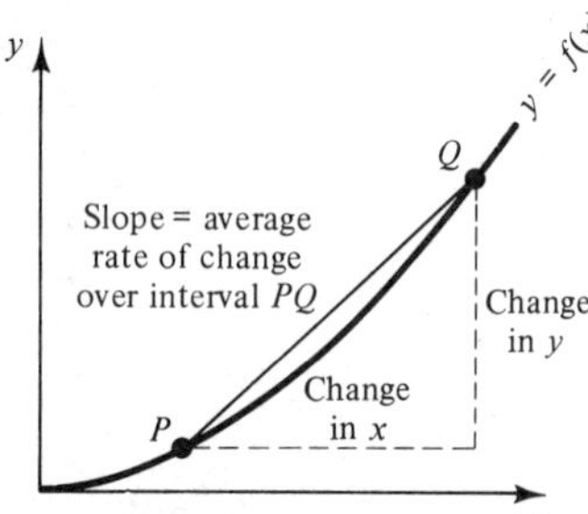

FIGURE 3-7. Average rate of change.

EXAMPLE 1: Find the average rate of change of the function $y = 3x^2 + 1$ over the interval $x = 1$ to $x = 3$.

Solution: We compute y at $x = 1$ and $x = 3$,

$$y(1) = 3(1)^2 + 1 = 4$$

and

$$y(3) = 3(3)^2 + 1 = 28$$

The endpoints of the interval are thus $P(1, 4)$ and $Q(3, 28)$. Then

$$\text{average rate of change} = \text{slope of } PQ = \frac{28 - 4}{3 - 1} = 12$$

EXAMPLE 2: During one day of a certain cross-country journey, a motorist travels 500 mi in 10 h. The average velocity is

$$\frac{500 \text{ mi}}{10 \text{ h}} = 50 \text{ mi/h}$$

even though part of the time may have been spent going faster than 50 mi/h, or slower, or even stopped or backing up.

Instantaneous Rate of Change

As the name implies, the instantaneous rate of change of some quantity is the rate of change at a given instant. In the preceding example of the motorist, it would be given by the speedometer reading at the instant in question (except when backing up). Graphically, the instantaneous rate of change at a point P (Fig. 3-8) is equal to the slope of the tangent line PT at P.

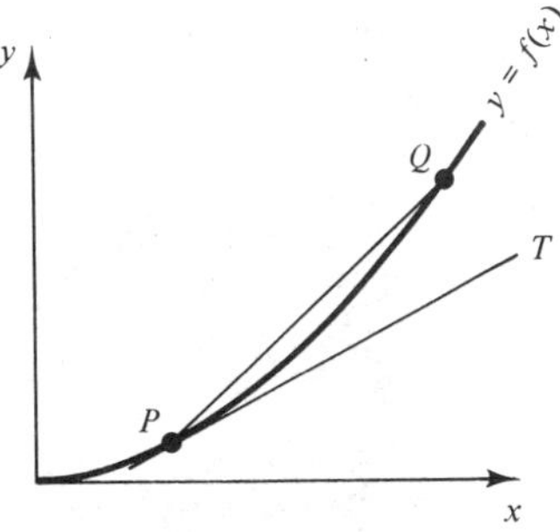

FIGURE 3-8

We come now to the heart of the problem. We were able to compute the average rate of change over the interval PQ because we had *two points* on line PQ from which to calculate its slope. But how can we find the slope of the tangent line PT when we have only the single point P?

In Sec. 1-2 we defined the tangent line at P as the limiting position of the secant line PQ, as Q approached P. But when it came to finding the *slope* of the tangent, we used a graphical method, actually measuring rise and run on the graph of the tangent line. We'll now derive an equation for finding that slope. We'll first write an expression for the slope of PQ. Then we'll let point Q approach P along the curve. Since the tangent PT is the limiting position of the secant PQ, then *the slope of PT will be the limit of the slope of PQ*.

The Derivative

Let us redraw Fig. 3-8 in more detail, as in Fig. 3-9. Say that we want the instantaneous rate of change (and hence the slope of the tangent PT) of the function $y = f(x)$ at the point $P(x, y)$. We place a second point Q on the curve, very close to P. The small horizontal distance between P and Q we call an *increment* in x, Δx, and the small vertical distance we call Δy, as we did in Sec. 1-1.

The *average* rate of change over the interval Δx is then equal to the slope of line PQ. Since we have two points, P and Q, on this line, we can write an expression for its slope. By Eq. 248,

$$\text{slope of } PQ = \frac{\text{rise}}{\text{run}} = \frac{\Delta y}{\Delta x} = \frac{f(x + \Delta x) - f(x)}{\Delta x}$$

We now let Δx approach zero, and point Q will approach P along the curve. Since the tangent PT is the limiting position of the secant PQ, the slope of PQ will approach the slope of PT.

$$\text{slope of } PT = \lim_{\Delta x \to 0} \frac{\Delta y}{\Delta x} = \lim_{\Delta x \to 0} \frac{f(x + \Delta x) - f(x)}{\Delta x}$$

We will see later that *dy* and *dx* in the symbol *dy/dx* have separate meanings of their own. But for now we will treat *dy/dx* as a single symbol.

This important quantity is called the *derivative* of y with respect to x, and is given the symbol $\dfrac{dy}{dx}$.

The Derivative	$\dfrac{dy}{dx} = \lim_{\Delta x \to 0} \dfrac{\Delta y}{\Delta x} = \lim_{\Delta x \to 0} \dfrac{f(x + \Delta x) - f(x)}{\Delta x}$	**302**

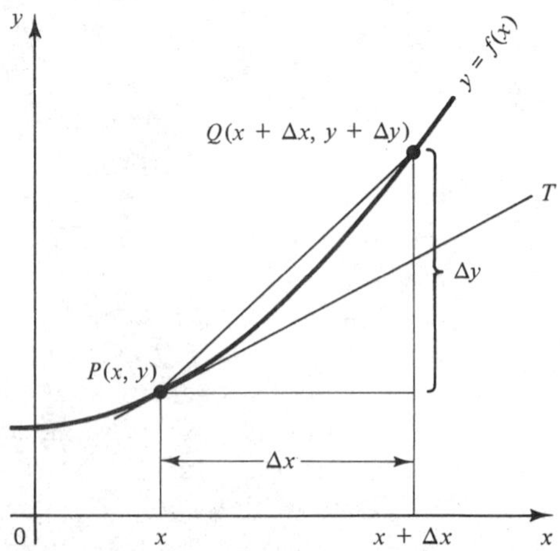

FIGURE 3-9. Derivation of the derivative.

EXAMPLE 3: Find the derivative of $f(x) = x^2$.

Solution: By Eq. 302,

$$\frac{dy}{dx} = \lim_{\Delta x \to 0} \frac{f(x + \Delta x) - f(x)}{\Delta x}$$

$$= \lim_{\Delta x \to 0} \frac{(x + \Delta x)^2 - x^2}{\Delta x}$$

$$= \lim_{\Delta x \to 0} \frac{x^2 + 2x\,\Delta x + (\Delta x)^2 - x^2}{\Delta x}$$

$$= \lim_{\Delta x \to 0} \frac{2x\,\Delta x + (\Delta x)^2}{\Delta x}$$

$$= \lim_{\Delta x \to 0} (2x + \Delta x) = 2x$$

Common Error	The symbol Δx is a *single symbol.* It is not Δ *times* x. Thus $$x \cdot \Delta x \neq \Delta x^2$$

Derivatives by the Delta Method

Our main use for the delta method will be to derive *rules* with which we can quickly find derivatives.

Instead of substituting directly into Eq. 302, some prefer to apply this equation in a series of steps. This is sometimes referred to as the *delta method, delta process,* or *four-step rule.*

EXAMPLE 4: (a) Find the derivative of the function $y = 3x^2$ by the delta method and (b) evaluate the derivative at $x = 1$.

Solution: (a) 1. First give x an increment Δx and compute the resulting increment in y, which we call Δy.

$$y + \Delta y = 3(x + \Delta x)^2$$
$$= 3[x^2 + 2x\,\Delta x + (\Delta x)^2]$$
$$= 3x^2 + 6x\,\Delta x + 3(\Delta x)^2$$

2. Subtract the original function, $y = 3x^2$.

$$(y + \Delta y) - y = 3x^2 + 6x\,\Delta x + 3(\Delta x)^2 - 3x^2$$
$$\Delta y = 6x\,\Delta x + 3(\Delta x)^2$$

3. Divide by Δx.

$$\frac{\Delta y}{\Delta x} = \frac{6x\,\Delta x + 3(\Delta x)^2}{\Delta x}$$

4. Let Δx approach zero. This causes Δy also to approach zero, and appears to make $\Delta y/\Delta x$ equal to the indeterminate expression 0/0. But recall from our study of limits in the preceding section that a fraction can often have a limit *even when both numerator and denominator approach zero*. To find it, we divide through by Δx and get $\Delta y/\Delta x = 6x + 3\,\Delta x$. Then

$$\frac{dy}{dx} = \lim_{\Delta x \to 0} \frac{\Delta y}{\Delta x} = 6x + 3(0) = 6x$$

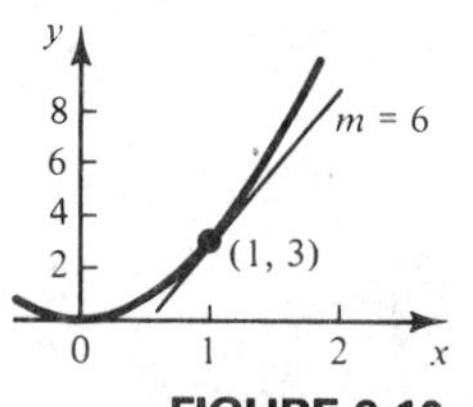

FIGURE 3-10

(b) When $x = 1$,

$$\left.\frac{dy}{dx}\right|_{x=1} = 6(1) = 6$$

The curve $y = 3x^2$ is graphed in Fig. 3-10 with a tangent line of slope 6 drawn at the point (1, 3).

If our expression is a fraction with x in the denominator, step 2 will require us to find a common denominator, as in the following example.

EXAMPLE 5: Find the derivative of $y = \dfrac{3}{x^2 + 1}$.

Solution: 1. Substitute $x + \Delta x$ for x and $y + \Delta y$ for y.

$$y + \Delta y = \frac{3}{(x + \Delta x)^2 + 1}$$

2. Subtracting,

$$y + \Delta y - y = \frac{3}{(x + \Delta x)^2 + 1} - \frac{3}{x^2 + 1}$$

Combining the fractions over a common denominator gives

$$\Delta y = \frac{3(x^2 + 1) - 3[(x + \Delta x)^2 + 1]}{[(x + \Delta x)^2 + 1](x^2 + 1)}$$

which reduces to

$$\Delta y = \frac{-6x\,\Delta x - 3(\Delta x)^2}{[(x + \Delta x)^2 + 1](x^2 + 1)}$$

3. Dividing by Δx,

$$\frac{\Delta y}{\Delta x} = \frac{-6x - 3\,\Delta x}{[(x + \Delta x)^2 + 1](x^2 + 1)}$$

4. Letting Δx approach zero gives

$$\frac{dy}{dx} = -\frac{6x}{(x^2 + 1)^2}$$

The following example shows how to differentiate an expression containing a radical.

EXAMPLE 6: Find the slope of the tangent to the curve $y = \sqrt{x}$ at the point (4, 2).

Solution: We first find the derivative by the delta method.

1. $$y + \Delta y = \sqrt{x + \Delta x}$$

2. $$\Delta y = \sqrt{x + \Delta x} - \sqrt{x}$$

When simplifying radicals, we used to rationalize the denominator. Here we rationalize the numerator instead!

The later steps will be easier if we now write this expression as a fraction with no radicals in the numerator. To do this, we multiply by the conjugate, and get

$$\Delta y = (\sqrt{x + \Delta x} - \sqrt{x}) \cdot \frac{\sqrt{x + \Delta x} + \sqrt{x}}{\sqrt{x + \Delta x} + \sqrt{x}}$$

$$= \frac{(x + \Delta x) - x}{\sqrt{x + \Delta x} + \sqrt{x}}$$

$$= \frac{\Delta x}{\sqrt{x + \Delta x} + \sqrt{x}}$$

3. Dividing by Δx yields

$$\frac{\Delta y}{\Delta x} = \frac{1}{\sqrt{x + \Delta x} + \sqrt{x}}$$

4. Letting Δx approach zero gives

$$\frac{dy}{dx} = \frac{1}{2\sqrt{x}}$$

When $x = 4$,

$$\left.\frac{dy}{dx}\right|_{x=4} = \frac{1}{4}$$

which is the slope of the tangent to $y = \sqrt{x}$ at the point (4, 2).

If the derivation of the derivative was less than convincing to you, you can at least convince yourself that the end result is correct. Make a good graph of this curve, carefully draw the tangent, and measure its slope. It should agree with the result found with the derivative.

Other Symbols for the Derivative

There are other symbols which are used instead of dy/dx to indicate a derivative.

A prime (′) symbol is often used. For example, if a certain function is equal to y, then the symbol y' can represent the derivative of that function. Further, if $f(x)$ is a certain function, then $f'(x)$ is the derivative of that function. Thus

$$\frac{dy}{dx} = y' = f'(x)$$

The y' or f' notation is handy for specifying the derivative at a particular value of x.

EXAMPLE 7: To specify a derivative evaluated at $x = 2$, we can write $y'(2)$, or $f'(2)$, instead of the clumsier

$$\left.\frac{dy}{dx}\right|_{x=2}$$

The Derivative as an Operator

We can think of the derivative as an *operator;* one that operates on a function to produce the derivative of that function. The symbol d/dx or D in front of an expression indicates that the expression is to be differentiated. For example, the symbols

$$\frac{d}{dx}(\ ) \qquad \text{or} \qquad D_x(\ ) \qquad \text{or} \qquad D(\ )$$

mean to find the derivative of the expression enclosed in parentheses.

EXAMPLE 8: We saw in Example 4(a) that if $y = 3x^2$, then $dy/dx = 6x$. This same result can be written

$$\frac{d(3x^2)}{dx} = 6x$$

$$D_x(3x^2) = 6x$$

$$D(3x^2) = 6x$$

Keep in mind that even though the notation is different, we find the derivative in *exactly the same way*.

Points Where the Derivative Does Not Exist

A curve is said to be *discontinuous* at a value of x where there is a break or gap. The derivative does not exist at such points. It also does not exist where the curve has a jump, corner, cusp, or any other feature at which it is not possible to draw a tangent line, as the points shown in Fig. 3-11. At such points, we say that the function is *not differentiable*.

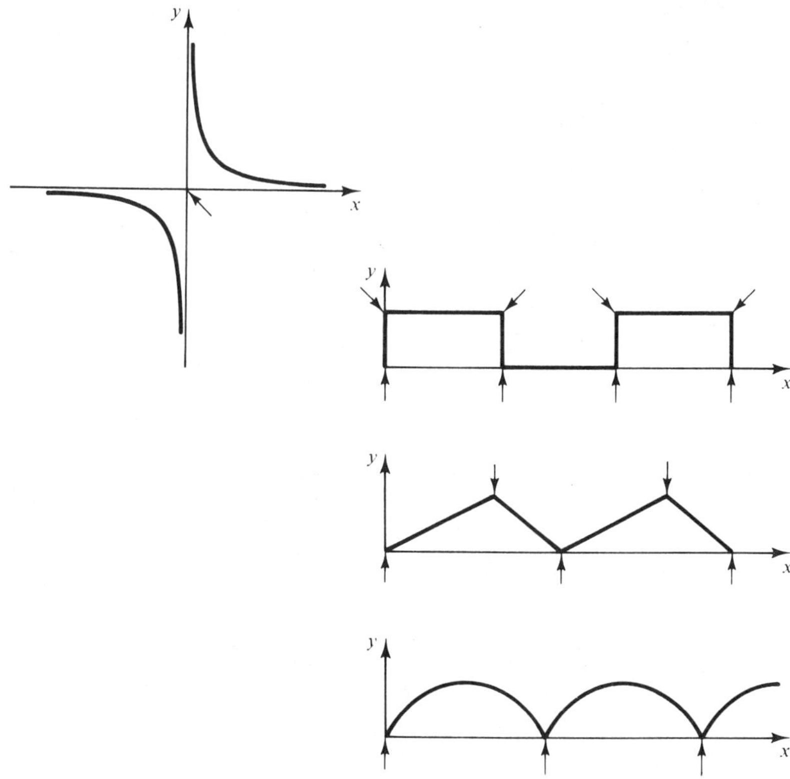

FIGURE 3-11. The arrows show the points where the derivative does not exist.

EXERCISE 2

Delta Method

Find the derivative by the delta method.

1. $y = 2x + 5$

2. $y = 7 - 4x$

3. $y = x^2 + 1$

4. $y = x^2 - 3x + 5$

5. $y = x^3$

6. $y = x^3 - x^2$

7. $y = \dfrac{3}{x}$

8. $y = \dfrac{x}{(x-1)^2}$

9. $y = \sqrt{3 - x}$

10. $y = \dfrac{1}{\sqrt{x}}$

Find the slope of the tangent at the given value of x.

11. $y = \dfrac{1}{x^2}$ at $x = 1$

12. $y = \dfrac{1}{x+1}$ at $x = 2$

13. $y = x + \dfrac{1}{x}$ at $x = 2$

14. $y = \dfrac{1}{x}$ at $x = 3$

Find the instantaneous rate of change of each function at the given value of x.

15. $y = 2x - 3$ at $x = 3$

16. $y = 16x^2$ at $x = 1$

17. $y = 2x^2 - 6$ at $x = 3$

18. $y = x^2 + 2x$ at $x = 1$

Other Symbols for the Derivative

19. If $y = 2x^2 - 3$, find y'.

20. In Problem 19, find $y'(3)$.

21. In Problem 19, find $y'(-1)$.

22. If $f(x) = 7 - 4x^2$, find $f'(x)$

23. In Problem 22, find $f'(2)$.

24. In Problem 22, find $f'(-3)$.

Operator Notation

Evaluate.

25. $\dfrac{d}{dx}(3x + 2)$

26. $\dfrac{d}{dx}(x^2 - 1)$

27. $D_x(7 - 5x)$

28. $D_x(x^2)$

29. $D(3x + 2)$

30. $D(x^2 - 1)$

Computer

31. Write a program or use a spreadsheet to compute the slope of the secant line drawn to the curve $y = x^2 - 2x$ at the point (3, 3). Start with $\Delta x = 1$, and halve it for each subsequent calculation. Have the computer print the following table.

DELTA X	X + DELTA X	Y + DELTA Y	DELTA Y	DELTA Y/DELTA X
1	4	8	5	5
0.5	3.5	5.25	2.25	4.5
0.25	3.25	4.06	1.06	4.24
0.125	3.125	3.516	0.516	4.13
.	.	.	.	.
.	.	.	.	.
.	.	.	.	.

Let the program run until $\Delta y/\Delta x$ stops changing in the second decimal place.

3-3. RULES FOR DERIVATIVES

We now apply the delta method to various kinds of simple functions, and derive a rule for each with which we can then find the derivative.

The Derivative of a Constant

To find the derivative of the function $y = c$, we substitute c for $f(x)$ and c for $f(x + \Delta x)$ in Eq. 302, and get

$$\frac{d(c)}{dx} = \lim_{\Delta x \to 0} \frac{c - c}{\Delta x} = 0$$

Derivative of a Constant	$\frac{d(c)}{dx} = 0$	**304**

The derivative of a constant is zero.

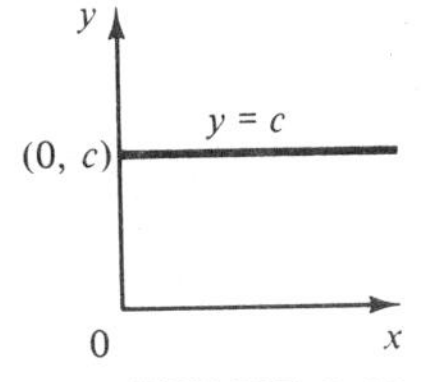

FIGURE 3-12

This is not surprising, because the graph of function $y = c$ is a straight line parallel to the x axis (Fig. 3-12) whose slope is, of course, zero.

EXAMPLE 1: If $y = 2\pi^2$ then

$$\frac{dy}{dx} = 0$$

Derivative of a Constant Times a Power of *x*

We let $y = cx^n$, where n is a positive integer and c is any constant. Using the delta method:

1. We substitute $x + \Delta x$ for x and $y + \Delta y$ for y,

$$y + \Delta y = c(x + \Delta x)^n$$

By the binomial formula (Eq. 212),

$$y + \Delta y = c\left[x^n + nx^{n-1}(\Delta x) + \frac{n(n-1)}{2}x^{n-2}(\Delta x)^2 + \cdots + (\Delta x)^n\right]$$

2. Subtracting $y = cx^n$, we get

The binomial formula shows the expression we get when a binomial $(a + b)$ is raised to any positive integral power n.

$$\Delta y = c\left[nx^{n-1}(\Delta x) + \frac{n(n-1)}{2}x^{n-2}(\Delta x)^2 + \cdots + (\Delta x)^n\right]$$

3. Dividing by Δx yields

$$\frac{\Delta y}{\Delta x} = c\left[nx^{n-1} + \frac{n(n-1)}{2}x^{n-2}\,\Delta x + \cdots + (\Delta x)^{n-1}\right]$$

4. Finally, letting Δx approach zero, all terms but the first vanish. Thus

$$\frac{dy}{dx} = cnx^{n-1}$$

or

Derivative of a Constant Times a Power of x	$\frac{d}{dx}cx^n = cnx^{n-1}$	**307**

The derivative of a constant times a power of x is equal to the product of the constant, the exponent, and x raised to the exponent reduced by one.

EXAMPLE 2

(a) If $y = x^3$, then by Eq. 307,

$$\frac{dy}{dx} = 3x^{3-1} = 3x^2$$

(b) If $y = 5x^2$, then

$$\frac{dy}{dx} = 5(2)x^{2-1} = 10x$$

Power Function with Negative Exponent

In our derivation of Eq. 307 we had required that the exponent n be a positive integer. We'll now show that the rule is also valid when n is a *negative* integer.

If n is a negative integer, then $m = -n$ is a positive integer. So

$$y = cx^n = cx^{-m} = \frac{c}{x^m}$$

We again use the delta method.

1. We substitute $x + \Delta x$ for x and $y + \Delta y$ for y.

$$y + \Delta y = \frac{c}{(x + \Delta x)^m}$$

2. Subtracting,

$$\Delta y = \frac{c}{(x + \Delta x)^m} - \frac{c}{x^m} = c\frac{x^m - (x + \Delta x)^m}{x^m(x + \Delta x)^m}$$

$$= c\frac{x^m - (x^m + mx^{m-1}\,\Delta x + kx^{m-2}\,\Delta x^2 + \cdots + \Delta x^m)}{x^m(x + \Delta x)^m}$$

3. Dividing by Δx gives

$$\frac{\Delta y}{\Delta x} = -c\frac{mx^{m-1} + kx^{m-2}\,\Delta x + \cdots + \Delta x^{m-1}}{x^m(x + \Delta x)^m}$$

4. Letting Δx go to zero, we get

$$\frac{dy}{dx} = -c\frac{mx^{m-1}}{x^{2m}} = -cmx^{m-1-2m} = -cmx^{-m-1} = cnx^{n-1}$$

This shows that Eq. 307 is valid when the exponent n is negative, as well as positive.

EXAMPLE 3

(a) If $y = x^{-4}$, then

$$\frac{dy}{dx} = -4x^{-5} = -\frac{4}{x^5}$$

(b) $\dfrac{d(3x^{-2})}{dx} = 3(-2)x^{-3} = -\dfrac{6}{x^3}$

(c) If $y = -3/x^2$, then $y = -3x^{-2}$ and

$$y' = -3(-2)x^{-3} = \frac{6}{x^3}$$

Power Function with Fractional Exponent

We have shown that the exponent n in Eq. 307 can be a positive or a negative integer. The rule is also valid when n is a positive or a negative rational number. We'll prove it in Sec. 3-4.

EXAMPLE 4: If $y = x^{-5/3}$, then

$$\frac{dy}{dx} = -\frac{5}{3}x^{-8/3}$$

To find the derivative of a radical, write it in exponential form, and use Eq. 307.

EXAMPLE 5: If $y = \sqrt[3]{x^2}$, then

$$\frac{dy}{dx} = \frac{d}{dx}x^{2/3} = \frac{2}{3}x^{-1/3} = \frac{2}{3\sqrt[3]{x}}$$

Derivative of a Sum

We want the derivative of the function

$$y = u + v + w$$

where u, v, and w are all functions of x. When x changes by an increment Δx, then u, v, and w will change by amounts Δu, Δv, and Δw, respectively. Using the delta method, we have

$$y + \Delta y = (u + \Delta u) + (v + \Delta v) + (w + \Delta w)$$

Subtracting yields

$$y + \Delta y - y = u + \Delta u + v + \Delta v + w + \Delta w - u - v - w$$

$$\Delta y = \Delta u + \Delta v + \Delta w$$

Dividing by Δx gives us

$$\frac{\Delta y}{\Delta x} = \frac{\Delta u + \Delta v + \Delta w}{\Delta x}$$

One of the theorems on limits (which we will not prove) is that the limit of the sum of several functions is equal to the sum of their individual limits.

So

$$\frac{dy}{dx} = \lim_{\Delta x \to 0}\left(\frac{\Delta u}{\Delta x} + \frac{\Delta v}{\Delta x} + \frac{\Delta w}{\Delta x}\right)$$

$$= \lim_{\Delta x \to 0}\frac{\Delta u}{\Delta x} + \lim_{\Delta x \to 0}\frac{\Delta v}{\Delta x} + \lim_{\Delta x \to 0}\frac{\Delta w}{\Delta x}$$

Derivative of a Sum	$\frac{d}{dx}(u + v + w) = \frac{du}{dx} + \frac{dv}{dx} + \frac{dw}{dx}$	**308**

The derivative of the sum of several functions is equal to the sum of the derivatives of those functions.

In addition to using the rule for sums here, we also need the rules for a power function and for a constant. We usually have to apply several rules in one problem.

EXAMPLE 6

$$\frac{d}{dx}(2x^3 - 3x^2 + 5x + 4) = 6x^2 - 6x + 5$$

EXAMPLE 7: Find the derivative of $y = \dfrac{x^2 + 3}{x}$.

Solution: At first glance it looks as if none of the rules learned so far apply here. But if we rewrite the function as

$$y = \frac{x^2 + 3}{x} = x + \frac{3}{x} = x + 3x^{-1}$$

we may write

$$\frac{dy}{dx} = 1 + 3(-1)x^{-2}$$

$$= 1 - \frac{3}{x^2}$$

Common Error	Students often forget to simplify an expression as much as possible **before** taking the derivative.

Other Variables

So far we have been using x for the independent variable and y for the dependent variable, but now let us get some practice using other variables.

EXAMPLE 8

(a) If $s = 3t^2 - 4t + 5$, then

$$\frac{ds}{dt} = 6t - 4$$

(b) If $y = 7u - 5u^3$, then

$$\frac{dy}{du} = 7 - 15u^2$$

(c) If $u = 9 + x^2 - 2x^4$, then

$$\frac{du}{dx} = 2x - 8x^3$$

EXAMPLE 9: Find the derivative of $T = 4z^2 - 2z + 5$.

Solution: What are we to find here, dT/dz, or dz/dT, or dT/dx, or something else?

> When we say "derivative," we mean the derivative of the given function **with respect to the independent variable,** unless otherwise noted.

So here we find the derivative of the given function T with respect to the independent variable z.

$$\frac{dT}{dz} = 8z - 2$$

EXERCISE 3

Here, as usual, the letters *a, b, c, . . . ,* represent constants.

Find the derivative of each function.

1. $y = a^2$

2. $y = 3b + 7c$

3. $y = x^7$

4. $y = x^4$

5. $y = 3x^2$

6. $y = 5.4x^3$

7. $y = x^{-5}$

8. $y = 2x^{-3}$

9. $y = \dfrac{1}{x}$

10. $y = \dfrac{1}{x^2}$

11. $y = \frac{3}{x^3}$

12. $y = \frac{3}{2x^2}$

13. $y = 7.5x^{1/3}$

14. $y = 4x^{5/3}$

15. $y = 4\sqrt{x}$

16. $y = 3\sqrt[3]{x}$

17. $y = -17\sqrt{x^3}$

18. $y = -2\sqrt[5]{x^4}$

19. $y = 3 - 2x$

20. $y = 4x^2 + 2x^3$

21. $y = 3x - x^3$

22. $y = x^4 + 3x^2 + 2$

23. $y = 3x^3 + 7x^2 - 2x + 5$

24. $y = x^3 - x^{3/2} + 3x$

25. $y = ax + b$

26. $y = ax^5 - 5bx^3$

27. $y = \frac{x^2}{2} - \frac{x^7}{7}$

28. $y = \frac{x^3}{1.75} + \frac{x^2}{2.84}$

29. $y = 2x^{3/4} + 4x^{-1/4}$

30. $y = \frac{2}{x} - \frac{3}{x^2}$

31. $y = 2x^{4/3} - 3x^{2/3}$

32. $y = x^{2/3} - a^{2/3}$

33. $y = \frac{x + 4}{x}$

34. $y = \frac{x^3 + 1}{x}$

35. If $y = 2x^3 - 3$, find y'.

36. In Problem 35, find $y'(2)$.

37. In Problem 35, find $y'(-3)$.

38. If $f(x) = 7 - 4x^2$, find $f'(x)$.

39. In Problem 38, find $f'(1)$.

40. In Problem 38, find $f'(-3)$.

Evaluate.

41. $\frac{d}{dx}(3x^5 + 2x)$

42. $\frac{d}{dx}(2.5x^2 - 1)$

43. $D_x(7.8 - 5.2x^{-2})$

44. $D_x(4x^2 - 1)$

45. $D(3x^2 + 2x)$

46. $D(1.75x^{-2} - 1)$

47. Find the slope of the tangent to the curve $y = x^2 - 2$, where x equals 1.

48. Find the instantaneous rate of change of the function $y = x - x^2$ at $x = 2$.

49. If $y = x^3 - 5$, find $y'(1)$.

50. If $f(x) = 1/x^2$, find $f'(2)$.

51. If $f(x) = 2.75x^2 - 5.02x$, find $f'(3.36)$.

52. If $y = \sqrt{83.2x^3}$, find $y'(1.74)$.

Other Variables

Find the derivative with respect to the independent variable.

53. $v = 5t^2 - 3t + 4$

54. $z = 9 - 8w + w^2$

55. $s = 58.3t^3 - 63.8t$

56. $x = 3.82y + 6.25y^4$

57. $y = \sqrt{5w^3}$

58. $w = \frac{5}{x} - \frac{3}{x^2}$

59. $v = \frac{85.3}{t^4}$

60. $T = 3.55\sqrt{1.06w^5}$

3-4. DERIVATIVE OF A FUNCTION RAISED TO A POWER

Composite Functions

In Sec. 3-3 we derived rule 307 for finding the derivative of x raised to a power. But rule 307 does not apply when we have an *expression* raised to a power, such as

$$y = (2x + 7)^5 \tag{1}$$

We may consider this function as being made up of two parts. One part is the function, which we shall call u, that is being raised to the power.

$$u = 2x + 7 \tag{2}$$

Then y, which is a function of u, is seen to be

$$y = u^5 \tag{3}$$

Our original function (1), which can be obtained by combining (2) and (3), is called a *composite function.*

The Chain Rule

The chain rule will enable us to take the derivative of a composite function, such as (1) above. Consider the situation where y is a function of u,

$$y = f(u)$$

and u, in turn, is a function of x,

$$u = g(x)$$

The chain rule is true even for those rare cases where Δu may be zero, but it takes a more complicated proof to show this.

Let x change by an increment Δx. This causes u to change by some amount which we shall call Δu, and this change, in turn, causes a change in y which we call Δy. The derivative dy/dx will be the limit of the quotient $\Delta y/\Delta x$ as Δx approaches zero. But before taking the limit, let us multiply our quotient by $\Delta u/\Delta u$, assuming here that Δu is not zero.

$$\frac{\Delta y}{\Delta x} = \frac{\Delta y}{\Delta x} \cdot \frac{\Delta u}{\Delta u}$$

Rearranging,

$$\frac{\Delta y}{\Delta x} = \frac{\Delta y}{\Delta u} \cdot \frac{\Delta u}{\Delta x}$$

Now as we let Δx approach zero, Δu will also approach zero, so

$$\lim_{\Delta x \to 0} \frac{\Delta y}{\Delta x} = \lim_{\Delta u \to 0} \frac{\Delta y}{\Delta u} \cdot \lim_{\Delta x \to 0} \frac{\Delta u}{\Delta x}$$

But by Eq. 302,

$$\lim_{\Delta x \to 0} \frac{\Delta y}{\Delta x} = \frac{dy}{dx}$$

Similarly,

$$\lim_{\Delta u \to 0} \frac{\Delta y}{\Delta u} = \frac{dy}{du} \qquad \text{and} \qquad \lim_{\Delta x \to 0} \frac{\Delta u}{\Delta x} = \frac{du}{dx}$$

So we get

Chain Rule	$\frac{dy}{dx} = \frac{dy}{du} \cdot \frac{du}{dx}$	303

The Power Rule

We now use the chain rule to find the derivative of a function raised to a power, $y = cu^n$. If, in Eq. 307 we replace x by u, we get

$$\frac{dy}{du} = \frac{d}{du}(cu^n) = cnu^{n-1}$$

Then by the chain rule, we get dy/dx by multiplying by du/dx,

Derivative of a Function Raised to a Power	$\frac{d(cu^n)}{dx} = cnu^{n-1}\frac{du}{dx}$	309

The derivative of a constant times a function raised to a power is equal to the product of the constant, the function raised to the power less 1, and the derivative of the function.

EXAMPLE 1: Take the derivative of $y = (x^3 + 1)^5$.

Solution: We use the rule for a function raised to a power, with

$$n = 5 \qquad \text{and} \qquad u = x^3 + 1$$

Then

$$\frac{du}{dx} = 3x^2$$

so

$$\frac{dy}{dx} = nu^{n-1}\frac{du}{dx}$$

$$= 5(x^3 + 1)^4(3x^2)$$

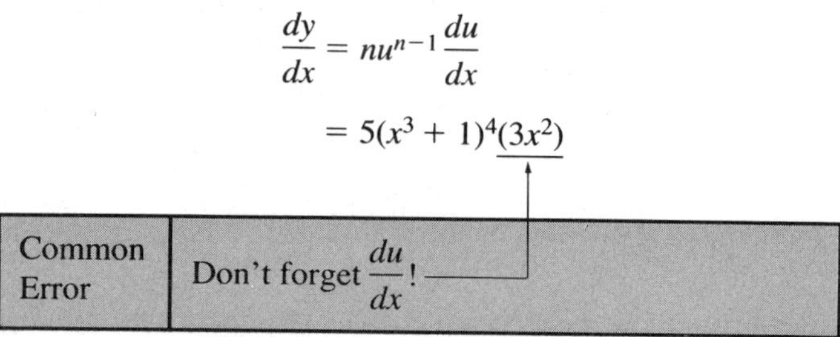

Now simplifying our answer, we get

$$\frac{dy}{dx} = 15x^2(x^3 + 1)^4$$

EXAMPLE 2: Take the derivative of $y = \dfrac{3}{x^2 + 2}$.

Solution: Rewriting our function as $y = 3(x^2 + 2)^{-1}$ and applying rule 309, we get

$$\frac{dy}{dx} = 3(-1)(x^2 + 2)^{-2}(2x)$$

$$= -\frac{6x}{(x^2 + 2)^2}$$

Power Function with Fractional Exponent

We'll now prove that Eq. 307, and hence Eq. 309, is valid when the exponent n is a fraction.

Let $n = p/q$, where p and q are both integers, positive or negative. Then

$$y = x^n = x^{p/q}$$

Raising both sides to the qth power,

$$y^q = x^p$$

Using Eq. 307, we take the derivative of each side,

$$qy^{q-1}\frac{dy}{dx} = px^{p-1}\frac{dx}{dx}$$

Solving for dy/dx,

$$\frac{dy}{dx} = \frac{p}{q}\frac{x^{p-1}}{y^{q-1}} = n\frac{x^{p-1}}{x^{(p/q)(q-1)}}$$

$$= n\frac{x^{p-1}}{x^{p-p/q}} = nx^{p-1-p+n} = nx^{n-1}$$

We have now shown that the power rule works for any rational exponent, positive or negative. It is also valid for an irrational exponent (such as π), as we will show in Sec. 6-5.

EXAMPLE 3: Differentiate $y = \sqrt[3]{1 + x^2}$.

Solution: We rewrite the radical in exponential form,

$$y = (1 + x^2)^{1/3}$$

Then by rule 309, with $u = 1 + x^2$, and $du/dx = 2x$,

$$\frac{dy}{dx} = \frac{1}{3}(1 + x^2)^{-2/3}(2x)$$

Or, returning to radical form,

$$\frac{dy}{dx} = \frac{2x}{3\sqrt[3]{(1 + x^2)^2}}$$

EXAMPLE 4: If $f(x) = \dfrac{5}{\sqrt{x^2 + 3}}$, find $f'(1)$.

Solution: Rewriting our function without radicals gives us

$$f(x) = 5(x^2 + 3)^{-1/2}$$

Taking the derivative yields

$$f'(x) = 5\left(-\frac{1}{2}\right)(x^2 + 3)^{-3/2}(2x)$$

$$= -\frac{5x}{(x^2 + 3)^{3/2}}$$

Substituting $x = 1$, we obtain

$$f'(1) = -\frac{5}{(1 + 3)^{3/2}} = -\frac{5}{8}$$

EXERCISE 4

Function Raised to a Power

As before, the letters *a*, *b*, *c*, . . . , represent constants.

Find the derivative of each function.

1. $y = (2x + 1)^5$

2. $y = (2 - 3x^2)^3$

3. $y = (3x^2 + 2)^4 - 2x$

4. $y = (x^3 + 5x^2 + 7)^2$

5. $y = (2 - 5x)^{3/5}$

6. $y = -\dfrac{2}{x + 1}$

7. $y = \dfrac{2.15}{x^2 + a^2}$

8. $y = \dfrac{31.6}{1 - 2x}$

9. $y = \dfrac{3}{x^2 + 2}$

10. $y = \dfrac{5}{(x^2 - 1)^2}$

11. $y = \left(a - \dfrac{b}{x}\right)^2$

12. $y = \left(a + \dfrac{b}{x}\right)^3$

13. $y = \sqrt{1 - 3x^2}$

14. $y = \sqrt{2x^2 - 7x}$

15. $y = \sqrt{1 - 2x}$

16. $y = \dfrac{b}{a}\sqrt{a^2 - x^2}$

17. $y = \sqrt[3]{4 - 9x}$

18. $y = \sqrt[3]{a^3 - x^3}$

19. $y = \dfrac{1}{\sqrt{x + 1}}$

20. $y = \dfrac{1}{\sqrt{x^2 - x}}$

Evaluate.

21. $\dfrac{d}{dx}(3x^5 + 2x)^2$

22. $\dfrac{d}{dx}(1.5x^2 - 3)^3$

23. $D_x(4.8 - 7.2x^{-2})^2$

24. $D(3x^3 - 5)^3$

Find the derivative with respect to the independent variable.

25. $v = (5t^2 - 3t + 4)^2$

26. $z = (9 - 8w)^3$

27. $s = (8.3t^3 - 3.8t)^{-2}$

28. $x = \sqrt{3.2y + 6.2y^4}$

29. Find the rate of change of the function $y = (4.82x^2 - 8.25x)^3$ when $x = 3.77$.

30. Find the slope of the tangent to the curve $y = 1/(x + 1)$ at $x = 2$.

31. If $y = (x^2 - x)^3$, find $y'(3)$.

32. If $f(x) = \sqrt[3]{2x} + (2x)^{2/3}$, find $f'(4)$.

3-5. DERIVATIVES OF PRODUCTS AND QUOTIENTS

Derivative of a Product

We often need the derivative of the product of two expressions, such as $y = (x^2 + 2)\sqrt{x - 5}$ where each of the expressions is itself a function of x. Let us label these expressions u and v. So our function is

$$y = uv$$

Using the delta method, we give an increment Δx to x which causes increments Δu in u, Δv in v, and Δy in y. So

$$\begin{aligned} y + \Delta y &= (u + \Delta u)(v + \Delta v) \\ &= uv + u\,\Delta v + v\,\Delta u + \Delta u\,\Delta v \end{aligned}$$

Subtracting $y = uv$ gives us

$$\Delta y = u\,\Delta v + v\,\Delta u + \Delta u\,\Delta v$$

Dividing by Δx yields

$$\frac{\Delta y}{\Delta x} = u\frac{\Delta v}{\Delta x} + v\frac{\Delta u}{\Delta x} + \Delta u\frac{\Delta v}{\Delta x}$$

As Δx now approaches 0, Δy, Δu, and Δv also approach 0, and

$$\lim_{\Delta x \to 0} \frac{\Delta y}{\Delta x} = \frac{dy}{dx}, \qquad \lim_{\Delta x \to 0} \frac{\Delta u}{\Delta x} = \frac{du}{dx}, \qquad \lim_{\Delta x \to 0} \frac{\Delta v}{\Delta x} = \frac{dv}{dx}$$

so

$$\frac{dy}{dx} = u\frac{dv}{dx} + v\frac{du}{dx} + 0\frac{dv}{dx}$$

or

Derivative of a Product of Two Factors	$\frac{d(uv)}{dx} = u\frac{dv}{dx} + v\frac{du}{dx}$	310

The derivative of a product of two factors is equal to the first factor times the derivative of the second factor, plus the second factor times the derivative of the first.

EXAMPLE 1: Find the derivative of $y = (x^2 + 2)(x - 5)$.

Solution: We let the first factor be u, and the second be v.

$$y = \underbrace{(x^2 + 2)}_{u}\underbrace{(x - 5)}_{v}$$

So

$$\frac{du}{dx} = 2x \quad \text{and} \quad \frac{dv}{dx} = 1$$

Using the product rule, we obtain

We could have multiplied the two factors together and taken the derivative term by term. Try it and see if you get the same result. But sometimes this is not possible, as in Example 2.

$$\frac{dy}{dx} = (x^2 + 2)\frac{d}{dx}(x - 5) + (x - 5)\frac{d}{dx}(x^2 + 2)$$

$$= (x^2 + 2)(1) + (x - 5)(2x)$$

$$= x^2 + 2 + 2x^2 - 10x$$

$$= 3x^2 - 10x + 2$$

EXAMPLE 2: Differentiate $y = (x + 5)\sqrt{x - 3}$.

Solution: By the product rule,

$$\frac{dy}{dx} = (x + 5) \cdot \frac{1}{2}(x - 3)^{-1/2}(1) + \sqrt{x - 3}(1)$$

$$= \frac{x + 5}{2\sqrt{x - 3}} + \sqrt{x - 3}$$

Products with More Than Two Factors

Our rule for the derivative of a product having two factors can be easily extended. Take an expression with three factors, for example,

$$y = uvw = (uv)w$$

By the product rule,

$$\frac{dy}{dx} = uv\frac{dw}{dx} + w\frac{d(uv)}{dx}$$

$$= uv\frac{dw}{dx} + w\left(u\frac{dv}{dx} + v\frac{du}{dx}\right)$$

so

Derivative of a Product of Three Factors	$\frac{d(uvw)}{dx} = uv\frac{dw}{dx} + uw\frac{dv}{dx} + vw\frac{du}{dx}$	**311**

The derivative of the product of three factors is an expression of three terms, each term being the product of two of the factors and the derivative of the third factor.

EXAMPLE 3: Differentiate $y = x^2(x-2)^5\sqrt{x+3}$.

Solution: By Eq. 311,

$$\frac{dy}{dx} = x^2(x-2)^5\left[\frac{1}{2}(x+3)^{-1/2}\right] + x^2(x+3)^{1/2}[5(x-2)^4]$$

$$+ (x-2)_5(x+3)^{1/2}(2x)$$

$$= \frac{x^2(x-2)^2}{2\sqrt{x+3}} + 5x^2(x-2)^4\sqrt{x+3} + 2x(x-2)^5\sqrt{x+3}$$

We now generalize this result (without proof) to cover any number of factors:

Derivative of a Product of n Factors	The derivative of the product of n factors is an expression of n terms, each term being the product of $n - 1$ of the factors and the derivative of the other factor.	**312**

Derivative of a Constant Times a Function

Let us use the product rule for the product cu, where c is a constant and u is a function of x.

$$\frac{d}{dx}cu = c\frac{du}{dx} + u\frac{dc}{dx} = c\frac{du}{dx}$$

since the derivative dc/dx of a constant is zero. Thus

Derivative of a Constant Times a Function	$\frac{d(cu)}{dx} = c\frac{du}{dx}$	**306**

The derivative of the product of a constant and a function is equal to the constant times the derivative of the function.

EXAMPLE 4: If $y = 3(x^2 - 3x)^5$, then

$$\frac{dy}{dx} = 3\frac{d}{dx}(x^2 - 3x)^5 = 3(5)(x^2 - 3x)^4(2x - 3) = 15(x^2 - 3x)^4(2x - 3)$$

Common Error	If one of the factors is a constant, it is much easier to use rule 306 for a constant times a function, rather than the product rule.

Derivative of a Quotient

To find the derivative of the function

$$y = \frac{u}{v}$$

where u and v are functions of x, we first rewrite it as a product,

$$y = uv^{-1}$$

Now using the rule for products and the rule for a power function,

$$\frac{dy}{dx} = u(-1)v^{-2}\frac{dv}{dx} + v^{-1}\frac{du}{dx}$$

$$= -\frac{u}{v^2}\frac{dv}{dx} + \frac{1}{v}\frac{du}{dx}$$

We combine the two fractions over the LCD, v^2,

$$\frac{dy}{dx} = \frac{v}{v^2}\frac{du}{dx} - \frac{u}{v^2}\frac{dv}{dx}$$

or

Derivative of a Quotient	$\frac{d}{dx}\left(\frac{u}{v}\right) = \frac{v\frac{du}{dx} - u\frac{dv}{dx}}{v^2}$	**313**

The derivative of a quotient equals the denominator times the derivative of the numerator minus the numerator times the derivative of the denominator, all divided by the square of the denominator.

Some prefer to use the product rule to do quotients, treating the quotient u/v as the product uv^{-1}.

EXAMPLE 5: Take the derivative of $y = 2x^3/(4x + 1)$.

Solution: The numerator is $u = 2x^3$, so

$$\frac{du}{dx} = 6x^2$$

and the denominator is $v = 4x + 1$, so

$$\frac{dv}{dx} = 4$$

Common Error	It is very easy, in Eq. 313, to interchange u and v, by mistake.

Applying the quotient rule yields

$$\frac{dy}{dx} = \frac{(4x + 1)(6x^2) - (2x^3)(4)}{(4x + 1)^2}$$

Simplifying, we get

$$\frac{dy}{dx} = \frac{24x^3 + 6x^2 - 8x^3}{(4x + 1)^2}$$

$$= \frac{16x^3 + 6x^2}{(4x + 1)^2}$$

Tip	If the numerator or denominator is a constant, rewrite the expression as a product and use the product rule.

EXAMPLE 6: Find the derivative of $y = \dfrac{2x^3}{3} + \dfrac{4}{x^2}$.

Solution: Rather than use the quotient rule, we rewrite the expression as

$$y = \frac{2}{3}x^3 + 4x^{-2}$$

Using the product rule and the sum rule,

$$\frac{dy}{dx} = \frac{2}{3}(3)x^2 + 4(-2)x^{-3} = 2x^2 - \frac{8}{x^3}$$

EXAMPLE 7: Find $s'(3)$ if $s = \dfrac{(t^3 - 3)^2}{\sqrt{t + 1}}$.

Solution: By the quotient rule,

$$s' = \frac{\sqrt{t+1}\,(2)(t^3 - 3)(3t^2) - (t^3 - 3)^2(\tfrac{1}{2})(t + 1)^{-1/2}}{t + 1}$$

$$= \frac{6t^2(t^3 - 3)\sqrt{t+1} - \dfrac{(t^3 - 3)^2}{2\sqrt{t+1}}}{t + 1}$$

Then

$$s'(3) = \frac{6(9)(27 - 3)\sqrt{3+1} - \dfrac{(27 - 3)^2}{2\sqrt{3+1}}}{3 + 1} = 612$$

EXERCISE 5

Products

Do not multiply out, where possible, but use the product rule for the practice.

Find the derivative of each function.

1. $y = x(x^2 - 3)$

2. $y = x^3(5 - 2x)$

3. $y = (5 + 3x)(3 + 7x)$

4. $y = (7 - 2x)(x + 4)$

5. $y = x(x^2 - 2)^2$

6. $y = x^3(8.24x - 6.24x^3)$

7. $y = (x^2 - 5x)^3(8x - 7)^2$

8. $y = (1 - x^2)(1 + x^2)^2$

9. $y = x\sqrt{1 + 2x}$

10. $y = 3x\sqrt{5 + x^2}$

11. $y = x^2\sqrt{3 - 4x}$

12. $y = 2x^2 - 2x\sqrt{x^2 - 5}$

13. $y = x\sqrt{a + bx}$

14. $y = \sqrt{x}\,(3x^2 + 2x - 3)$

15. $y = (3x + 1)^3 \sqrt{4x - 2}$

16. $y = (2x^2 - 3)\sqrt[3]{3x + 5}$

17. $v = (5t^2 - 3t)(t + 4)^2$

18. $z = (6 - 5w)^3(w^2 - 1)$

19. $s = (81.3t^3 - 73.8t)(t - 47.2)^2$

20. $x = (y - 49.3)\sqrt{23.2y + 16.2y^4}$

Evaluate.

21. $\dfrac{d}{dx}(2x^5 + 5x)^2(x - 3)$

22. $D_x(11.5x + 49.3)(14.8 - 27.2x^{-2})^2$

Products with More Than Two Factors

23. $y = x(x + 1)^2(x - 2)^3$

24. $y = x\sqrt{x + 1}\sqrt[3]{x}$

25. $y = (x - 1)^{1/2}(x + 1)^{3/2}(x + 2)$

26. $y = x\sqrt{x + 2}\sqrt[3]{x - 1}\sqrt[4]{x + 1}$

Quotients

Find the derivative of each function.

27. $y = \dfrac{x}{x + 2}$

28. $y = \dfrac{x}{x^2 + 1}$

29. $y = \dfrac{x^2}{4 - x^2}$

30. $y = \dfrac{x - 1}{x + 1}$

31. $y = \dfrac{x + 2}{x - 3}$

32. $y = \dfrac{2x - 1}{(x - 1)^2}$

33. $y = \dfrac{2x^2 - 1}{(x - 1)^2}$

34. $y = \dfrac{a - x}{n + x}$

35. $y = \dfrac{x^{1/2}}{x^{1/2} + 1}$

36. $s = \sqrt{\dfrac{t - 1}{t + 1}}$

37. $w = \dfrac{z}{\sqrt{z^2 - a^2}}$

38. $v = \sqrt{\dfrac{1 + 2t}{1 - 2t}}$

39. Find the slope of the tangent to the curve $y = \sqrt{16 + 3x}/x$ at $x = 3$.

40. Find the rate of change of the function $y = x/(7.42x^2 - 2.75x)$ when $x = 1.47$.

41. If $y = x\sqrt{8 - x^2}$, find $y'(2)$.

42. If $f(x) = x^2/\sqrt{1 + x^3}$, find $f'(2)$.

3-6. DERIVATIVES OF IMPLICIT RELATIONS

Derivatives Not with Respect to the Independent Variable

Our rule for the derivative of a power function is

$$\frac{d(u^n)}{dx} = nu^{n-1}\frac{du}{dx} \tag{309}$$

Up to now, we have used this rule only when the variable in the function u has been the *same variable* that we take the derivative with respect to.

EXAMPLE 1

(a) $\dfrac{d}{dx}(x^3) = 3x^2\dfrac{dx}{dx} = 3x^2$ (same)

(b) $\dfrac{d}{dt}(t^5) = 5t^4\dfrac{dt}{dt} = 5t^4$ (same)

Since $dx/dx = dt/dt = 1$, we have not bothered to write these in. Of course, rule 309 is just as valid when our independent variable is *different* from the variable that we are taking the derivative with respect to.

Mathematical ideas do not, of course, depend upon which letters of the alphabet we happen to have chosen when doing a derivation. Thus in any of our rules you can replace any letter, say *x*, with any other letter, such as *z* or *t*.

EXAMPLE 2

(a) $\dfrac{d}{dx}(u^5) = 5u^4\dfrac{du}{dx}$ (different)

(b) $\dfrac{d}{dt}(w^4) = 4w^3\dfrac{dw}{dt}$ (different)

(c) $\dfrac{d}{dx}(y^6) = 6y^5\dfrac{dy}{dx}$ (different)

(d) $\dfrac{d}{dx}y = 1y^0\dfrac{dy}{dx} = \dfrac{dy}{dx}$ (different)

Common Error	It is very easy to forget to include the *dy/dx* in problems such as the preceding example. $\dfrac{d}{dx}(y^6) = 6y^5\dfrac{dy}{dx}$ ← don't forget!

Our other rules for derivatives (for sums, products, quotients, etc.) also work when the independent variable(s) are different from the variable we are taking the derivative with respect to.

EXAMPLE 3

(a) $\dfrac{d}{dx}(2y + z^3) = 2\dfrac{dy}{dx} + 3z^2\dfrac{dz}{dx}$

(b) $\frac{d}{dt}(wz) = w\frac{dz}{dt} + z\frac{dw}{dt}$

(c) $\frac{d}{dx}(x^2y^3) = x^2(3y^2)\frac{dy}{dx} + y^3(2x)\frac{dx}{dx}$

$$= 3x^2y^2\frac{dy}{dx} + 2xy^3$$

Implicit Relations

Recall that in an *implicit relation* neither variable is isolated on one side of the equal sign.

EXAMPLE 4: The relation $x^2 + y^2 = y^3 - x$ is implicit.

We need to be able to differentiate implicit relations because we cannot always solve for one of the variables.

To find the derivative of an implicit relation, take the derivative of both sides of the equation, and then solve for dy/dx. When taking the derivatives, keep in mind that the derivative of x with respect to x is 1, and that the derivative of y with respect to x is dy/dx.

EXAMPLE 5: Given the implicit relation in Example 4, find dy/dx.

Solution: We take the derivative term by term,

$$2x\frac{dx}{dx} + 2y\frac{dy}{dx} = 3y^2\frac{dy}{dx} - \frac{dx}{dx}$$

or

$$2x + 2y\frac{dy}{dx} = 3y^2\frac{dy}{dx} - 1$$

Solving for dy/dx gives us

$$2y\frac{dy}{dx} - 3y^2\frac{dy}{dx} = -2x - 1$$

$$(2y - 3y^2)\frac{dy}{dx} = -2x - 1$$

Note that the derivative, unlike those for explicit functions, contains y.

$$\frac{dy}{dx} = \frac{2x + 1}{3y^2 - 2y}$$

When taking implicit derivatives, it is convenient to use the y' notation instead of dy/dx.

EXAMPLE 6: Find the derivative dy/dx for the relation

$$x^2y^3 = 5$$

Solution: Using the product rule we obtain

$$x^2(3y^2)y' + y^3(2x) = 0$$

$$3x^2y^2y' = -2xy^3$$

$$y' = -\frac{2xy^3}{3x^2y^2} = -\frac{2y}{3x}$$

EXERCISE 6

Derivatives with Respect to Other Variables

1. If $y = 2u^3$, find dy/dw.
2. If $z = (w + 3)^2$, find dz/dy.
3. If $w = y^2 + u^3$, find dw/du.
4. If $y = 3x^2$, find dy/du.

Evaluate.

5. $\dfrac{d}{dx}(x^3y^2)$
6. $\dfrac{d}{dx}(w^2 - 3w - 1)$
7. $\dfrac{d}{dt}\sqrt{3z^2 + 5}$
8. $\dfrac{d}{dz}(y - 3)\sqrt{y - 2}$

Derivatives of Implicit Relations

Find dy/dx.

9. $5x - 2y = 7$
10. $2x + 3y^2 = 4$
11. $xy = 5$
12. $x^2 + 3xy = 2y$
13. $y^2 = 4ax$
14. $y^2 - 2xy = a^2$
15. $x^3 + y^3 - 3axy = 0$
16. $x^2 + y^2 = r^2$
17. $y + y^3 = x + x^3$
18. $x + 2x^2y = 7$
19. $y^3 - 4x^2y^2 + y^4 = 9$
20. $y^{3/2} + x^{3/2} = 16$

Find the slope of the tangent to each curve at the given point.

21. $x^2 + y^2 = 25$ at $x = 2$ in the first quadrant
22. $x^2 + y^2 = 25$ at $(3, 4)$
23. $2x^2 + 2y^3 - 9xy = 0$ at $(1, 2)$
24. $x^2 + xy + y^2 - 3 = 0$ at $(1, 1)$

3-7. HIGHER-ORDER DERIVATIVES

After taking the derivative of a function, we may then take the derivative of the derivative. That is called the *second derivative*. Our original derivative we now call the *first derivative*. The symbols used for the second derivative are

We will have lots of uses for the second derivative in the next few chapters, but derivatives higher than second order are rarely needed.

$$\frac{d^2y}{dx^2} \quad \text{or} \quad y'' \quad \text{or} \quad f''(x) \quad \text{or} \quad D^2y$$

We can then go on to find third, fourth, and higher derivatives.

EXAMPLE 1: Given the function

$$y = x^4 + 2x^3 - 3x^2 + x - 5$$

we get

$$y' = 4x^3 + 6x^2 - 6x + 1$$

and

$$y'' = 12x^2 + 12x - 6$$

$$y''' = 24x + 12$$

$$y^{(iv)} = 24$$

$$y^{(v)} = 0$$

and all higher derivatives will also be zero.

EXAMPLE 2: Find the second derivative of $y = (x + 2)\sqrt{x - 3}$.

Solution: Using the product rule,

$$y' = (x + 2)\left(\frac{1}{2}\right)(x - 3)^{-1/2} + \sqrt{x - 3}\,(1)$$

and

$$y'' = \frac{1}{2}\left[(x + 2)\left(-\frac{1}{2}\right)(x - 3)^{-3/2} + (x - 3)^{-1/2}(1)\right] + \frac{1}{2}(x - 3)^{-1/2}$$

$$= -\frac{x + 2}{4(x - 3)^{3/2}} + \frac{1}{\sqrt{x - 3}}$$

Common Error	Be sure to *simplify* the first derivative before taking the second.

EXERCISE 7

Find the second derivative of each function.

1. $y = 3x^4 - x^3 + 5x$
2. $y = x^3 - 3x^2 + 6$
3. $y = \dfrac{x^2}{x + 2}$
4. $y = \dfrac{3 + x}{3 - x}$
5. $y = \sqrt{5 - 4x^2}$
6. $y = \sqrt{x + 2}$
7. $y = (x - 7)(x - 3)^3$
8. $y = x^2\sqrt{2.3x - 5.82}$
9. If $y = x(9 + x^2)^{1/2}$, find $y''(4)$.
10. If $f(x) = 1/\sqrt{x} + \sqrt{x}$, *find* $f''(1)$.

CHAPTER TEST

1. Differentiate $y = \sqrt{\dfrac{x^2 - 1}{x^2 + 1}}$.
2. Evaluate $\lim\limits_{x \to 1} \dfrac{x^2 + 2 - 3x}{x - 1}$.
3. Find dy/dx if $x^{2/3} + y^{2/3} = 9$.
4. Find y'' if $y = \dfrac{x}{\sqrt{2x + 1}}$.
5. If $f(x) = \sqrt{4x^2 + 9}$, find $f'(2)$.
6. Evaluate $\lim\limits_{x \to \infty} \dfrac{5x + 3x^2}{x^2 - 1 - 3x}$.
7. Find dy/dx if $\dfrac{x^2}{a^2} + \dfrac{y^2}{b^2} = 1$.
8. Find dy/dx if $y = \sqrt[3]{\dfrac{3x + 2}{2 - 3x}}$.
9. If $f(x) = \sqrt{25 - 3x}$, find $f''(3)$.
10. Find dy/dx if $y = \dfrac{x^2 + a^2}{a^2 - x^2}$.
11. Find dy/dx by the delta method if $y = 5x - 3x^2$.
12. Evaluate $\lim\limits_{x \to 0} \dfrac{\sin x}{x + 1}$.
13. Find the slope of the tangent to the curve $y = \dfrac{1}{\sqrt{25 - x^2}}$ at $x = 3$.
14. Evaluate $\lim\limits_{x \to 0} \dfrac{e^x}{x}$.
15. If $y = 2.15x^3 - 6.23$, find $y'(5.25)$.
16. Evaluate $\lim\limits_{x \to -7} \dfrac{x^2 + 6x - 7}{x + 7}$.
17. Evaluate $\dfrac{d}{dx}(3x + 2)$.
18. Evaluate $D(3x^4 + 2)^2$.
19. Evaluate $D_x(x^2 - 1)(x + 3)^{-4}$.
20. If $v = 5t^2 - 3t + 4$, find dv/dt.
21. If $z = 9 - 8w + w^2$, find dz/dx.
22. Evaluate $\lim\limits_{x \to 5} \dfrac{25 - x^2}{x - 5}$.
23. If $f(x) = 7x - 4x^3$, find $f''(x)$.
24. Evaluate $D_x(21.7x + 19.1)(64.2 - 17.9x^{-2})^2$.
25. If $s = 58.3t^3 - 63.8t$, find ds/dt.

Graphical Applications of the Derivative

We saw in Chapter 3 that the first derivative gives us the slope of the tangent to a curve at any point. This fact leads to several graphical applications of the derivative. In addition to the obvious one of finding the equations of the tangent and normal to a curve at a given point, we will find the "peaks" and "valleys" on a curve by locating the points at which the tangent is parallel to the x axis.

The second derivative is also used here to locate points on a curve where the curvature changes direction, and also to tell whether a point at which the tangent is horizontal is a peak or a valley.

The derivative is further used in Newton's method for finding roots of equations, a technique that works very well on the computer. Finally, the chapter closes with a summary of steps to aid in curve sketching.

4-1. TANGENTS AND NORMALS

Tangent to a Curve

We saw in Chapter 3 that the slope m_t of the tangent to a curve at some point (x_1, y_1) is given by the derivative of the equation of the curve evaluated at that point.

$$\text{slope of tangent} = m_t = y'(x_1)$$

Knowing the slope and the coordinates of the given point, we then use the point-slope form of the straight-line equation (Eq. 255) to write the equation of the tangent.

EXAMPLE 1: Write the equation of the tangent to the curve $y = x^2$, where $x = 2$ (Fig. 4-1).

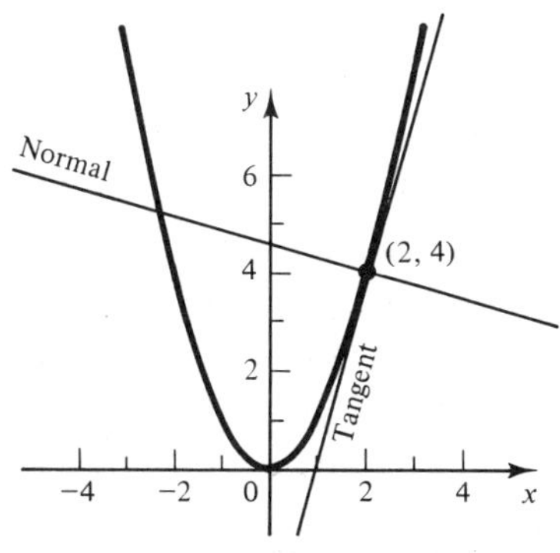

FIGURE 4-1

Solution: The derivative is $y' = 2x$. When $x = 2$, $y(2) = 2^2 = 4$ and

$$y'(2) = 2(2) = 4 = m_t$$

From Eq. 255,

$$\frac{y-4}{x-2} = 4$$

or

$$y - 4 = 4x - 8$$

so $4x - y - 4 = 0$ is the equation of the tangent.

Normal to a Curve

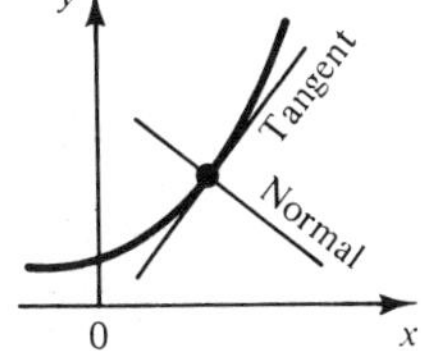

FIGURE 4-2. Tangent and normal to a curve.

The *normal* to a curve at a given point is the line perpendicular to the tangent at that point, as in Fig. 4-2. From Eq. 259, we know that the slope m_n of the normal will be the negative reciprocal of the slope of the tangent.

EXAMPLE 2: Find the equation of the normal at the same point as in the preceding example.

Solution: By Eq. 259,

$$m_n = -\frac{1}{m_t} = -\frac{1}{4}$$

By Eq. 255,

$$\frac{y-4}{x-2} = -\frac{1}{4}$$

$$4y - 16 = -x + 2$$

so $x + 4y - 18 = 0$ is the equation of the normal.

Implicit Relations

When the equation of the curve is an implicit relation, you may choose to solve for y, when possible, before taking the derivative. Often, though, it is easier to take the derivative implicitly, as in the following example.

EXAMPLE 3: Find (a) the equation of the tangent to the ellipse (Fig. 4-3) $4x^2 + 9y^2 = 40$ at the point $(1, -2)$ and (b) the x intercept of the tangent.

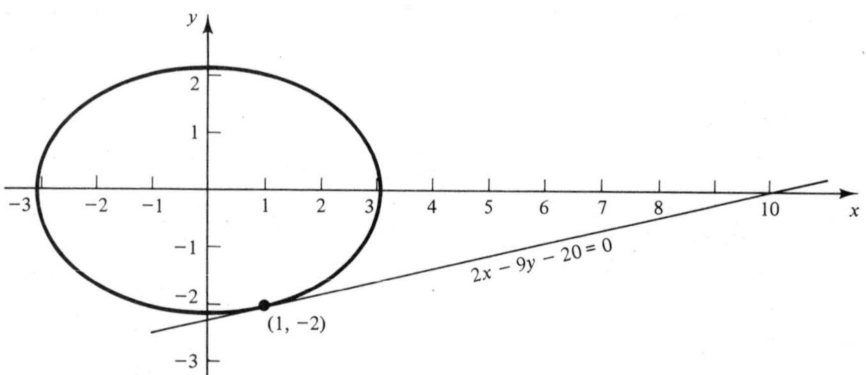

FIGURE 4-3

Solution: (a) Taking the derivative implicitly, we have $8x + 18yy' = 0$, or

$$y' = -\frac{4x}{9y}$$

At $(1, -2)$,

$$y'(1, -2) = -\frac{4(1)}{9(-2)} = \frac{2}{9} = m_t$$

From Eq. 255,

$$\frac{y - (-2)}{x - 1} = \frac{2}{9}$$

$$9y + 18 = 2x - 2$$

so $2x - 9y - 20 = 0$ is the equation of the tangent.
(b) Setting y equal to zero in the equation of the tangent gives $2x - 20 = 0$, or an x intercept of

$$x = 10$$

Angle of Intersection of Two Curves

If the point(s) of intersection are not known, solve the two equations simultaneously.

The angle between two curves is defined as the angle between their tangents at the point of intersection, which is found from Eq. 260.

EXAMPLE 4: Find the angle of intersection between the parabolas

$$(1)\ y^2 = x \qquad \text{and} \qquad (2)\ y = x^2$$

at the point of intersection (1, 1) (Fig. 4-4).

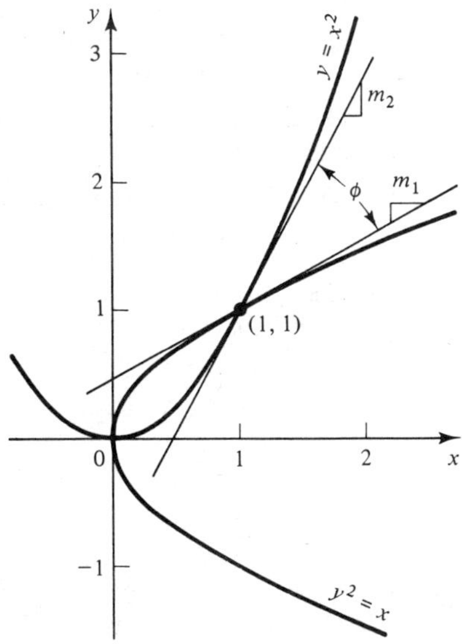

FIGURE 4-4. Angle of intersection of two curves.

Solution: Taking the derivative of (1), we have $2yy' = 1$, or

$$y' = \frac{1}{2y}$$

At (1, 1),

$$y'(1, 1) = \tfrac{1}{2} = m_1$$

Now taking the derivative of (2) gives us $y' = 2x$. At (1, 1),

$$y'(1, 1) = 2 = m_2$$

From Eq. 260,

$$\tan \phi = \frac{m_2 - m_1}{1 + m_1 m_2} = \frac{2 - \frac{1}{2}}{1 + \frac{1}{2}(2)} = \frac{1.5}{2} = 0.75$$

$$\phi = 36.9°$$

EXERCISE 1

Tangents and Normals

Write the equation of the tangent and normal at the given point.

1. $y = x^2 + 2$ when $x = 1$

2. $y = x^3 - 3x$ at (2, 2)

3. $y = 3x^2 - 1$ when $x = 2$

4. $y = x^2 - 4x + 5$ at (1, 2)

5. $x^2 + y^2 = 25$ at (3, 4)

6. $16x^2 + 9y^2 = 144$ at (2, 2.98)

7. Find the first quadrant point on the curve $y = x^3 - 3x^2$ at which the tangent to the curve is parallel to $y = 9x + 7$.

8. Write the equation of the tangent to the parabola $y^2 = 4x$ that makes an angle of 45° with the x axis.

9. The curve $y = 2x^3 - 6x^2 - 2x + 1$ has two tangents each of which is parallel to the line $2x + y = 12$. Find their equations.

10. Each of two tangents to the circle $x^2 + y^2 = 25$ has a slope $\frac{3}{4}$. Find the points of contact.

11. Find the equation of the line tangent to the upper branch of the hyperbola $4x^2 - 9y^2 + 36 = 0$, perpendicular to the line $5x - 2y - 10 = 0$.

Intercepts

12. Find the x intercept of the tangent to $y^2 = 18x$ at (2, 6).

13. Find the x intercept of the tangent to the curve $x^2 + y^2 = 50$ at $(-5, -5)$.

Angles between Curves

Find the angle(s) of intersection between the given curves.

14. $y = x^2 + x - 2$ and $y = x^2 - 5x + 4$ at (1, 0)

15. $y = -2x$ and $y = x^2(1 - x)$ at (0, 0), (2, −4), and (−1, 2)

16. $y = 2x + 2$ and $x^2 - xy + y^2 = 4$ at (0, 2) and (−2, −2)

4-2. MAXIMUM AND MINIMUM POINTS

Some Definitions

We will limit this discussion to *smooth curves*, without cusps or corners at which there is no derivative.

Figure 4-5 shows a path over the mountains from A to H. It goes over three peaks, B, D, and F. These are called *maximum points*. The highest, peak D, is called the *absolute maximum* in the section of trail from A to H, while peaks B and F are called *relative maximum* points. Similarly, valley G is called an *absolute minimum*, while valleys C and E are *relative minimums*. All the peaks and valleys are referred to as *extreme values*.

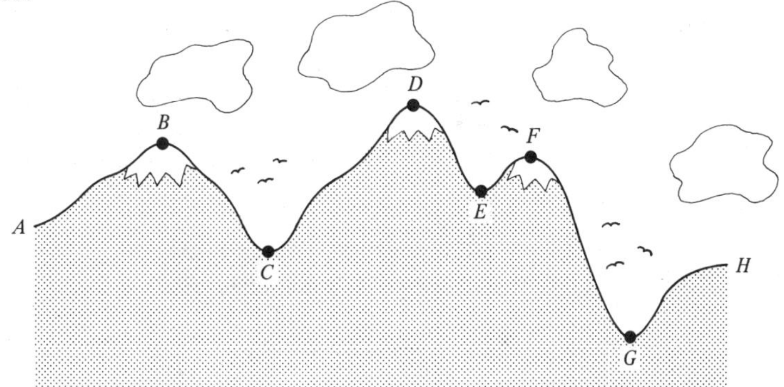

FIGURE 4-5. Path over the mountains.

Increasing and Decreasing Functions

Within the interval from A to K in Fig. 4-6, the function is said to be *increasing* from A to D, from F to H, and from J to K. It is called a *decreasing function* from D to F, and from H to J.

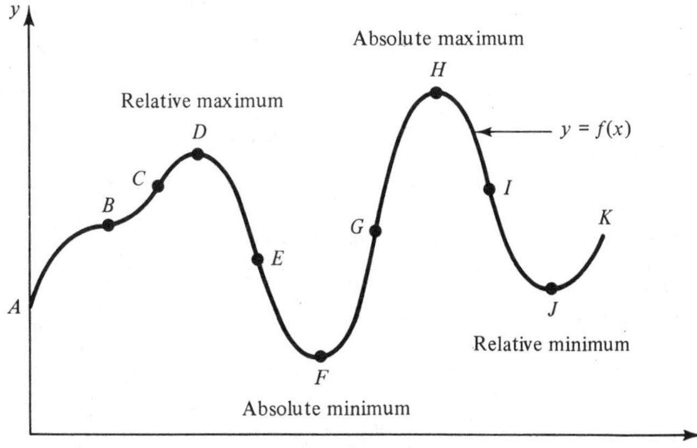

FIGURE 4-6. Maximum, minimum, and inflection points.

The slope of the tangent to the curve, and hence the first derivative, is *positive* for an *increasing* function. Conversely, the first derivative is *negative* for a *decreasing* function.

As with the mountains, point D is a relative maximum and H is an absolute maximum. Further, F is an absolute minimum and J is a relative minimum.

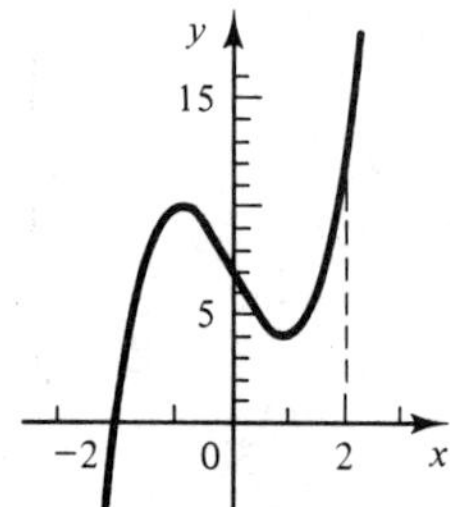

FIGURE 4-7. Graph of $y = 2x^3 - 5x + 7$.

EXAMPLE 1: Is the function $y = 2x^3 - 5x + 7$ increasing or decreasing at $x = 2$?

Solution: The derivative is $y' = 6x^2 - 5$. At $x = 2$,

$$y'(2) = 6(4) - 5 = 19$$

A positive derivative means that the given function is *increasing* at $x = 2$ (Fig. 4-7).

EXAMPLE 2: For what values of x is the curve $y = 3x^2 - 12x - 2$ rising, and for what values of x is it falling?

Solution: The first derivative is $y' = 6x - 12$. This derivative will be equal to zero when $6x - 12 = 0$, or $x = 2$. We see that y' is negative for values of x less than 2. A *negative* derivative tells us that our original function is *falling* in that region. Further, the derivative is positive for $x > 2$, so the given function is rising in that region (see Fig. 4-8).

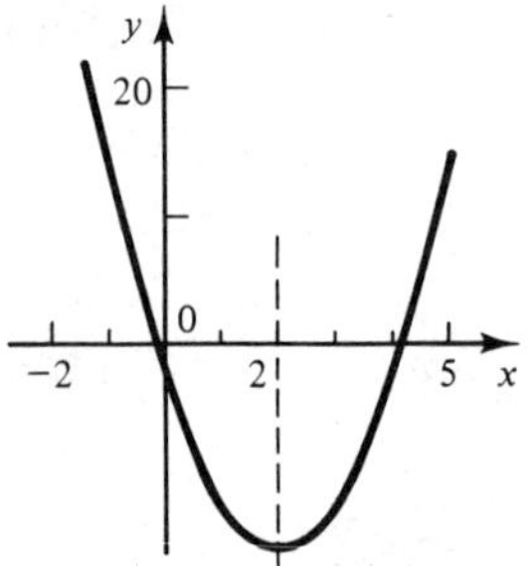

FIGURE 4-8. Graph of $y = 3x^2 - 12x - 2$.

Finding Maximum and Minimum Points

The idea of maximum and minimum values will lead to some interesting applications in Chapter 5.

A *stationary point* on a curve is one at which the tangent is horizontal. In Fig. 4-6 they include the maximums, D and H, the minimums F and J, and point B, which is neither a maximum nor a minimum. At all such stationary points *the first derivative is equal to zero.* Since the first derivative is zero at maximum and minimum points (as well as at other stationary points such as point B in Fig. 4-6), we find such points by taking the first derivative of the function and finding the value(s) of x that make the first derivative equal to zero.

To find maximum and minimum points (and other stationary points) set the first derivative equal to zero and solve for x.	**330**

EXAMPLE 3: Find the maximum and minimum values for the function $y = x^3 - 3x$.

Solution: We take the first derivative,

$$y' = 3x^2 - 3$$

We can find the value of x that makes the derivative zero by *setting the derivative equal to zero and solving for x.*

$$3x^2 - 3 = 0$$

$$x = \pm 1$$

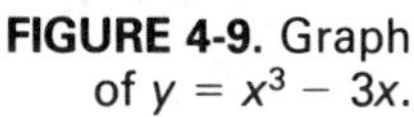
FIGURE 4-9. Graph of $y = x^3 - 3x$.

Solving for y,

$$y(1) = 1 - 3 = -2$$

and

$$y(-1) = -1 + 3 = 2$$

So the points of zero slope are $(1, -2)$ and $(-1, 2)$, as in Fig. 4-9.

We see from the graph that $(1, -2)$ is a minimum and that $(-1, 2)$ is a maximum. However, we can *test* for whether a point is a maximum or minimum, without having to graph the function, as shown in the following section.

Testing for Maximum or Minimum

The simplest way to tell whether a stationary point is a maximum, a minimum, or neither is to graph the curve. There are also a few tests that will identify such points.

If we look at the slope of the tangent on either side of a minimum point B (Fig. 4-10), we see that it is negative to the left (at A) and positive to the right (at C) of that point. The reverse is true for a maximum point. Since the slope is given by the first derivative, we have the following:

First-Derivative Test	The first derivative is negative to the left of, and positive to the right of, a minimum point. The reverse is true for a maximum point.	**331**

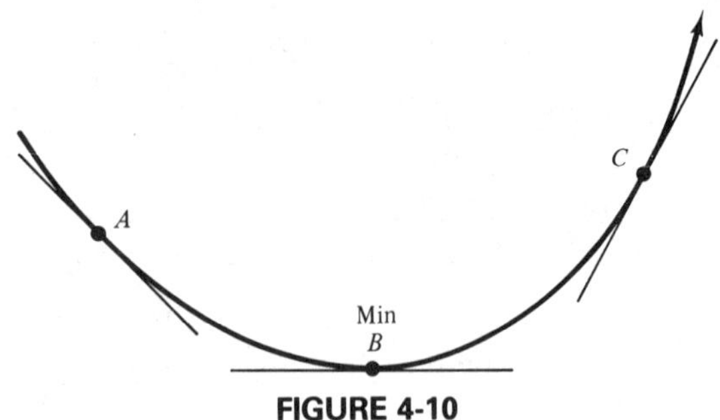

FIGURE 4-10

We use this test only on points *close* to the suspected maximum or minimum. If our test points are too far away, the curve might have already changed direction.

(a) Concave upward

(b) Concave downward

FIGURE 4-11

Further, a minimum point occurs in a region of a curve (Fig. 4-11a) that is said to be *concave upward,* while a maximum point occurs where a curve is *concave downward* (Fig. 4-11b). Thus a test for concavity at a stationary point will tell whether that point is a maximum or a minimum.

If we look at the slope of the tangent at each of several points on the concave upward curve, we see that the slope is increasing as we proceed in the positive x direction. At point A in Fig. 4-10 the slope is negative, at B it is zero, and at C it is positive. And since the slope of the tangent is increasing, it means that the first derivative, which gives the value of the slope, is also increasing. Therefore, the *rate of change of the first derivative must be positive,* because a positive rate of change means an increase.

But the rate of change of the first derivative is given by the *second derivative.* We thus see that *the second derivative is positive where a curve is concave upward.* The opposite is also true, that *the second derivative is negative where a curve is concave downward.* We can also reason that the second derivative is zero when the curve is neither concave upward or downward, such as point B in Fig. 4-6.

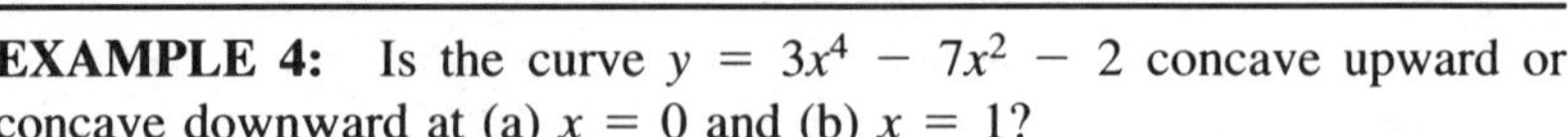

EXAMPLE 4: Is the curve $y = 3x^4 - 7x^2 - 2$ concave upward or concave downward at (a) $x = 0$ and (b) $x = 1$?

Solution: The first derivative is $y' = 12x^3 - 14x$, and the second derivative is $y'' = 36x^2 - 14$.
(a) At $x = 0$,

$$y''(0) = -14$$

A negative second derivative tells that the given curve is *concave downward* at that point, as in Fig. 4-12.
(b) At $x = 1$,

$$y''(1) = 36 - 14 = 22$$

or positive, so the curve is concave upward at that point.

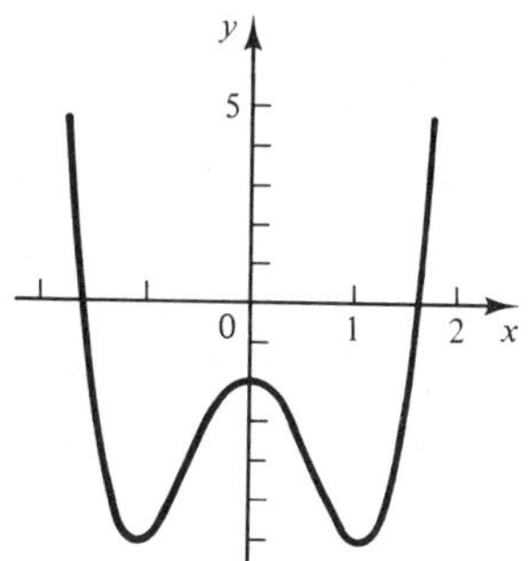

FIGURE 4-12. Graph of $y = 3x^4 - 7x^2 - 2$.

Thus the second derivative tells us whether a curve is concave upward or downward at a given point, and provides us with the following test.

This test will not work on rare occasions. For example, the function $y = x^4$ has a minimum point at the origin, but the second derivative there is zero, not a positive number as expected.

Second-Derivative Test	If the first derivative at some point is zero, then, if the second derivative is 1. Positive, the point is a minimum 2. Negative, the point is a maximum 3. Zero, the test fails	**332**

EXAMPLE 5: From Example 3, the function $y = x^3 - 3x$ had extreme values of $(1, -2)$ and $(-1, 2)$. Use the second derivative test to decide which is a maximum and which a minimum.

Solution: The first derivative is $y' = 3x^2 - 3$ and the second derivative is

$$y'' = 6x$$

At the point $(1, -2)$,

$$y''(1) = 6(1) = 6$$

A positive second derivative tells us that $(1, -2)$ is a minimum. Now at $(-1, 2)$,

It might help to think of the curve as a bowl. When it is concave upward it will hold (+) water, and when concave downward it will spill (−) water.

$$y''(-1) = 6(-1) = -6$$

which is negative, so $(-1, 2)$ is a maximum, as in Fig. 4-9.

Common Error	It is tempting to group *maximum with positive,* and *minimum with negative.* Remember that they are just the *reverse* of this.

A third test for distinguishing between a maximum point and a minimum point is called the *ordinate* test.

Ordinate Test	Find y a small distance to either side of the point to be tested. If y is greater there, we have a minimum; if less, we have a maximum.

EXAMPLE 6: Testing the point $(1, -2)$ in Example 5, we compute y at, say, $x = 0.9$ and $x = 1.1$.

$$y(0.9) = (0.9)^3 - 3(0.9) = -1.97$$

$$y(1.1) = (1.1)^3 - 3(1.1) = -1.97$$

Thus the curve is higher a small distance to either side of the extreme value $(1, -2)$, so we conclude that $(1, -2)$ is a minimum point.

The procedure for finding maximum and minimum points is no different for an implicit relation, although it is usually more work.

EXAMPLE 7: Find any maximum and minimum points on the curve $x^2 + 4y^2 - 6x + 2y + 3 = 0$.

Solution: Taking the derivative implicitly gives

$$2x + 8yy' - 6 + 2y' = 0$$

$$y'(8y + 2) = 6 - 2x$$

$$y' = \frac{6 - 2x}{8y + 2}$$

Setting this derivative equal to zero gives $6 - 2x = 0$, or

$$x = 3$$

Substituting $x = 3$ into the original equation, we get

$$9 + 4y^2 - 18 + 2y + 3 = 0$$

Collecting terms,

$$2y^2 + y - 3 = 0$$

Factoring,

$$(y - 1)(2y + 3) = 0$$

$$y = 1, \qquad y = -\frac{3}{2}$$

So the points of zero slope are (3, 1) and $(3, -\frac{3}{2})$. We now apply the second-derivative test. Using the quotient rule gives

$$y'' = \frac{(8y + 2)(-2) - (6 - 2x)8y'}{(8y + 2)^2}$$

Replacing y' by $(6 - 2x)/(8y + 2)$ and simplifying,

$$y'' = \frac{(-16y - 4)(8y + 2) - 8(6 - 2x)^2}{(8y + 2)^3}$$

At (3, 1), $y'' = -0.640$. The negative second derivative tells that (3, 1) is a maximum point. At $(3, -\frac{3}{2})$, $y'' = 0.2$, telling that we have a minimum, as shown in Fig. 4-13.

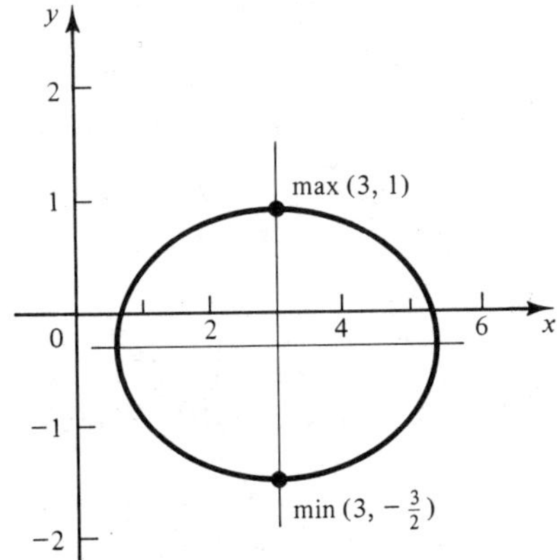

FIGURE 4-13. Graph of $x^2 + 4y^2 - 6x + 2y + 3 = 0$.

EXERCISE 2

Increasing and Decreasing Functions

For what values of x is each curve rising, and for what values is each falling?

1. $y = 3x + 5$

2. $y = 4x^2 + 16x - 7$

Is each function increasing or decreasing at the value indicated?

3. $y = 3x^2 - 4$ at $x = 2$

4. $y = x^2 - x - 3$ at $x = 0$

5. $y = 4x^2 - x$ at $x = -2$

6. $y = x^3 + 2x - 4$ at $x = -1$

State whether each curve is concave upward or concave downward at the given value of x.

7. $y = x^4 + x^2$ at $x = 2$

8. $y = 4x^5 - 5x^4$ at $x = 1$

9. $y = -2x^3 - 2\sqrt{x + 2}$ at $x = \frac{1}{4}$

10. $y = \sqrt{x^2 + 3x}$ at $x = 2$

Maxima and Minima

Find the maximum and minimum points for each function.

11. $y = x^2$

12. $y = x^3 + 3x^2 - 2$

13. $y = 6x - x^2 + 4$

14. $y = x^3 - 3x + 4$

15. $y = x^3 - 7x^2 + 36$

16. $y = x^4 - 4x^3$

17. $y = 2x^2 - x^4$

18. $16y = x^2 - 32x$

19. $y = x^4 - 4x$

20. $2y = x^2 - 4x + 6$

21. $y = 3x^4 - 4x^3 - 12x^2$

22. $y = 2x^3 - 9x^2 + 12x - 3$

23. $y = x^3 + 3x^2 - 9x + 5$

24. $y = \dfrac{x^2 - 7x + 6}{x - 10}$

25. $y = (x - 1)^4(x + 2)^3$

26. $y = (x - 2)^2(2x + 1)$

Find the maximum and minimum points for each implicit relation.

27. $4x^2 + 9y^2 = 36$

28. $x^2 + y^2 - 2x + 4y = 4$

29. $x^2 - x - 2y^2 + 36 = 0$

30. $x^2 + y^2 - 8x - 6y = 0$

4-3. INFLECTION POINTS

A point where the curvature changes from concave upward to concave downward (or vice versa) is called an *inflection point,* or *point of inflection.* In Fig. 4-6 they are points B, C, E, G, and I.

We saw that the second derivative was positive where a curve was concave upward and negative where concave downward. In going from positive to negative, the second derivative must somewhere be zero, and this is at the point of inflection.

Unfortunately, a curve may have a point of inflection where the second derivative does not exist, such as the curve $y = x^{1/3}$ at $x = 0$. Fortunately, these cases are rare.

Points of Inflection	To find points of inflection, set the second derivative to zero and solve for x. Test by seeing if the second derivative changes sign a small distance to either side of the point.	333

EXAMPLE 1: Find any points of inflection on the curve $y = x^3 - 3x^2 - 5x + 7$.

Solution: We take the derivative twice,

$$y' = 3x^2 - 6x - 5$$

$$y'' = 6x - 6$$

We now set y'' to 0 and solve for x.

$$6x - 6 = 0$$

$$x = 1$$

and

$$y(1) = 1 - 3 - 5 + 7 = 0$$

A graph of the given function (Fig. 4-14) shows the point (1, 0) clearly to be a point of inflection. If there were any doubt, we would test it by seeing if the second derivative has opposite signs on either side of the point.

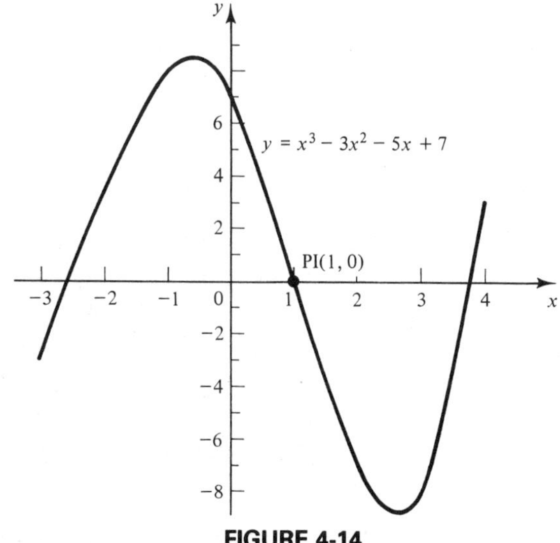

FIGURE 4-14

EXERCISE 3

Find the points of inflection for each curve.

1. $y = x^3$
2. $y = x^3 - 3x^2$
3. $y = x^5$
4. $y = 3x^4 - 4x^3 + 1$

5. $y = x^3 - 9x^2 - 24x$

6. $y = 3x - x^3$

7. $y = x^4 - 6x^2$

8. $y = x^2(x^2 + 1)$

9. $y = x^4 + 2x^3 + 10x - 17$

10. $y = (x + 2)(x - 2)(x - 3)$

We have had so many laws and methods named "Newton" that you might think there were twenty Newtons all working in different fields. But no. All are from the same Isaac Newton (1642–1727).

APPROXIMATE SOLUTION OF EQUATIONS BY NEWTON'S METHOD

The ability to find the slope of a tangent to a curve simply by taking the derivative is made use of in Newton's method. With this method we can quickly find the root of an equation of the form $y = f(x)$, even though we cannot solve the equation for this root. The result is *approximate,* but this is hardly a drawback since we get as many significant digits as we want.

Figure 4-15 shows a graph of the function $y = f(x)$ and the zero where the curve crosses the x axis. The solution to the equation $f(x) = 0$ is then the value a.

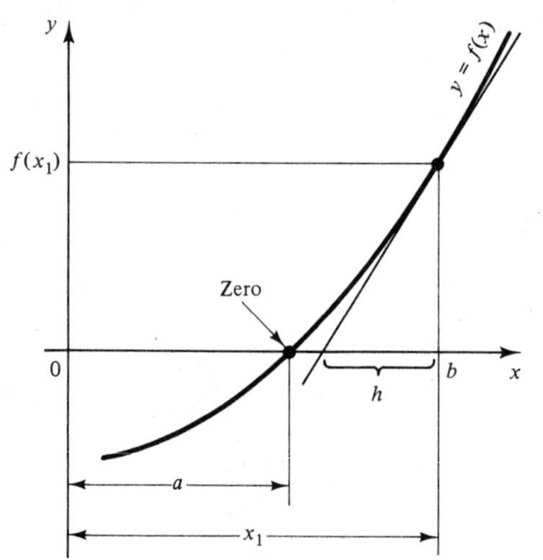

FIGURE 4-15. Newton's method.

Suppose that our first approximation to a is the value x_1 (our first guess is too high). We can *correct* our first guess by subtracting from it the amount h, taking for our second guess the point where the tangent line crosses the x axis. The slope of the tangent line at x_1 is

$$m = f'(x_1) = \frac{\text{rise}}{\text{run}} = \frac{f(x_1)}{h}$$

So

$$h = \frac{f(x_1)}{f'(x_1)}$$

and

$$(\text{second guess}) = (\text{first guess}) - h$$

or

$$x_2 = x_1 - \frac{f(x_1)}{f'(x_1)}$$

You can stop at this point, or repeat the calculation several times to get the accuracy needed, with each new value x_{n+1} obtained from the old value x_n by

Newton's Method	$x_{n+1} = x_n - \dfrac{f(x_n)}{f'(x_n)}$	**334**

A good way to get a first guess is to graph the function. Note that the equations are the same whether your first guess is to the left or to the right of the zero. Also, the method is the same regardless if the curve is rising or falling where it crosses the x axis.

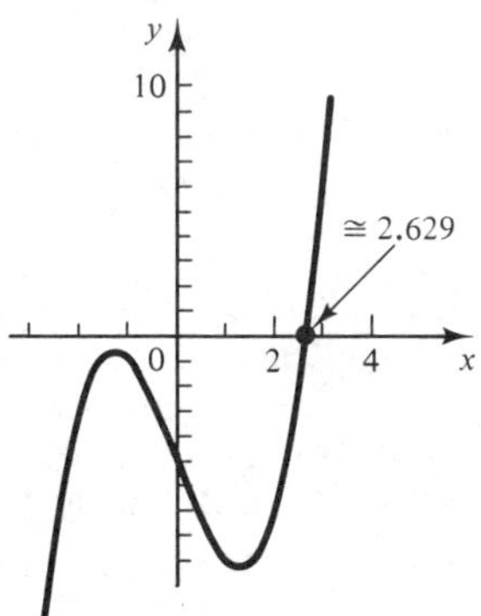

FIGURE 4-16. A root found by Newton's method.

EXAMPLE 1: Find one root of the equation $x^3 - 5x = 5$ (Fig. 4-16) to one decimal place. Take $x = 3$ as a first guess.

Solution: We transpose all terms to one side of the equation and set them equal to $f(x)$:

$$f(x) = x^3 - 5x - 5$$

Taking the derivative yields

$$f'(x) = 3x^2 - 5$$

At the value 3,

$$f'(3) = 3(9) - 5 = 22$$

and

$$f(3) = 27 - 15 - 5 = 7$$

The first correction is then

$$h = \frac{7}{22} = 0.32$$

Our second guess is then

$$3 - 0.32 = 2.68$$

We now repeat the entire calculation, using 2.68 as our guess.

Newton's method will usually fail when the root you seek is near a maximum, minimum, or inflection point, or near a discontinuity. If *h* does not shrink rapidly as you repeat the computation, this may be the cause.

$$f(2.68) = 0.8488$$

$$f'(2.68) = 16.55$$

$$h = \frac{0.8488}{16.55} = 0.0513$$

$$\text{third guess} = 2.68 - 0.0513 = 2.629$$

Since h is 0.0513 and getting smaller, we know that it will have no further affect on the first decimal place, so we stop here.

Can you imagine trying to solve this equation by an exact method?

EXAMPLE 2: Find the root of the equation $1.07\sqrt{x^2 + 7.25} = x^3 + 9.84$ to five decimal places.

Solution: Transposing, we have

$$f(x) = x^3 - 1.07\sqrt{x^2 + 7.25} + 9.84$$

$$f'(x) = 3x^2 + \frac{1.07x}{\sqrt{x^2 + 7.25}}$$

A plot of our function (Fig. 4-17) shows a zero near $x = -2$, so we use this for our first guess. The calculated values are given in the following table.

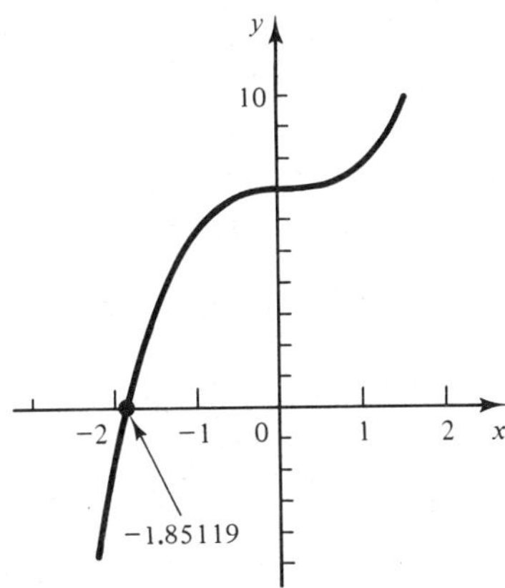

FIGURE 4-17. A root found by Newton's method.

x	$f(x)$	$f'(x)$	*Correction*	*New x*
−2.00000	0.05540	9.61893	−0.15392	−1.84608
−1.84608	−0.00716	9.68154	0.00576	−1.85184
−1.85184	0.00089	9.67349	−0.00074	−1.85110
−1.85110	−0.00011	9.67450	0.00009	−1.85119

Our root then, to five decimal places, is −1.85119.

EXERCISE 4

Each equation has at least one root between $x = -10$ to 10. Find one such root to two decimal places using Newton's method.

1. $x^3 - 40 = 0$

2. $x^3 + 3x^2 - 10 = 0$

3. $x^3 - 3x^2 + 8x - 25 = 0$

4. $x^3 - 6x + 12 = 0$

5. $x^3 + 2x - 8 = 0$

6. $3x^3 - 4x = 1$

7. $x^3 + 4x + 12 = 0$

8. $x^4 + 8x - 12 = 0$

9. A square of side x is cut from each corner of a square sheet of metal 16 in. on a side, and the edges are turned up to form a tray whose volume is 282 in^3. Write an equation for the volume as a function of x, and use Newton's method to find x to two decimal places.

10. A solid sphere of radius r and specific gravity S will sink in water to a depth x, where

$$x^3 - 3rx^2 + 4r^3S = 0$$

Use Newton's method to find x to two decimal places, if $r = 4$ in. and $S = 0.700$.

11. A certain silo consists of a cylindrical base of radius r and height h, topped by a hemisphere (Fig. 4-18). It contains a volume V equal to

$$V = \pi r^2 h + \frac{2}{3}\pi r^3$$

If $h = 20.0$ m and $V = 800$ m^3, use Newton's method to find r to two decimal places.

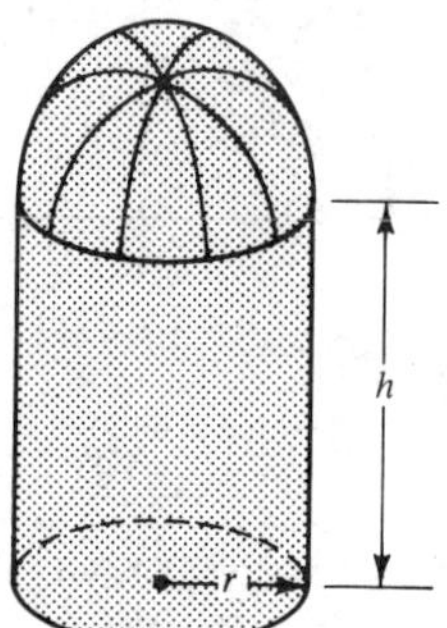

FIGURE 4-18. Silo.

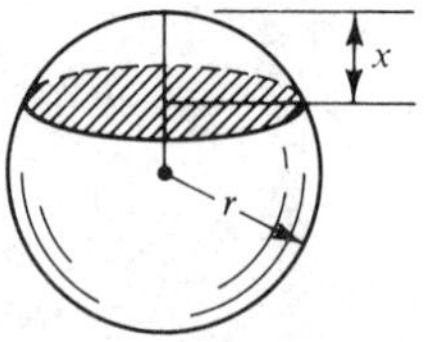

FIGURE 4-19. Segment cut from a sphere.

12. A segment of height x cut from a sphere of radius r (Fig. 4-19) has a volume

$$V = \pi\left(rx^2 - \frac{x^3}{3}\right)$$

Find x to two decimal places if $r = 4.0$ cm and $V = 150$ cm^3.

Computer

13. The labor of Newton's method can be reduced by the computer. Write a program or use a spreadsheet that will accept as input the given equation, its derivative, the approximate location of a root, and the degree of accuracy you require. The computer should then locate and print the value of the root. Test your program on any of the problems in this exercise.

4-5. CURVE SKETCHING

When we first studied curve sketching we simply obtained a table of point pairs by substituting into the given function or relation, and then plotted each point. Since then, we have learned methods for finding particular features of interest, such as intercepts or inflection points, which can make curve sketching faster and easier. We summarize them here.

1. *Recognition of type:* Look at the equation. From our work in analytic geometry you should recognize the equation of a straight line or a conic. For a polynomial, the *degree* will tell the maximum number of extreme points you can expect (two for third degree, three for fourth degree, and so on).

2. *Intercepts:* Find the x intercept(s), or roots, by setting y equal to zero and solving the resulting equation for x. Find the y intercept(s) by setting x equal to zero and solving for y.

A function whose graph is symmetrical about the y axis is called an *even function*. One symmetrical about the origin is called an *odd function*.

3. *Symmetry:* A curve is symmetrical about the
   (a) x axis, if the equation does not change when we substitute $-y$ for y
   (b) y axis, if the equation does not change when we substitute $-x$ for x
   (c) Origin, if the equation does not change when we substitute $-x$ for x and $-y$ for y

4. *Extent:* Look for values of the variables that give division by zero, or that result in negative numbers under a radical sign. The curve will not exist at these values.

5. *Asymptotes:* Look for some value of x which, when *approached* by x (from above, or from below) will cause y to become infinite. You will then have found a *vertical asymptote*. Then if y approaches some particular value as x becomes infinite, we will have found a *horizontal* asymptote. Similarly, check what happens to y when x becomes infinite in the negative direction.

We look for any of these items, only if it can be easily found. These steps are intended to make curve sketching faster and easier, not to increase our work. These steps can, of course, be done in any order.

6. *Increasing or decreasing:* The first derivative will be positive in regions where the function is increasing, and negative where the function is decreasing. Thus inspection of the first derivative can show where the curve is rising, and where falling.

7. *Maximum and minimum points:* Find the points at which the first derivative is zero. Then test each of these points to decide if it is a maximum, minimum, or neither of these.

   In some cases it may be useful to find the values of y that make dx/dy equal to zero, thus finding maximum and minimum points in the *horizontal* direction. Usually, though, we will have enough other information without this step.

8. *Inflection points:* Find any inflection points by setting the second derivative equal to zero, and solving for x. Test by seeing if the second derivative changes sign a small distance to either side of the point.

9. *Large values:* What happens to y as x gets very large (both in the positive and negative directions)? Does y continue to grow without bound, or approach some asymptote, or is it possible that the curve will turn and perhaps cross the x axis again? Similarly, can you say what happens to x as y gets very large?

In graphing any function, we use whichever of these nine steps seem useful and appropriate (seldom all of them), and fill in any gaps in our graph by plotting extra points.

EXAMPLE 1: Graph the function $y = \dfrac{x^3}{x-1}$.

Solution: We make a table of point pairs in the usual way:

x	-4	-3	-2	-1	0	1	2	3	4
y	12.8	6.75	2.67	0.500	0	—	8	13.5	21.3

These points are plotted in Fig. 4-20. Now if we were not too curious about why we get no value for y at $x = 1$, we might be tempted to connect these points with the smooth curve shown dashed in Fig. 4-20 which seems to fit the points very well. But we know that something is wrong. The dashed curve shows $y = 3$, approximately, when $x = 1$, while the equation gives division by zero at $x = 1$. Let us now go through the steps for curve sketching, but not necessarily in the order given above, to see what the graph really looks like.

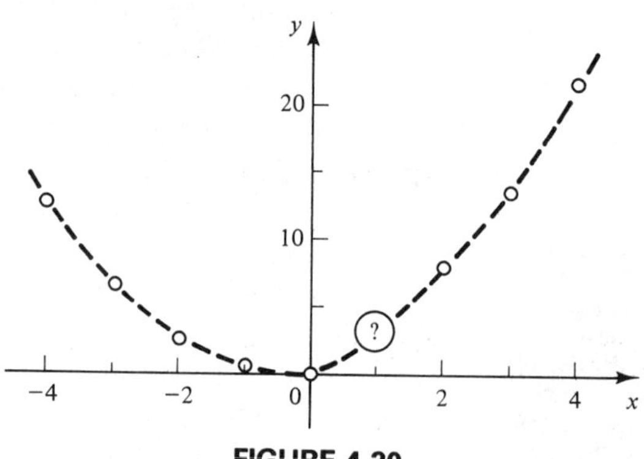

FIGURE 4-20

Extent: We get division by zero when $x = 1$, so the curve cannot exist at that value. On our graph we draw a vertical line at $x = 1$. Our curve cannot cross that line.

Asymptotes: As x approaches 1 from above, the denominator of our function gets smaller, but remains positive. Thus y takes on larger and larger positive values. As x approaches 1 from below, the denominator gets smaller but is now negative. Thus y gets very large in the negative direction. Thus the line $x = 1$ is a vertical asymptote.

Large values: As x gets large, the -1 in the denominator becomes negligible compared to x, and our equation is approximately equal to

$y = x^3/x$, or $y = x^2$. So, far from the origin, the curve will look like a parabola with vertex at the origin, which opens upward.

Maximum and minimum points: Looking at the information already obtained, we know there must be a minimum point somewhere between $x = 1$ and $x = 2$, and perhaps a maximum between $x = 0$ and $x = 1$. To find these points, we take the first derivative,

$$y' = \frac{(x-1)(3x^2) - x^3(1)}{(x-1)^2}$$

$$= \frac{x^2(2x-3)}{(x-1)^2}$$

Setting the first derivative equal to zero, we get $x^2(2x - 3) = 0$, so

$$x = 0 \qquad \text{and} \qquad x = \frac{3}{2}$$

Solving for y yields $y(0) = 0$, and

$$y\left(\frac{3}{2}\right) = \frac{\left(\frac{3}{2}\right)^3}{\frac{3}{2} - 1} = 6.75$$

So the first derivative is zero at the points (0, 0) and (1.5, 6.75). We now need the second derivative to determine what type of point each of these is.

$$y'' = \frac{(x-1)^2(6x^2 - 6x) - x^2(2x-3)(2)(x-1)}{(x-1)^4}$$

which eventually reduces to

$$y'' = \frac{2x(x^3 - 4x^2 + 6x - 3)}{(x-1)^4}$$

Now testing the points found above yields $y''(0) = 0$. So (0, 0) is not a maximum or minimum, but is a stationary point, and also an *inflection point*.

From our graph, we see that the point (1.5, 6.75) can only be a minimum. We verify this with the second derivative test,

$$y''(1.5) = \frac{2(1.5)[(1.5)^3 - 4(1.5)^2 + 6(1.5) - 3]}{(1.5-1)^4} = 18$$

The positive second derivative indicates a minimum point, as expected.

Increasing or decreasing: Looking again at the first derivative,

$$y' = \frac{x^2(2x - 3)}{(x - 1)^2}$$

we see that the denominator cannot be negative, and x^2 in the numerator cannot be negative, but that y' can be negative when $2x$ is less than 3, or, when x is less than $\frac{3}{2}$. We conclude that the function is decreasing when x is less than $\frac{3}{2}$ and increasing when x is greater than $\frac{3}{2}$.

Intercepts: Setting $x = 0$ in our original equation gives $y = 0$. Setting $y = 0$ and solving for x, we get $x = 0$. Thus the origin is the only intercept.

Symmetry: Substituting $-y$ for y,

$$-y = \frac{x^3}{x - 1}$$

we get an equation that is not equivalent to the original equation, so there is no symmetry about the x axis. Substituting $-x$ for x,

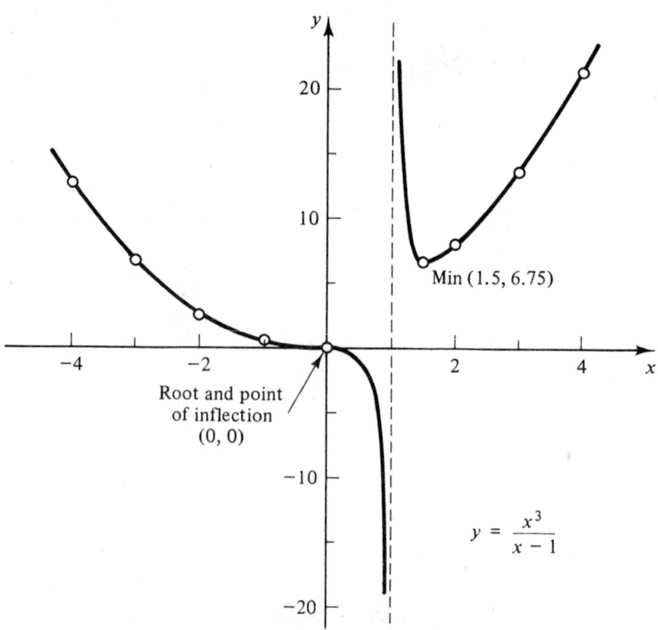

FIGURE 4-21

$$y = \frac{(-x)^3}{-x-1} = \frac{-x^3}{-x-1} = \frac{x^3}{x+1}$$

we get a changed equation, which rules out symmetry about the y axis. Finally, substituting $-x$ for x and $-y$ for y,

$$-y = \frac{(-x)^3}{-x-1} = \frac{x^3}{x+1}$$

which is also different than the original. Thus there is no symmetry about the origin.

Our final graph is shown in Fig. 4-21.

Graphing Regions

In later applications we will have to identify *regions* that are bounded by two or more curves. Here we get some practice in doing that.

EXAMPLE 2: Locate the region bounded by the y axis and the curves $y = x^2$, $y = 1/x$, and $y = 4$.

Solution: The three curves given are, respectively, a parabola, a hyperbola, and a straight line, and are graphed in Fig. 4-22. We see that three different closed areas are formed, but the shaded area is the only one *bounded by each one of the given curves.*

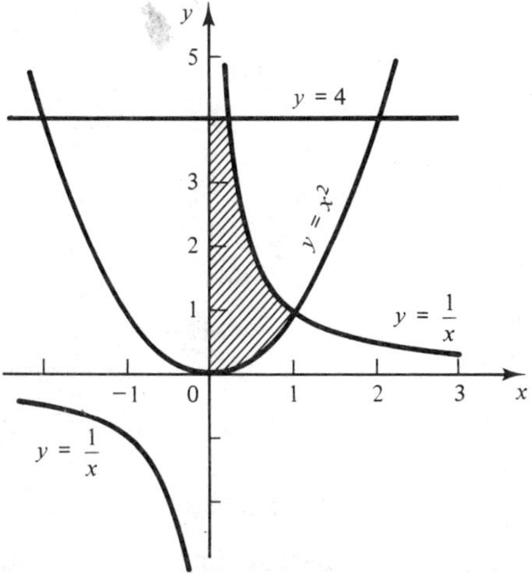

FIGURE 4-22. Graphing a region bounded by several curves.

EXERCISE 5

Graph each function.

1. $y = 4x^2 - 5$

2. $y = 3x - 2x^2$

3. $y = 5 - \dfrac{1}{x}$

4. $y = \dfrac{3}{x} + x^2$

5. $y = x^4 - 8x^2$

6. $y = \dfrac{1}{x^2 - 1}$

7. $y = x^3 - 9x^2 + 24x - 7$

8. $y = x\sqrt{1 - x}$

9. $y = 5x - x^5$

10. $y = \dfrac{9}{x^2 + 9}$

11. $y = \dfrac{6x}{3 + x^2}$

12. $y = \dfrac{x}{\sqrt{4 - x^2}}$

13. $y = x^3 - 6x^2 + 9x + 3$

14. $y = x^2\sqrt{6 - x^2}$

15. $y = \dfrac{96x - 288}{x^2 + 2x + 1}$

16. $y = \dfrac{\sqrt{x}}{x - 1}$

17. $y = \dfrac{x}{\sqrt{x^2 + 1}}$

18. $y = \dfrac{x^3}{(1 + x^2)^2}$

19. $y = x^2 + 2x$

20. $y = x^2 - 3x + 2$

21. $y = x^3 + 4x^2 - 5$

22. $y = x^4 - x^2$

Graph the region bounded by the given curves.

23. $y = 3x^2$ and $y = 2x$

24. $y^2 = 4x$, $x = 5$, and the x axis, in the first quadrant

25. $y = 5x^2 - 2x$, the y axis, and $y = 4$, in the second quadrant

26. $y = 4/x$, $y = x$, $x = 6$, and the x axis

CHAPTER TEST

1. Find the roots of $x^3 - 4x + 2 = 0$ to two decimal places.

2. Find any points of inflection on the curve $y = x^3(1 + x^2)$.

3. Find the maximum and minimum points on the curve $y = 3\sqrt[3]{x} - x$.

4. Find any maximum points, minimum points, and points of inflection for the function

$$3y = x^3 - 3x^2 - 9x + 11.$$

5. Write the equations of the tangent and normal to the curve $y = \dfrac{1 + 2x}{3 - x}$ at the point (2, 5).

6. Find the coordinates of the point on the curve $y = \sqrt{13 - x^2}$ at which the slope of the tangent is $-\frac{2}{3}$.

7. Find the x intercept of the tangent to the curve $y = \sqrt{x^2 + 7}$ at the point (3, 4).

8. Graph the function $y = \dfrac{4 - x^2}{\sqrt{1 - x^2}}$ and locate any features of interest.

9. For which values of x is the curve $y = \sqrt{4x}$ rising, and for what values is it falling?

10. Is the function $y = \sqrt{5 - 3x}$ increasing or decreasing at $x = 1$?

11. Is the curve $y = x^2 - x^5$ concave upward or concave downward at $x = 1$?

12. Graph the region bounded by the curves $y = 3x^3$, $x = 1$, and the x axis.

13. Write the equations of the tangent and normal to the curve $y = 3x^3 - 2x + 4$ at $x = 2$.

14. Find the x intercept of the tangent and normal in Problem 13.

15. Find the angle of intersection between the curves $y = x^2/4$, and $y = 2/x$.

16. Graph the function $y = 3x^3 \sqrt{9 - x^2}$, and locate any features of interest.

More Applications of the Derivative

What is this stuff good for? This chapter gives the main answer to that question, at least for the derivative.

Most of the applications hinge on two ideas. First, that the derivative gives the *rate of change* of the function of which we took the derivative. With this we can find out *how fast* something (say, the pressure in an engine cylinder) is changing. If the changing quantity is a *distance,* the rate of change with respect to time is called *speed,* or *velocity.* In this chapter we find velocities of moving objects, and also the rate of change of velocity, or *acceleration.*

For electrical applications, we use the rate of change of charge to find current, the rate of change of current to find the voltage across an inductor, and the rate of change of voltage to find the current in a capacitor.

The other important idea for these applications is that the rate of change, and hence the derivative, is zero at a maximum or minimum point. We used this fact in Chapter 4 to find the peaks and valleys on a curve, and we use the same procedure here to find maximum and minimum values of a varying quantity.

Finally, we will use the idea of a *differential* to make fast estimations of some quantities that would otherwise be difficult to find.

5-1. RATE OF CHANGE

Rate of Change

When two or more related quantities are changing, we often speak about the *rate of change* of one quantity with respect to one of the other quantities. For example, if a steel rod is placed in a furnace, its temperature and its length both increase. Since the length varies with the temperature, we can speak about the *rate of change of length with respect to temperature*. But the length of the bar is also varying with time, so we can speak about the *rate of change of length with respect to time*. Time rates are the most common rates of change with which we have to deal.

Average and Instantaneous Rates of Change

The *average* rate of change of some quantity over an *interval* is the amount of change of the quantity divided by the interval.

The *instantaneous rate of change*, on the other hand, gives the rate at a *particular instant*. We saw in Chapter 4 that *the derivative of a function gives us the instantaneous rate of change of the function.*

EXAMPLE 1: Find the instantaneous rate of change of the function $y = 3x^2 + 5$ when $x = 2$.

Solution: Taking the derivative,

$$y' = 6x$$

When $x = 2$,

$$y'(2) = 6(2) = 12 = \text{instantaneous rate of change}$$

EXAMPLE 2: The temperature T (°F) in a certain furnace varies with time $t(s)$ according to the function $T = 4.85t^3 + 2.96t$. Find the rate of change of temperature at $t = 3.75$ s.

Solution: Taking the derivative of T with respect to t,

$$\frac{dT}{dt} = 14.55t^2 + 2.96 \quad \text{°F/s}$$

At $t = 3.75$ s,

$$\frac{dT}{dt} = 14.55(3.75)^2 + 2.96 = 208 \quad \text{°F/s}$$

Electric Current

The *coulomb* (C) is the unit of electrical *charge*. The *current* in amperes (A) is the number of coulombs passing a point in a circuit in 1 second. The *instantaneous current* is given by

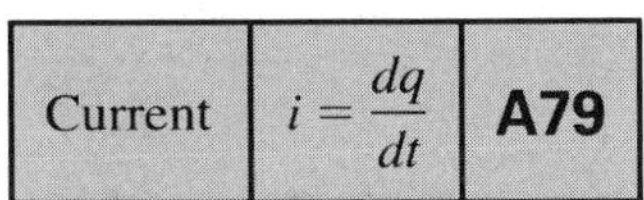

Current	$i = \dfrac{dq}{dt}$	**A79**

Current is the rate of change of charge.

EXAMPLE 3: The charge through a 2.85-Ω resistor is given by $q = 1.08t^3 - 3.82t$ coulomb. Write an expression for (a) the instantaneous current through the resistor, (b) the voltage across the resistor, and (c) the instantaneous power in the resistor. (d) Evaluate each at 2.00 s.

Solution: (a) $i = dq/dt = 3.24t^2 - 3.82$ A
(b) By Ohm's law,

$$\begin{aligned} v = Ri &= 2.85(3.24t^2 - 3.82) \\ &= 9.23t^2 - 10.9 \text{ V} \end{aligned}$$

(c) By Eq. A65,

$$\begin{aligned} P &= vi \\ &= (9.23t^2 - 10.9)(3.24t^2 - 3.82) \\ &= 29.9t^4 - 70.5t^2 + 41.6 \text{ W} \end{aligned}$$

(d) At $t = 2.00$ *s*,

$$i = 3.24(4.00) - 3.82 = 9.14 \text{ A}$$

$$v = 9.23(4.00) - 10.9 = 26.0 \text{ V}$$

$$P = 29.9(16.0) - 70.5(4.00) + 41.6 = 238 \text{ W}$$

Current in a Capacitor

If a steady voltage is applied across a capacitor, no current will flow into the capacitor (after the initial transient currents have died down). But if the applied voltage varies with time, the instantaneous current i to the capacitor will be proportional to the rate of change of the voltage. The constant of proportionality is called the *capacitance C*.

Current in a Capacitor	$i = C\dfrac{dv}{dt}$	**A81**

The current to a capacitor equals the capacitance times the rate of change of the voltage.

The units here are volts for v, seconds for t, farads for C, and amperes for i.

EXAMPLE 4: The voltage applied to a 2.85-microfarad (μF) capacitor is $v = 1.47t^2 + 48.3t - 38.2$ V. Find the current at $t = 2.50$ s.

A microfarad equals 10^{-6} farad

Solution: The derivative of the voltage equation is $dv/dt = 2.94t + 48.3$. Then from Eq. A81,

$$i = C\frac{dv}{dt} = (2.85 \times 10^{-6})(2.94t + 48.3) \quad \text{A}$$

At $t = 2.50$ s,

$$i = (2.85 \times 10^{-6})[2.94(2.50) + 48.3] = 159 \times 10^{-6} \quad \text{A}$$
$$= 0.159\ \mu\text{A}$$

Voltage across an Inductor

If the current through an inductor (such as a coil of wire) is steady, there will be no voltage drop across the inductor. But if the current varies, a voltage will be induced that is proportional to the rate of change of the current. The constant of proportionality L is called the *inductance* and is measured in henrys (H).

Voltage across an Inductor	$v = L\dfrac{di}{dt}$	**A86**

The voltage across an inductor equals the inductance times the rate of change of the current.

EXAMPLE 5: The current in a 8.75-H inductor is given by $i = \sqrt{t^2 + 5.83t}$. Find the voltage across the inductor at $t = 5.00$ s.

Solution: By Eq. A86,

$$v = 8.75\frac{di}{dt} = 8.75\left(\frac{1}{2}\right)(t^2 + 5.83t)^{-1/2}(2t + 5.83)$$

At $t = 5.00$ s,

$$v = 8.75\left(\frac{1}{2}\right)[25.0 + 5.83(5.00)]^{-1/2}(10.0 + 5.83) = 9.41 \text{ V}$$

EXERCISE 1

Use Boyle's law, $pv = k$.

1. The air in a certain cylinder is at a pressure of 25.5 lb/in.^{2} when its volume is 146 in.^{3}. Find the rate of change of the pressure with respect to volume as the piston descends farther.

Use the inverse square law, $I = k/d^2$.

2. A certain light source produces an illumination of 655 lux on a surface at a distance of 2.75 m. Find the rate of change of illumination with respect to distance, and evaluate it at 2.75 m.

3. A spherical balloon starts to shrink as the gas escapes. Find the rate of change of its volume with respect to its radius when the radius is 1.00 m.

Use Eq. A67, $P = I^2R$

4. The power dissipated in a certain resistor is 865 W at a current of 2.48 A. What is the rate of change of the power with respect to the current as the current starts to increase?

5. The period (in seconds) for a pendulum of length L inches to complete one oscillation is equal to $P = 0.324\sqrt{L}$. Find the rate of change of the period with respect to length when the length is 9.00 in.

The rate of change dT/dx is called the *temperature gradient*.

6. The temperature T at a distance x inches from the end of a certain heated bar is given by $T = 2.24x^3 + 1.85x + 95.4$ (°F). Find the rate of change of temperature with respect to distance at a point 3.75 in. from the end.

The shape taken by the axis of a bent beam is called the *elastic curve*.

7. The cantilever beam (Fig. 5-1) has a deflection y at a distance x from the built-in end, where

$$y = \frac{wx^2}{24EI}(x^2 + 6L^2 - 4Lx)$$

where E is the modulus of elasticity and I is the moment of inertia. Write an expression for the rate of change of deflection with respect to the distance x. Regard E, I, w, and L as constants.

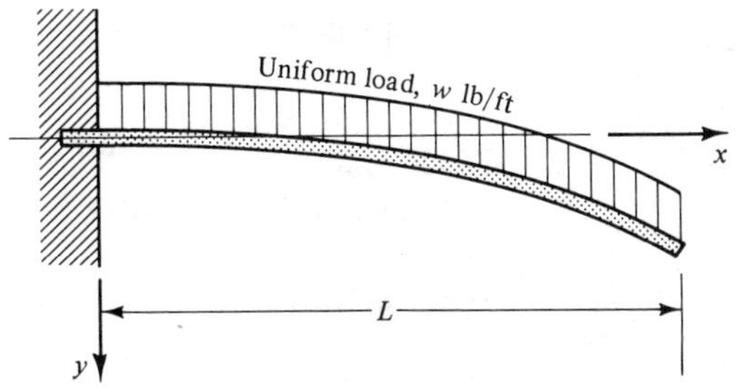

FIGURE 5-1. Cantilever beam with uniform load.

8. The equation of the elastic curve for the beam of Fig. 5-2 is

$$y = \frac{wx}{24EI}(L^3 - 2Lx^2 + x^3)$$

Write an expression for the rate of change of deflection (the slope) of the elastic curve at $x = L/4$.

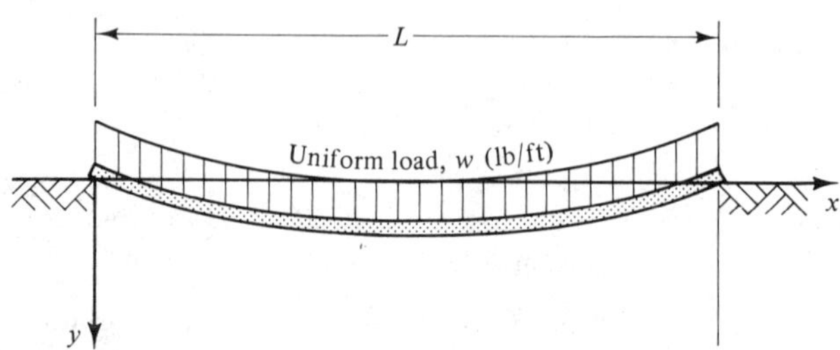

FIGURE 5-2. Simply supported beam with uniform load.

9. The charge through a 4.82-Ω resistor varies with time according to the function $q = 3.48t^2 - 1.64t$ coulomb. Write an expression for the instantaneous current through the resistor.

10. Evaluate the current in Problem 9 at $t = 5.92$ s.

11. Find the voltage across the resistor of Problem 9. Evaluate it at $t = 1.75$ s.

12. Find the instantaneous power in the resistor of Problem 9. Evaluate it at $t = 4.88$ s.

13. The charge at a resistor varies with time according to the function $q = 22.4t + 41.6t^3$ coulomb. Write an expression for the instantaneous current through the resistor and evaluate it at 2.50 s.

14. The voltage applied to a 33.5-μF capacitor is $v = 6.27t^2 - 15.3t + 52.2$ V. Find the current at $t = 5.50$ s.

15. The voltage applied to a 1.25-μF capacitor is $v = 3.17 + 28.3t + 29.4t^2$ V. Find the current at $t = 33.2$ s.

16. The current in a 1.44-H inductor is given by $i = 5.22t^2 - 4.02t$. Find the voltage across the inductor at $t = 2.00$ s.

17. The current in a 8.75-H inductor is given by $i = 8.22 + 5.83t^3$. Find the voltage across the inductor at $t = 25.0$ s.

5-2. MOTION OF A POINT

Velocity in Straight-Line Motion

Let us first consider straight-line, or rectilinear motion, and later we will study curvilinear motion. We first distinguish between speed and velocity. As an object moves along some path, the distance traveled *along the path* per unit time is called the *speed*. No account is taken of any change in direction; hence speed is a *scalar* quantity.

Velocity, on the other hand, is a *vector* quantity, having both *magnitude* and *direction*. For an object moving along a curved path, the magnitude of the velocity along the path is equal to the speed, and the *direction* of the velocity is the same as that of the *tangent to the curve* at that point. We will also speak of the components of that velocity in directions other than along the path, usually in the x and y directions.

As with average and instantaneous rates of change, we also can have average speed and average velocity, or instantaneous speed or instantaneous velocity. These terms have the same meaning as in Sec. 3-2.

Velocity or speed is the rate of change of displacement, and hence *is given by the derivative* of the displacement. If we give displacement the symbol s, then

Instantaneous Velocity	$v = \dfrac{ds}{dt}$	**A23**

The velocity is the rate of change of the displacement.

EXAMPLE 1: The displacement of an object is given by $s = 2t^2 + 5t + 4$ (in.), where t is the time in seconds. Find the velocity at 1 s.

Solution: We take the derivative

$$\frac{ds}{dt} = 4t + 5$$

At $t = 1$ s,

$$\left.\frac{ds}{dt}\right|_{t=1} = 4(1) + 5 = 9 \text{ in./s}$$

Acceleration in Straight-Line Motion

The acceleration is defined as the time rate of change of velocity. It is also a vector quantity. Since the velocity is itself the derivative of the displacement, the acceleration is the derivative of the derivative of the displacement, or the *second derivative* of displacement, with respect to time.

Instantaneous Acceleration	$a = \dfrac{dv}{dt} = \dfrac{d^2s}{dt^2}$	**A25**

The acceleration is the rate of change of the velocity.

EXAMPLE 2: One point in a certain mechanism moves according to the equation $s = 3t^3 + 5t - 3$ (cm), where t is in seconds. Find the instantaneous velocity and acceleration at $t = 2$ s.

Later we will do the reverse of differentiation, called *integration*, to find the velocity from the acceleration, and then the displacement from the velocity.

Solution: We take the derivative twice, with respect to t.

$$v = \frac{ds}{dt} = 9t^2 + 5$$

and

$$a = \frac{dv}{dt} = 18t$$

At $t = 2$ s,

$$v(2) = 9(2)^2 + 5 = 41 \text{ cm/s}$$

and

$$a(2) = 18(2) = 36 \text{ cm/s}^2$$

Velocity in Curvilinear Motion

FIGURE 5-3

At any instant we may think of a point as moving in a direction *tangent* to the path, as in Fig. 5-3. Thus if the speed is known and the direction of the tangent can be found, the instantaneous velocity (a vector having both magnitude and direction) can be found.

A more useful way of giving the instantaneous velocity, however, is by its x and y components (Fig. 5-4). If the magnitude and direction of the velocity are known, the components can be found by resolving the velocity vector into its x and y components.

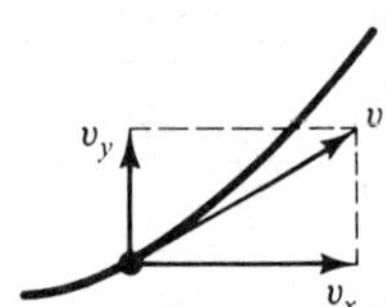

FIGURE 5-4. x and y components of velocity.

EXAMPLE 3: A point moves along the curve $y = 2x^3 - 5x^2 - 1$. (a) Find the direction of travel at $x = 2.00$ cm. (b) If the speed of the point along the curve is 3.00 cm/s, find the x and y components of the velocity when $x = 2.00$ cm.

Solution: (a) Taking the derivative of the given function gives $dy/dx = 6x^2 - 10x$. When $x = 2.00$,

$$y'(2) = 6(2)^2 - 10(2) = 4$$

The slope of the curve at that point is thus 4, and the direction of travel is $\tan^{-1} 4 = 76.0°$. (b) Resolving the velocity vector into x and y components,

$$v_x = 3.00 \cos 76.0° = 0.726 \text{ cm/s}$$

$$v_y = 3.00 \sin 76.0° = 2.91 \text{ cm/s}$$

Displacement Given by Parametric Equations

If the x displacement and y displacement are each given by a separate function of time (parametric equations) we may find the x and y components directly by taking the derivative of each equation.

x and y Components of Velocity	(a) $v_x = \frac{dx}{dt}$	(b) $v_y = \frac{dy}{dt}$	**A33**

Once we have expressions for the x and y components of velocity, we simply have to take the derivative again to get the x and y components of acceleration.

x and y Components of Acceleration	(a) $a_x = \frac{dv_x}{dt} = \frac{d^2x}{dt^2}$	(b) $a_y = \frac{dv_y}{dt} = \frac{d^2y}{dt^2}$	**A35**

EXAMPLE 4: A point moves along a curve such that its horizontal displacement is

$$x = 2t^3 - 15t \quad \text{cm}$$

and its vertical displacement is

$$y = 3 + t^2 \quad \text{cm}$$

Find (a) the horizontal and vertical components of the instantaneous velocity at $t = 2.00$ s, (b) the magnitude and direction of the instantaneous velocity at $t = 2.00$ s, (c) the x and y components of the instantaneous acceleration at $t = 2.00$ s, and (d) the magnitude and direction of the instantaneous acceleration at the same instant.

Solution: (a) Taking derivatives, we get

$$v_x = 6t^2 - 15 \qquad \text{and} \qquad v_y = 2t$$

At $t = 2.00$ s,

$$v_x = 6(2.00)^2 - 15 = 9.00 \text{ cm/s}$$

and

$$v_y = 2(2.00) = 4.00 \text{ cm/s}$$

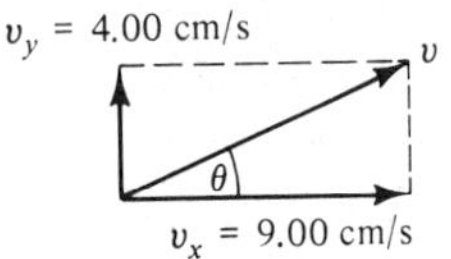

FIGURE 5-5

(b) The x and y components of the velocity are shown in Fig. 5-5. We find the resultant by vector addition.

$$v = \sqrt{v_x^2 + v_y^2} = \sqrt{81.0 + 16.0} = 9.85 \text{ cm/s}$$

Now finding the angle,

$$\tan\theta = \frac{v_y}{v_x} = \frac{4.00}{9.00} = 0.444$$

$$\theta = 24.0°$$

(c) Again taking derivatives we have

$$a_x = \frac{dv_x}{dt} = 12t \qquad \text{and} \qquad a_y = \frac{dv_y}{dt} = 2$$

At $t = 2.00$ s,

$$a_x = 12(2.00) = 24.0 \text{ cm/s}^2$$

and

$$a_y = 2.00 \text{ cm/s}^2$$

(d) Finding the resultant (Fig. 5-6) gives us

Note that the direction of the acceleration is *different* from that of the velocity. The velocity is always in a direction tangent to the path, while the acceleration vector is turned more toward the inside of the curve.

$$a = \sqrt{(24.0)^2 + (2.00)^2} = 24.1 \text{ cm/s}^2$$

and the angle

$$\tan\phi = \frac{a_y}{a_x} = \frac{2.00}{24.0} = 0.0833$$

$$\phi = 4.76°$$

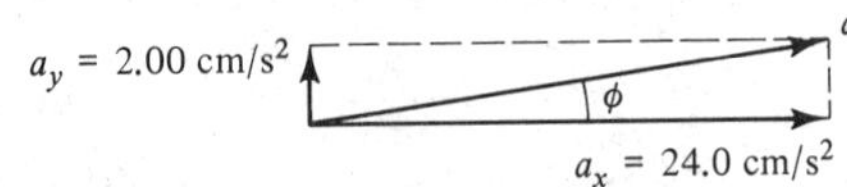

FIGURE 5-6. x and y components of acceleration.

If the *equation of the path* which the point follows is known, the derivative of that equation will give the slope of the tangent to the curve, and hence the direction of the velocity vector.

EXAMPLE 5: A point moves along the curve $y = x^2 + 2$ so that its horizontal displacement is $x = 3t^2 - 2t$ in. Find the magnitude and direction of the velocity when $t = 1.00$ s.

Solution: The velocity in the x direction is

$$v_x = \frac{dx}{dt} = 6t - 2 \qquad \text{in./s}$$

and at $t = 1.00$ s, the horizontal velocity is

$$v_x = 6(1.00) - 2 = 4.00 \text{ in./s}$$

and the horizontal displacement is

$$x = 3(1.00)^2 - 2(1.00) = 1.00 \text{ in.}$$

We now find the slope of the curve:

$$\text{slope} = \frac{dy}{dx} = 2x$$

and at $t = 1.00$, $x = 1.00$ in., so

$$\text{slope} = 2(1.00) = 2.00$$

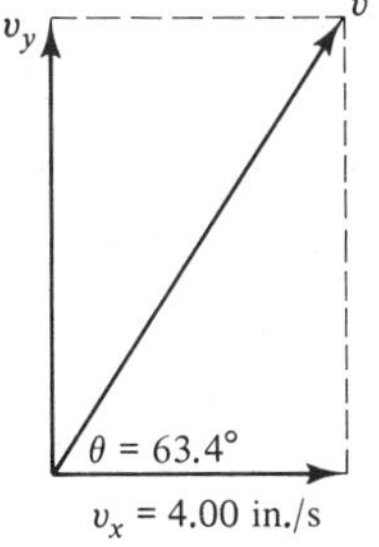

FIGURE 5-7

Thus the angle θ between v_x and v (Fig. 5-7) is

$$\tan \theta = 2.00$$
$$\theta = 63.4°$$

The magnitude of the velocity vector is, then,

$$v = \frac{4.00}{\cos 63.4°} = 8.94 \text{ in./s}$$

Rotation

We saw for straight-line motion that velocity is the rate of change of displacement. Similarly, for rotation, the *angular velocity,* ω, is the rate of change of *angular displacement,* θ.

Angular Velocity	$\omega = \dfrac{d\theta}{dt}$	**A29**

The angular velocity is the rate of change of the angular displacement.

Similarly for acceleration,

Angular Acceleration	$\alpha = \dfrac{d\omega}{dt} = \dfrac{d^2\theta}{dt^2}$	**A31**

The angular acceleration is the rate of change of the angular velocity.

EXAMPLE 6: The angular displacement of a rotating body is given by $\theta = 1.75t^3 + 2.88t^2 + 4.88$ rad. Find (a) the angular velocity and (b) the angular acceleration, at $t = 2.00$ s.

Solution: (a) From Eq. A29,

$$\omega = \frac{d\theta}{dt} = 5.25t^2 + 5.76t$$

At 2.00 s, $\omega = 5.25(4.00) + 5.76(2.00) = 32.5$ rad/s. (b) From Eq. A31,

$$\alpha = \frac{d\omega}{dt} = 10.5t + 5.76$$

At 2.00 s, $\alpha = 10.5(2.00) + 5.76 = 26.8$ rad/s^2.

EXERCISE 2

Straight-Line Motion

Find the instantaneous velocity and acceleration at the given time for the straight-line motion described by each equation, where s is in centimeters and t is in seconds.

1. $s = 32t - 8t^2$ at $t = 2$

2. $s = 6t^2 - 2t^3$ at $t = 1$

3. $s = t^2 + t^{-1} + 3$ at $t = \frac{1}{2}$

4. $s = (t + 1)^4 - 3(t + 1)^3$ at $t = -1$

5. $s = 120t - 16t^2$ at $t = 4$

6. $s = 3t - t^4 - 8$ at $t = 1$

7. The distance in feet traveled in time t seconds by a point moving in a straight line is given by the formula $s = 40t + 16t^2$. Find the velocity and the acceleration at the end of 2.0 s.

8. A car moves according to the equation $s = 250t^2 - \frac{5}{4}t^4$, where t is measured in minutes and s in feet. (a) How far does the car go in the first 10 min? (b) What is the maximum speed? (c) How far has the car moved when its maximum speed is reached?

9. If the distance traveled by a ball rolling down an incline in t seconds is s feet, where $s = 6t^2$, find its speed when $t = 5$ s.

10. The height s in feet reached by a ball t seconds after being thrown vertically upward at 320 ft/s is given by $s = 320t - 16t^2$. Find the greatest height reached by the ball, and the velocity with which it reaches the ground.

11. A bullet was fired straight upward so that its height in feet after t seconds was $s = 2000t - 16t^2$. What was its initial velocity? What was its greatest height? What was its velocity at the end of 10 s?

12. If the height h kilometers to which a balloon will rise in t minutes is given by the formula

$$h = \frac{10t}{\sqrt{4000 + t^2}}$$

at what rate is the balloon rising at the end of 30 min?

13. If the equation of motion of a point is $s = 16t^2 - 64t + 64$, find the position and acceleration at which the point first comes to rest.

Motion along a Curve

14. A point moves along the curve $y = 2x^3 - 3x^2 - 3$ with a speed of 14.00 in./s. Find the direction of travel at $x = 1.50$ in. and the x and y components of the velocity.

15. A point has a horizontal and vertical displacements (in cm) of $x = 3t^2 + 5t$ and $y = 13 - 3t^2$, respectively. Find the x and y components of the velocity and acceleration at $t = 4.55$ s.

16. A point moves along a curve having the parametric equations $x = 2t$, $y = 2t^2 - 4$. Find the coordinates and the velocity of the point when $t = 1$ s.

17. A point moves on the curve $xy = 2$, such that $x = 2t^2$, where x and y are in feet and t in seconds. Find the magnitude and direction of the velocity of the point when $t = 2$.

18. A point moves along the curve $y^2 = 8x$ (x and y in feet) in such a way that its y coordinate is $y = t^2 - 4t$, where y is in feet and t in seconds. Find its velocity when $t = 5$ s.

19. A point moves on the path $xy = 8$ such that its abscissa is $x = 4t^2$. Find the x and y components of the velocity and of the acceleration when $t = 2$, and find the total acceleration.

20. A point moves on the curve $y^2 = x$ such that its ordinate is $y = 4t$. Find the magnitude and direction of the acceleration of the point when $t = 1$ s.

21. A point moves on a parabola $y = x^2$ so that its abscissa changes at the constant rate of 2 units/s. How fast is its ordinate changing when the point passes through (3, 9)?

22. A point moves along the parabola $y^2 = 12x$. At what point do the abscissa and ordinate increase at the same rate?

23. A point moves along the parabola $y^2 = 8x$. What is the position of the point when $v_x = v_y$?

Rotation

24. The angular displacement of a rotating body is given by $\theta = 44.8t^3 + 29.3t^2 + 81.5$ rad. Find the angular velocity at $t = 4.25$ s.

25. Find the angular acceleration in Problem 24 at $t = 22.4$ s.

26. The angular displacement of a rotating body is given by $\theta = 184 + 271t^3$ rad. Find (a) the angular velocity and (b) the angular acceleration, at $t = 1.25$ s.

5-3. RELATED RATES

You would, of course, study the problem statement and make a diagram, as you would for other word problems.

In *related rate* problems, there are *two* quantities changing with time. The rate of change of one of the quantities is given and the other must be found. A procedure that can be followed is:

1. Locate the *given rate*. Since it is a rate, it can be *expressed as a derivative* with respect to time.
2. Determine the *unknown* rate. Express it also as a derivative with respect to time.
3. Find an *equation* linking the variable in the given rate with that in the unknown rate. If there are other variables in the equation they must be eliminated by means of other relationships.
4. Take the derivative of the equation *with respect to time.*
5. Substitute the given values and solve for the unknown rate.

EXAMPLE 1: A 20-ft ladder (Fig. 5-8) leans against a building. The foot of the ladder is pulled away from the building at a rate of 2.00 ft/s. How fast is the top of the ladder falling when the foot is 10.0 ft from the building?

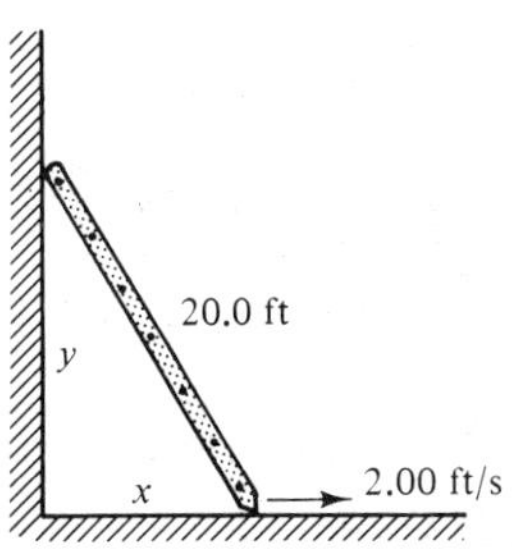

FIGURE 5-8. The ladder problem.

Solution: 1. If we let x be the distance from the foot of the ladder to the building, we have given

$$\frac{dx}{dt} = 2.00 \text{ ft/s}$$

2. If y is the distance from the ground to the top of the ladder, we are looking for dy/dt.

3. The equation linking x and y is the Pythagorean theorem,

$$x^2 + y^2 = 20.0^2$$

It will often be easiest to take the derivative implicitly rather than first to solve for one of the variables.

4. We could solve for y before taking the derivative, or do it implicitly,

$$2x\frac{dx}{dt} + 2y\frac{dy}{dt} = 0$$

5. When $x = 10.0$ ft, the height of the top of the ladder is

$$y = \sqrt{400 - (10.0)^2} = 17.3 \text{ ft}$$

We now substitute 2.00 for dx/dt, 10.0 for x, and 17.3 ft for y.

$$2(10.0)(2.00) + 2(17.3)\frac{dy}{dt} = 0$$

$$\frac{dy}{dt} = -1.16 \text{ ft/s}$$

The negative sign indicates that y is decreasing.

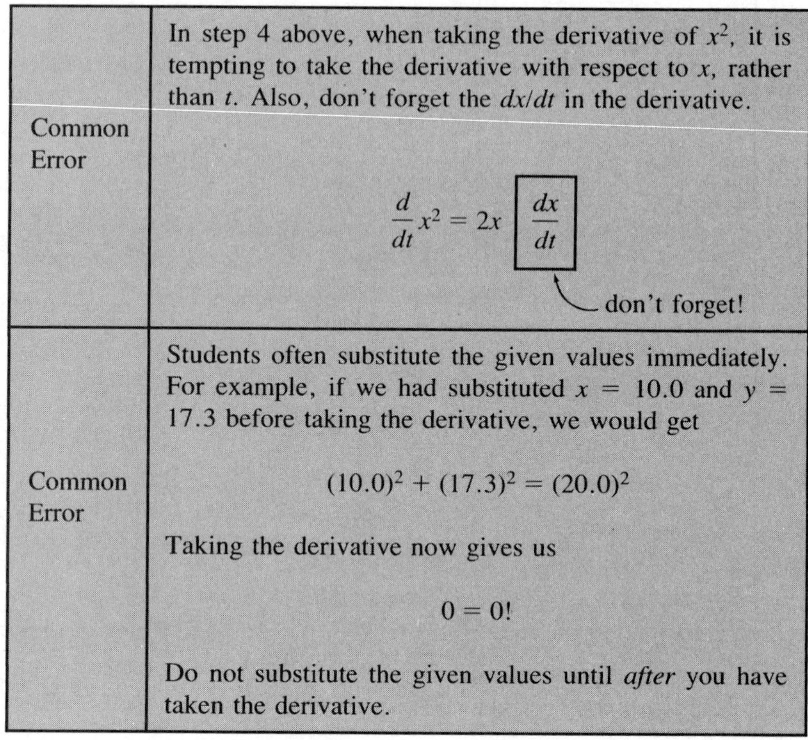

Common Error	In step 4 above, when taking the derivative of x^2, it is tempting to take the derivative with respect to x, rather than t. Also, don't forget the dx/dt in the derivative. $$\frac{d}{dt}x^2 = 2x \boxed{\frac{dx}{dt}}$$ don't forget!
Common Error	Students often substitute the given values immediately. For example, if we had substituted $x = 10.0$ and $y = 17.3$ before taking the derivative, we would get $$(10.0)^2 + (17.3)^2 = (20.0)^2$$ Taking the derivative now gives us $$0 = 0!$$ Do not substitute the given values until *after* you have taken the derivative.

In the next example when we find an equation linking the variables, it is seen to contain *three* variables. We then need a *second equation* with which we eliminate one variable.

EXAMPLE 2: A conical tank with vertex down has a base radius of 3.00 m and a height of 6.00 m (Fig. 5-9). Water flows in at a rate of 2.00 m^3/h. How fast is the water level rising when the depth is 3.00 m?

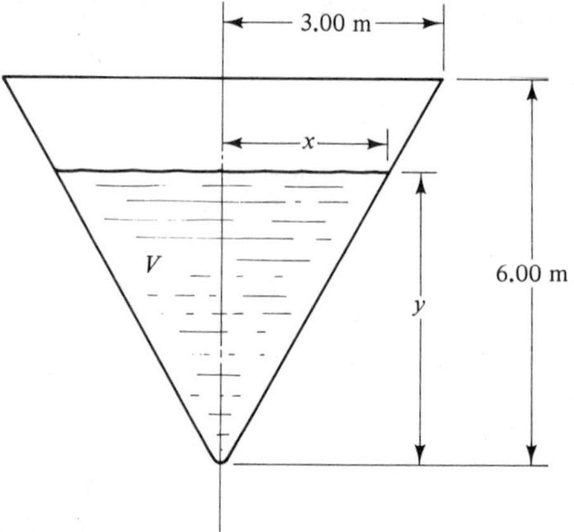

FIGURE 5-9. Conical tank.

Solution: We let y equal the depth, x the base radius, and V the volume of the water when the tank is partially filled. Then:

1. Given: $dV/dt = 2.00\ \text{m}^3/\text{h}$.
2. Unknown: dy/dt when $y = 3.00$ m.
3. The equation linking V and y is that for the volume of a cone, $V = (\pi/3)x^2y$. But in addition to the two variables in our derivatives, V and y, we have the third variable, x. We must eliminate x by means of another equation. By similar triangles,

$$\frac{x}{y} = \frac{3}{6}$$

from which

$$x = \frac{y}{2}$$

Substituting yields

$$V = \frac{\pi}{3} \cdot \frac{y^2}{4} \cdot y = \frac{\pi}{12} y^3$$

4. Now we have V as a function of y only. Taking the derivative,

$$\frac{dV}{dt} = \frac{\pi}{4} y^2 \frac{dy}{dt}$$

5. Substituting 3.00 for y and 2.00 for dV/dt, we obtain

$$2.00 = \frac{\pi}{4}(9.00)\frac{dy}{dt}$$

$$\frac{dy}{dt} = \frac{8.00}{9.00\pi} = 0.283\ \text{m/h}$$

Common Error	You cannot write an equation linking the variables until you have *defined* those variables. Indicate them right on your diagram. Draw axes if needed.

Some problems have two *independently* moving objects, as in the following example.

EXAMPLE 3: Ship A leaves a port P and travels west at 11.5 mi/h. After 2.25 h, ship B leaves P and travels north at 19.4 mi/h. How fast are the ships separating 5.00 h after the departure of A?

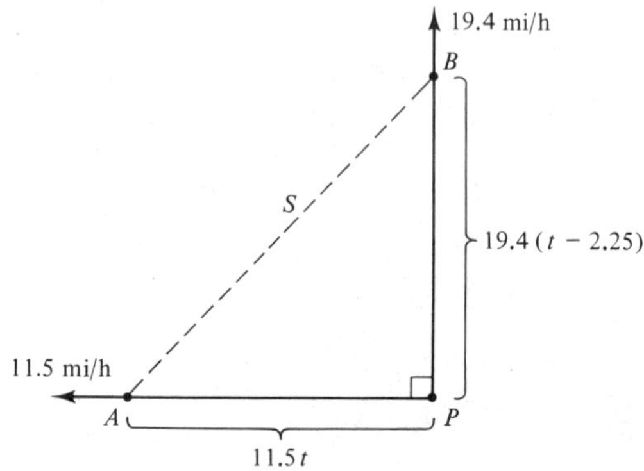

FIGURE 5-10. Note that we give the position of each ship after t hours have elapsed.

Solution: Figure 5-10 shows the ships t hours after A has left. Ship A has gone $11.5t$ miles and B has gone $19.4(t - 2.25)$ miles. The distance S between them is given by

$$S^2 = (11.5t)^2 + [19.4(t - 2.25)]^2$$

$$= 132t^2 + 376(t - 2.25)^2$$

Taking the derivative,

$$2S\frac{ds}{dt} = 2(132)t + 2(376)(t - 2.25)$$

$$\frac{ds}{dt} = \frac{132t + 376t - 846}{S} = \frac{508t - 846}{S}$$

At $t = 5.00$ h, A has gone $11.5(5.00) = 57.5$ mi, and B has gone $19.4(5.00 - 2.25) = 53.4$ mi. The distance S between them is then

$$S = \sqrt{(57.5)^2 + (53.4)^2} = 78.4 \text{ mi}$$

Substituting gives

$$\frac{ds}{dt} = \frac{508(5.00) - 846}{78.4} = 21.6 \text{ mi/h}$$

Common Error	Be sure to distinguish which quantities in a problem are *constants* and which are *variables*. Represent each variable by a letter, and do not substitute a given numerical value for a variable *until the very last step*.

EXERCISE 3

One Moving Object

1. An airplane flying horizontally at a height of 8000 m and at a rate of 100 m/s passes directly over a pond. How fast is its straight-line distance from the pond increasing 1 min later?

2. A ship moving 30 mi/h is 6.0 mi from a straight beach and moving parallel to the beach. How fast is the ship approaching a lighthouse on the beach when 10 mi (straight-line distance) from it?

3. A person is running at the rate of 8.0 mi/h on a horizontal street directly toward the foot of a tower 100 ft high. How fast is the person approaching the top of the tower when 50 ft from the foot?

Ropes and Cables

4. A boat is fastened to a rope that is wound about a winch 20 ft above the level at which the rope is attached to the boat. The boat is drifting away at the horizontal rate of 8 ft/s. How fast is the rope increasing in length when 30 ft of rope is out?

5. A boat with its anchor on the bottom at a depth of 40 m is drifting at 4.0 m/s, while the anchor cable slips from the boat at water level. At what rate is the cable leaving the boat when 50 m are out? Assume that the cable is straight.

6. A kite is at a constant height of 120 ft and moves horizontally, at 4.0 mi/h, in a straight line away from the person holding the cord. Assuming that the cord remains straight, how fast is the cord being paid out when its length is 130 ft?

7. A rope (Fig. 5-11) runs over a pulley at *A* and is attached at *B*. The rope is being wound in at the rate of 4.0 ft/s. How fast is *B* rising when *AB* is horizontal?

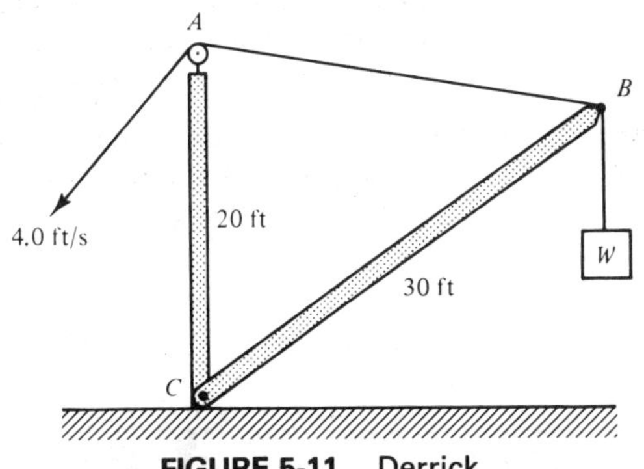

FIGURE 5-11. Derrick.

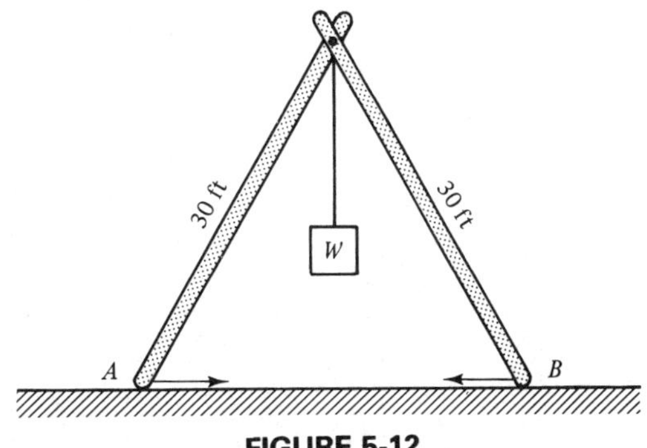

FIGURE 5-12

8. A weight *W* is being lifted between two poles (Fig. 5-12). How fast is *W* being raised when *A* and *B* are 20 ft apart if they are being drawn together, each moving at the rate of 9 in./s?

9. A bucket (Fig. 5-13) is raised by a person who walks away from the building at 1.0 ft/s. How fast is the bucket rising when $x = 80$ in.?

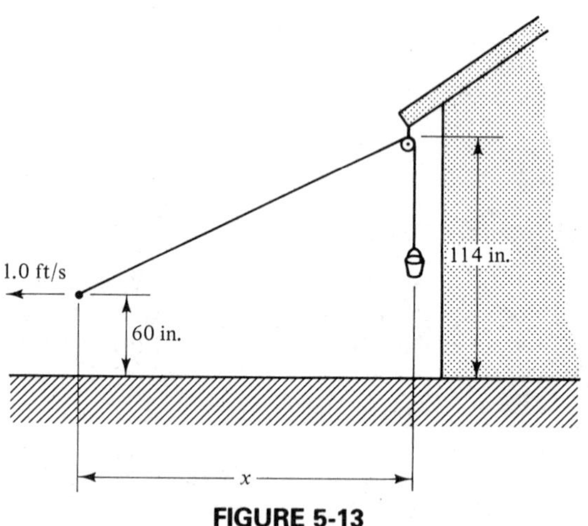

FIGURE 5-13

Two Moving Objects

10. Two trains start from the same point at the same time, one going east at a rate of 40 mi/h and the other going south at 60 mi/h. Find the rate at which they are separating after 1.0 h of travel.

11. An airplane leaves a field at noon and flies east at 100 km/h. A second airplane leaves the same field at 1 P.M. and flies south at 150 km/h. How fast are the airplanes separating at 2 P.M.?

12. An elevated train on a track 30 m above the ground crosses a street (which is at right angles to the track) at the rate of 20 m/s. At that instant, an automobile, approaching at the rate of 30 m/s is 40 m from a point directly beneath the track. Find how fast the train and the automobile are separating 2 s later.

13. As a person is jogging over a bridge at 5.0 ft/s, a boat, 30 ft beneath the bridge and moving perpendicular to the bridge, passes directly underneath at 10 ft/s. How fast are the person and the boat separating 3.0 s later?

Moving Shadows

14. A light is 100 ft from a wall (Fig. 5-14). A person runs at 13 ft/s away from the wall. Find the speed of the shadow on the wall when the person's distance from the wall is 50 ft.

15. A lamp is located on the ground 30 ft from a building. A person 6 ft tall walks from the light toward the building at a rate of 5 ft/s. Find the rate at which the person's shadow on the wall is shortening when the person is 15 ft from the building.

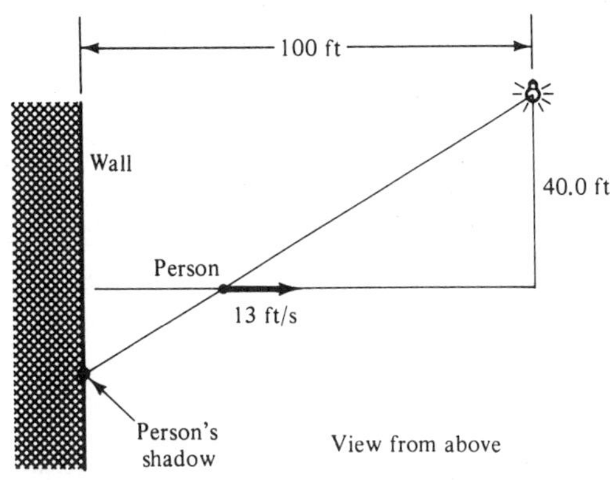

FIGURE 5-14

The height of the ball at t seconds is $s = 100 - \frac{1}{2}gt^2$.

16. A ball is dropped from a height of 100 ft above level ground when the sun is at an altitude of 40°. Find the rate at which the shadow of the ball is traveling along the ground when the ball has fallen 50 ft.

Expansion

17. A square sheet of metal 10 in. on a side is expanded by increasing its temperature so that each side of the square increases 0.005 in./s. At what rate is the area of the square increasing at 20 s?

18. A circular plate in a furnace is expanding so that its radius is changing 0.01 cm/s. How fast is the area of one face changing when the radius is 5 cm?

19. The volume of a cube is increasing at 10 in.^{3}/min. At the instant when its volume is 125 in.^{3}, what is the rate of change of its edge?

20. The edge of an expanding cube is changing at the rate of 0.003 in./s. Find the rate of change of its volume when its edge is 5 in. long.

21. The diameter and altitude of a right circular cylinder are found at a certain instant to be 10 in. and 20 in., respectively. If the diameter is increasing at the rate of 1 in./min, what change of rate in the altitude will keep the volume constant?

Fluid Flow

22. Water is running from a vertical cylindrical tank 3.0 m in diameter at the rate of $3\pi\sqrt{h}$ m^3/min, where h is the depth of the water in the tank. How fast is the surface of the water falling when $h = 9.0$ m?

23. Water is flowing into a conical reservoir 20 m deep and 10 m across the top, at a rate of 15 m^3/min. How fast is the surface rising when the water is 8 m deep?

24. Sand poured on the ground at the rate of 3.0 m^3/min forms a conical pile whose height is one-third the diameter of its base. How fast is the altitude of the pile increasing when the radius of its base is 2 m?

25. A horizontal trough 10.0 ft long has ends in the shape of an isosceles right triangle (Fig. 5-15). If water is poured into it at the rate of 8.0 ft^3/min, at what rate is the surface of the water rising when the water is 2 ft deep?

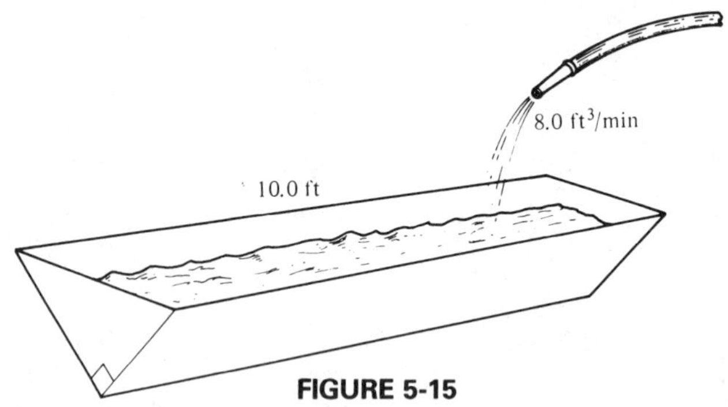

FIGURE 5-15

Gas Laws

Use Boyle's law, pv = constant.

26. A tank contains 1000 ft^3 of gas at a pressure of 5 lb/in.^{2}. If the pressure is decreasing at the rate of 0.05 lb/in.^{2} per hour, find the rate of increase of the volume.

27. The adiabatic law for the expansion of air is $pv^{1.4} = C$. If at a given time the volume is observed to be 10 ft^3, and the pressure is 50 lb/in.^{2}, at what rate is the pressure changing if the volume is decreasing 1 ft^3/s?

Miscellaneous

28. If the speed v ft/s of a certain bullet passing through wood is given by $v = 500\sqrt{1 - 3x}$, where x is the depth in feet, find the rate at which the speed is decreasing after the bullet has penetrated 3.0 in.

29. As a man walks a distance x (ft) along a board (Fig. 5-16), he sinks a distance $y = x^2(x + 2)/150$ in. If he moves at the rate of 2 ft/s, how fast is he sinking when $x = 8$ ft?

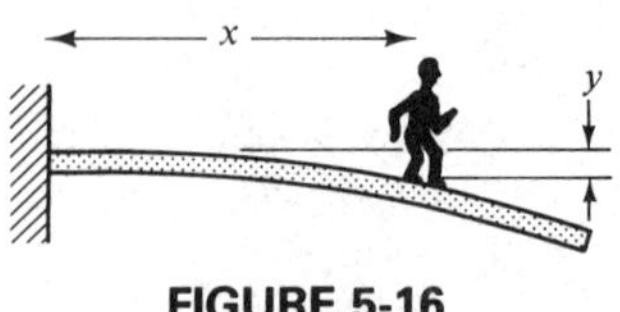

FIGURE 5-16

30. A stone dropped into a calm lake causes a series of circular ripples. The radius of the outer one increases at 2.0 ft/s. How rapidly is the disturbed area changing at the end of 3 s?

5-4. APPLIED MAXIMUM–MINIMUM PROBLEMS

In Chapter 4, we found extreme points, the peaks and valleys on a curve. We did this, you recall, by finding the points where the slope (and hence the first derivative) was zero. We now apply the same idea to problems in which we find, for example, the point of minimum cost, or the point of maximum efficiency, or the point of maximum carrying capacity.

Suggested Steps for Maximum–Minimum Problems

1. Locate the quantity (which we will call Q) to be maximized or minimized, and locate the independent variable, say x, which is varied in order to maximize or minimize Q.
2. Write an equation linking Q and x. If this equation contains another variable as well, that variable must be eliminated by means of a second equation. It is a good idea, although not essential, to sketch the function $Q = f(x)$. The sketch will give the approximate location of any maximum or minimum points.
3. Take the derivative dQ/dx.
4. Set the derivative equal to zero. Solve for x.
5. Check any extreme points found to see if they are maxima or minima. This can be done simply by looking at your graph of $Q = f(x)$, by your knowledge of the physical problem, or by the first- or second-derivative tests. Also check if the maximum or minimum value you seek is at one of the endpoints. An endpoint can be a maximum or minimum point in the given interval, even though the slope there is not zero.

A list of general suggestions such as these is usually so vague as to be useless without examples. Our first example is one in which the relation between the variables is given verbally in the problem statement.

EXAMPLE 1: What two positive numbers whose product is 100 have the least possible sum?

Solution: 1. We want to minimize the sum S of the two numbers by varying one of them, which we call x. Then

$$\frac{100}{x} = \text{other number}$$

2. The sum of the two numbers is

$$S = x + \frac{100}{x}$$

3. Taking the derivative yields

$$\frac{dS}{dx} = 1 - \frac{100}{x^2}$$

4. Setting the derivative to zero and solving for x gives us

$$x^2 = 100$$

$$x = \pm 10$$

The sum of two positive numbers whose product is a constant is always a minimum when the numbers are equal as in this example. Can you prove the general case?

Since we are asked for positive numbers, we discard the -10. The other number is $100/10 = 10$.

5. But have we found those numbers that will give a *minimum* sum, as requested, or a *maximum* sum? We can check this by means of the second derivative test. Taking the second derivative, we have

$$S'' = \frac{200}{x^3}$$

When $x = 10$,

$$S''(10) = \frac{200}{1000} = 0.2$$

which is *positive,* indicating a *minimum* point. This is verified by our graph of S versus x (Fig. 5-17) which shows a minimum at $x = 10$.

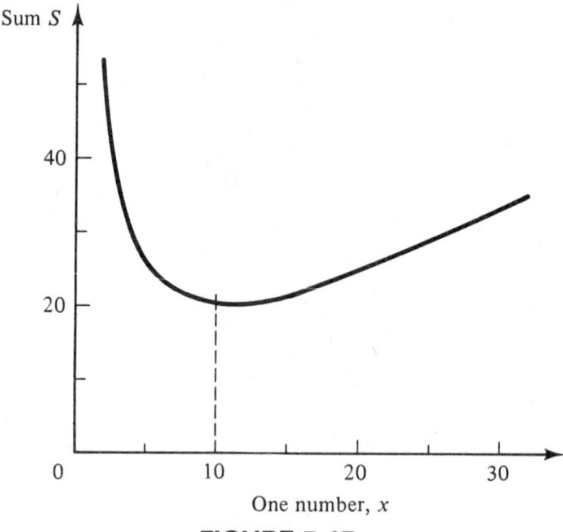

FIGURE 5-17

In the next example, the equation linking the variables is easily written from the geometrical relationships in the problem.

EXAMPLE 2: An open-top box is to be made from a square of sheet metal 40 cm on a side by cutting a square from each corner and bending up the sides along the dashed lines in Fig. 5-18. Find the dimension x of the cutout that will result in a box of the greatest volume.

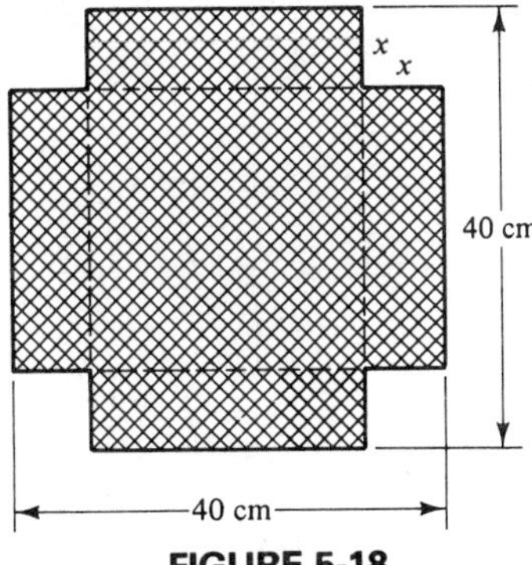

FIGURE 5-18

Solution: 1. We want to maximize the volume V by varying x.

2. The equation is

$$\begin{aligned} V &= \text{length}\cdot\text{width}\cdot\text{depth} \\ &= (40 - 2x)(40 - 2x)x \\ &= x(40 - 2x)^2 \end{aligned}$$

3. Taking the derivative, we obtain

$$\begin{aligned} \frac{dV}{dx} &= x(2)(40 - 2x)(-2) + (40 - 2x)^2 \\ &= -4x(40 - 2x) + (40 - 2x)^2 \\ &= (40 - 2x)(-4x + 40 - 2x) \\ &= (40 - 2x)(40 - 6x) \end{aligned}$$

4. Setting the derivative to zero and solving for x, we have

$$40 - 2x = 0 \qquad\qquad 40 - 6x = 0$$

$$x = 20 \text{ cm} \qquad\qquad x = \frac{20}{3} = 6.67 \text{ cm}$$

We discard $x = 20$ cm, for it is a minimum value, and results in the entire sheet of metal being cut away, and keep $x = 6.67$ cm as our answer.

5. The graph of volume versus x in Fig. 5-19 shows that our point is indeed a maximum, so no further test is needed.

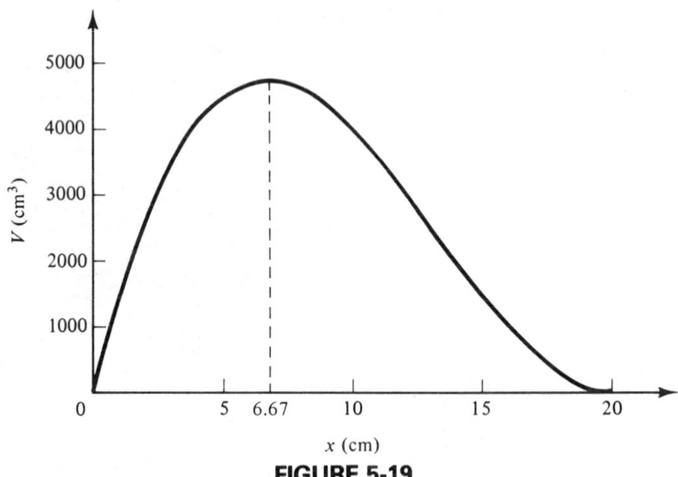

FIGURE 5-19

Our next example is one in which the equation is given.

EXAMPLE 3: The deflection y of a simply supported beam with a concentrated load P (Fig. 5-20) is given by

$$y = \frac{Pbx}{6LEI}(L^2 - x^2 - b^2)$$

where E is the modulus of elasticity, I is the moment of inertia of the beam's cross section, and x is the distance from the left end of the beam. Find the value of x at which the deflection is a maximum, for a 20.0-ft-long beam with a concentrated load 5.00 ft from the right end.

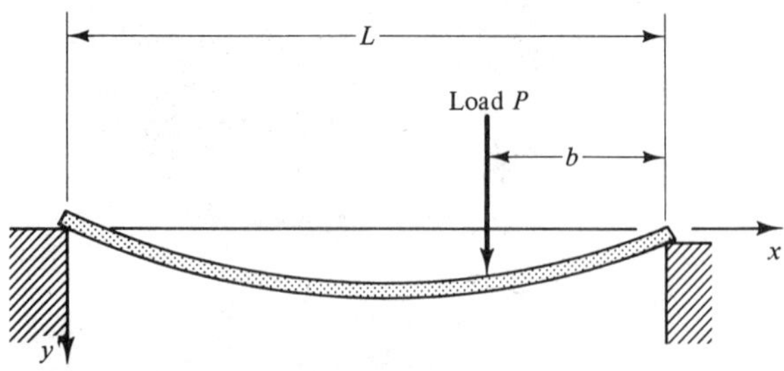

FIGURE 5-20. Simply supported beam with concentrated load.

Solution: 1. We want to find the distance x from the left end at which the deflection is a maximum.

2. The equation is given in the problem statement.

3. We take the derivative using the product rule, noting that every quantity but x or y is a constant.

$$\frac{dy}{dx} = \frac{Pb}{6LEI}[x(-2x) + (L^2 - x^2 - b^2)(1)]$$

4. We set this derivative equal to zero and solve for x:

$$2x^2 = L^2 - x^2 - b^2$$

$$3x^2 = L^2 - b^2$$

$$x = \pm\sqrt{\frac{L^2 - b^2}{3}}$$

We drop the negative value, since x cannot be negative in this problem. Now substituting $L = 20.0$ ft and $b = 5.00$ ft gives us

$$x = \sqrt{\frac{400 - 25}{3}} = 11.2 \text{ ft}$$

Thus the maximum deflection occurs between the load and the midpoint of the beam, as we might expect.

5. It is clear from the physical problem that our point is a maximum, so no test is needed.

The equation linking the variables in the preceding example had only two variables, x and y. In the following example, our equation will have *three* variables, one of which must be eliminated before we take the derivative. We eliminate the third variable by means of a second equation.

EXAMPLE 4: If we assume that the strength of a rectangular beam varies directly as its width and the square of its depth, find the dimensions of the strongest beam that can be cut from a round log 12.0 in. in diameter (Fig. 5-21).

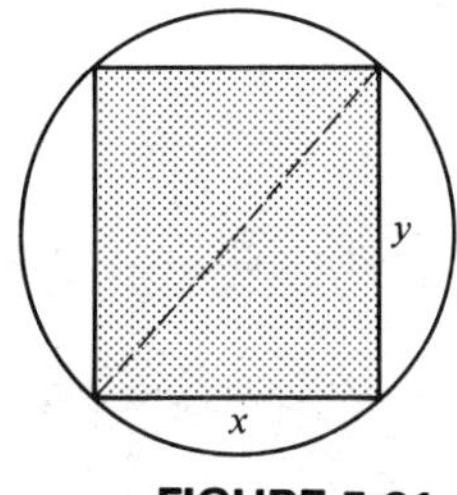

FIGURE 5-21. Beam cut from a log.

Solution: 1. We want to maximize the strength S by varying the width x.

2. The strength S is $S = kxy^2$, where k is a constant of proportionality. Note that we have three variables, S, x, and y. We must eliminate x or y. By the Pythagorean theorem,

$$x^2 + y^2 = 12.0^2$$

so

$$y^2 = 144 - x^2$$

Substituting, we obtain

$$S = kx(144 - x^2)$$

3. The derivative is

$$\frac{dS}{dx} = kx(-2x) + k(144 - x^2)$$

$$= -3kx^2 + 144k$$

4. Setting it equal to zero and solving for x gives us $3x^2 = 144$, so

$$x = \pm 6.93 \text{ in.}$$

We discard the negative value, of course. The depth is

$$y = \sqrt{144 - 48} = 9.80 \text{ in.}$$

5. But have we found the dimensions of the beam with *maximum* strength, or *minimum* strength? Let us use the second derivative test to tell us. The second derivative is

$$\frac{d^2S}{dx^2} = -6kx$$

When $x = 6.93$,

$$\frac{d^2S}{dx^2} = -41.6k$$

Since k is positive, the second derivative is negative, which tells us that we have found a *maximum*.

EXERCISE 4

Number Problems

1. What number added to half the square of its reciprocal gives the smallest sum?

2. Separate the number 10 into two parts such that their product will be a maximum.

3. Separate the number 20 into two parts such that the product of one part and the square of the other part is a maximum.

4. Separate the number 5 into two parts such that the square of one part times the cube of the other part shall be a maximum.

Minimum Perimeter

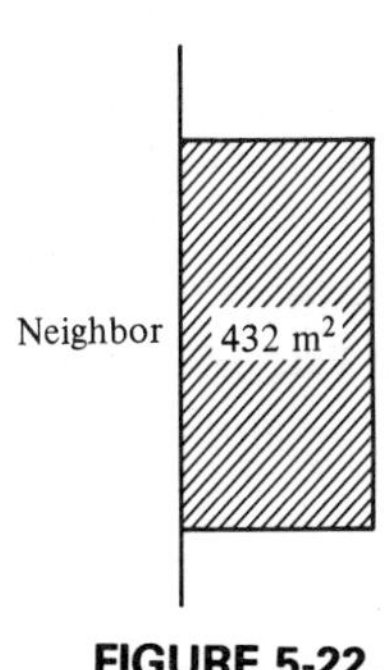

FIGURE 5-22. Rectangular garden.

5. A rectangular garden (Fig. 5-22) laid out along your neighbor's lot contains 432 m^2. It is to be fenced on all sides. If the neighbor pays for half the shared fence, what should be the dimensions of the garden so that your cost is a minimum?

6. It is required to enclose a rectangular field by a fence (Fig. 5-23), and then divide it into two lots by a fence parallel to the short sides. If the area of the field is 25,000 ft^2, find the lengths of the sides so that the total length of fence will be a minimum.

7. A rectangular pasture 162 yd^2 in area is built so that a long, straight wall serves as one side of it. If the length of the fence along the remaining three sides is the least possible, find the dimensions of the pasture.

Maximum Volume of Containers

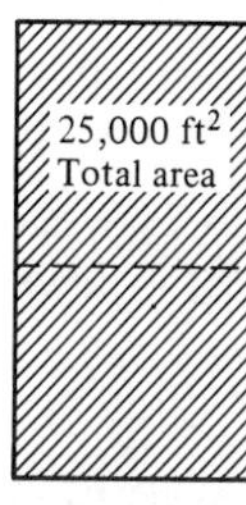

FIGURE 5-23. Rectangular field.

8. Find the volume of the largest box that can be made from a rectangular sheet of metal 6.00 in. by 16.0 in. (Fig. 5-24) by cutting a square from each corner and turning up the sides.

9. Find the height and base diameter of a cylindrical, topless tin cup of maximum volume if its area (sides and bottom) is 100 cm^2.

10. The slant height of a certain cone is 50.0 cm. What cone height will make the volume a maximum?

Maximum Area of Plane Figures

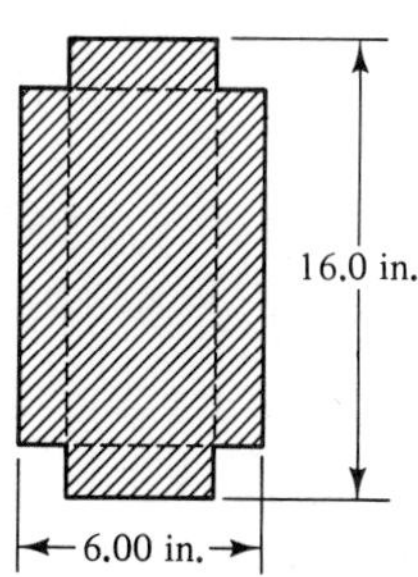

FIGURE 5-24. Sheet metal for box.

11. Find the area of the greatest rectangle that has a perimeter of 20 in.

12. A window composed of a rectangle surmounted by an equilateral triangle is 15 ft in perimeter (Fig. 5-25). Find the dimensions that will make its total area a maximum.

FIGURE 5-25. Window.

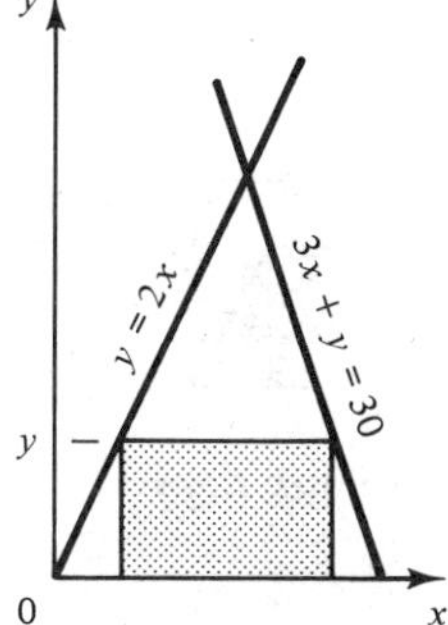

FIGURE 5-26

13. Two corners of a rectangle are on the x axis between $x = 0$ and $x = 10$ (Fig. 5-26). The other two corners are on the lines whose equations are $y = 2x$

and $3x + y = 30$. For what value of y will the area of the rectangle be a maximum?

Maximum Cross-Sectional Area

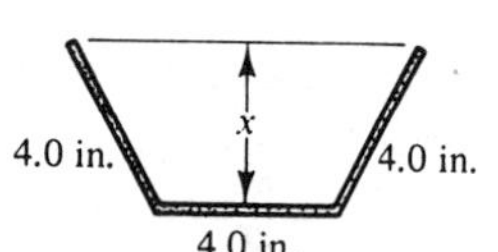

FIGURE 5-27

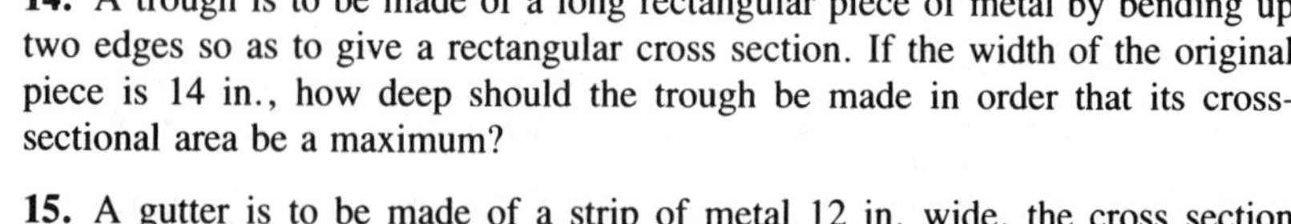

14. A trough is to be made of a long rectangular piece of metal by bending up two edges so as to give a rectangular cross section. If the width of the original piece is 14 in., how deep should the trough be made in order that its cross-sectional area be a maximum?

15. A gutter is to be made of a strip of metal 12 in. wide, the cross section having the form shown in Fig. 5-27. What depth x gives a maximum cross-sectional area?

FIGURE 5-28

Minimum Distance

16. Two railroad tracks intersect at right angles. There is a train on each track approaching the crossing at 40 mi/h, one being 10 mi, the other 20 mi from the intersection. What will be their minimum distance apart?

17. Find the point Q on the curve $y = \frac{x^2}{2}$ which is nearest the point (4, 1) (Fig. 5-28).

18. Given one branch of the parabola $y^2 = 8x$ and the point $P(6, 0)$ on the x axis (Fig. 5-29), find the coordinates of point Q so that PQ is a minimum.

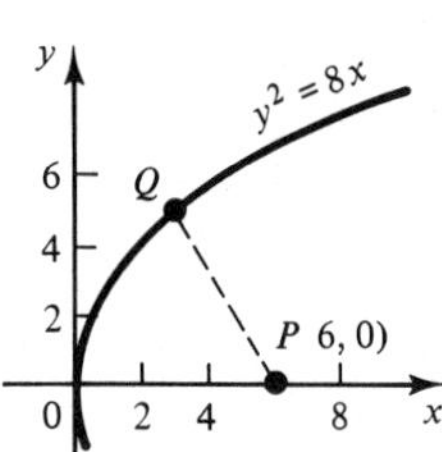

FIGURE 5-29

Inscribed Plane Figures

19. Find the dimensions of the rectangle of greatest area that can be inscribed in an equilateral triangle each of whose sides is 10.0 in., if one of the sides of the rectangle is on a side of the triangle (Fig. 5-30). (*Hint:* Let the independent variable be the height x of the rectangle.)

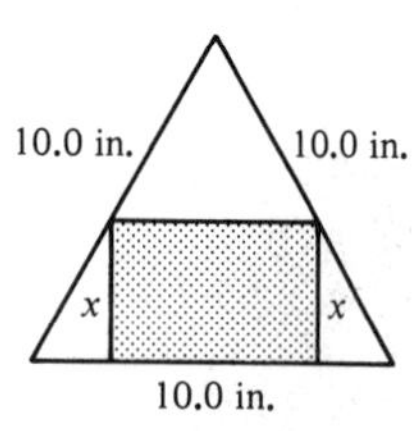

FIGURE 5-30

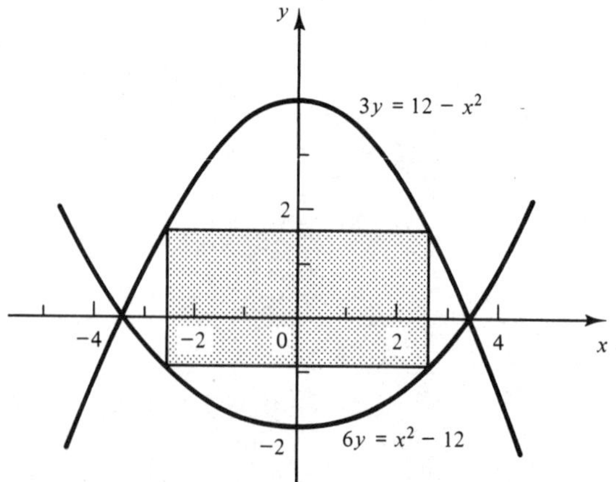

FIGURE 5-31

20. Find the area of the largest rectangle with sides parallel to the coordinate axes which can be inscribed in the figure bounded by the two parabolas $3y = 12 - x^2$ and $6y = x^2 - 12$ (Fig. 5-31).

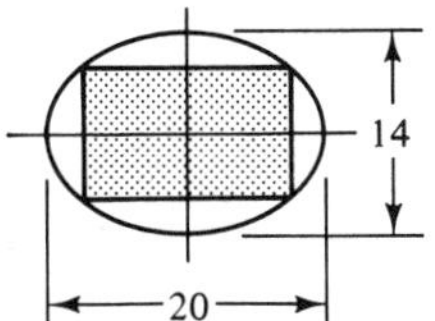

FIGURE 5-32. Rectangle inscribed in ellipse.

21. Find the dimensions of the largest rectangle that can be inscribed in an ellipse whose major axis is 20 units and whose minor axis is 14 units (Fig. 5-32).

Inscribed Volumes

22. Find the dimensions of the largest rectangular parallelepiped with a square base which can be cut from a solid sphere 18.0 in. in diameter (Fig. 5-33).

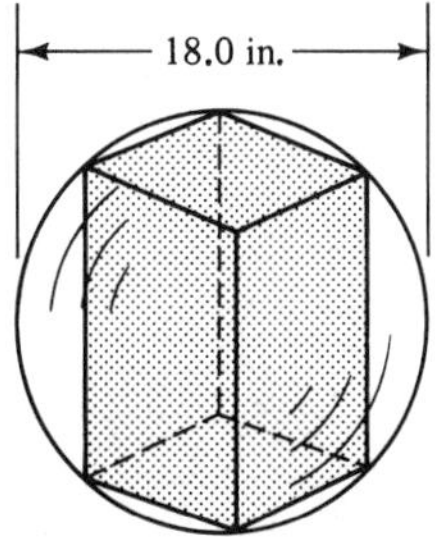

FIGURE 5-33. Parallelepiped inscribed in sphere.

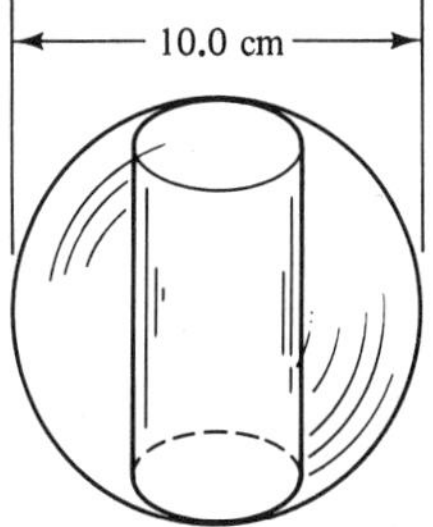

FIGURE 5-34. Cylinder inscribed in sphere.

23. Find the dimensions of the largest right circular cylinder that can be inscribed in a sphere with a diameter of 10.0 cm (Fig. 5-34).

24. Find the height of the cone of minimum volume circumscribed about a sphere of radius 10.0 m (Fig. 5-35).

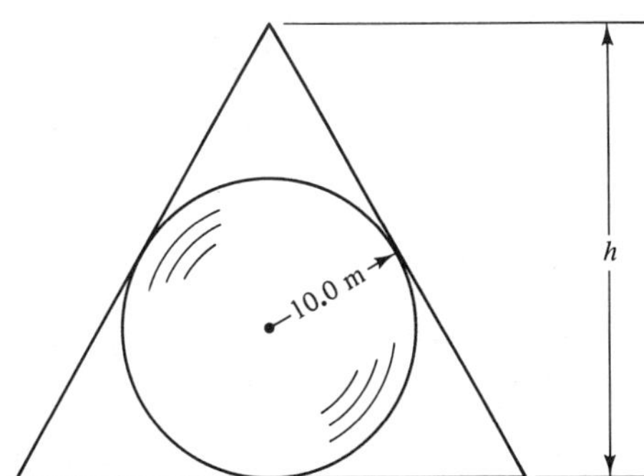

FIGURE 5-35. Cone circumscribed about sphere.

25. Find the altitude of the cone of maximum volume that can be inscribed in a sphere of radius 9.0 ft.

Most Economical Dimensions of Containers

26. What should be the diameter of a can holding 1 qt (58 in.^{3}) and requiring the least amount of metal, if the can is open at the top?

27. A silo (Fig. 4-18) has a hemispherical roof, cylindrical sides, and circular floor, all made of steel. Find the dimensions for a silo having a volume of 755 m^3 (including the dome) that needs the least steel.

Minimum Travel Time

28. A man in a rowboat at P (Fig. 5-36) 6.0 mi from shore desires to reach point Q on the shore at a straight-line distance of 10.0 mi from his present position. If he can walk 4 mi/h and row 3 mi/h, at what point L should he land in order to reach Q in the shortest time?

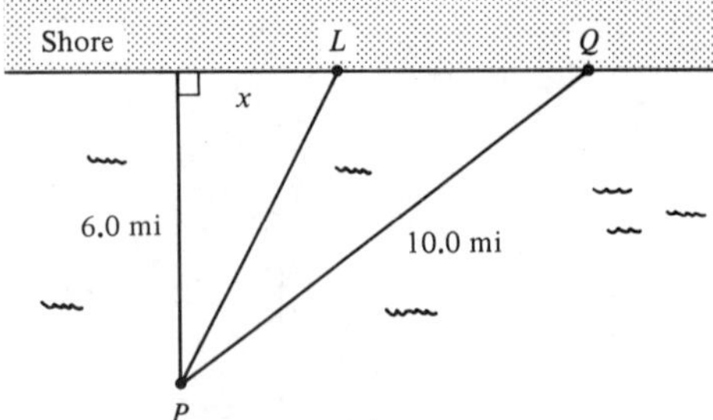

FIGURE 5-36. Find the fastest route from boat P to shore at Q.

Beam Problems

29. The strength S of the beam in Fig. 5-21 is given by $S = kxy^2$, where k is a constant. Find x and y for the strongest rectangular beam that can be cut from an 18.0-in.-diameter cylindrical log.

30. The stiffness Q of the beam in Fig. 5-21 is given by $Q = kxy^3$, where k is a constant. Find x and y for the stiffest rectangular beam that can be cut from an 18.0-in.-diameter cylindrical log.

Light

31. The intensity E of illumination at a point due to a light at a distance x from the point is given by $E = kI/x^2$, where k is a constant and I is the intensity of the source. A light M has an intensity three times that of N (Fig. 5-37). At what distance from M is the illumination a minimum?

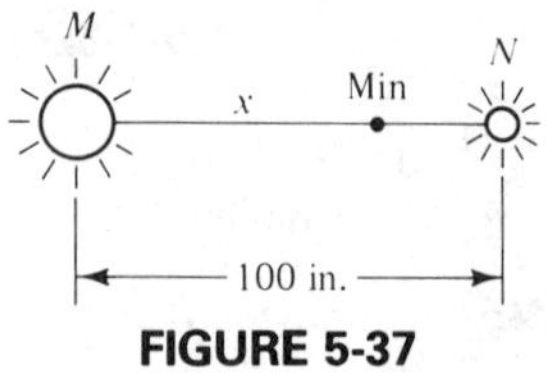

M x Min N

100 in.

FIGURE 5-37

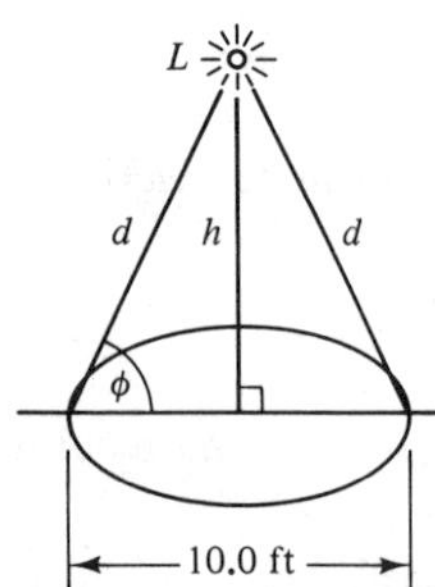

FIGURE 5-38

32. The intensity of illumination from a given source of light L varies as the sine of the angle ϕ at which the light rays strike the illuminated surface, divided by the square of the distance d from the light. Find the height h of a light directly over the center of a given circle 10.0 ft in diameter (Fig. 5-38) so that it shall give a maximum illumination to the circumference.

Electrical

33. The power delivered to a load by a 30-V source of internal resistance 2 Ω is $30i - 2i^2$ watts, where i is the current in amperes. For what current will this source deliver the maximum power?

34. When 12 cells, each having an EMF of e and an internal resistance r are connected to a load R as in Fig. 5-39 the current in R is

$$i = \frac{3e}{\frac{3r}{4} + R}$$

Show that the maximum power (i^2R) delivered to the load is a maximum when the load R is equal to the equivalent internal resistance of the source, $3r/4$.

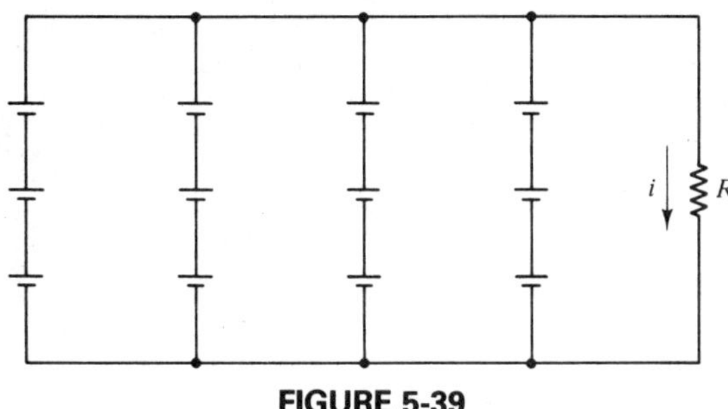

FIGURE 5-39

35. A certain transformer has an efficiency E when delivering a current i, where

$$E = \frac{115i - 25 - i^2}{115i}$$

At what current is the efficiency of the transformer a maximum?

Mechanisms

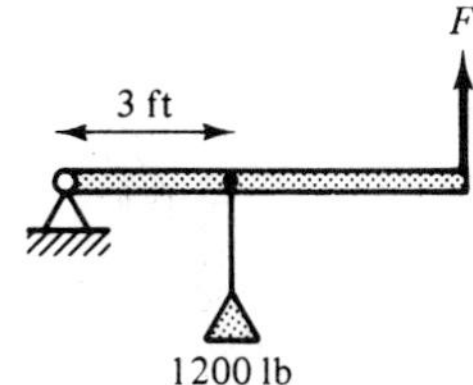

FIGURE 5-40

36. If the lever in Fig. 5-40 weighs 12 lb per foot, find its length so as to make the lifting force F a minimum.

37. The efficiency E of a screw is given by

$$E = \frac{x - \mu x^2}{x + \mu}$$

where μ is the coefficient of friction and x the tangent of the pitch angle of the screw. Find x for maximum efficiency if $\mu = 0.45$.

5-5. DIFFERENTIALS

Definition

Up to now, we have treated the symbol dy/dx as a *whole*, and not as the quotient of two quantities dy and dx. Here, we give dy and dx separate names and meanings of their own. The quantity dy is called the *differential of y*, and dx is called the *differential of x*.

These two differentials, dx and dy, have a simple geometric interpretation. Figure 5-41 shows a tangent drawn to a curve $y = f(x)$ at some point P. The slope of the curve is found by evaluating dy/dx at P. *The differential dy is then the rise of the tangent line, in some arbitrary run dx.* Since the rise of a line is equal to the slope of the line times the run:

Differential of y	$dy = f'(x)\,dx$	**335**

where we have represented the slope by $f'(x)$ instead of dy/dx, to avoid confusion.

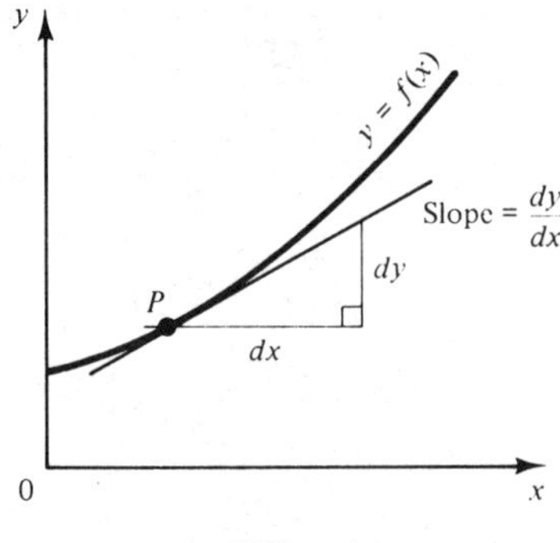

FIGURE 5-41. Differentials.

Differentials and Increments Compared

You may be thinking that the differentials dx and dy look something like the increments Δx and Δy back in Fig. 3-8. In fact, it is easy to confuse differentials and increments.

Recall from Sec. 3-2 that when we gave a small increment Δx to x, then the function $f(x)$ changed by an increment Δy, as in Fig. 5-42. We see then that the difference between differentials and increments is that:

1. An increment Δx is a *small* change in x, while the differential dx can be of any size, large or small (although we will soon

see that, for applications, we will choose only small values for differentials).

2. The increment Δy is the change in y for the *curve* itself, corresponding to an increment Δx, whereas the differential dy is the change in y for the *tangent line,* corresponding to a change dx.

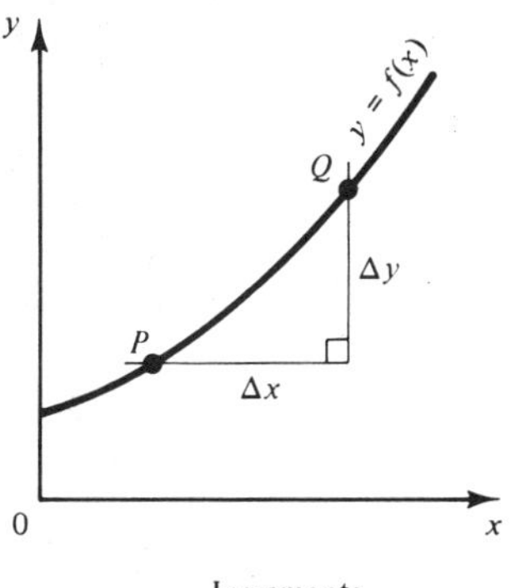

Increments

FIGURE 5-42. Increments.

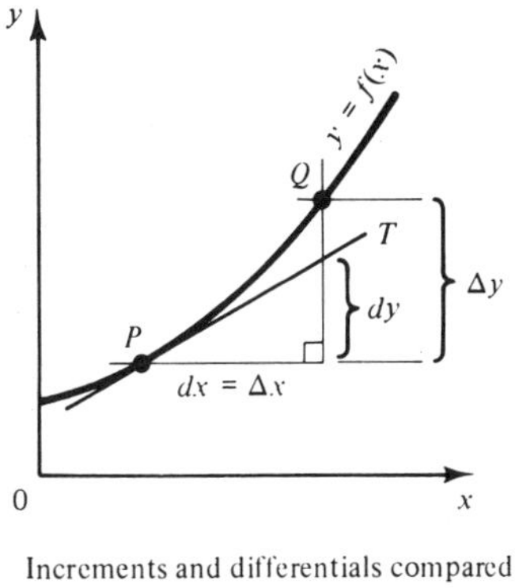

Increments and differentials compared

FIGURE 5-43. Increments and differentials compared.

These quantities can be clearly seen in Fig. 5-43 where we have chosen the differential dx to be equal to the increment Δx.

Also notice that if we choose a small enough dx (or Δx) then dy *will be approximately equal to* Δy. We make use of this fact shortly to find approximate changes in one quantity due to small changes in another quantity.

Differential Form

If we take the derivative of some function, say, $y = x^3$ we get

$$\frac{dy}{dx} = 3x^2$$

Since we may now think of dy and dx as separate quantities (differentials), we can multiply both sides by dx,

An equation containing a derivative (called a *differential equation*) is usually written in differential form before it is solved.

$$dy = 3x^2\, dx$$

This expression is said to be in *differential form.*

Thus to find dy, the differential of y, given some function of x, simply take the derivative of the function and multiply by dx.

EXAMPLE 1: If $y = 3x^2 - 2x + 5$, find the differential dy.

Solution: Taking the derivative gives us

$$\frac{dy}{dx} = 6x - 2$$

Multiplying by dx, we get the differential of y

$$dy = (6x - 2)\,dx$$

Using Differentials to Estimate Small Changes in a Function

Let us look again at Fig. 5-43. There we took the run equal to the increment Δx and also equal to the differential dx, and we saw that the tangent line had a rise of dy, while the curve itself rose by an amount Δy.

Now it is clear from the figure that dy and Δy are not equal. *But,* if Δx is small, *the differential dy can be a close enough approximation to the increment Δy for many applications.*

We obtain an expression for Δy from our definition of the derivative (Eq. 302),

$$\lim_{\Delta x \to 0} \frac{\Delta y}{\Delta x} = \frac{dy}{dx}$$

That is, that $\Delta y/\Delta x$ approaches dy/dx as Δx goes to zero. But when Δx has some small but finite value, we see that

$$\frac{\Delta y}{\Delta x} \cong \frac{dy}{dx}$$

Multiplying by Δx yields

Approximations Using Differentials	$\Delta y \cong \dfrac{dy}{dx}\Delta x$	**336**

The approximate change Δy in some function due to a small change Δx is the change Δx multiplied by the derivative of the function.

EXAMPLE 2: Using differentials, estimate the change in the function $y = 2x^3$ when x changes from 5 to 5.1. Compare this estimate with the exact value.

Solution: The change Δx in x is 0.1, and the derivative is

$$\frac{dy}{dx} = 6x^2$$

When $x = 5$,

$$\frac{dy}{dx} = 6(5)^2 = 150$$

Then by Eq. 336,

$$\Delta y \cong \frac{dy}{dx}\,\Delta x = 150(0.1) = 15$$

The exact change is

$$2(5.1)^3 - 2(5.0)^3 = 265.302 - 250 = 15.302$$

or about 2% different than our estimate.

Approximate Volumes of Shells and Rings

We can use differentials to derive approximate formulas for the volumes of thin-walled shells, such as the volume of rubber in a spherical balloon of given radius. We do this by thinking of the volume of the wall of the shell as the increase in volume of the sphere as the radius increases by an amount equal to the wall thickness.

EXAMPLE 3: Derive a formula for the volume of a hollow sphere of inside radius r, having a wall thickness t.

Solution: The volume of a sphere is, by Eq. 128, $V = \frac{4}{3}\pi r^3$. Taking the derivative, we have

$$\frac{dV}{dr} = 4\pi r^2$$

The increment Δr as r increases from the inside of the wall to the outside is equal to the wall thickness t, so by Eq. 336,

$$\Delta V \cong \frac{dV}{dr}\,\Delta r \cong 4\pi r^2 t$$

Estimating the Effect of Small Errors in Measurement

Equation 336 tells us how much y will depart from its original value when x departs from its original value by an amount Δx. In this section we let Δx represent some small error in measurement or manufacture, and compute the resulting error in some quantity y which is related to x.

EXAMPLE 4: In order to calculate the power dissipated in a certain 1565-Ω resistor, the current through the resistor is measured and found to be 2.84 A. What would be the error in the calculated value of the power if the error in the current measurement is 5%?

Solution: The power to a resistor is, by Eq. A67, $P = I^2R$. Taking the derivative,

$$\frac{dP}{dI} = 2IR$$

The increment in I is equal to 5% of the measurement,

$$\Delta I = 0.05(2.84) = 0.142 \text{ A}$$

Then by Eq. 336,

$$\Delta P \cong \frac{dP}{dI}\,\Delta I = 2IR(0.142)$$

Substituting $I = 2.84$ A and $R = 1565$ Ω gives us

$$\Delta P \cong 2(2.84)(1565)(0.142)$$
$$= 1260 \text{ W}$$

EXERCISE 5

Write the differential dy for each function.

1. $y = x^3$

2. $y = x^2 + 2x$

3. $y = \dfrac{x-1}{x+1}$

4. $y = (2 - 3x^2)^3$

5. $y = (x+1)^2(2x+3)^3$

6. $y = \sqrt{1 - 2x}$

7. $y = \dfrac{\sqrt{x-4}}{3-2x}$

Write the differential dy in terms of x, y, and dx for each implicit relation.

8. $3x^2 - 2xy + 2y^2 = 3$

9. $x^3 + 2y^3 = 5$

10. $2x^2 + 3xy + 4y^2 = 20$

11. $2\sqrt{x} + 3\sqrt{y} = 4$

Estimating Small Changes

Find the approximate change in y when x changes as indicated.

12. $y = 5x^2$, and x changes from 8 to 8.01

13. $y = \sqrt{x}$, and x changes from 10 to 10.1

14. $y = \sqrt[3]{3x}$, and x changes from 4 to 3.99

15. $y = 2\sqrt{x^2 - 3x}$, and x changes from 15 to 15.3

16. The height h of a certain projectile at time t is given by $h = 100 + 160t - 16t^2$ (ft). Find the approximate change in height during the interval when t changes from 15 s to 15.1 s.

17. The approximate time t for one oscillation of a pendulum of length L ft is $t = 2\pi\sqrt{\frac{L}{g}}$ (s) where $g = 32.2$ ft/s approximately. By how much will t change if a 13 ft long pendulum is lengthened by 0.01 ft?

18. A cube of metal 10.00 cm on a side is heated until each side has increased by 0.01 cm. Find the approximate increase in volume.

Shells and Rings

Use differentials to derive an approximate formula for the volume of each figure.

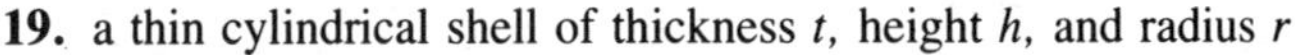

19. a thin cylindrical shell of thickness t, height h, and radius r

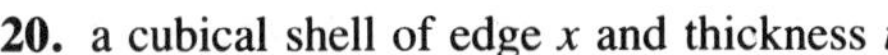

20. a cubical shell of edge x and thickness t

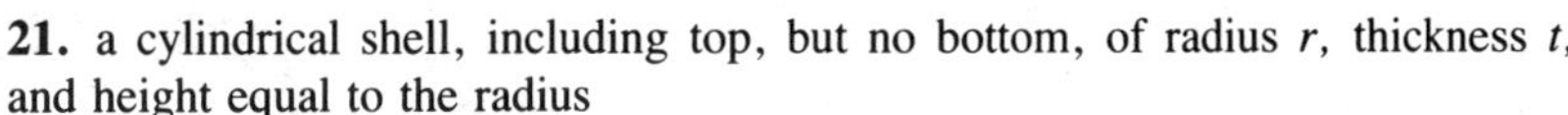

21. a cylindrical shell, including top, but no bottom, of radius r, thickness t, and height equal to the radius

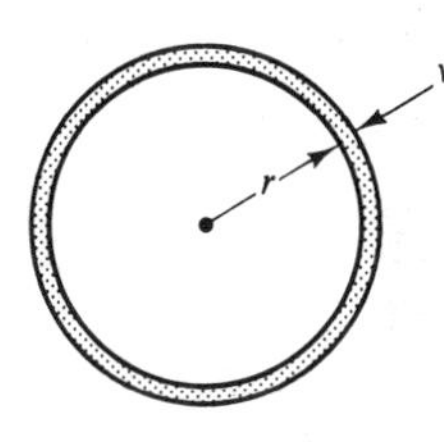

FIGURE 5-44

22. Write an approximate formula for the area of a circular ring of radius r and width w (Fig. 5-44).

Effects of Errors in Measurement

23. If x and y are related by the function $y = x^{2/3}$, what is the approximate error in y when x is measured at 27.9 instead of the correct value of 27?

24. What is the approximate error in the volume of a cube of edge 6 cm if an error of 0.2 mm is made when measuring the edge?

25. What is the approximate error in the surface area of a sphere of radius 3 ft if an error of 0.01 ft is made when measuring the radius?

26. How accurately must the edge of a cubical box be made so that the volume will not differ from the intended value of 1000 ft^3 by more than 3 ft^3?

27. To what percent accuracy must the diameter of a circle be measured so that the calculated area will be correct to 1%?

CHAPTER TEST

1. Airplane A is flying south at a speed of 120 ft/s. It passes over a bridge 12 min before another airplane, B, which is flying east at the same height at a speed of 160 ft/s. How fast are the airplanes separating 12 min after B passes over the bridge?

2. Find the approximate change in the function $y = 3x^2 - 2x + 5$, when x changes from 5 to 5.01.

3. A person walks towards the base of a 60-m-high tower at the rate of 5.0 km/h. At what rate does the person approach the top of the tower when 80 m from the base?

4. A point moves along the hyperbola $x^2 - y^2 = 144$ with a horizontal velocity $v_x = 15$ cm/s. Find the total velocity when the point is at (13, 5).

5. A conical tank with vertex down has a vertex angle of 60°. Water flows from the tank at a rate of 5.0 cm^3/min. At what rate is the inner surface of the tank being exposed when the water is 6.0 cm deep?

6. Find the instantaneous velocity and acceleration at $t = 2.0$ s for a point moving in a straight line according to the equation $s = 4t^2 - 6t$.

7. A turbine blade (Fig. 5-45) is driven by a jet of water having a speed s. The power output from the turbine is given by $P = k(sv - v^2)$, where v is the blade speed and k is a constant. Find the blade speed v for maximum power output.

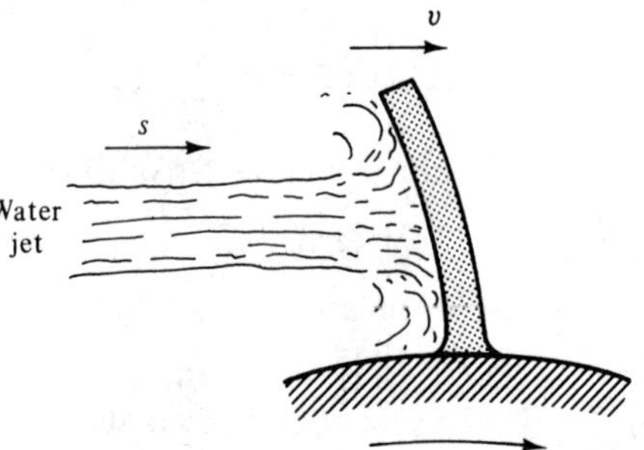

FIGURE 5-45. Turbine blade.

8. A pole (Fig. 5-46) is braced by a cable 24.0 ft long. Find the distance x from the foot of the pole to the cable anchor so that the moment produced by the tension in the cable about the foot of the pole is a maximum. Assume that the tension in the cable does not change as the anchor point is changed.

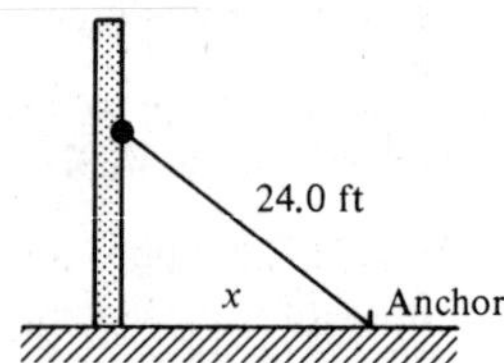

FIGURE 5-46. Pole braced by a cable.

9. Find the height of a right circular cylinder of maximum volume that can be inscribed in a sphere of radius 6.

10. Three sides of a trapezoid each have a length of 10 units. What length must the fourth side be to make the area a maximum?

11. The air in a certain balloon has a pressure of 40.0 lb/in.^{2}, a volume of 5.0 ft^3, and is expanding at the rate of 0.20 ft^3/s. If the pressure and volume are related by the equation $pv^{1.41} =$ constant, find the rate at which the pressure is changing.

12. A certain item costs \$10 to make, and the number that can be sold is estimated to be inversely proportional to the cube of the selling price. What selling price will give the greatest net profit?

13. The distance s of a point moving in a straight line is given by

$$s = -t^3 + 3t^2 + 24t + 28$$

At what times and at what distances is the point at rest?

14. A stone dropped into water produces a circular wave which increases in radius at the rate of 5.0 ft/s. How fast is the area within the ripple increasing when its diameter is 20.0 ft?

15. What is the area of the largest rectangle that can be drawn with one side on the x axis and with two corners on the curve $y = 8/(x^2 + 4)$?

16. The power P delivered to a load by a 120-V source having an internal resistance of 5 Ω is $P = 120I - 5I^2$, where I is the current to the load. At what current will the power be a maximum?

17. Separate the number 10 into two parts so that the product of the square of one part and the cube of the other part is a maximum.

18. The radius of a circular metal plate is increasing at the rate of 0.01 m/s. At what rate is the area increasing when the radius is 2.0 m?

19. Find the dimensions of the largest rectangular box with square base and open top that can be made from 300 in.^{2} of metal.

20. Use differentials to show that a 1% increase in the side of a square gives a 2% increase in its area.

21. Use differentials to show that a 1% increase in the edge of a cube gives a 3% increase in its volume.

22. The charge through a 8.24-Ω resistor varies with time according to the function $q = 2.26t^3 - 8.28$ coulomb. Write an expression for the instantaneous current through the resistor.

23. The charge at a resistor varies with time according to the function $q = 2.84t^2 + 6.25t^3$ coulomb. Write an expression for the instantaneous current through the resistor and evaluate it at 1.25 s.

24. The voltage applied to a 3.25μF capacitor is $v = 1.03t^2 + 1.33t + 2.52$ V. Find the current at $t = 15.0$ s.

25. The angular displacement of a rotating body is given by $\theta = 18.5t^2 + 12.8t + 14.8$ rad. Find (a) the angular velocity and (b) the angular acceleration, at $t = 3.50$ s.

Derivatives of Trigonometric, Logarithmic, & Exponential Functions

In this chapter we extend our ability to take derivatives to include the trigonometric, logarithmic, and exponential functions. This will enable us to solve a larger range of problems than was possible before. After learning the rules for derivatives of these functions, we apply them to problems quite similar to those in Chapters 4 and 5, that is, tangents, related rates, maximum–minimum, and the rest.

If you have not used the trigonometric, logarithmic, or exponential functions for a while, you might want to thumb quickly through that material before starting here.

6-1. DERIVATIVE OF THE SINE AND COSINE FUNCTIONS

Derivative of sin *u*

You might want to glance back at the delta method in Sec. 3-2.

We want the derivative of the function $y = \sin u$ where u is a function of x [such as $y = \sin(x^2 + 3x)$]. We will use the delta method to derive a rule for finding this derivative. We give an increment Δu to u, and y thus changes by an amount Δy,

$$y + \Delta y = \sin(u + \Delta u)$$

Subtracting the original equation,

$$\Delta y = \sin(u + \Delta u) - \sin u$$

We now make use of the identity, Eq. 177,

$$\sin\alpha - \sin\beta = 2\cos\frac{\alpha + \beta}{2}\sin\frac{\alpha - \beta}{2}$$

to transform our equation for Δy into a more useful form. We let

$$\alpha = u + \Delta u \qquad \text{and} \qquad \beta = u$$

so

$$\begin{aligned}\Delta y &= 2\cos\left(\frac{u + \Delta u + u}{2}\right)\sin\left(\frac{u + \Delta u - u}{2}\right)\\ &= 2\cos\left(u + \frac{\Delta u}{2}\right)\sin\frac{\Delta u}{2}\end{aligned}$$

Now dividing by Δu,

$$\begin{aligned}\frac{\Delta y}{\Delta u} &= 2\left[\cos\left(u + \frac{\Delta u}{2}\right)\right]\frac{\sin(\Delta u/2)}{\Delta u}\\ &= \left[\cos\left(u + \frac{\Delta u}{2}\right)\right]\frac{\sin(\Delta u/2)}{\Delta u/2}\end{aligned}$$

Remember that when we evaluated this limit in Sec. 3-1 we required the angle ($\Delta u/2$ in this case) to be *in radians*. Thus the formulas we derive here also require the angle to be in radians.

If we now let Δu approach zero, the quantity

$$\frac{\sin(\Delta u/2)}{\Delta u/2}$$

approaches 1, as we saw in Sec. 3-1. Also, the quantity $\Delta u/2$ will approach zero, leaving us with

$$\frac{dy}{du} = \cos u$$

Now by the chain rule

$$\frac{dy}{dx} = \frac{dy}{du}\cdot\frac{du}{dx}$$

So

Derivative of the Sine	$\dfrac{d(\sin u)}{dx} = \cos u \dfrac{du}{dx}$	**314**

The derivative of the sine of some function is the cosine of that function, multiplied by the derivative of that function.

EXAMPLE 1:

(a) If $y = \sin 3x$, then

$$\frac{dy}{dx} = (\cos 3x) \cdot \frac{d}{dx} 3x$$
$$= 3 \cos 3x$$

(b) If $y = \sin (x^3 + 2x^2)$, then

$$y' = \cos (x^3 + 2x^2) \cdot \frac{d}{dx} (x^3 + 2x^2)$$
$$= (3x^2 + 4x) \cos (x^3 + 2x^2)$$

Recall that $\sin^3 x$ is the same as $(\sin x)^3$.

(c) If $y = \sin^3 x$, then by the power rule

$$y' = 3(\sin x)^2 \cdot \frac{d}{dx} (\sin x)$$
$$= 3 \sin^2 x \cos x$$

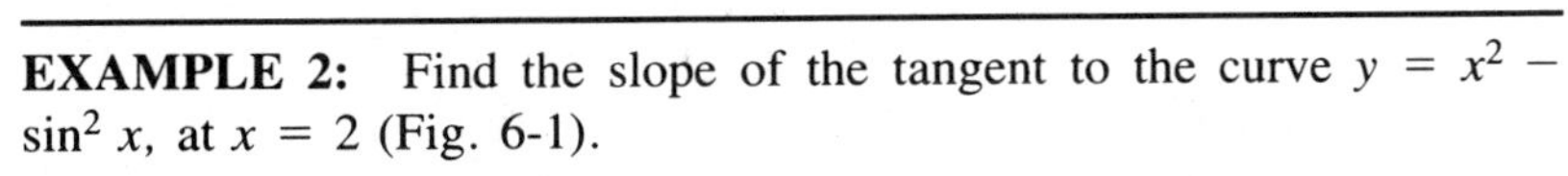

EXAMPLE 2: Find the slope of the tangent to the curve $y = x^2 - \sin^2 x$, at $x = 2$ (Fig. 6-1).

Solution: Taking the derivative,

$$y' = 2x - 2 \sin x \cos x$$

At $x = 2$ rad,

$$y'(2) = 4 - 2 \sin 2 \cos 2$$
$$= 4 - 2(0.909)(-0.416) = 4.757$$

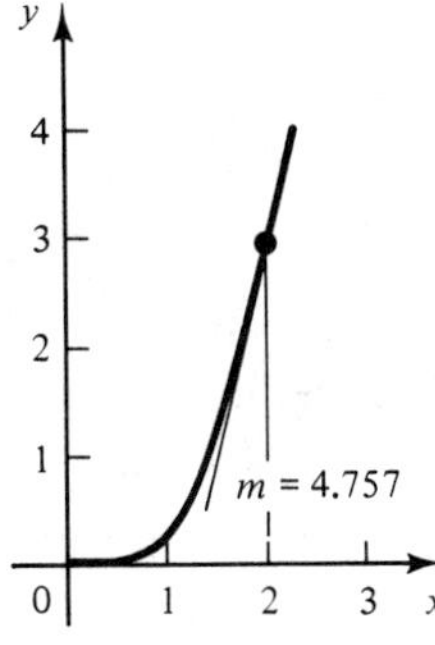

FIGURE 6-1. Graph of $y = x^2 - \sin^2 x$.

Common Error	Remember that x is in *radians*, unless otherwise specified. Be sure that your calculator is in radian mode.

Derivative of cos *u*

We now take the derivative of $y = \cos u$.

We do not need the delta method again, because we can relate the cosine to the sine with Eq. 154b, $\cos A = \sin B$, where A and B are complementary angles ($B = \pi/2 - A$).

So

$$y = \cos u = \sin\left(\frac{\pi}{2} - u\right)$$

Then by Eq. 314,

$$\frac{dy}{dx} = \cos\left(\frac{\pi}{2} - u\right)\left(-\frac{du}{dx}\right)$$

But $\cos(\pi/2 - u) = \sin u$, so

Derivative of the Cosine	$\dfrac{d(\cos u)}{dx} = -\sin u \dfrac{du}{dx}$	**315**

The derivative of the cosine of some function is the negative of the sine of that function, multiplied by the derivative of that function.

EXAMPLE 3: If $y = \cos 3x^2$, then

$$y' = (-\sin 3x^2)\frac{d}{dx}(3x^2)$$

$$= (-\sin 3x^2)6x = -6x \sin 3x^2$$

Common Error	$\cos 3x^2$ *is not the same as* $(\cos 3x)^2$.

EXAMPLE 4: Differentiate $y = \sin 3x \cos 5x$.

Solution: Using the product rule, we have

$$\frac{dy}{dx} = (\sin 3x)(-\sin 5x)(5) + (\cos 5x)(\cos 3x)(3)$$

$$= -5 \sin 3x \sin 5x + 3 \cos 5x \cos 3x$$

EXAMPLE 5: Differentiate $y = \dfrac{2\cos x}{\sin 3x}$.

Solution: Using the quotient rule gives us

$$y' = \frac{(\sin 3x)(-2\sin x) - (2\cos x)(3\cos 3x)}{(\sin 3x)^2}$$

$$= \frac{-2\sin 3x \sin x - 6\cos x \cos 3x}{\sin^2 3x}$$

EXAMPLE 6: Find the maximum and minimum points on the curve $y = 3\cos x$.

Solution: We take the derivative and set it equal to zero.

$$y' = -3\sin x = 0$$

$$\sin x = 0$$

$$x = 0, \pm\pi, \pm 2\pi, \pm 3\pi, \ldots$$

The second derivative is

$$y'' = -3\cos x$$

The second derivative is negative when x equals $0, \pm 2\pi, \pm 4\pi, \ldots$, so these are the locations of the maximum points. The others, where the second derivative is positive, are minimum points. This, of course, agrees with our plot of the cosine curve in Fig. 6-2.

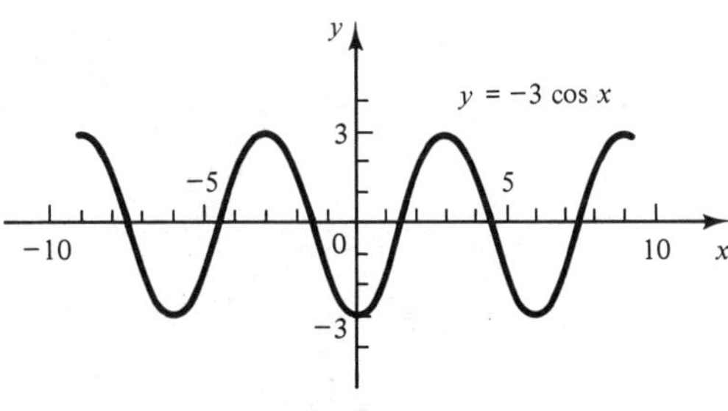

FIGURE 6-2

EXAMPLE 7: Find dy/dx, if $y \sin x = x^2$

Solution: We proceed as with other implicit derivatives. Using the product rule,

$$y\cos x + (\sin x)y' = 2x$$

Rearranging gives

$$y' \sin x = 2x - y \cos x$$

$$y' = \frac{2x - y \cos x}{\sin x}$$

EXERCISE 1

First Derivatives

Find the derivative.

1. $y = \sin x$

2. $y = 3 \cos 2x$

3. $y = \cos^3 x$

4. $y = \sin x^2$

5. $y = \sin 3x$

6. $y = \cos 6x$

7. $y = \sin x \cos x$

8. $y = 15.4 \cos^5 x$

9. $y = 3.75\, x \cos x$

10. $y = \sin (\theta + \pi) \cos (\theta - \pi)$

11. $y = \sin^2 (\pi - x)$

12. $y = \dfrac{\sin \theta}{\theta}$

13. $y = \sin 2x \cos x$

14. $y = \sin^5 x$

15. $y = \sin^2 x \cos x$

16. $y = \sin 2x \cos 3x$

17. $y = 1.23 \sin^2 x \cos 3x$

18. $y = \frac{1}{2} \sin^2 x$

19. $y = \sqrt{\cos 2t}$

20. $y = 42.7 \sin^2 x$

Second Derivatives

Find the second derivative of each function.

21. $y = \sin x$

22. $y = \frac{1}{4} \cos 2\theta$

23. $y = x \cos x$

24. If $f(x) = x^2 \cos^3 x$, find $f''(0)$.

25. If $f(x) = x \sin (\pi/2)x$, find $f''(1)$.

Implicit Functions

Find dy/dx for each implicit function.

26. $y \sin x = 1$

27. $xy - y \sin x - x \cos y = 0$

28. $y = \cos (x - y)$

29. $x = \sin (x + y)$

30. $x \sin y - y \sin x = 0$

Tangents

Find the slope of the tangent at the given value of x.

31. $y = \sin x$ at $x = 2$ rad

32. $y = x - \cos x$ at $x = 1$ rad

33. $y = x \sin \frac{x}{2}$ at $x = 2$ rad

34. $y = \sin x \cos 2x$ at $x = 1$ rad

Extreme Values and Inflection Points

Find the maximum, minimum, and inflection points for each curve between $x = 0$ and $x = 2\pi$.

35. $y = \sin x$

36. $y = \frac{x}{2} - \sin x$

37. $y = 3 \sin x - 4 \cos x$

Newton's Method

Sketch each function from $x = 0$ to 10. Calculate a root to three decimal places by Newton's method. If there is more than one root, find only the smallest positive root.

38. $\cos x - x = 0$

39. $\cos 2x - x = 0$

40. $3 \sin x - x = 0$

41. $2 \sin x - x^2 = 0$

42. $\cos x - 2x^2 = 0$

6-2. DERIVATIVE OF THE TANGENT, COTANGENT, SECANT, AND COSECANT

Derivative of tan *u*

We seek the derivative of $y = \tan u$. By Eq. 162,

$$y = \frac{\sin u}{\cos u}$$

Using the rule for the derivative of a quotient, Eq. 313,

$$\frac{dy}{dx} = \frac{\cos^2 u \dfrac{du}{dx} + \sin^2 u \dfrac{du}{dx}}{\cos^2 u}$$

$$= \frac{(\sin^2 u + \cos^2 u)\dfrac{du}{dx}}{\cos^2 u}$$

and by Eq. 164,

$$= \frac{1}{\cos^2 u} \frac{du}{dx}$$

and by Eq. 152b,

$$= \sec^2 u \frac{du}{dx}$$

So

Derivative of the Tangent	$\frac{d(\tan u)}{dx} = \sec^2 u \frac{du}{dx}$	**316**

The derivative of the tangent of some function is the secant squared of that function, multiplied by the derivative of that function.

EXAMPLE 1

(a) If $y = 3 \tan x^2$, then

$$y' = 3(\sec^2 x^2)(2x) = 6x \sec^2 x^2$$

(b) If $y = 2 \sin 3x \tan 3x$, then, by the product rule,

$$\begin{aligned} y' &= 2[\sin 3x(\sec^2 3x)(3) + \tan 3x(\cos 3x)(3)] \\ &= 6 \sin 3x \sec^2 3x + 6 \cos 3x \tan 3x \end{aligned}$$

Derivative of cot *u*, sec *u*, and csc *u*

Each of these derivatives can be obtained using the rules for the sine, cosine, and tangent already derived, and the identities

$$\cot u = \frac{\cos u}{\sin u}, \qquad \sec u = \frac{1}{\cos u}, \qquad \csc u = \frac{1}{\sin u}$$

You should memorize at least the first three of these. Notice how the signs alternate. The derivative of each cofunction is negative.

We list them here, together with those already found.

Derivatives of the Trigonometric Functions	$\frac{d(\sin u)}{dx} = \cos u \frac{du}{dx}$	**314**
	$\frac{d(\cos u)}{dx} = -\sin u \frac{du}{dx}$	**315**
	$\frac{d(\tan u)}{dx} = \sec^2 u \frac{du}{dx}$	**316**
	$\frac{d(\cot u)}{dx} = -\csc^2 u \frac{du}{dx}$	**317**
	$\frac{d(\sec u)}{dx} = \sec u \tan u \frac{du}{dx}$	**318**
	$\frac{d(\csc u)}{dx} = -\csc u \cot u \frac{du}{dx}$	**319**

EXAMPLE 2

(a) If $y = \sec(x^3 - 2x)$, then

$$y' = [\sec(x^3 - 2x)\tan(x^3 - 2x)](3x^2 - 2)$$
$$= (3x^2 - 2)\sec(x^3 - 2x)\tan(x^3 - 2x)$$

(b) If $y = \cot^3 5x$, then by the power rule,

$$y' = 3(\cot 5x)^2(-\csc^2 5x)(5)$$
$$= -15\cot^2 5x\csc^2 5x$$

EXERCISE 2

First Derivative

Find the derivative.

1. $y = \tan 2x$

2. $y = \sec 4x$

3. $y = 5\csc 3x$

4. $y = 9\cot 8x$

5. $y = 3.25\tan x^2$

6. $y = 5.14\sec 2.11x^2$

7. $y = 7\csc x^3$

8. $y = 9\cot 3x^3$

9. $y = x\tan x$

10. $y = x\sec x^2$

11. $y = 5x\csc 6x$

12. $y = 9x^2\cot 2x$

13. $w = \sin\theta\tan 2\theta$

14. $s = \cos t\sec 4t$

15. $v = 5\tan t\csc 3t$

16. $z = 2\sin 2\theta\cot 8\theta$

17. If $y = 5.83\tan^2 2x$, find $y'(1)$.

18. If $f(x) = \sec^3 x$, find $f'(3)$.

19. If $f(x) = 3\csc^3 3x$, find $f'(3)$.

20. If $y = 9.55x\cot^2 8x$, find $y'(1)$.

Second Derivative

Find the second derivative.

21. $y = 3\tan x$

22. $y = 2\sec 5\theta$

23. If $y = 3\csc 2\theta$, find $y''(1)$.

24. If $f(x) = 6\cot 4x$, find $f''(3)$.

Implicit Functions

Find dy/dx for each implicit function.

25. $y\tan x = 2$

26. $xy + y\cot x = 0$

27. $\sec(x + y) = 7$

28. $x\cot y = y\sec x$

Tangents

29. Find the tangent to the curve $y = \tan x$, at $x = 1$ rad.

30. Find the tangent to the curve $y = \sec 2x$, at $x = 2$ rad.

Extreme Values and Inflection Points

Find any maximum, minimum, or inflection points between 0 and π for each function.

31. $y = 2x - \tan x$ **32.** $y = \tan x - 4x$

Rate of Change

33. An object moves with simple harmonic motion so that its displacement y at time t is $y = 6 \sin 4t$ cm. Find the velocity and acceleration of the object when $t = 0.05$ s.

Related Rates

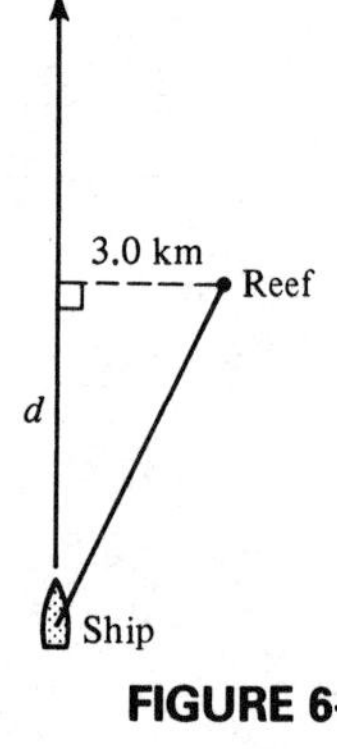

FIGURE 6-3

34. A ship is sailing at 10.0 km/h in a straight line (Fig. 6-3). It keeps its searchlight trained on a reef which is 3.0 km off the path of the ship, measured perpendicularly. How fast (rad/h) is the light turning when the distance d is 5.0 km?

35. Two cables pass over fixed pulleys A and B (Fig. 6-4) forming an isosceles triangle ABP. Point P is being raised at the rate of 3.0 in./min. How fast is θ changing when h is 4.0 ft?

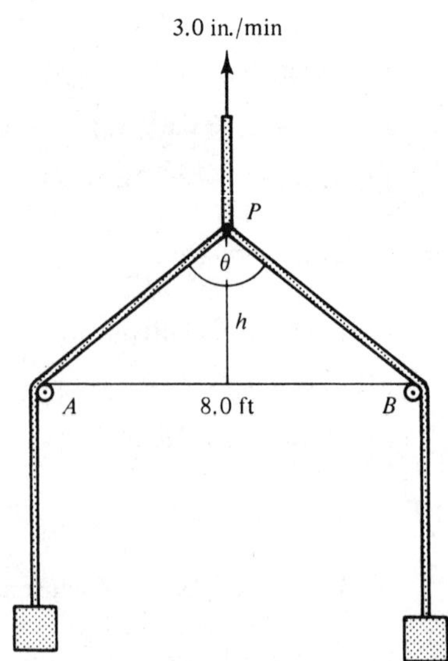

FIGURE 6-4

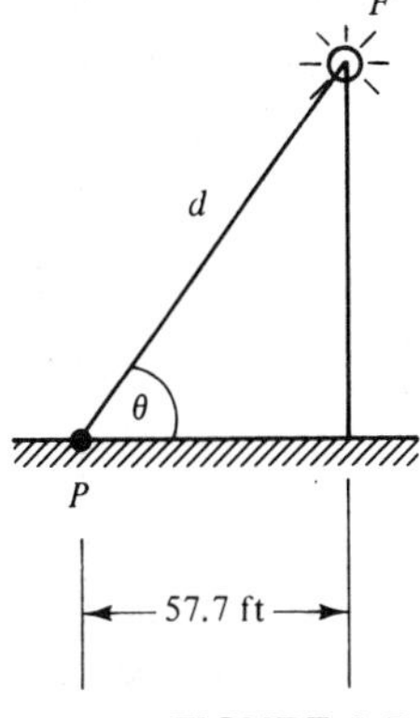

FIGURE 6-5. Falling flare.

36. The illumination at a point P on the ground (Fig. 6-5) due to a flare F is $I = k \sin \theta/d^2$ lux, where k is a constant. Find the rate of change of I when the flare is 100 ft above the ground and falling at a rate of 1.0 ft/s, if the illumination at P at that instant is 65 lux.

Maximum–Minimum Problems

37. Find the length of the shortest ladder (Fig. 6-6) that will touch the ground, the wall, and the house.

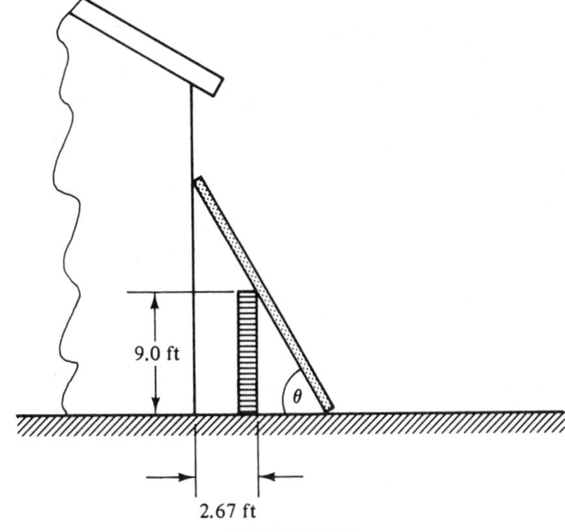

FIGURE 6-6

38. The range x of a projectile fired at an angle θ with the horizontal at a velocity v is $x = (v^2/g) \sin 2\theta$ where g is the acceleration due to gravity. Find θ for maximum range.

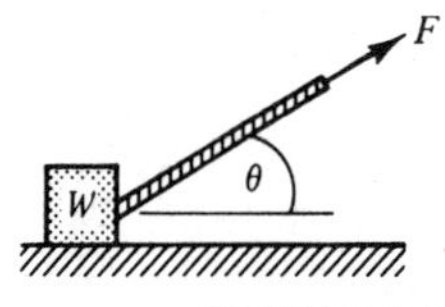

FIGURE 6-7

39. A force F (Fig. 6-7) pulls the weight along a horizontal surface. If f is the coefficient of friction, then $F = \dfrac{fW}{f \sin \theta + \cos \theta}$. Find θ for a minimum force if $f = 0.60$.

40. A 20-ft-long steel girder is dragged along a corridor 10.0 ft wide, and then around a corner into another corridor at right angles to the first (Fig. 6-8). Neglecting the thickness of the girder, what must be the width of the second corridor to allow the girder to turn the corner?

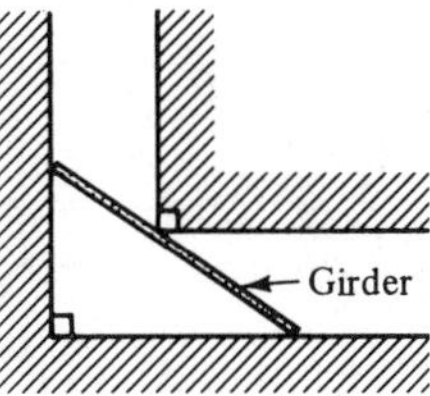

FIGURE 6-8. Top view of a girder dragged along a corridor.

41. If a girder (Fig. 6-8) is to be dragged from a 12.8-ft-wide corridor into another 5.4-ft-wide corridor, find the length of the longest girder that will fit around the corner. (Neglect the thickness of the girder.)

6-3. DERIVATIVES OF THE INVERSE TRIGONOMETRIC FUNCTIONS

Derivative of Arcsin u

We now seek the derivative of $y = \text{Sin}^{-1} u$, where y is some angle whose sine is u, as in Fig. 6-9, whose value we restrict to the range $-\pi/2$ to $\pi/2$. We can then write

$$\sin y = u$$

Taking the derivative,

$$\cos y \frac{dy}{dx} = \frac{du}{dx}$$

so

$$\frac{dy}{dx} = \frac{1}{\cos y}\frac{du}{dx} \tag{1}$$

But from Eq. 164,

$$\cos^2 y = 1 - \sin^2 y$$

$$\cos y = \pm\sqrt{1 - \sin^2 y}$$

But since y is restricted to values between $-\pi/2$ and $\pi/2$, $\cos y$ cannot be negative. So

$$\cos y = +\sqrt{1 - \sin^2 y} = \sqrt{1 - u^2}$$

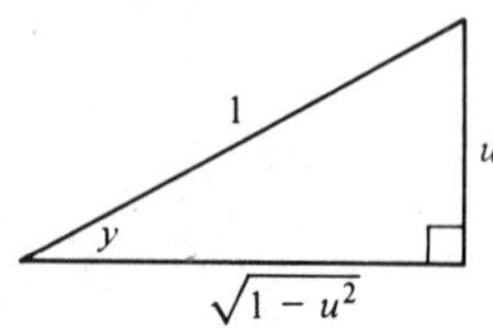

FIGURE 6-9

Substituting into (1) we have

$$\frac{d(\text{Sin}^{-1} u)}{dx} = \frac{1}{\sqrt{1-u^2}}\frac{du}{dx}$$ $-1 < u < 1$	**320**

EXAMPLE 1: If $y = \text{Sin}^{-1} 3x$, then

$$y' = \frac{1}{\sqrt{1-(3x)^2}}(3)$$

$$= \frac{3}{\sqrt{1-9x^2}}$$

Derivatives of Arccos, Arctan, Arccot, Arcsec, and Arccsc

Try to derive one or more of these. Follow steps similar to those we used for the Arcsin.

The rules for taking derivatives of the remaining inverse trigonometric functions, and the Arcsin as well, are

Derivatives of the Inverse Trigonometric Functions	$$\frac{d(\text{Sin}^{-1} u)}{dx} = \frac{1}{\sqrt{1-u^2}}\frac{du}{dx}$$ $-1 < u < 1$	**320**
	$$\frac{d(\text{Cos}^{-1} u)}{dx} = \frac{-1}{\sqrt{1-u^2}}\frac{du}{dx}$$ $-1 < u < 1$	**321**
	$$\frac{d(\text{Tan}^{-1} u)}{dx} = \frac{1}{1+u^2}\frac{du}{dx}$$	**322**
	$$\frac{d(\text{Cot}^{-1} u)}{dx} = \frac{-1}{1+u^2}\frac{du}{dx}$$	**323**
	$$\frac{d(\text{Sec}^{-1} u)}{dx} = \frac{1}{u\sqrt{u^2-1}}\frac{du}{dx}$$ $\lvert u \rvert > 1$	**324**
	$$\frac{d(\text{Csc}^{-1} u)}{dx} = \frac{-1}{u\sqrt{u^2-1}}\frac{du}{dx}$$ $\lvert u \rvert > 1$	**325**

EXAMPLE 2

(a) If $y = \mathrm{Cot}^{-1}(x^2 + 1)$, then

$$y' = \frac{-1}{1 + (x^2 + 1)^2}(2x)$$
$$= \frac{-2x}{2 + 2x^2 + x^4}$$

(b) If $y = \mathrm{Cos}^{-1}\sqrt{1 - x}$, then

$$y' = \frac{-1}{\sqrt{1 - (1 - x)}} \cdot \frac{-1}{2\sqrt{1 - x}}$$
$$= \frac{1}{2\sqrt{x - x^2}}$$

EXERCISE 3

Find the derivative.

1. $y = x \,\mathrm{Sin}^{-1} x$

2. $y = \mathrm{Sin}^{-1}\dfrac{x}{a}$

3. $y = \mathrm{Cos}^{-1}\dfrac{x}{a}$

4. $y = \mathrm{Tan}^{-1}(\sec x + \tan x)$

5. $y = \mathrm{Sin}^{-1}\dfrac{\sin x - \cos x}{\sqrt{2}}$

6. $y = \sqrt{2ax - x^2} + a\,\mathrm{Cos}^{-1}\dfrac{\sqrt{2ax - x^2}}{a}$

7. $y = t^2\,\mathrm{Cos}^{-1} t$

8. $y = \mathrm{Arcsin}\ 2x$

9. $y = \mathrm{Arctan}\ (1 + 2x)$

10. $y = \mathrm{Arccot}\ (2x + 5)^2$

11. $y = \mathrm{Arccot}\dfrac{x}{a}$

12. $y = \mathrm{Arcsec}\dfrac{1}{x}$

13. $y = \mathrm{Arccsc}\ 2x$

14. $y = \mathrm{Arcsin}\sqrt{x}$

15. $y = t^2\,\mathrm{Arcsin}\dfrac{t}{2}$

16. $y = \mathrm{Sin}^{-1}\dfrac{x}{\sqrt{1 + x^2}}$

17. $y = \mathrm{Sec}^{-1}\dfrac{a}{\sqrt{a^2 - x^2}}$

Find the slope of the tangent to each curve.

18. $y = x\,\mathrm{Arcsin}\,x \quad$ at $x = \frac{1}{2}$

19. $y = \dfrac{\mathrm{Arctan}\,x}{x} \quad$ at $x = 1$

20. $y = x^2 \text{ Arccsc } \sqrt{x}$ at $x = 2$

21. $y = \sqrt{x} \text{ Arccot } \dfrac{x}{4}$ at $x = 4$

22. Find the equations of the tangents to the curve $y = \text{Arctan } x$ having a slope of $\frac{1}{4}$.

23. Use differentials to find the approximate amount by which Arcsin x will change (in degrees) when x changes from 0.50 to 0.52.

6-4. DERIVATIVES OF LOGARITHMIC FUNCTIONS

Derivative of $\log_b u$

Let us now use the delta method again to find the derivative of the logarithmic function $y = \log_b u$. We first let u take an increment Δu and y an increment Δy,

$$y + \Delta y = \log_b (u + \Delta u)$$

Subtracting,

$$\Delta y = \log_b (u + \Delta u) - \log_b u = \log_b \left(\frac{u + \Delta u}{u}\right)$$

by the law of logarithms for quotients. Now dividing by Δu yields

$$\frac{\Delta y}{\Delta u} = \frac{1}{\Delta u} \log_b \left(\frac{u + \Delta u}{u}\right)$$

We now do some manipulation that will get our expression into a form that will be easier to evaluate. We start by multiplying the right side by u/u.

$$\frac{\Delta y}{\Delta u} = \frac{u}{u} \cdot \frac{1}{\Delta u} \log_b \left(\frac{u + \Delta u}{u}\right)$$

$$= \frac{1}{u} \cdot \frac{u}{\Delta u} \log_b \left(\frac{u + \Delta u}{u}\right)$$

Then using the law of logarithms for powers (Eq. 188),

$$\frac{\Delta y}{\Delta u} = \frac{1}{u} \log_b \left(\frac{u + \Delta u}{u}\right)^{u/\Delta u}$$

We now let Δu approach zero,

$$\lim_{\Delta u \to 0} \frac{\Delta y}{\Delta u} = \frac{dy}{du} = \lim_{\Delta u \to 0} \frac{1}{u} \log_b \left(\frac{u + \Delta u}{u}\right)^{u/\Delta u}$$

$$= \frac{1}{u} \log_b \left[\lim_{\Delta u \to 0} \left(1 + \frac{\Delta u}{u}\right)^{u/\Delta u}\right]$$

Let us simplify the expression inside the brackets by making the substitution

$$k = \frac{u}{\Delta u}$$

Then k will approach infinity as Δu approaches zero, and the expression inside the brackets becomes

$$\lim_{\Delta u \to 0} \left(1 + \frac{\Delta u}{u}\right)^{u/\Delta u} = \lim_{k \to \infty} \left(1 + \frac{1}{k}\right)^k$$

The number e is named after a Swiss mathematician, Leonhard Euler (1707–1783). Its value has been computed to thousands of decimal places. In Chapter 16 we show how to compute e using series.

This limit defines the number e, the familiar base of natural logarithms.

$$e \equiv \lim_{k \to \infty} \left(1 + \frac{1}{k}\right)^k$$

We can get an approximate numerical value for this limit by calculator. We let k get large and compute $(1 + 1/k)^k$.

k	$\left(1 + \frac{1}{k}\right)^k$
1	2
10	2.5937 . . .
100	2.7048 . . .
1,000	2.7169 . . .
10,000	2.7181 . . .
100,000	2.7183 . . .
1,000,000	2.7183 . . .

Thus $e \cong 2.7183$. Our derivative thus becomes

$$\frac{dy}{du} = \frac{1}{u} \log_b e$$

We are nearly finished now. Using the chain rule we get dy/dx by multiplying dy/du by du/dx, so

Derivative of $\log_b u$	$\dfrac{d(\log_b u)}{dx} = \dfrac{1}{u} \log_b e \dfrac{du}{dx}$	**326a**

or, since $\log_b e = 1/\ln b$,

This form is more useful when the base b is a number other than 10.

Derivative of $\log_b u$	$\dfrac{d(\log_b u)}{dx} = \dfrac{1}{u \ln b}\dfrac{du}{dx}$	**326b**

EXAMPLE 1: Take the derivative of $y = \log (x^2 - 3x)$.

Solution: The base b, when not indicated, is taken as 10. By Eq. 326a,

$$\frac{dy}{dx} = \frac{1}{x^2 - 3x} \cdot \log e \cdot (2x - 3)$$

$$= \left(\frac{2x - 3}{x^2 - 3x}\right) \log e$$

Or, since $\log e \cong 0.4343$,

$$\frac{dy}{dx} = 0.4343 \left(\frac{2x - 3}{x^2 - 3x}\right)$$

EXAMPLE 2: If $y = x \log_3 x^2$, then by the product rule,

$$y' = x\left(\frac{1}{x^2 \ln 3}\right)(2x) + \log_3 x^2$$

$$= \frac{2}{\ln 3} + \log_3 x^2 \cong 1.82 + \log_3 x^2$$

Derivative of ln *u*

Our efforts in deriving Eq. 326 will now pay off, because we can use that result to find the derivative of the natural logarithm of a function, as well as derivatives of exponential functions in the following sections. To find the derivative of $y = \ln u$ we use Eq. 326a:

$$\frac{dy}{dx} = \frac{1}{u} \ln e \frac{du}{dx}$$

But by Eq. 191, $\ln e = 1$, so

Derivative of $\ln u$	$\dfrac{d(\ln u)}{dx} = \dfrac{1}{u}\dfrac{du}{dx}$	**327**

The derivative of the natural logarithm of a function is the reciprocal of that function multiplied by its derivative.

EXAMPLE 3: Differentiate $y = \ln(2x^3 + 5x)$.

Solution: By Eq. 327,

$$\frac{dy}{dx} = \frac{1}{2x^3 + 5x}(6x^2 + 5)$$

$$= \frac{6x^2 + 5}{2x^3 + 5x}$$

The rule for derivatives of the logarithmic function is often used with our former rules for derivatives.

EXAMPLE 4: Take the derivative of $y = x^3 \ln(5x + 2)$.

Solution: Using the product rule together with our rule for logarithms gives us

$$\frac{dy}{dx} = x^3 \cdot \frac{1}{5x + 2} \cdot 5 + \ln(5x + 2) \cdot 3x^2$$

$$= \frac{5x^3}{5x + 2} + 3x^2 \ln(5x + 2)$$

Our work is sometimes made easier if we first use the laws of logarithms to simplify a given expression.

EXAMPLE 5: Take the derivative of $y = \ln \frac{x\sqrt{2x - 3}}{\sqrt[3]{4x + 1}}$.

Solution: By the laws of logarithms,

$$y = \ln x + \frac{1}{2}\ln(2x - 3) - \frac{1}{3}\ln(4x + 1)$$

We now take the derivative term by term,

$$\frac{dy}{dx} = \frac{1}{x} + \frac{1}{2} \cdot \frac{1}{2x - 3} \cdot 2 - \frac{1}{3} \cdot \frac{1}{4x + 1} \cdot 4$$

$$= \frac{1}{x} + \frac{1}{2x - 3} - \frac{4}{3(4x + 1)}$$

Using Logarithms to Aid in Differentiating Nonlogarithmic Expressions

This procedure is called *logarithmic differentiation.*

Derivatives of some complicated expressions can be found more easily if we first take the logarithm of both sides of the given expression, simplify by means of the laws of logarithms, and then take the derivative.

EXAMPLE 6: Differentiate $y = \dfrac{\sqrt{x-2}\,\sqrt[3]{x+3}}{\sqrt[4]{x+1}}$.

We could instead take the common log of both sides, but the natural log has a simpler derivative.

Solution: Instead of proceeding in the usual way, we first take the natural log of both sides, and apply the laws of logarithms:

$$\ln y = \frac{1}{2}\ln(x-2) + \frac{1}{3}\ln(x+3) - \frac{1}{4}\ln(x+1)$$

Taking the derivative, we have

$$\frac{1}{y}\frac{dy}{dx} = \frac{1}{2}\frac{1}{x-2} + \frac{1}{3}\frac{1}{x+3} - \frac{1}{4}\frac{1}{x+1}$$

Finally, multiplying by y gives us

$$\frac{dy}{dx} = y\left[\frac{1}{2(x-2)} + \frac{1}{3(x+3)} - \frac{1}{4(x+1)}\right]$$

In other cases, this method will allow us to take derivatives not possible with our other rules.

EXAMPLE 7: Find the derivative of $y = x^{2x}$.

Solution: This is not a power function, because the exponent is not a constant, nor is it an exponential function, because the base is not a constant. So neither rule 309 nor 328 applies. But let us take the log of both sides:

$$\ln y = \ln x^{2x} = 2x \ln x$$

Now taking the derivative by means of the product rule yields

$$\frac{1}{y}\frac{dy}{dx} = 2x\left(\frac{1}{x}\right) + (\ln x)2$$
$$= 2 + 2\ln x$$

Multiplying by y, we get

$$\frac{dy}{dx} = 2(1 + \ln x)y$$
$$= 2(1 + \ln x)x^{2x}$$

since $y = x^{2x}$.

EXERCISE 4

Derivative of $\log_b u$

Differentiate.

1. $y = \log 7x$
2. $y = \log x^{-2}$
3. $y = \log_b x^3$
4. $y = \log_a(x^2 - 3x)$
5. $y = \log (x\sqrt{5 + 6x})$
6. $y = \log_a\left(\dfrac{1}{2x + 5}\right)$
7. $y = x \log \dfrac{2}{x}$
8. $y = \log \dfrac{(1 + 3x)}{x^2}$

Derivative of $\ln u$

Differentiate.

9. $y = \ln 3x$
10. $y = \ln x^3$
11. $y = \ln (x^2 - 3x)$
12. $y = \ln (4x - x^3)$
13. $y = 2.75x \ln 1.02x^3$
14. $w = z^2 \ln (1 - z^2)$
15. $y = \dfrac{\ln (x + 5)}{x^2}$
16. $y = \dfrac{\ln x^2}{3 \ln (x - 4)}$
17. $s = \ln \sqrt{t - 5}$
18. $y = 5.06 \ln \sqrt{x^2 - 3.25x}$

With Trigonometric Functions

Differentiate.

19. $y = \ln \sin x$
20. $y = \ln \sec x$
21. $y = \sin x \ln \sin x$
22. $y = \ln (\sec x + \tan x)$

Implicit Relations

Find dy/dx.

23. $y \ln y + \cos x = 0$
24. $\ln x^2 - 2x \sin y = 0$
25. $x - y = \ln (x + y)$
26. $xy = a^2 \ln \dfrac{x}{a}$

27. $\ln y + x = 10$

Logarithmic Differentiation

Remember to start these by taking the logarithm of both sides.

Differentiate.

28. $y = \dfrac{\sqrt{x+2}}{\sqrt[3]{2-x}}$

29. $y = \dfrac{\sqrt{a^2 - x^2}}{x}$

30. $y = x^x$

31. $y = x^{\sin x}$

32. $y = (\cot x)^{\sin x}$

33. $y = (\cos^{-1} x)^x$

Tangent to a Curve

Find the slope of the tangent at the point indicated.

34. $y = \log x$ at $x = 1$

35. $y = \ln x$ where $y = 0$

36. $y = \ln (x^2 + 2)$ at $x = 4$

37. $y = \log (4x - 3)$ at $x = 2$

38. Find the equation of the tangent to the curve $y = \ln x$ at $y = 0$.

Angle of Intersection

Find the angles of intersection of each pair of curves.

39. $y = \ln (x + 1)$ and $y = \ln (7 - 2x)$ at $x = 2$

40. $y = x \ln x$ and $y = x \ln (1 - x)$ at $x = \frac{1}{2}$

Extreme Values and Points of Inflection

Find the maximum, minimum, and inflection points for each curve.

41. $y = x \ln x$

42. $y = x^3 \ln x$

43. $y = \dfrac{x}{\ln x}$

44. $y = \ln (8x - x^2)$

Newton's Method

Find the smallest positive root between $x = 0$ and $x = 10$, to two decimal places.

45. $x - 10 \log x = 0$

46. $\tan x - \log x = 0$

Applications

47. A certain underwater cable has a core of copper wires covered by insulation. The speed of transmission of a signal along the cable is

$$S = x^2 \ln \frac{1}{x}$$

where x is the ratio of the radius of the core to the thickness of the insulation. What value of x gives the greatest signal speed?

48. The heat loss q per foot of cylindrical pipe insulation (Fig. 6-10) having an inside radius r_1 and outside radius r_2 is given by the logarithmic equation

$$q = \frac{2\pi k(t_1 - t_2)}{\ln (r_2/r_1)} \qquad \text{Btu/h}$$

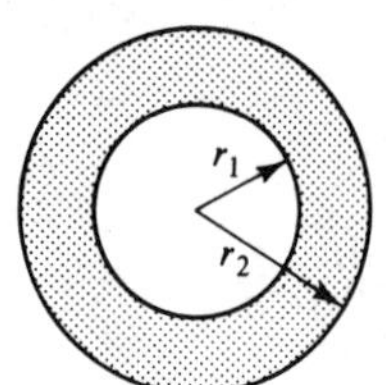

FIGURE 6-10. Insulated pipe.

where t_1 and t_2 are the inside and outside temperatures (°F) and k is the conductivity of the insulation. A 4-in.-thick insulation having a conductivity of 0.036 is wrapped around a 9-in.-diameter pipe at 550°F, and the surroundings are at 90°F. Find the rate of change of heat loss q if the insulation thickness is decreasing at the rate of 0.1 in./h.

49. The pH value of a solution having a concentration C of hydrogen ions is $\text{pH} = -\log_{10} C$. Use differentials to estimate the pH change if the concentration of hydrogen ions is increased from 2.0×10^{-5} to 2.1×10^{-5}.

Hint: Treat B_2 as a ***constant.***

50. The difference in elevation, in feet, between two locations having barometric readings of B_1 and B_2 inches of mercury is given by $h = 60{,}470 \log \dfrac{B_2}{B_1}$, where B_1 is the pressure at the upper location. At what rate is an airplane changing in elevation if the barometric pressure outside the airplane is 21.5 in. of mercury and decreasing at the rate of 0.50 in. per minute?

51. The power input to a certain amplifier is 2.0 W and the power output is 400 W, but due to a defective component is dropping at the rate of 0.5 W per day. Use Eq. A101 to find the rate (decibels per day) at which the decibels are decreasing.

6-5. DERIVATIVE OF THE EXPONENTIAL FUNCTION

Derivative of b^u

We seek the derivative of the exponential function $y = b^u$, where b is a constant and u is a function of x. We can get the derivative without having to use the delta method by using the rule we derived for the logarithmic function (Eq. 327). We first take the natural logarithm of both sides,

$$\ln y = \ln b^u = u \ln b$$

by Eq. 188. We now take the derivative of both sides, remembering that $\ln b$ is a constant,

$$\frac{1}{y}\frac{dy}{dx} = \frac{du}{dx}\ln b$$

Multiplying by y, we have

$$\frac{dy}{dx} = y\frac{du}{dx}\ln b$$

Finally, replacing y by b^u gives us

Derivative of b^u	$\frac{d(b^u)}{dx} = b^u\frac{du}{dx}\ln b$	**328**

The derivative of a base b *raised to a function* u *is the product of* b^u, *the derivative of the function, and the natural log of the base.*

EXAMPLE 1: Find the derivative of $y = 10^{x^2+2}$.

Solution: By Eq. 328,

$$\frac{dy}{dx} = 10^{x^2+2}(2x)\ln 10 = 2x(\ln 10)10^{x^2+2}$$

Derivative of e^u

We will use Eq. 328 mostly when the base b is the base of natural logarithms, e.

$$y = e^u$$

Taking the derivative by Eq. 328 gives us

$$\frac{dy}{dx} = e^u\frac{du}{dx}\ln e$$

But since $\ln e = 1$, we get

Derivative of e^u	$\frac{d}{dx}e^u = e^u\frac{du}{dx}$	**329**

The derivative of an exponential function e^u *is the same exponential function, multiplied by the derivative of the exponent.*

Note that $y = c_1e^{mx}$ and its derivatives $y' = c_2e^{mx}$, $y'' = c_3e^{mx}$, etc., are all *like terms*. We use this fact when solving differential equations in Chapter 14.

EXAMPLE 2: Find the first, second, and third derivatives of $y = 2e^{3x}$.

Solution: By rule 329,

$$y' = 6e^{3x}, \qquad y'' = 18e^{3x}, \qquad \text{and } y''' = 54e^{3x}$$

EXAMPLE 3: The derivative of $y = e^{x3+5x2}$ is

$$\frac{dy}{dx} = e^{x3+5x2}(3x^2 + 10x)$$

EXAMPLE 4: Find the derivative of $y = x^3e^{x^2}$.

Solution: Using the product rule together with Eq. 329, we have

$$\frac{dy}{dx} = x^3(e^{x^2})(2x) + e^{x^2}(3x^2)$$

$$= 2x^4e^{x^2} + 3x^2e^{x^2} = x^2e^{x^2}(2x^2 + 3)$$

Derivative of x^n, Where n Is Any Real Number

We have already shown that the derivative of x^n is nx^{n-1}, when n is any rational number, positive or negative. Now we show that n can be an irrational number, such as e or π. We let

$$x^n = e^{\ln x^n} = e^{n\ln x}$$

Then

$$\frac{d}{dx}x^n = \frac{d}{dx}e^{n\ln x} = \frac{n}{x}e^{n\ln x} = \frac{n}{x}x^n = nx^{n-1}$$

Thus, the rule for the derivative of x^n holds when n is an irrational number.

EXAMPLE 5: The derivative of $3x^\pi$ is

$$\frac{d}{dx}3x^\pi = 3\pi x^{\pi-1}$$

EXERCISE 5

Derivative of b^u

Differentiate.

1. $y = 3^{2x}$ **2.** $y = 10^{2x+3}$ **3.** $y = (x)(10^{2x+3})$

4. $y = 10^{3x}$ **5.** $y = 2^{x^2}$ **6.** $y = 7^{2x}$

Derivative of e^u

Differentiate.

7. $y = e^{2x}$ **8.** $y = e^{x^2}$ **9.** $y = e^{e^x}$

10. $y = e^{3x^2+4}$ **11.** $y = e^{\sqrt{1-x^2}}$ **13.** $y = \dfrac{2}{e^x}$

12. $y = xe^x$ **14.** $y = (3x + 2)e^{-x^2}$ **15.** $y = x^2e^{3x}$

16. $y = xe^{-x}$ **17.** $y = \dfrac{e^x}{x}$ **18.** $y = \dfrac{x^2 - 2}{e^{3x}}$

19. $y = \dfrac{e^x - e^{-x}}{x^2}$ **20.** $y = \dfrac{e^x - x}{e^{-x} + x^2}$ **21.** $y = (x + e^x)^2$

22. $y = (e^x + 2x)^3$ **23.** $y = \dfrac{(1 + e^x)^2}{x}$ **24.** $y = \left(\dfrac{e^x + 1}{e^x - 1}\right)^2$

Implicit Relations

Find dy/dx.

25. $e^x + e^y = 1$ **26.** $e^x \sin y = 0$ **27.** $e^y = \sin(x + y)$

With Trigonometric Functions

Differentiate.

28. $y = e^x \sin x$ **29.** $y = \sin^3 e^x$

30. $y = e^\theta \cos 2\theta$ **31.** $y = e^x(\cos bx + \sin bx)$

Evaluate each expression.

32. $f'(2)$ where $f(x) = e^{\sin(\pi x/2)}$

33. $f''(0)$ where $f(t) = e^{\sin t} \cos t$

34. $f'(1)$ and $f''(1)$ where $f(x) = e^{x-1} \sin \pi x$

With Logarithmic Functions

Differentiate.

35. $y = e^x \ln x$ **36.** $y = \ln(x^2e^x)$

37. $y = \ln x^{e^2}$ **38.** $y = \ln e^{2x}$

Find the second derivative.

39. $y = e^t \cos t$ **40.** $y = e^{-t} \sin 2t$

41. $y = e^x \sin x$ **42.** $y = \frac{1}{2}(e^x + e^{-x})$

Newton's Method

Find the smallest root that is greater than zero, to two decimal places.

43. $e^x + x - 3 = 0$

44. $xe^{-0.02x} = 1$

45. $5e^{-x} + x - 5 = 0$

46. $e^x = \tan x$

Maximum, Minimum, and Inflection Points

47. Find the minimum point of $y = e^{2x} + 5e^{-2x}$.

48. Find the maximum point and the points of inflection of $y = e^{-x^2}$.

49. Find the maximum and minimum points for one cycle of $y = 10e^{-x} \sin x$.

Applications

50. If \$10,000 is invested for t years at an annual interest rate of 10%, compounded continuously, it will accumulate to an amount y, where $y = 10{,}000e^{0.1t}$. At what rate, in dollars per year, is the balance growing when (a) $t = 0$ yr? (b) $t = 10$ yr?

51. If we assume that the price of an automobile is increasing or ''inflating'' exponentially at an annual rate of 8%, at what rate in dollars per year is the price of a car that initially cost \$9000 increasing after 3 years?

52. When a certain object is placed in an oven at 1000°F, its temperature T rises according to the equation $T = 1000(1 - e^{-0.1t})$, where t is the elapsed time (minutes). How fast is the temperature rising (in degrees per minute) when (a) $t = 0$ and (b) $t = 10$ min?

53. A catenary has the equation $y = \frac{1}{2}(e^x + e^{-x})$. Find the slope of the catenary when $x = 5$.

54. Verify that the minimum point on the catenary occurs at $x = 0$.

55. The speed N of a certain flywheel is decaying exponentially according to the equation $N = 1855e^{-0.5t}$ (rev/min), where t is the time (minutes) after the power is disconnected. Find the angular acceleration (the rate of change of N) when $t = 1$ min.

56. The height y of a certain pendulum released from a height of 50.0 cm is $y = 50.0e^{-0.75t}$ cm, where t is the time after release, in seconds. Find the vertical component of the velocity of the pendulum when $t = 1.00$ s.

57. The number of bacteria in a certain culture is growing exponentially. The number N of bacteria at any time t (hours) is $N = 10{,}000e^{0.1t}$. At what rate (number of bacteria per hour) is the population increasing (a) when $t = 0$ and (b) when $t = 100$ h?

58. A capacitor (Fig. 6-11) is charged to 300 V. When the switch is closed, the voltage across R is initially 300 V but then drops according to the equation $V_1 = 300e^{-t/RC}$, where t is the time (in seconds), R is the resistance, and C is the capacitance. If the voltage V_2 also starts to rise at the instant of switch closure so that $V_2 = 100t$, the total voltage V will be $V = 300e^{-t/RC} + 100t$. Find t when V is a minimum.

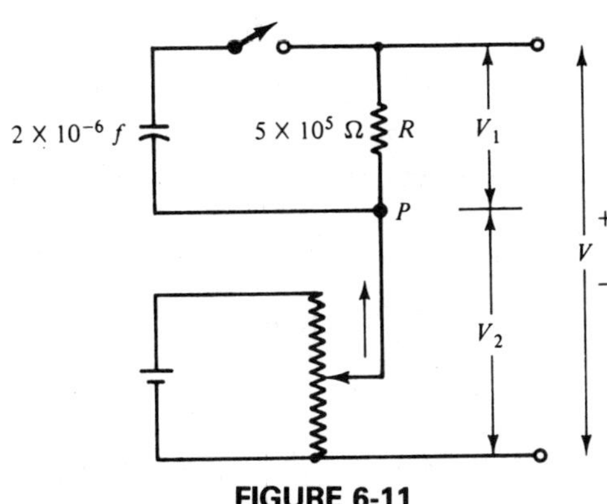

FIGURE 6-11

59. The atmospheric pressure at a height of h miles above the earth's surface is given by $p = 29.92e^{-h/5}$ inches of mercury. Find the rate of change of the pressure on a rocket which is at 18 mi and climbing at a rate of 1500 mi/h.

60. The equation in Problem 59 becomes $p = 2121e^{-0.000037h}$ when h is in feet and p is in pounds per square foot. Find the rate of change of pressure on an aircraft at 5000 ft, climbing at a rate of 10 ft/s.

61. The approximate density of seawater at a depth of h miles is $d = 64.0e^{0.00676h}$ lb/ft^3. Use differentials to estimate the change in density when going from a depth of 1.0 mi to 1.1 mi.

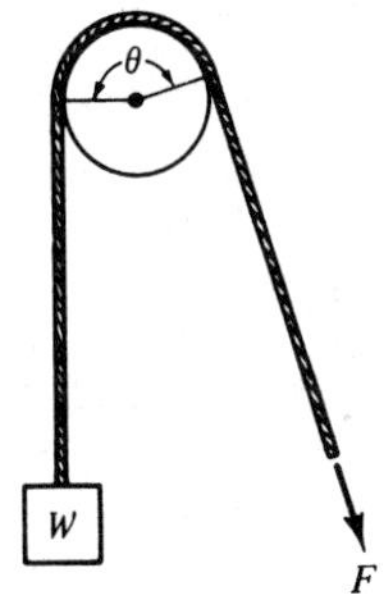

FIGURE 6-12

62. The force F needed to hold a weight W (Fig. 6-12) is $F = We^{-\mu\theta}$. For a certain beam with $\mu = 0.15$, an angle of wrap of 4.62 rad is needed to hold a weight of 200 lb with a force of 100 lb. Find the rate of change of F if the rope is unwrapping at a rate of 15°/s.

CHAPTER TEST

Find dy/dx.

1. $y = \dfrac{a}{2}(e^{x/a} - e^{-x/a})$

2. $y = 5^{2x+3}$

3. $y = 8 \tan \sqrt{x}$

4. $y = \sec^2 x$

5. $y = x \text{ Arctan } 4x$

6. $y = \dfrac{1}{\sqrt{\text{Arcsin } 2x}}$

7. $y^2 = \sin 2x$

8. $y = xe^{2x}$

9. $y = x \sin x$

10. $y = x^2 \sin x$

11. $y = x^3 \cos x$

12. $y = \ln \sin (x^2 + 3x)$

13. $y = \dfrac{\sin x}{x}$

14. $y = (\log x)^2$

15. $y = \log x(1 + x^2)$

16. $\cos (x - y) = 2x$

17. $y = \ln (x + \sqrt{x^2 + a^2})$

18. $y = \ln(x + 10)$

19. $y = \csc 3x$

20. $y = \ln(x^2 + 3x)$

21. $y = \dfrac{\sin x}{\cos x}$

22. $y = \dfrac{1}{\cos^2 x}$

23. $y = \ln(2x^3 + x)$

24. $y = x \text{ Arcsin } 2x$

25. $y = x^2 \text{ Arccos } x$

26. Find the minimum point of the curve $y = \ln(x^2 - 2x + 3)$.

27. Find the points of inflection of the curve $xy = 4 \log(x/2)$.

28. Find a minimum point and a point of inflection on the curve $y \ln x = x$. Write the equation of the tangent at the point of inflection.

29. At what x between $-\pi/2$ and $\pi/2$ is there a maximum on the curve $y = 2 \tan x - \tan^2 x$?

Find the value of dy/dx for the given value of x.

30. $y = x \text{ Arccos } x \quad$ at $x = -\frac{1}{2}$

31. $y = \dfrac{\text{Arcsec } 2x}{\sqrt{x}} \quad$ at $x = 1$

32. If $x^2 + y^2 = \ln y + 2$, find y' and y'' at the point $(1, 1)$.

Find the equation of the tangent to each curve.

33. $y = \sin x \quad$ at $x = \pi/6$

34. $y = x \ln x$ parallel to the line $3x - 2y = 5$

35. At what x is the tangent to the curve $y = \tan x$ parallel to the line $y = 2x + 5$?

Find the smallest positive root between x = 0 and x = 10, to three decimal places.

36. $\sin 3x - \cos 2x = 0$

37. $2 \sin \frac{1}{2}x - \cos 2x = 0$

38. Find the angle of intersection between $y = \ln\left(\dfrac{x^3}{8} - 1\right)$ and $y = \ln\left(3x - \dfrac{x^2}{4} - 1\right)$ at $x = 4$.

39. A casting is taken from one oven at 1500°F and placed in another oven whose temperature is 0°F and rising at a linear rate of 100° per hour. The temperature T of the casting after t hours is then $T = 100t + 1500e^{-0.2t}$. Find the minimum temperature reached by the casting and the time at which it occurs.

40. A statue which is 11 ft tall is on a pedestal, so that the bottom of the statue is 25 ft above eye level. How far from the statue (measured horizontally) should an observer stand so that the statue will subtend the greatest angle at the observer's eye?

Integration

Every operation in mathematics has its inverse; we reverse the squaring operation by taking the square root, the arcsin is the inverse of the sine, and so on. In this chapter we learn how to reverse the process of differentiation with the process of *integration*.

The material in this and the following three chapters usually falls under the heading of *integral calculus,* as opposed to the differential calculus already introduced. We will have some applications (accelerated motion and electric circuits) in this chapter, but most applications of integration will be in Chapters 8 through 10.

7-1. THE INDEFINITE INTEGRAL

Reversing the Process of Differentiation

The *antiderivative* of an expression is a new expression which, if differentiated, gives the original expression.

EXAMPLE 1: The derivative of x^3 is $3x^2$, so the antiderivative of $3x^2$ is x^3.

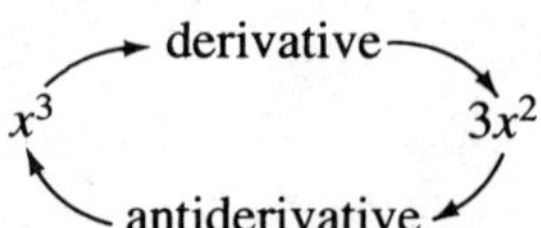

The derivative of x^3 is $3x^2$, but the derivative $x^3 + 6$ is also $3x^2$. The derivatives of $x^3 - 99$, and x^3 + *any constant* are also $3x^2$. This constant, called the *constant of integration,* must be included when we find the antiderivative of a function.

We will learn how to evaluate the constant of integration in Sec. 7-3.

EXAMPLE 2: The derivative of x^3 + (any constant) is $3x^2$, so the antiderivative of $3x^2$ is x^3 + (a constant).

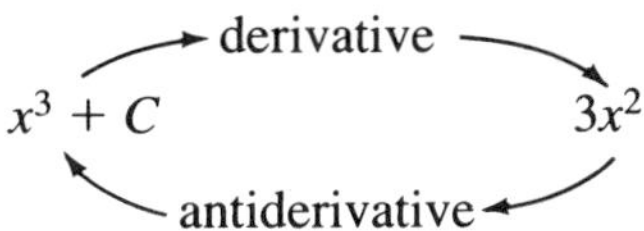

The antiderivative is also called the *indefinite integral,* and we will use both names. The process of finding the integral or antiderivative is called *integration.*

The Integral Sign

We have seen above that the derivative of $x^3 + C$ is $3x^2$, so the antiderivative of $3x^2$ is $x^3 + C$. Let us now state this same idea more formally.

Let there be a function $F(x) + C$. If the derivative of $F(x) + C$ is denoted $F'(x)$, then

$$\text{antiderivative of } F'(x) = F(x) + C$$

It is no accident that the integral sign looks like "S." We will see in Chapter 8 that it stands for *summation*.

The integral is called indefinite **because of the unknown constant. In Chapter 8 we study the definite integral, which has no unknown constant.**

Instead of writing "antiderivative of" we use the *integral* sign $\int$ to indicate the antiderivative.

Indefinite Integral or Antiderivative	$\int F'(x)\,dx = F(x) + C$	**337**

We read this as "the integral of $F'(x)$ with respect to x is $F(x) + C$." The expression $F'(x)$ to be integrated is called the *integrand,* and C is the constant of integration. The dx tells that the *variable of integration* (the variable that we take the integral with respect to) is x. In the following section we'll see where the dx comes from.

We use both capital and lowercase *F* in this section. Be careful not to mix them up.

If we let $F'(x)$ be denoted by $f(x)$, Eq. 337 becomes

Indefinite Integral or Antiderivative	$\int f(x)\,dx = F(x) + C$	**338**

EXAMPLE 3: Using the same expressions as before, we can write

$$\int 3x^2\, dx = x^3 + C$$

The Integral of a Differential

Let there be a function $u = F(x)$ whose derivative is

$$\frac{du}{dx} = F'(x)$$

or

$$du = F'(x)\, dx$$

Taking the integral of both sides gives

$$\int du = \int F'(x)\, dx = F(x) + C$$

by Eq. 337. Substituting $F(x) = u$ we get our first rule for finding integrals.

$$1. \quad \int du = u + C$$

The integral of the differential of a function is equal to the function itself, plus a constant.

EXAMPLE 4:

(a) $\int dy = y + C$

(b) $\int dz = z + C$

Finding the Original Function from Its Derivative

This equation, as well as any other that contains a derivative, is called a *differential equation*. In this example we are solving a differential equation.

Suppose that we have the derivative of a function. Let's use our familiar

$$\frac{dy}{dx} = 3x^2$$

We now seek the equation, $y = f(x)$, of which $3x^2$ is the derivative. Proceed as follows. First write the given differential equation in differential form by multiplying both sides by dx.

$$dy = 3x^2\, dx$$

We now take the integral of both sides of the equation,

$$\int dy = \int 3x^2\,dx$$

so

$$y + C_1 = x^3 + C_2$$

We now combine the two arbitrary constants, C_1 and C_2,

$$y = x^3 + C_1 - C_2 = x^3 + C$$

From now on, we will not bother writing C_1 and C_2, but will combine them immediately into a single constant C.

This is the function whose derivative is $3x^2$. This function is also called the *solution* to the differential equation $dy/dx = 3x^2$.

Rules for Finding Integrals

We can obtain rules for integration by reversing the rules we had previously derived for differentiation. Such a list of rules, called a *table of integrals,* is given in Appendix C. This is a very short table of integrals; some fill entire books.

Our second rule is for the integral of a constant times a function.

$$2.\ \int a\,f(x)\,dx = a\int f(x)\,dx = a\,F(x) + C$$

The integral of a constant times a function is equal to the constant times the integral of the function, plus a constant.

Our third rule is for the sum of several functions.

$$3.\ \int [f(x) + g(x) + h(x) \cdots]\,dx = \int f(x)\,dx + \int g(x)\,dx + \int h(x)\,dx + \cdots + C$$

The integral of the sum of several functions equals the sum of the integrals of those functions, plus a constant.

Our fourth rule is for a power of x. Since the derivative of $x^{n+1}/(n + 1)$ is

$$\frac{d}{dx}\left(\frac{x^{n+1}}{n+1}\right) = (n + 1)\frac{x^n}{n+1} = x^n$$

going to differential form and switching sides gives us

$$x^n\,dx = d\left(\frac{x^{n+1}}{n+1}\right)$$

Taking the integral of both sides gives rule 4,

$$4. \quad \int x^n\,dx = \frac{x^{n+1}}{n+1} + C \qquad (n \neq -1)$$

The integral of x raised to a power is x raised to that power increased by one, divided by the new power, plus a constant.

EXAMPLE 5: Integrate $\int x^3\,dx$.

Solution: We use rule 4, with $u = x$ and $du = dx$,

$$\int x^3\,dx = \frac{x^{3+1}}{3+1} + C = \frac{x^4}{4} + C$$

EXAMPLE 6: Integrate $\int x^5\,dx$.

Solution: This is similar to Example 5, except that the exponent is 5. By rule 4,

$$\int x^5\,dx = \frac{x^{5+1}}{5+1} + C = \frac{x^6}{6} + C$$

Rule 2 says that we may move constants to the left of the integral sign, as in the following example.

EXAMPLE 7: Integrate $\int 7x^3\,dx$.

Solution: This is similar to the first example, except that our function is multiplied by 7. By rule 2,

$$\int 7x^3\,dx = 7\int x^3\,dx = 7\left(\frac{x^4}{4} + C_1\right)$$

$$= \frac{7x^4}{4} + 7C_1 = \frac{7x^4}{4} + C$$

where $C = 7C_1$. Since C_1 is an unknown constant, $7C_1$ is also an unknown constant which can be more simply represented by C. From now on we will not even bother with C_1 but will write the final constant C directly.

Rule 3 says that when integrating an expression having several terms, we may integrate each term separately.

EXAMPLE 8: Integrate $\int (x^3 + x^5)\,dx$.

Solution: By rule 3,

$$\int (x^3 + x^5)dx = \int x^3\,dx + \int x^5\,dx$$
$$= \frac{x^4}{4} + \frac{x^6}{6} + C$$

Even though each of the two integrals has produced its own constant of integration, we have combined them immediately into the single constant C.

EXAMPLE 9: Find the function $y = f(x)$ that has the derivative

$$\frac{dy}{dx} = 5x^2 + 2x - 3$$

Solution: We put the given equation into differential form by multiplying by dx,

$$dy = (5x^2 + 2x - 3)dx$$

and take the integral of both sides,

$$\int dy = \int (5x^2 + 2x - 3)dx$$

Integrating yields

$$y = \frac{5x^3}{3} + x^2 - 3x + C$$

Rule 4 is also used when the exponent n is not an integer.

EXAMPLE 10: Integrate $\int \sqrt[3]{x}\,dx$.

Solution:

$$\int \sqrt[3]{x}\,dx = \int x^{1/3}\,dx$$
$$= \frac{x^{4/3}}{4/3} + C$$
$$= \frac{3x^{4/3}}{4} + C$$

The exponent n can also be negative (with the exception of -1, which would result in division by zero).

The variable of integration can, of course, be any letter other than x, as in this example.

EXAMPLE 11: Integrate $\int \frac{1}{t^3}\,dt.$

Solution:

$$\int \frac{1}{t^3}\,dt = \int t^{-3}\,dt$$

$$= \frac{t^{-2}}{-2} + C = -\frac{1}{2t^2} + C$$

Checking an Integral

Many rules for integration are presented without derivation or proof, so you would be correct in being suspicious of them. However, you can convince yourself that a rule works (and that you have used it correctly) simply by taking the derivative of your result. You should get back the original expression.

EXAMPLE 12: Taking the derivative of the expression obtained in Example 11,

$$\frac{d}{dt}\left(-\frac{t^{-2}}{2} + C\right) = -\frac{1}{2}(-2t^{-3}) + 0 = t^{-3} = \frac{1}{t^3}$$

which is the expression we started with, so our integration was correct.

Simplify before Integrating

If an expression does not seem to fit any given rule at first, try performing the indicated operations (squaring, removing parentheses, and so on.)

EXAMPLE 13: Integrate $\int (x^2 + 3)^2\,dx$.

Solution: None of our rules (so far) seem to fit. Rule 3, for example, is for x raised to a power, *not* for $(x^2 + 3)$ raised to a power. However, if we square $x^2 + 3$, we get

$$\int (x^2 + 3)^2\,dx = \int (x^4 + 6x^2 + 9)\,dx$$

$$= \frac{x^5}{5} + \frac{6x^3}{3} + 9x + C$$

$$= \frac{x^5}{5} + 2x^3 + 9x + C$$

EXAMPLE 14: Integrate $\int \frac{x^5 - 2x^3 + 5x}{x}\,dx.$

Solution: This looks complicated at first, but let us perform the division:

$$\int \frac{x^5 - 2x^3 + 5x}{x}\,dx = \int (x^4 - 2x^2 + 5)dx$$
$$= \frac{x^5}{5} - \frac{2x^3}{3} + 5x + C$$

---

* **EXAMPLE 15:** Evaluate $\int \frac{x^3 - x^2 + 5x - 5}{x - 1}\,dx.$

Solution: Again, no rule seems to fit, so we try to simplify the given expression by long division.

$$\begin{array}{r|l} & x^2 + 5 \\ \hline x - 1 & x^3 - x^2 + 5x - 5 \\ & \underline{x^3 - x^2} \\ & 0 + 5x - 5 \\ & \underline{5x - 5} \end{array}$$

So our quotient is $x^2 + 5$, which we now integrate,

$$\int \frac{x^3 - x^2 + 5x - 5}{x - 1}\,dx = \int (x^2 + 5)\,dx = \frac{x^3}{3} + 5x + C$$

EXERCISE 1

Find the indefinite integral.

1. $\int dx$	**2.** $\int dy$	**3.** $\int x\,dx$
4. $\int x^4\,dx$	**5.** $\int \frac{dx}{x^2}$	**6.** $\int$L $x^{2/3}\,dx$
7. $\int 3x\,dx$	**8.** $\int 6x\,dx$	**9.** $\int 3x^3\,dx$
10. $\int x^5\,dx$	**11.** $\int x^n\,dx$	**12.** $\int x^{1/2}\,dx$
13. $\int \frac{dx}{\sqrt{x}}$	**14.** $\int 3x^2\,dx$	**15.** $\int 4x^3\,dx$
16. $\int \left(\frac{x^2}{2} - \frac{2}{x^2}\right) dx$	**17.** $\int \frac{7}{2}x^{5/2}\,dx$	**18.** $\int 5x^4\,dx$
19. $\int 2(x + 1)dx$	**20.** $\int x^{5/3}\,dx$	**21.** $\int \frac{4}{3}x^{1/3}\,dx$

22. $\int (x^{3/2} - 2x^{2/3} + 5\sqrt{x} - 3)dx$ **23.** $\int 2u\,du$

24. $\int \sqrt[3]{t}\,dt$ **25.** $\int 4s^{1/2}\,ds$ **26.** $\int 3\sqrt[3]{y}\,dy$

Simplify and integrate.

27. $\int \sqrt{x}\,(3x - 2)dx$ **28.** $\int (x + 1)^2\,dx$ **29.** $\int \frac{4x^2 - 2\sqrt{x}}{x}\,dx$

30. $\int (t + 2)(t - 3)\,dt$ **31.** $\int (1 - s)^3\,ds$ **32.** $\int \frac{v^3 + 2v^2 - 3v - 6}{v + 2}\,dv$

Find the function whose derivative is given (solve each differential equation).

33. $\frac{dy}{dx} = 4x^2$ **34.** $\frac{dy}{dx} = 2x(x^2 + 6)$ **35.** $\frac{dy}{dx} = x^{-3}$

36. $\frac{ds}{dt} = 10t^{-6}$ **37.** $\frac{ds}{dt} = \frac{1}{2}t^{-2/3}$ **38.** $\frac{dv}{dt} = 6t^3 - 3t^{-2}$

7-2. INTEGRAL OF A POWER FUNCTION

Our next rule is for a function u raised to a power n. If u is some function of x, the derivative of $u^{n+1}/(n + 1)$ is

$$\frac{d}{dx}\left(\frac{u^{n+1}}{n + 1}\right) = (n + 1)\frac{u^n}{n + 1}\frac{du}{dx} = u^n\frac{du}{dx}$$

or

$$d\left(\frac{n^{n+1}}{n + 1}\right) = u^n\,du$$

Reversing the process, we get for the integral of $u^n\,du$,

$$5.\quad \int u^n\,du = \frac{u^{n+1}}{n + 1} + C \qquad (n \neq -1)$$

The expression u in rule 5 can be any function of x, say, $x^3 + 3$. However, in order to use rule 5 *the quantity u^n must be followed by the derivative of u.*

EXAMPLE 1: Find the integral $\int (x^3 + 3)^6(3x^2)\,dx$.

Solution: If we let $u = x^3 + 3$, we see that the derivative of u is

$$\frac{du}{dx} = 3x^2$$

or, in differential form, $du = 3x^2\,dx$. Notice now that our given integral exactly matches rule 5.

$$\int \underbrace{(x^3 + 3)}_{u}{}^{\textcircled{n}\,6}\underbrace{(3x^2)\,dx}_{du}$$

Applying rule 5,

$$\int (x^3 + 3)^6(3x^2)\,dx = \frac{(x^3 + 3)^7}{7} + C$$

Students are often puzzled at the disappearance of the $3x^2\,dx$ in Example 1.

$$\int (x^3 + 3)^6 \boxed{(3x^2)\,dx} = \frac{(x^3 + 3)^7}{7} + C$$

Where did this go?

The $3x^2\,dx$ is the differential of x^3, and does not remain after integration. Do not be alarmed when it vanishes.

Very often an integral will not exactly match the form of rule 5 or any other rule in the table. However, if all we lack is a constant factor, it can usually be supplied as in the following example.

EXAMPLE 2: Integrate $\int (x^3 + 3)^6 x^2 dx$.

Solution: This is almost identical to Example 1, except that the factor 3 is missing. Realize that we cannot use rule 5 *yet,* because if $u = x^2 + 3$ then

$$du = 3x^2\,dx$$

and our integral contains $x^2\,dx$ but not $3x^2\,dx$. But we can insert a factor of 3 into our integrand, as long as we compensate for it by multiplying the whole integral by $\frac{1}{3}$.

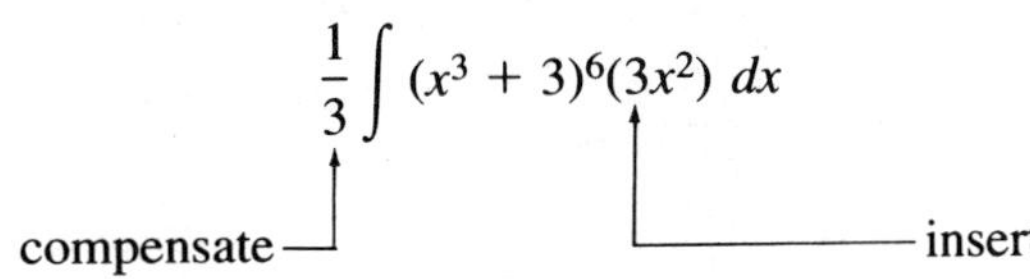
$$\frac{1}{3}\int (x^3 + 3)^6(3x^2)\,dx$$

Integrating gives us

$$\frac{1}{3}\int (x^3 + 3)^6(3x^2)\,dx = \frac{1}{3}\frac{(x^3 + 3)^7}{7} + C$$

$$= \frac{(x^3 + 3)^7}{21} + C$$

EXAMPLE 3: Evaluate $\int \frac{5z\,dz}{\sqrt{3z^2 - 7}}$.

Solution: We rewrite the given expression in exponential form and move the 5 outside the integral.

$$\int \frac{5z\,dz}{\sqrt{3z^2 - 7}} = 5\int (3z^2 - 7)^{-1/2}\,z\,dz$$

Since u is $3z^2 - 7$, du is $6z\,dz$. We thus insert a 6 before the z and compensate with $\frac{1}{6}$ outside the integral sign.

$$5\int (3z^2 - 7)^{-1/2}\,z\,dz = 5\left(\frac{1}{6}\right)\int (3z^2 - 7)^{-1/2}(6z\,dz)$$

$$= \frac{5}{6}\frac{(3z^2 - 7)^{1/2}}{\frac{1}{2}} + C = \frac{5}{3}\sqrt{3z^2 - 7} + C$$

Common Error	Rule 2 allows us to move *only constants* to the left of the integral sign. You cannot use this same procedure with variables.

Sometimes rule 5 can be used in unexpected places, as in the following example.

EXAMPLE 4: Evaluate $\int \sin^3 2x \cos 2x\,dx$.

Solution: If we let $u = \sin 2x$, then

$$du = 2\cos 2x\,dx$$

We insert the factor 2, and compensate with $\frac{1}{2}$ to the left of the integral sign:

$$\frac{1}{2}\int (\sin 2x)^3(2\cos 2x\,dx) = \frac{1}{2}\frac{\sin^4 2x}{4} + C$$

$$= \frac{1}{8}\sin^4 2x + C$$

EXERCISE 2

Integrate.

1. $\int (x^4 + 1)^3\, 4x^3\, dx$

2. $\int (2x^2 - 6)^3 4x\, dx$

3. $\int 2(x^2 + 2x)(2x + 2)dx$

4. $\int (1 - 2x)^5\, dx$

5. $\int \frac{dx}{(1 - x)^2}$

6. $\int 3(x^2 - 1)^2 2x\, dx$

7. $\int 3(x^3 + 1)^2 3x^2\, dx$

8. $\int 3x \sqrt{x^2 - 1}\, dx$

9. $\int \frac{y^2\, dy}{\sqrt{1 - y^3}}$

10. $\int (x + 1)(x^2 + 2x + 6)^2\, dx$

11. $\int \frac{4x\, dx}{(9 - x^2)^2}$

12. $\int z \sqrt{z^2 - 2}\, dz$

13. $\int \tan^2 x \sec^2 x\, dx$

14. $\int \cos^3 2x \sin 2x\, dx$

Find the function whose derivative is given.

15. $\frac{dy}{dx} = x(1 - x^2)^3$

16. $\frac{dy}{dx} = x^2(x^3 - 2)^{1/2}$

17. $\frac{dy}{dx} = (3x^2 + x)(2x^3 + x^2)^3$

18. $\frac{ds}{dt} = \frac{t^2}{\sqrt{5t^3 + 7}}$

19. $\frac{dv}{dt} = 5t \sqrt{7 - t^2}$

20. $\frac{dy}{dx} = \frac{x}{(9 - 4x^2)^2}$

7-3. CONSTANT OF INTEGRATION

Families of Curves

We have seen in the preceding sections that we get a constant of integration whenever we find an indefinite integral. We now give a geometric meaning to this constant.

EXAMPLE 1: Find the function whose derivative is $\frac{dy}{dx} = 2x$, and make a graph of the function.

Solution: Proceeding as in Sec. 7-1, we write the given differential equation in differential form, $dy = 2x\, dx$. Integrating, we get

$$y = \int 2x\, dx = 2\left(\frac{x^2}{2}\right) + C$$

$$y = x^2 + C$$

From Sec. 3-2 we recognize this as a parabola opening upward. Further, its vertex is on the y axis at a distance of C units from the origin. When we graph this function (Fig. 7-1) we get not a single curve, but a *family of curves,* each of which has a different value for C.

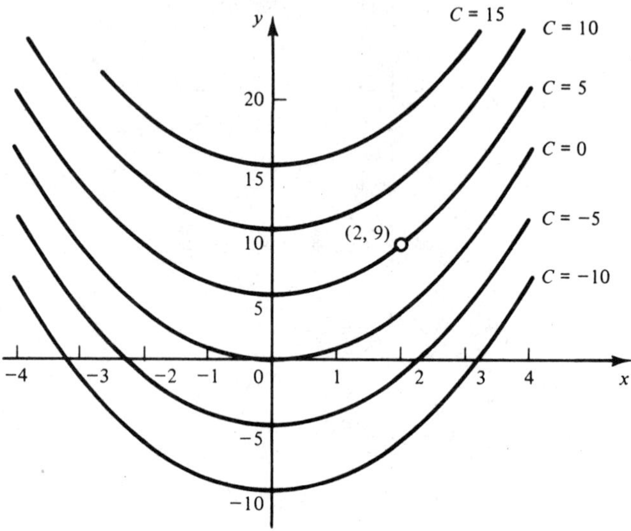

FIGURE 7-1. Family of curves, $y = x^2 + C$.

Boundary Conditions

The term *initial* is more appropriate when our variable is *time.*

In order for us to evaluate the constant of integration, more information must be given. Such additional information is called a *boundary condition* or *initial condition.*

EXAMPLE 2: Find the constant of integration in the preceding example if the curve whose derivative is $2x$ is to pass through the point (2, 9).

Solution: The equation of the curve was found to be $y = x^2 + C$. Letting $x = 2$ and $y = 9$, we have

$$C = 9 - 2^2 = 5$$

as can be verified from Fig. 7-1.

Successive Integration

If we are given the *second* derivative of some function and wish to find the function itself, we simply have to *integrate twice.* However, each time we integrate we get another constant of integration. These constants can be found if we are given enough additional information, as in the following example.

EXAMPLE 3: Find the equation of a curve that has a second derivative $y'' = 4x$, and that has a slope of 10 at the point (2, 9).

Solution: We can write the second derivative as

$$\frac{d}{dx}(y') = 4x$$

or, in differential form,

$$d(y') = 4x\,dx$$

Integrating gives us

$$y' = \int 4x\,dx = 2x^2 + C_1 \tag{1}$$

But the slope, and hence y', is 10 when $x = 2$, so

$$C_1 = 10 - 2(2)^2 = 2$$

So (1) becomes $y' = 2x^2 + 2$ or, in differential form,

$$dy = (2x^2 + 2)\,dx$$

Integrating again, we obtain

$$y = \int (2x^2 + 2)dx = \frac{2x^3}{3} + 2x + C_2 \tag{2}$$

But $y = 9$ when $x = 2$, so

$$C_2 = 9 - \frac{2(2)^3}{3} - 2(2) = -\frac{1}{3}$$

Substituting into (2), we get

$$y = \frac{2x^3}{3} + 2x - \frac{1}{3}$$

as our final equation, with all constants evaluated.

EXERCISE 3

Families of Curves

Write the function that has the given derivative. Then graph that function for $C = -1$, $C = 0$, and $C = 1$.

1. $\dfrac{dy}{dx} = 3$ **2.** $\dfrac{dy}{dx} = 5x$ **3.** $\dfrac{dy}{dx} = 3x^2$

Constant of Integration

Write the function that has the given derivative and passes through the given point.

4. $y' = 3x$, passes through (2, 6)

5. $y' = x^2$, passes through (1, 1)

6. $y' = \sqrt{x}$, passes through (2, 4)

7. If $dy/dx = 2x + 1$, and $y = 7$ when $x = 1$, find the value of y when $x = 3$.

8. If $dy/dx = \sqrt{2x}$ and $y = \frac{1}{3}$ when $x = \frac{1}{2}$, find the value of y when $x = 2$.

Successive Integration

9. Find the equation of a curve that passes through the point (3, 0), has the slope $\frac{7}{2}$ at that point, and has a second derivative $y'' = x$.

10. Find the equation of the curve for which $y'' = 4$ at every point if the curve is tangent to the line $y = 3x$ at (2, 6).

11. Find the equation of a curve that passes through the point (1, 0), is tangent to the line $6x + y = 6$ at that point, and has a second derivative $y'' = 12/x^3$.

12. The second derivative of a curve is $y'' = 12x^2 - 6$. The tangent to this curve at (2, 4) is perpendicular to the line $x + 20y - 40 = 0$. Find the equation of the curve.

7-4. USE OF A TABLE OF INTEGRALS

This is a good time to review the rules already covered.

A very short table of integrals is given in Appendix C, but much longer ones are available. With an extensive table we can often find the integral of a given expression without having to modify it a lot first. It will usually be necessary, however, to adjust the values of the constants, as we did before, and place a suitable compensating factor to the left of the integral sign.

Integral of du/u

The derivative of $y = \ln u$ (Fig. 7-2) is

$$\frac{d(\ln u)}{dx} = \frac{1}{u}\frac{du}{dx} \tag{327}$$

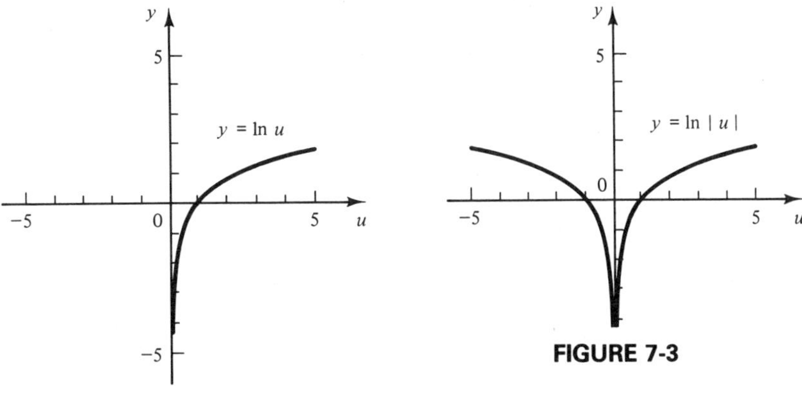

FIGURE 7-3

FIGURE 7-2

Since we can take logs of positive numbers only, $y = \ln u$ exists for positive u only. However, the function $y = \ln |u|$ (Fig. 7-3) exists for positive *and negative u,* since $|u|$ is never negative. Its derivative is the same as for $y = \ln u$.

$$\frac{d(\ln |u|)}{dx} = \frac{1}{u}\frac{du}{dx}$$

or, in differential form, $d(\ln |u|) = du/u$. Thus the integral of du/u is $\ln |u|$.

When we took derivatives of ln *u*, we didn't care about negative values because *u* had to be positive for the logarithm to exist. But we sometimes want the integral of *y* = 1/*u* at negative values of *u*.

$$7. \quad \int \frac{du}{u} = \ln |u| + C \qquad (u \neq 0)$$

EXAMPLE 1: Integrate $\int 5x^{-1}\,dx$.

Solution: $\displaystyle\int 5x^{-1}\,dx = 5\int \frac{dx}{x} = 5 \ln |x| + C$, by rule 7.

EXAMPLE 2: Integrate $\displaystyle\int \frac{x\,dx}{3 - x^2}$.

Solution: The derivative of $3 - x^2$ is $-2x$, so we insert a factor of -2 and compensate with a factor of $-\frac{1}{2}$,

$$\int \frac{x\,dx}{3 - x^2} = -\frac{1}{2}\int \frac{-2x\,dx}{3 - x^2} = -\frac{1}{2}\ln|3 - x^2| + C$$

by rule 7.

EXAMPLE 3: Find the equation of the curve whose derivative is $6x/(x^2 - 4)$ and which passes through the point (0, 5).

Solution: Our derivative, in differential form, is

$$dy = \frac{6x\,dx}{x^2 - 4}$$

Integrating,

$$y = 3\int \frac{2x\,dx}{x^2 - 4} = 3 \ln |x^2 - 4| + C$$

At (0, 5) we get $C = 5 - 3 \ln |0 - 4| = 0.841$, so

$$y = 3 \ln |x^2 - 4| + 0.841$$

This function is graphed as a solid line in Fig. 7-4. Its domain is from $x = -2$ to $x = 2$.

Do not try to use our function $y = 3 \ln |x^2 - 4| + 0.841$ for $x < -2$ or $x > 2$. Those parts of the curve (shown dashed) *are not continuous* with the section for which we had the known value, so the function we found may not apply there.

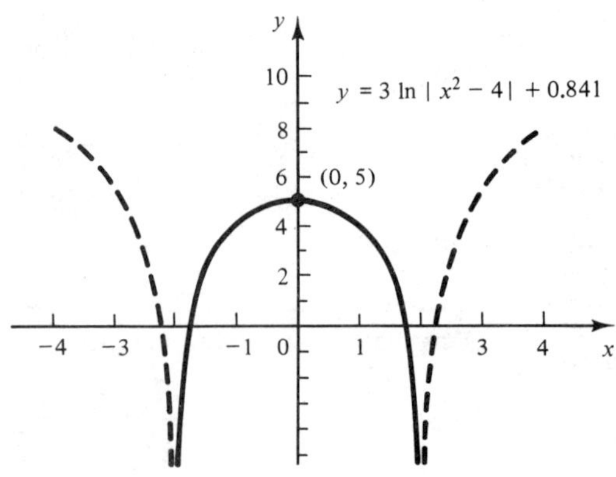

FIGURE 7-4

Integral of $e^u\,du$

Since the derivative of e^u is e^u, then the integral of e^u is e^u.

$$8. \int e^u\,du = e^u + C$$

EXAMPLE 4: Integrate $\int e^{6x}\,dx$.

Solution: We insert a factor of 6, and compensate:

$$\int e^{6x}\,dx = \frac{1}{6}\int e^{6x}(6\,dx) = \frac{1}{6}e^{6x} + C$$

by rule 8.

EXAMPLE 5: Integrate $\int \frac{6e^{\sqrt{4x}}}{\sqrt{4x}}\,dx$.

Solution: Since the derivative of $\sqrt{4x}$ is $2(4x)^{-1/2}$, we insert a factor of 2 and compensate:

$$\begin{aligned}\int \frac{6e^{\sqrt{4x}}}{\sqrt{4x}}\,dx &= 6\int e^{\sqrt{4x}}(4x)^{-1/2}\,dx \\ &= 6\left(\frac{1}{2}\right)\int e^{\sqrt{4x}}\,[2(4x)^{-1/2}\,dx] \\ &= 3e^{\sqrt{4x}} + C\end{aligned}$$

by rule 8.

Integral of $b^u\,du$

The derivative of $b^u/\ln b$ is

$$\frac{d}{dx}\left(\frac{b^u}{\ln b}\right) = \frac{1}{\ln b}(b^u)\,(\ln b)\frac{du}{dx} = b^u\frac{du}{dx}$$

or, in differential form, $d(b^u/\ln b) = b^u\,du$. Thus the integral of $b^u\,du$ is

$$9. \int b^u\,du = \frac{b^u}{\ln b} + C \qquad (b > 0,\ b \neq 1)$$

EXAMPLE 6: Integrate $\int 3xa^{2x^2}\,dx$.

Solution:

$$\begin{aligned}\int 3xa^{2x^2}\,dx &= 3\int a^{2x^2}x\,dx \\ &= 3\left(\frac{1}{4}\right)\int a^{2x^2}(4x\,dx)\end{aligned}$$

$$= \frac{3}{4} \frac{a^{2x^2}}{\ln a} + C$$

by rule 9.

Integrals of the Trigonometric Functions

To our growing list of rules we add those for the six trigonometric functions. Since, by Eq. 315,

$$\frac{d(-\cos u)}{dx} = \sin u$$

or $d(-\cos u) = \sin u \, du$. Taking the integral of both sides gives

$$\int \sin u \, du = -\cos u + C$$

The integrals of the other trigonometric functions are found in the same way. We thus get the following rules,

10. $\int \sin u \, du = -\cos u + C$
11. $\int \cos u \, du = \sin u + C$
12. $\int \tan u \, du = -\ln|\cos u| + C$
13. $\int \cot u \, du = \ln|\sin u| + C$
14. $\int \sec u \, du = \ln|\sec u + \tan u| + C$
15. $\int \csc u \, du = \ln|\csc u - \cot u| + C$

We use these rules just as we did the preceding ones. Match the given integral *exactly* with one of the rules, inserting a factor and compensating when necessary, and then copy off the integral in the rule.

EXAMPLE 7: Integrate $\int x \sin x^2 \, dx$.

Solution:

$$\int x \sin x^2 \, dx = \int \sin x^2 (x \, dx)$$

$$= \frac{1}{2}\int \sin x^2(2x\,dx) = -\frac{1}{2}\cos x^2 + C$$

from rule 10.

Sometimes the trigonometric identities can be used to simplify an expression before integrating.

EXAMPLE 8: Integrate $\int \frac{\cot 5x}{\cos 5x}\,dx$.

Solution: We replace cot 5*x* by cos 5*x*/sin 5*x*:

$$\int \frac{\cot 5x}{\cos 5x}\,dx = \int \frac{\cos 5x}{\sin 5x \cos 5x}\,dx = \int \frac{1}{\sin 5x}\,dx$$

$$= \int \csc 5x\,dx$$

$$= \frac{1}{5}\int \csc 5x(5\,dx) = \frac{1}{5}\ln|\csc 5x - \cot 5x| + C$$

by rule 15.

Miscellaneous Rules from the Table

Now that you can use rules 1 to 15, you should find it no harder to use any rule from the table of integrals.

EXAMPLE 9: Integrate $\int e^{3x}\cos 2x\,dx$.

Solution: We search the table for a similar form and find

$$42.\quad \int e^{au}\cos bu\,du = \frac{e^{au}}{a^2 + b^2}(a\cos bu + b\sin bu) + C$$

This matches our integral if we set

$$u = x, \qquad a = 3, \qquad b = 2, \qquad du = dx$$

so

$$\int e^{3x}\cos 2x\,dx = \frac{e^{3x}}{3^2 + 2^2}(3\cos 2x + 2\sin 2x) + C$$

$$= \frac{e^{3x}}{13}(3\cos 2x + 2\sin 2x) + C$$

EXAMPLE 10: Integrate $\int \frac{dx}{4x^2 + 25}$.

Solution: From the table we find

$$56. \quad \int \frac{du}{a^2 + b^2u^2} = \frac{1}{ab} \tan^{-1} \frac{bu}{a} + C$$

Letting $a = 5$, $b = 2$, $u = x$, and $du = dx$,

$$\int \frac{dx}{4x^2 + 25} = \frac{1}{10} \tan^{-1} \frac{2x}{5} + C$$

EXAMPLE 11: Integrate $\int \frac{dx}{(4x^2 + 9)2x}$.

Solution: We match this with

$$60. \quad \int \frac{du}{u(u^2 + a^2)} = \frac{1}{2a^2} \ln \left| \frac{u^2}{u^2 + a^2} \right| + C$$

with $a = 3$, $u = 2x$, and $du = 2dx$, so

$$\int \frac{dx}{(4x^2 + 9)2x} = \frac{1}{2} \int \frac{2\,dx}{(2x)[(2x)^2 + 3^2]}$$

$$= \frac{1}{36} \ln \left| \frac{4x^2}{4x^2 + 9} \right| + C$$

Common Error	When using the table of integrals, be sure that *all of du is present* before writing the result.

EXAMPLE 12: Integrate $\int \frac{dx}{x\sqrt{x^2 + 16}}$.

Solution: We use

$$64. \quad \int \frac{du}{u\sqrt{u^2 + a^2}} = \frac{1}{a} \ln \left| \frac{u}{a + \sqrt{u^2 + a^2}} \right| + C$$

We're not done learning how to find integrals. All of Chapter 11 is devoted to other methods of integration.

with $u = x$, $a = 4$, and $du = dx$, so

$$\int \frac{dx}{x\sqrt{x^2 + 16}} = \frac{1}{4} \ln \left| \frac{x}{4 + \sqrt{x^2 + 16}} \right| + C$$

EXERCISE 4

Evaluate each integral.

Integral of du/u

1. $\int \frac{3}{x}\, dx$ **2.** $\int \frac{dx}{x - 1}$ **3.** $\int \frac{x + 1}{x}\, dx$

4. $\int \frac{t\, dt}{6 - t^2}$ **5.** $\int \frac{5z^2}{z^3 - 3}\, dz$ **6.** $\int \frac{w^2 + 5}{w}\, dw$

Exponential Functions

7. $\int a^{5x}\, dx$ **8.** $\int a^{9x}\, dx$ **9.** $\int 5^{7x}\, dx$

10. $\int 10^x\, dx$ **11.** $\int a^{3y}\, dy$ **12.** $\int a^{3x^2} x\, dx$

13. $\int 4e^x\, dx$ **14.** $\int e^{2x}\, dx$ **15.** $\int xe^{x^2}\, dx$

16. $\int e^{x^3} x^2\, dx$ **17.** $\int e^{3x^2}\, x\, dx$ **18.** $\int \sqrt{e^t}\, dt$

19. $\int \frac{e^{\sqrt{x}}\, dx}{\sqrt{x}}$ **20.** $\int e^{x^2+6x-2}(x + 3)\, dx$

21. $\int (e^x - 1)^2\, dx$ **22.** $\int e^{-x^2} x\, dx$

23. $\int \frac{e^{\sqrt{x-2}}}{\sqrt{x - 2}}\, dx$ **24.** $\int \frac{(e^{x/2} - e^{-x/2})^2}{4}\, dx$

25. $\int (e^{x/a} + e^{-x/a})\, dx$ **26.** $\int (e^{x/a} - e^{-x/a})^2\, dx$

Trigonometric Functions

27. $\int \sin 3x\, dx$ **28.** $\int \cos 7x\, dx$

29. $\int \tan 5\theta\, d\theta$ **30.** $\int \sec 2\theta\, d\theta$

31. $\int \sec 4x\, dx$

32. $\int \cot 8x\, dx$

33. $\int 3 \tan 9\theta\, d\theta$

34. $\int 7 \sec 3\theta\, d\theta$

35. $\int x \sin x^2\, dx$

36. $\int 5x \cos 2x^2\, dx$

37. $\int \theta^2 \tan \theta^3\, d\theta$

38. $\int \theta \sec 2\theta^2\, d\theta$

39. $\int \sin (x + 1)\, dx$

40. $\int \cos (7x - 3)\, dx$

41. $\int \tan (4 - 5\theta)\, d\theta$

42. $\int \sec (2\theta + 3)\, d\theta$

43. $\int x \sec (4x^2 - 3)\, dx$

44. $\int 3x^2 \cot (8x^3 + 3)\, dx$

Miscellaneous Integrals from the Table

45. $\int \frac{(y + 2)\, dy}{y^2 + 4y}$

46. $\int \frac{ds}{\sqrt{s^2 - 16}}$

47. $\int \sqrt{25 - 9x^2}\, dx$

48. $\int \sqrt{4x^2 + 9}\, dx$

49. $\int \frac{dx}{x^2 + 9}$

50. $\int \frac{dx}{x^2 + 2x}$

51. $\int \frac{dx}{16x^2 + 9}$

52. $\int \frac{dx}{2 + 3x}$

53. $\int \frac{x^2\, dx}{2 + x^3}$

54. $\int \overline{1 + 9x^2}\, dx$

55. $\int \frac{dx}{x^2 - 4}$

56. $\int \frac{dy}{\sqrt{25 - y^2}}$

57. $\int \sqrt{\frac{x^2}{4} - 1}\, dx$

58. $\int \frac{x^2\, dx}{\sqrt{3x + 5}}$

59. $\int \frac{5x\, dx}{\sqrt{1 - x^4}}$

60. $\int x\sqrt{1 + 3x}\, dx$

61. $\int \frac{dt}{4 - 9t^2}$

62. $\int \frac{x\, dx}{3 - 4x}$

7-5. APPLICATIONS TO MOTION

Our main applications for the integral come in the next several chapters, but here we give two applications for the indefinite integral; motion and electric circuits.

Displacement and Velocity

In Sec. 6-2 we saw that the velocity v of a moving point was defined as the rate of change of the displacement s of the point. The velocity was thus equal to the derivative of the displacement, or $v = ds/dt$. We now reverse the process and find the displacement when given the velocity. Since $ds = v\,dt$, integrating gives

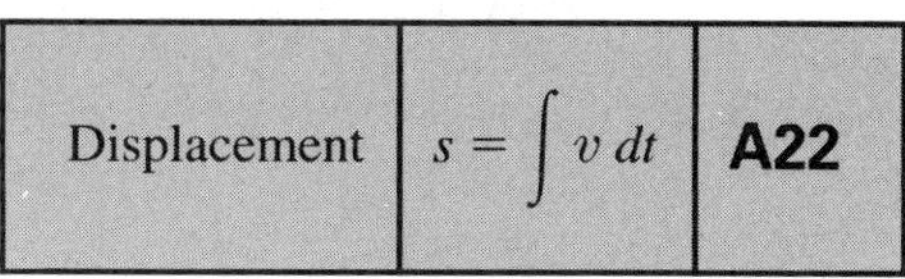

Similarly, the acceleration a is given by dv/dt, so $dv = a\,dt$. Integrating we get

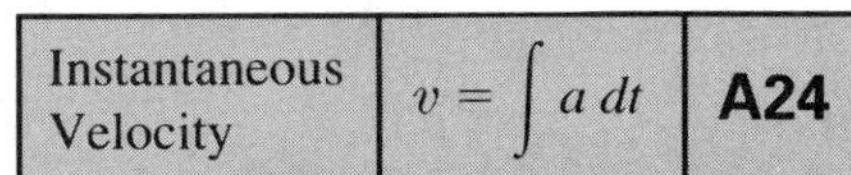

Thus, if given an equation for acceleration, we can integrate to get an equation for velocity, and integrate again to get displacement. We evaluate the constants of integration by substituting boundary conditions, as shown in the following example.

EXAMPLE 1: A particle moves with a constant acceleration of 4 ft/s^2. It has an initial velocity of 6 ft/s and an initial displacement of 2 ft. Find the equations for velocity and displacement, and graph the displacement, velocity, and acceleration for $t = 0$ to 10 s.

Solution: We are given $a = dv/dt = 4$, so

$$v = \int 4\,dt = 4t + C_1$$

Since $v = 6$ ft/s when $t = 0$, we get $C_1 = 6$, so

$$v = 4t + 6$$

is our equation for the velocity of the particle. Now since $v = ds/dt$,

$$ds = (4t + 6)\,dt$$

Integrating again gives us

$$s = \frac{4t^2}{2} + 6t + C_2$$

Since the initial displacement is 2 ft when $t = 0$, we get $C_2 = 2$. The complete equation for the displacement of the particle is then

$$s = 2t^2 + 6t + 2$$

The three curves are graphed one above the other in Fig. 7-5 with the same scale used for the horizontal axis of each. Note that each curve is the derivative of the one below it, and, conversely, each curve is the integral of the one above it.

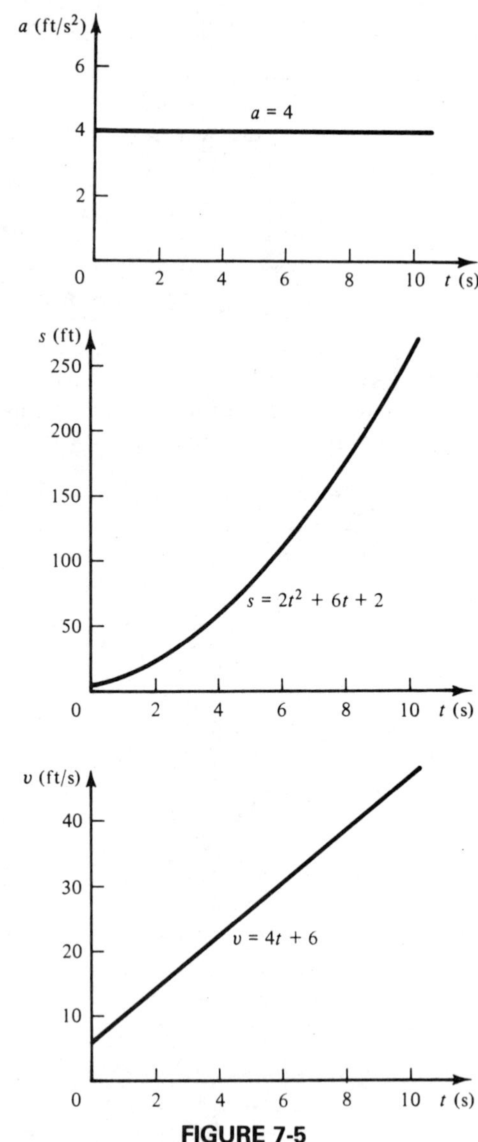

FIGURE 7-5

Freely Falling Body

Integration provides us with a slick way to derive the equations for the displacement and velocity of a freely falling body.

EXAMPLE 2: An object falls with constant acceleration due to gravity of g. Write the equations for the displacement and velocity of the body at any time.

Solution: We are given that $a = dv/dt = g$, so

$$dv = g\,dt$$

Integrating, we find that

$$v = \int g\,dt = gt + C_1$$

These two formulas are slightly different from Eqs. A18 and A19 given in Appendix A. There *a* is used instead of *g* to stand for any constant acceleration, and in Eq. A18, the initial displacement s_0 is assumed to be zero.

When $t = 0$, $v = C_1$, so C_1 is the initial velocity. Let us relabel it v_0.

$v = v_0 + gt$	**A19**

But $v = ds/dt$, so

$$ds = (v_0 + gt)\,dt$$

$$s = \int (v_0 + gt)\,dt = v_0t + \frac{gt^2}{2} + C_2$$

When $t = 0$, $s = C_2$, so we interpret C_2 as the initial displacement. Let us call it s_0. So the displacement is

$s = s_0 + v_0t + \dfrac{gt^2}{2}$	**A18**

Motion along a Curve

In Sec. 6-2 the motion of a point along a curve was described by parametric equations, with the x and y displacements each given as a separate function of time. We saw that dx/dt gave the velocity v_x in the x direction and that dy/dt gave the velocity v_y in the y direction. Now, given the velocities, we integrate to get the displacements.

Displacement in x and y Directions	(a) $x = \int v_x \, dt$	(b) $y = \int v_y \, dt$	**A32**

Similarly, if we have parametric equations for the accelerations in the x and y directions, we integrate to get the velocities.

Velocity in x and y Directions	(a) $v_x = \int a_x \, dt$	(b) $v_y = \int a_y \, dt$	**A34**

EXAMPLE 3: A point starts from (2, 4) with initial velocities $v_x = 7$ cm/s and $v_y = 5$ cm/s, and moves along a curved path. It has x and y accelerations of $a_x = 3t$ cm/s^2 and $a_y = 5$ cm/s^2. Write expressions for the x and y components of velocity and displacement.

Solution: We integrate to find the velocities,

$$v_x = \int 3t \, dt = \frac{3t^2}{2} + C_1 \qquad \text{and} \qquad v_y = \int 5 \, dt = 5t + C_2$$

At $t = 0$, $v_x = 7$ and $v_y = 5$, so

$$v_x = \frac{3t^2}{2} + 7 \text{ cm/s} \qquad \text{and} \qquad v_y = 5t + 5 \text{ cm/s}$$

Integrating again gives the displacements.

$$x = \int \left(\frac{3t^2}{2} + 7\right) dt \qquad \text{and} \qquad y = \int (5t + 5) \, dt$$

$$= \frac{t^3}{2} + 7t + C_3 \qquad \qquad = \frac{5t^2}{2} + 5t + C_4$$

At $t = 0$, $x = 2$ and $y = 4$, so our complete equations for the displacements are

$$x = \frac{t^3}{2} + 7t + 2 \text{ cm} \qquad \text{and} \qquad y = \frac{5t^2}{2} + 5t + 4 \text{ cm}$$

Rotation

In Sec. 6-2 we saw that the angular velocity ω of a rotating body was given by the derivative $d\theta/dt$ of the angular displacement θ. Thus θ is the integral of the angular velocity,

Angular Displacement	$\theta = \int \omega \, dt$	**A28**

Similarly, the angular velocity is the integral of the angular acceleration α,

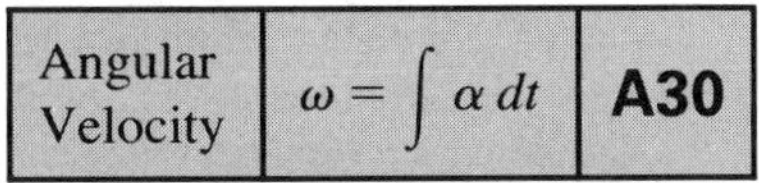

EXAMPLE 4: A flywheel in a machine starts from rest and accelerates at $3.85t$ rad/s^2. Find the angular velocity and the total number of revolutions after 10.0 s.

Solution: We integrate to get the angular velocity,

$$\omega = \int 3.85t \, dt = \frac{3.85t^2}{2} + C_1 \qquad \text{rad/s}$$

Since the flywheel starts from rest, $\omega = 0$ at $t = 0$, so $C_1 = 0$. Integrating again gives the angular displacement.

$$\theta = \int \frac{3.85t^2}{2} \, dt = \frac{3.85t^3}{6} + C_2 \qquad \text{rad}$$

Since θ is 0 at $t = 0$, we get $C_2 = 0$. Now evaluating ω and θ at $t = 10.0$ s,

$$\omega = \frac{3.85(10.0)^2}{2} = 193 \text{ rad/s}$$

and

$$\theta = \frac{3.85(10.0)^3}{6} = 642 \text{ rad} = 102 \text{ revolutions}$$

EXERCISE 5

Velocity and Acceleration

1. Find the relation between v and t if $v = 2$ m/s when $t = 3$s, if the acceleration $a = 4 - t^2$ m/s^2.

2. A car starts from rest and continues at a rate of $v = \frac{1}{8}t^2$ ft/s. Find the formula for the distance s from the rest position, when it has moved t seconds. How far will it go in 4 s?

3. A particle starts from the origin. Its x component of velocity is $t^2 - 4$ and its y component is $4t$ (cm/s). (a) Write equations for x and y and (b) find the distance between the particle and the origin when $t = 2$ s.

4. A body is moving at the rate $v = \frac{3}{2}t^2$ (m/s). Find the distance it will move in t seconds if $s = 0$ when $t = 0$.

5. Find the relation between s and t if $s = 0$ and $v = 20$ ft/s when $t = 0$ if the acceleration is $a = -32$ ft/s^2.

Motion along a Curve

6. A point starts from rest at the origin and moves along a curved path with x and y accelerations of $a_x = 2$ cm/s^2 and $a_y = 8t$ cm/s^2. Write expressions for the x and y components of velocity.

7. A point starts from rest at the origin and moves along a curved path with x and y accelerations of $a_x = 5t^2$ cm/s^2 and $a_y = 2t$ cm/s^2. Find the x and y components of velocity at $t = 10$ s.

8. A point starts from (5, 2) with initial velocities of $v_x = 2$ cm/s and $v_y = 4$ cm/s and moves along a curved path. It has x and y accelerations of $a_x = 7t$ and $a_y = 2$. Find the x and y displacements at $t = 5$ s.

9. A point starts from (9, 1) with initial velocities of $v_x = 6$ cm/s and $v_y = 2$ cm/s and moves along a curved path. It has x and y accelerations of $a_x = 3t$ and $a_y = 2t^2$. Find the x and y components of velocity at $t = 15$ s.

Rotation

10. A wheel starts from rest and accelerates at 3.00 rad/s^2. Find the angular velocity after 12.0 s.

11. A certain gear starts from rest and accelerates at $8.5t^2$ rad/s^2. Find the total number of revolutions after 20.0 s.

12. A link in a mechanism rotating with an angular velocity of 3.00 rad/s is given an acceleration of 5.00 rad/s^2 at $t = 0$. Find the angular velocity after 20.0 s.

13. A pulley in a magnetic tape drive is rotating at 1.25 rad/s when it is given an acceleration of 7.24 rad/s^2 at $t = 0$. Find the angular velocity at 2.00 s.

7-6. APPLICATION TO ELECTRIC CIRCUITS

Charge

We stated in Sec. 6-1 that the current i (amperes) at some point in a conductor was equal to the time rate of change of the charge q (coulombs) passing that point, or $i = dq/dt$. We can now solve this equation for q. Multiplying by dt gives $dq = i\,dt$. Integrating, we get

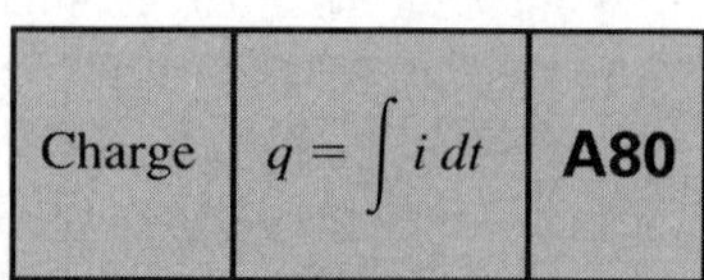

EXAMPLE 1: The current to a certain capacitor is given by $i = 2t^3 + t^2 + 3$. The initial charge on the capacitor is 6.83 coulombs (C). Find (a) an expression for the charge on the capacitor and (b) the charge when $t = 5.00$ s.

Solution: (a) Integrating the expression for current,

$$q = \int i\,dt = \int (2t^3 + t^2 + 3)\,dt$$

$$= \frac{t^4}{2} + \frac{t^3}{3} + 3t + C \qquad \text{coulombs}$$

We find the constant of integration by substituting the initial conditions, $q = 6.83$ C at $t = 0$. So $C = 6.83$ C. Our complete equation is then

$$q = \frac{t^4}{2} + \frac{t^3}{3} + 3t + 6.83 \qquad \text{coulombs}$$

(b) When $t = 5.00$ s,

$$q = \frac{(5.00)^4}{2} + \frac{(5.00)^3}{3} + 3(5.00) + 6.83 = 376 \text{ coulombs}$$

Voltage across a Capacitor

The current in a capacitor has already been given by Eq. A81, $i = C\,dv/dt$, where i is in amperes, C in farads, v in volts, and t in seconds. We now integrate to find the voltage across the capacitor.

$$dv = \frac{1}{C}\,i\,dt$$

Voltage across a Capacitor	$v = \dfrac{1}{C}\displaystyle\int i\,dt$ volts	**A82**

EXAMPLE 2: A 1.25-F capacitor that has an initial voltage of 25.0 V is charged with a current that varies with time according to the equation $i = t\sqrt{t^2 + 6.83}$. Find the voltage across the capacitor at 1.00 s.

Solution: By Eq. A82,

$$v = \frac{1}{1.25}\int t\sqrt{t^2 + 6.83}\,dt = 0.80\left(\frac{1}{2}\right)\int (t^2 + 6.83)^{1/2}(2t\,dt)$$

$$v = \frac{0.40(t^2 + 6.83)^{3/2}}{3/2} + k = 0.267(t^2 + 6.83)^{3/2} + k$$

where we have used k for the constant of integration to avoid confusion with the symbol for capacitance. Since $v = 25.0$ V when $t = 0$, we get

$$k = 25.0 - 0.267\,(6.83)^{3/2} = 20.2 \text{ V}$$

When $t = 1.00$ s,

$$v = 0.267(1.00^2 + 6.83)^{3/2} + 20.2 = 26.0 \text{ V}$$

Current in an Inductor

The voltage across an inductor was given by Eq. A86 as $v = L\,di/dt$, where L is the inductance in henrys. From this we get

Current in an Inductor	$i = \dfrac{1}{L}\displaystyle\int v\,dt$ amperes	**A85**

EXAMPLE 3: The voltage across a 10.6-H inductor is $v = \sqrt{3t + 25.4}$ V. Find the current in the inductor at 5.25 s if the initial current is 6.15 A.

Solution: From Eq. A85,

$$i = \frac{1}{10.6}\int \sqrt{3t + 25.4}\,dt = 0.0943\left(\frac{1}{3}\right)\int (3t + 25.4)^{1/2}(3\,dt)$$

$$= \frac{0.0314\,(3t + 25.4)^{3/2}}{\frac{3}{2}} + C = 0.0210(3t + 25.4)^{3/2} + C$$

When $t = 0$, $i = 6.15$ A, so

$$C = 6.15 - 0.0210\,(25.4)^{3/2} = 3.46 \text{ A}$$

When $t = 5.25$ s,

$$i = 0.0314[3(5.25) + 25.4]^{3/2} + 3.46 = 11.8 \text{ A}$$

EXERCISE 6

1. The current to a capacitor is given by $i = 2t + 3$. The initial charge on the capacitor is 8.13 C. Find the charge when $t = 1.00$ s.

2. The current to a certain circuit is given by $i = t^2 + 4$. If the initial charge is zero, find the charge at 2.50 s.

3. The current to a certain capacitor is $i = 3.25 + t^3$. If the initial charge on the capacitor is 16.8 C, find the charge when $t = 3.75$ s.

4. A 21.5-F capacitor with zero initial voltage has a charging current of $i = \sqrt{t}$. Find the voltage across the capacitor at 2.00 s.

5. A 15.2-F capacitor has an initial voltage of 2.00 V. It is charged with a current given by $i = t\sqrt{5 + t^2}$. Find the voltage across the capacitor at 1.75 s.

6. A 75.0-μF capacitor has an initial voltage of 125 V and is charged with a current equal to $i = \sqrt{t + 16.3}$. Find the voltage across the capacitor at 4.00 s.

7. The voltage across a 1.05-H inductor is $v = \sqrt{23t}$ V. Find the current in the inductor at 1.25 s if the initial current is zero.

8. The voltage across a 52.0-H inductor is $v = t^2 - 3t$ V. If the initial current is 2.00 A, find the current in the inductor at 1.00 s.

9. The voltage across a 15.0-H inductor is given by $v = 28.5 + \sqrt{6t}$ V. Find the current in the inductor at 2.50 s if the initial current is 15.0 A.

CHAPTER TEST

Integrate.

1. $\int \tan 3\theta \, d\theta$

2. $\int \cot 5\theta \, d\theta$

3. $\int a^{2x} \, dx$

4. $\int \frac{dx}{\sqrt[3]{x}}$

5. $\int (e^{5x} + a^{5x})dx$

6. $\int (e^x + 4)e^{-x} \, dx$

7. $\int (e^{2x+1} + x)dx$

8. $\int 6e^{3x} \, dx$

9. $\int \frac{x^4 + x^3 + 1}{x^3} \, dx$

10. $\int e^{5x} \, dx$

11. $\int \frac{dx}{e^x}$

12. $\int \csc^2(3x + 2) \, dx$

13. $\int \csc^2 3x \, dx$

14. $\int 3.1 \, y^2 \, dy$

15. $\int \frac{2dt}{t^2}$

16. $\int \sqrt{4x} \, dx$

17. $\int \sec 2\theta \, d\theta$

18. $\int \tan^2 5\theta \, d\theta$

19. $\int x^2(x^3 - 4)^2 \, dx$

20. $\int (x^4 - 2x^3)(2x^3 - 3x^2) \, dx$

21. $\int \frac{x \, dx}{x^2 + 3}$

22. $\int \frac{dx}{x + 5}$

23. Find the equation of a curve that passes through the point (3,0), has a slope of 9 at that point, and has a second derivative $y'' = x$.

24. The rate of growth of the number N of bacteria in a culture is $dN/dt = 0.5N$. If $N = 100$ when $t = 0$, derive the formula for N at any time.

25. The acceleration of an object that starts from rest is given by $a = 3t$. Write equations for the velocity and displacement of the object.

26. The voltage across a 25.0-H inductor is given by $v = 8.9 + \sqrt{3t}$ V. Find the current in the inductor at 5.00 s if the initial current is 1.00 A.

27. A flywheel starts from rest and accelerates at $7.25t^2$ rad/s^2. Find the angular velocity and the total number of revolutions after 20.0 s.

28. The current to a certain capacitor is $i = t^3 + 18.5$. If the initial charge on the capacitor is 6.84 C, find the charge when $t = 5.25$ s.

29. A point starts from (1, 1) with initial velocities of $v_x = 4$ cm/s and $v_y = 15$ cm/s, and moves along a curved path. It has x and y accelerations of $a_x = t$ and $a_y = 5t$. Write expressions for the x and y components of velocity and displacement.

30. A 15.0-F capacitor has an initial voltage of 25 V and is charged with a current equal to $i = \sqrt{4t + 21.6}$. Find the voltage across the capacitor at 14.00 s.

The Definite Integral

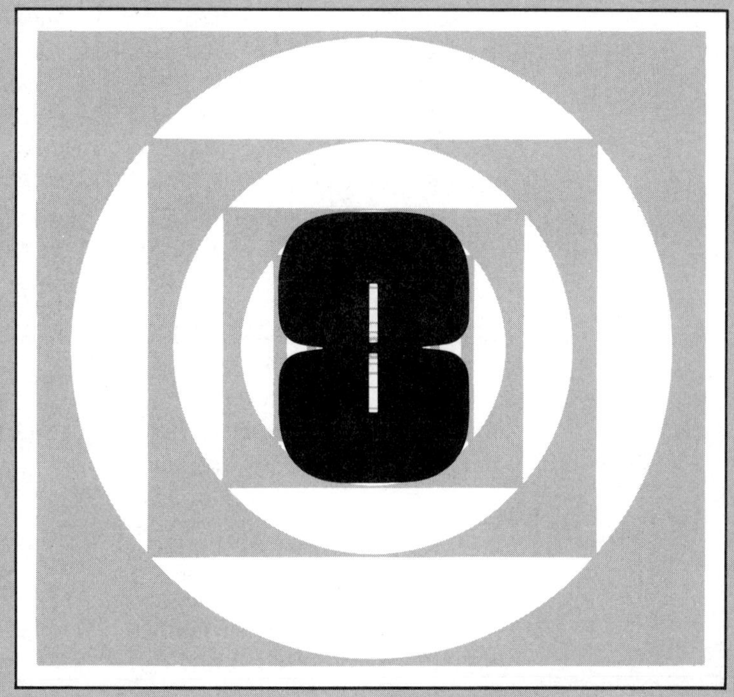

In this chapter we first define the *definite integral,* and then show how to evaluate it. Next we discuss the problem of finding the area bounded by a curve and the x axis, between two given values of x. We find such areas, first approximately by the *midpoint method,* and then exactly by means of the definite integral. In the process we develop the *fundamental theorem of calculus,* which ties together the derivative, the integral, and the area under a curve. Finally, learn a fast way to set up the integral for finding areas, which will be of great use in later chapters for finding other quantities, such as volumes, by integration.

8-1. THE DEFINITE INTEGRAL

In Chapter 7 we learned how to find the indefinite integral, or antiderivative, of a function. For example,

$$\int x^2\,dx = \frac{x^3}{3} + C \tag{1}$$

We can, of course, evaluate the antiderivative at some particular value, say, $x = 6$. Substituting into (1),

$$\int x^2\,dx\bigg|_{x=6} = \frac{6^3}{3} + C = 72 + C$$

Similarly, we can evaluate the same integral at, say, $x = 3$. Again substituting into (1),

$$\int x^2\,dx\bigg|_{x=3} = \frac{3^3}{3} + C = 9 + C$$

Suppose, now, that we subtract the second integral from the first. We get

$$72 + C - 9 - C = 63$$

Although we don't know the value of the constant C, we do know that it has the same value in both integrals, since both were obtained from (1), so C will drop out when we subtract.

We now introduce new notation to replace the left side of this equation. Thus

$$\int x^2\,dx\bigg|_{x=6} - \int x^2\,dx\bigg|_{x=3} = \int_3^6 x^2\,dx$$

But what is this number, and what is it good for? We'll soon see that it gives us the area under the curve $y = x^2$, from $x = 3$ to $x = 6$, and has lots of applications.

is called a *definite integral.* Here 6 is called the *upper limit* and 3 is the *lower limit.* This notation tells us to evaluate the antiderivative at the upper limit, and from that, subtract the antiderivative evaluated at the lower limit. Notice that a definite integral (unlike the indefinite integral) has a *numerical* value, in this case, 63.

In general, if

$$\int f(x)\,dx = F(x) + C$$

then

$$\int_a^b f(x)\,dx = F(b) + C - F(a) - C$$

We require, as usual, that the function $f(x)$ be continuous in the interval under consideration.

The constants drop out, leaving

Definite Integral	$\int_a^b f(x)\,dx = F(b) - F(a)$	**339**

The definite integral of a function is equal to the antiderivative of that function evaluated at the upper limit b minus the antiderivative evaluated at the lower limit a.

Evaluating a Definite Integral

To *evaluate* a definite integral, first integrate the expression (omitting the constant of integration), and write the upper and lower limits on a vertical bar or bracket to the right of the integral. Then substitute the upper limit, then the lower limit, and subtract.

EXAMPLE 1: Evaluate $\int_2^4 x^2\,dx$.

Solution:

$$\int_2^4 x^2\,dx = \frac{x^3}{3}\bigg|_2^4$$

$$= \frac{4^3}{3} - \frac{2^3}{3} = \frac{56}{3}$$

EXAMPLE 2: Evaluate $\int_0^{\pi/2} \sin 2x\,dx$.

Solution:

$$\int_0^{\pi/2} \sin 2x\,dx = \frac{1}{2}\int_0^{\pi/2} \sin 2x(2\,dx)$$

By rule 10,

$$= -\frac{1}{2}\cos 2x\bigg|_0^{\pi/2}$$

$$= -\frac{1}{2}\left[\cos 2\left(\frac{\pi}{2}\right) - \cos 0\right] = 1$$

Common Error	Don't assume that an integral is zero when the limit is zero. Work it out.

EXAMPLE 3: Evaluate $\int_1^2 e^{3x}\,dx$.

Solution:

$$\int_1^2 e^{3x}\,dx = \frac{1}{3}\int_1^2 e^{3x}\,(3\,dx)$$

$$= \frac{1}{3}e^{3x}\bigg|_1^2$$

$$= \frac{1}{3}[e^6 - e^3] \cong 128$$

By rule 8.

Integrals with Absolute Value Signs

Integrals will often contain absolute value signs, as in the following example.

EXAMPLE 4

$$\int_{-3}^{-2} \frac{dx}{x} = \ln|x| \Big|_{-3}^{-2} = \ln|-2| - \ln|-3|$$

The logarithm of a negative number is not defined. But here we are taking the logarithm of the *absolute value* of a negative number. Thus

$$\ln|-2| - \ln|-3| = \ln 2 - \ln 3 \cong -0.405$$

Continuity

If a function is *discontinuous* between two limits a and b, the definite integral is not defined over that interval.

EXAMPLE 5: The integral $\int_{-3}^{2} \frac{dx}{x}$ is not defined, because the function $y = 1/x$ is discontinuous at $x = 0$.

Common Error	Be sure that your function is continuous between the given limits before evaluating a definite integral.

EXERCISE 1

Evaluate each definite integral.

1. $\int_{1}^{2} x\,dx$
2. $\int_{-2}^{2} x^2\,dx$
3. $\int_{1}^{3} 7x^2\,dx$
4. $\int_{-2}^{2} 3x^4\,dx$
5. $\int_{0}^{4} (x^2 + 2x)\,dx$
6. $\int_{-2}^{2} x^2(x + 2)\,dx$
7. $\int_{2}^{4} (x + 3)^2\,dx$
8. $\int_{0}^{a} (a^2x - x^3)\,dx$
9. $\int_{1}^{10} \frac{dx}{x}$
10. $\int_{1}^{e} \frac{dx}{x}$
11. $\int_{0}^{1} \frac{x\,dx}{4 + x^2}$
12. $\int_{-2}^{-3} \frac{2t\,dt}{1 + t^2}$
13. $\int_{0}^{1} \frac{dx}{\sqrt{3 - 2x}}$
14. $\int_{0}^{1} \frac{x\,dx}{\sqrt{2 - x^2}}$

15. $\int_0^1 xe^{x^2}\,dx$

16. $\int_0^1 \frac{dx}{e^{3x}}$

17. $\int_0^{\pi} \sin\phi\,d\phi$

18. $\int_0^{\pi/2} \cos\phi\,d\phi$

19. $\int_0^{\pi} \cos\frac{\theta}{2}\,d\theta$

20. $\int_{\pi/3}^{\pi/2} \sin^2 x\cos x\,dx$

8-2. AREA UNDER A CURVE

Summation Notation

Before we start deriving an expression for the area under a curve, we must learn some new notation to express the sum of a string of terms. We use the capital Greek sigma Σ to stand for summation, or adding up. Thus

$$\Sigma\, n$$

means to sum a string of n's. Of course, we must indicate a starting and ending value for n, and these values are placed on the sigma symbol. Thus

$$\sum_{n=1}^{5} n$$

means to add up the n's starting with $n = 1$ and ending with $n = 5$.

$$\sum_{n=1}^{5} n = 1 + 2 + 3 + 4 + 5 = 15$$

EXAMPLE 1: Evaluate $\sum_{n=1}^{4} (n^2 - 1)$.

Solution: $\sum_{n=1}^{4} (n^2 - 1) = (1^2 - 1) + (2^2 - 1) + (3^2 - 1) + (4^2 - 1)$

$$= 0 + 3 + 8 + 15 = 26$$

EXAMPLE 2

(a) $\sum_{k=2}^{5} k^2 = 2^2 + 3^2 + 4^2 + 5^2 = 54$

(b) $\sum_{x=1}^{4} f(x) = f(1) + f(2) + f(3) + f(4)$

(c) $\sum_{i=1}^{n} f(x_i) = f(x_1) + f(x_2) + f(x_3) + \cdots + f(x_n)$

In the following section, we use the sigma notation for expressions similar to Example 2(c).

Approximate Area under a Curve

Figure 8-1 shows a graph of some function $f(x)$. Our problem is to find the area (shown lightly shaded) bounded by that curve, the x axis, and the lines $x = a$ and $x = b$.

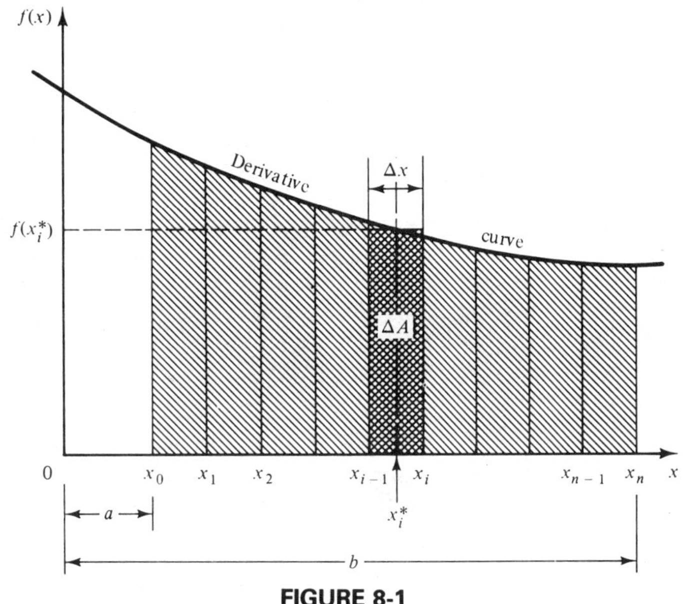

FIGURE 8-1

We start by subdividing that area into n vertical strips, called *panels*, by drawing vertical lines at $x_0, x_1, x_2, \ldots, x_n$. The panels do not have to be of equal width, but we make them equal for simplicity. Let the width of each panel be Δx.

Now look at one particular panel, the one lying between x_{i-1} and x_i (shown shaded darkly). Anywhere within this panel we choose a point x_i^*. The height of the curve at this value of x is then $f(x_i^*)$. The area ΔA of the dark panel is then *approximately* equal to the area of a rectangle of width Δx and height $f(x_i^*)$.

$$\Delta A \cong f(x_i^*)\,\Delta x$$

Our approximate area may be greater than or less than the actual area ΔA, depending on where we chose x_i^*. (Later we show that x_i^* is chosen in a particular place, but for now, consider it to be anywhere between x_{i-1} and x_i.)

The area of the first panel is, similarly, $f(x_i^*)\ \Delta x$; of the second panel, $f(x_2^*)\ \Delta x$; and so on. To get an approximate value for the total area, we add up the areas of each panel.

$$A \cong f(x_i^*)\ \Delta x + f(x_2^*)\ \Delta x + f(x_3^*)\ \Delta x + \cdots + f(x_n^*)\ \Delta x$$

These are called *Riemann sums,* after Georg Friedrich Bernhard Riemann (1826–1866).

Rewriting this expression using our sigma notation gives

$$A \cong \sum_{i=1}^{n} f(x_i^*)\ \Delta x$$

Midpoint Method

If we approximate the area under a curve by rectangular panels of width Δx and choose our values of x^* at the *midpoints* of these panels, we are using what is often called the *midpoint method.*

In this chapter we use the midpoint method mainly as a lead-in to finding areas by integration. But it is also a valuable method for finding areas for functions that cannot be exactly integrated. Other standard approximation methods are given in Sec. 11-6.

Midpoint Method	$A = \sum_{i=1}^{n} f(x_i^*)\ \Delta x$ where $f(x_i^*)$ is the height of the ith panel at its midpoint.	**345**

EXAMPLE 3: Use the midpoint method to calculate the approximate area under the curve $f(x) = 3x^2$, from $x = 0$ to $x = 10$, taking panels of width 2.

Solution: Our graph (Fig. 8-2) shows the panels, with midpoints at 1, 3, 5, 7, and 9. At each midpoint x^* we compute the height $f(x^*)$ of the curve.

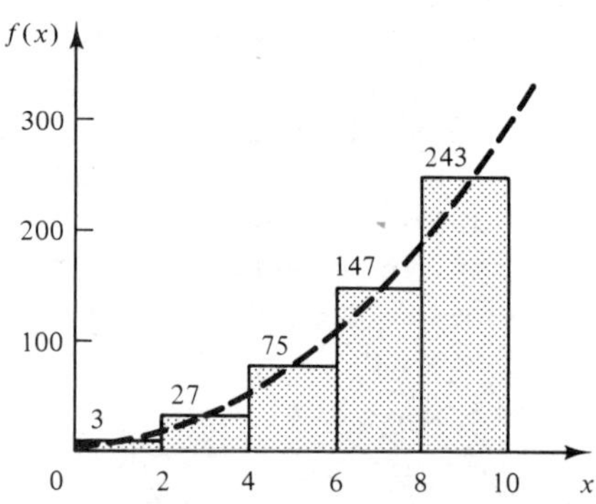

FIGURE 8-2. Midpoint method.

x^*	1	3	5	7	9
$f(x^*)$	3	27	75	147	243

We'll see later that the exact area is 1000 square units.

The approximate area is then the sum of the areas of each panel.

$$A \cong 3(2) + 27(2) + 75(2) + 147(2) + 243(2)$$
$$= 495(2) = 990 \text{ square units}$$

Exact Area under a Curve

We can obtain greater accuracy in computing the area under a curve simply by reducing the width of each rectangular panel. Clearly, the panels in Fig. 8-3b are a better fit to the curve than those in Fig. 8-3a. Thus, as the panel width Δx approaches zero (and the number of panels approaches infinity) the sum of the areas of the panels approaches the exact area A under the curve.

Exact Area under a Curve	$A = \lim_{\Delta x \to 0} \sum_{i=1}^{n} f(x_i^*) \, \Delta x$	**340**

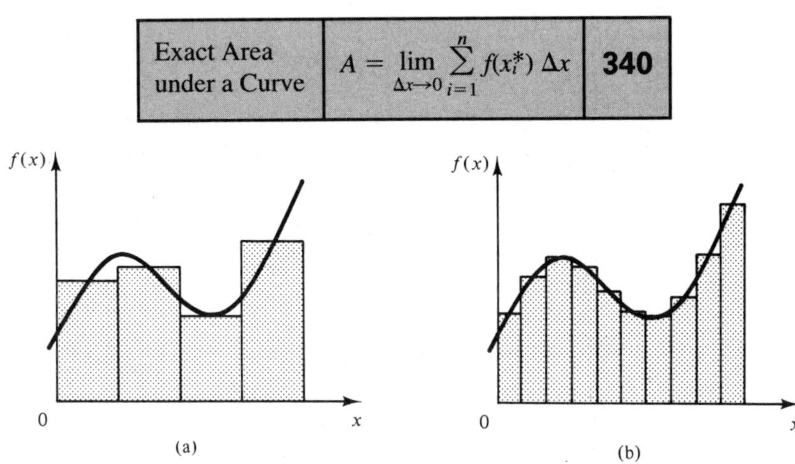

FIGURE 8-3. More panels give greater accuracy.

EXAMPLE 4: Compute the area under the curve in Example 3 by the midpoint method, using panel widths of 2, 1, $\frac{1}{2}$, $\frac{1}{4}$, and so on.

Solution: We compute the approximate area just as in Example 3. We omit the tedious computations (which were done by computer) and show only the results.

Width	*Area*
2.0000	990.0000
1.0000	997.5000
0.5000	999.3750
0.2500	999.8438
0.1250	999.9609
0.0625	999.9902
0.0313	999.9968
0.0156	999.9996

Notice that as the panel width decreases, the computed area seems to be approaching a limit of 1000. We'll see in the next section that 1000 is the exact area under the curve.

The Fundamental Theorem

How, then, do we find the exact area under a curve? Must we compute the areas of hundreds of rectangular panels and add them up, as in Example 4? No. We can let the process of integration add them up for us, as we'll now see.

Let us return to Fig. 8-1, but now we draw above it the curve $F(x)$ (Fig. 8-4). Thus the lower curve $f(x)$ is the derivative of the upper curve $F(x)$, and conversely, the upper curve $F(x)$ is the integral of the lower curve $f(x)$.

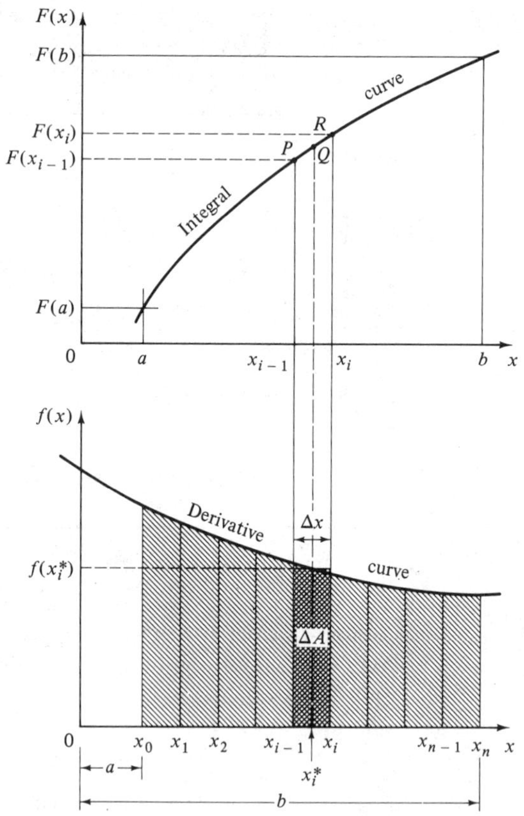

FIGURE 8-4. Area as the limit of a sum.

FIGURE 8-5

For the midpoint method, we arbitrarily selected x_i^* at the midpoint of each panel. We now do it differently. *We select x_i^* so that the slope at point Q on the integral (upper) curve is equal to the slope of the straight line PR* (Fig. 8-5).

There is a theorem, called the *mean value theorem*, that says there must be at least one point *Q*, between *P* and *R*, at which the slope is equal to the slope of *PR*. We won't prove it, but can you see intuitively that it must be so?

The slope at Q is equal to $f(x_i^*)$, and the slope of PR is equal to

$$f(x_i^*) = \frac{\text{rise}}{\text{run}} = \frac{F(x_i) - F(x_{i-1})}{\Delta x}$$

or

$$f(x_i^*)\,\Delta x = F(x_i) - F(x_{i-1})$$

If we write this expression for each panel, we get

$$f(x_1^*)\,\Delta x = F(x_1) - F(a)$$
$$f(x_2^*)\,\Delta x = F(x_2) - F(x_1)$$
$$f(x_3^*)\,\Delta x = F(x_3) - F(x_2)$$
$$\vdots \qquad\qquad \vdots$$
$$f(x_n^*\,\Delta x = F(b) - F(x_{n-1})$$

If we add all these equations, every term on the right drops out except $F(a)$ and $F(b)$.

$$f(x_1^*)\,\Delta x + f(x_2^*)\,\Delta x + f(x_3^*)\,\Delta x + \cdots + f(x_n^*)\,\Delta x = F(b) - F(a)$$

$$\sum_{i=1}^{n} f(x_i^*)\,\Delta x = F(b) - F(a)$$

As before, we let x approach zero.

$$\lim_{\Delta x \to 0} \sum_{i=1}^{n} f(x_i^*)\,\Delta x = F(b) - F(a)$$

The left side of this equation is equal to the exact area A under the curve. The right side is equal to the definite integral from a to b of the function $f(x)$. Thus we get

Fundamental Theorem of Calculus	$A = \int_a^b f(x)\,dx = F(b) - F(a)$	**339**

This result is so important that it is called the *Fundamental Theorem of Calculus*. It gives the amazingly simple result that the area under a curve is equal to the change in the integral between the same limits, as shown graphically in Fig. 8-6.

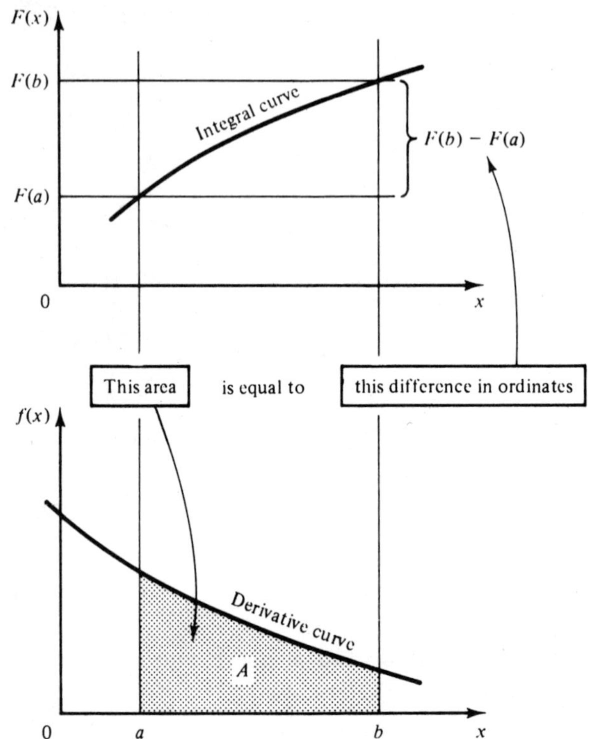

FIGURE 8-6. Illustration of the fundamental theorem.

EXAMPLE 5: Find the area bounded by the curve $y = 3x^2 + x + 1$, the x axis, and the lines $x = 2$ and $x = 5$.

Solution: By Eq. 339,

$$A = \int_2^5 (3x^2 + x + 1)\,dx = x^3 + \frac{x^2}{2} + x \Big|_2^5$$

$$= \left(5^3 + \frac{5^2}{2} + 5\right) - \left(2^3 + \frac{2^2}{2} + 2\right)$$

$$\cong 130.5 \text{ square units}$$

EXERCISE 2

Sigma Notation

Evaluate each expression.

1. $\sum_{n=1}^{5} n$

2. $\sum_{r=1}^{9} r^2$

3. $\sum_{n=1}^{7} 3n$

4. $\sum_{m=1}^{4} \frac{1}{m}$

5. $\sum_{n=1}^{5} n(n-1)$

6. $\sum_{q=1}^{6} \frac{q}{q+1}$

Approximate Areas by Midpoint Method

Find the approximate area under each curve by the midpoint method, using panels 2 units wide.

7. $y = x^2 + 1$ from $x = 0$ to 8.

8. $y = x^2 + 3$ from $x = -4$ to 4.

9. $y = \frac{1}{x}$ from $x = 2$ to 10.

10. $y = 2 + x^4$ from $x = -10$ to 0.

Exact Areas by Integration

Find the area bounded by each curve, the given lines, and the x axis.

11. $y = 2x$ from $x = 0$ to $x = 10$

12. $y = x^2 + 1$ from $x = 1$ to 20

13. $y = 3 + x^2$ from $x = -5$ to 5

14. $y = x^4 + 4$ from $x = -10$ to -2

15. $y = x^3$ from $x = 0$ to $x = 4$

16. $y = 9 - x^2$ from $x = 0$ to $x = 3$

17. $y = 1/\sqrt{x}$ from $x = \frac{1}{2}$ to $x = 8$

18. $y = x^3 + 3x^2 + 2x + 10$ from $x = -3$ to $x = 3$

19. $y = x^2 + x + 1$ from $x = 2$ to $x = 3$

20. $y = \sqrt{3x}$ from $x = 2$ to $x = 8$

21. $y = 2x + \frac{1}{x^2}$ from $x = 1$ to $x = 4$

22. $y = \frac{10}{\sqrt{x + 4}}$ from $x = 0$ to $x = 5$

Computer

23. Write a program or use a spreadsheet to compute areas using the midpoint method, and test it on any of the problems in this exercise. Have the program accept any panel width as input, and print the panel width and the approximate area.

8-3. FINDING AREAS BY MEANS OF THE DEFINITE INTEGRAL

A Fast Way to Set Up the Integral

In Section 8-2 we have found areas bounded by simple curves and the x axis. In this section we go on to find areas bounded by the y, rather than the x axis, and areas *between* curves. Later, we find volumes, surface areas, and so forth, where Eq. 339 cannot be used directly. We will have to set up a *different integral each time*. Looking back at the work we had to do to get Eq. 339 makes us wish for an easier way to set up an integral. We have shown in Sec. 8-2 that the definite integral can be thought of as the sum of many small elements of area. This provides us with an intuitive shortcut for setting up an integral, without having to go through a long derivation each time.

Think of the integral sign as an "S," standing for *sum*. It indicates that we are to *add up* the elements that are written after the integral sign. Thus

$$A = \int_a^b f(x)\,dx$$

can be read: "The area A is the sum of all the elements having a height $f(x)$ and a width dx, between the limits of a and b," as shown in Fig. 8-7.

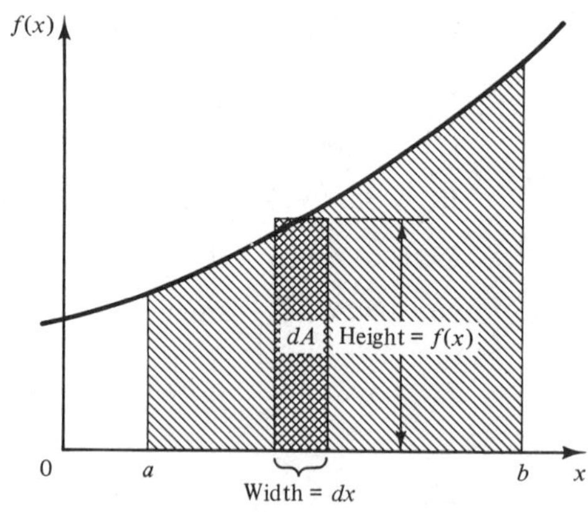

FIGURE 8-7

EXAMPLE 1: Find the area bounded by the curve $y = x^2 + 3$, the x axis, and the lines $x = 1$ and $x = 4$:

Solution: The usual steps are as follows:

1. Make a sketch showing the bounded area, as in Fig. 8-8. Locate a point (x, y) on the curve.

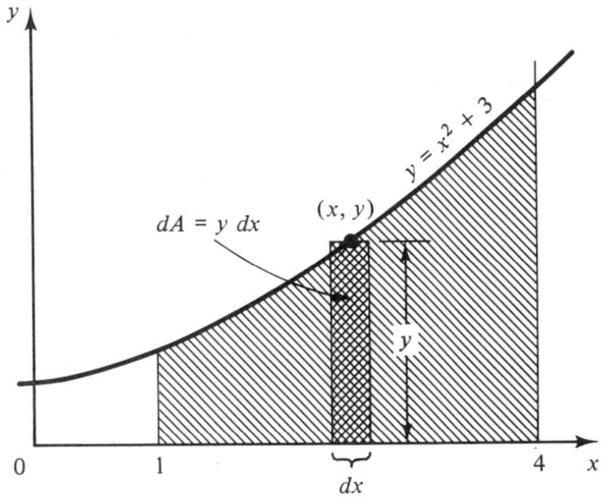

FIGURE 8-8

2. Through (x, y) draw a rectangular element of area, which we call dA. Give the rectangle dimensions. We call the width dx, because it is measured in the x direction, and we call the height y. The area of the element is thus

$$\begin{aligned} dA &= y\,dx \\ &= (x^2 + 3)\,dx \end{aligned}$$

3. We think of A as being the summation of all the small dA's, which we accomplish by integration.

$$A = \int dA = \int (x^2 + 3)\,dx$$

This looks like a long procedure, but it will save us a great deal of time later.

4. We locate the limits from the figure. Since we are summing the elements in the x direction, our limits must be on x. It is clear that we start the summing at $x = 1$ and end at $x = 4$. So

$$A = \int_1^4 (x^2 + 3)\,dx$$

5. Check that all parts of the integral, including the integrand, the differential, and the limits of integration, are in terms of the *same variable*. In our example, everything is in terms of x, and the limits are on x, so we can proceed. If, however, our integral contained both x and y, one of the variables would have to be eliminated.

6. Our integral is now set up and we evaluate it by Eq. 339.

$$\begin{aligned} A &= \int_1^4 (x^2 + 3)\,dx = \frac{x^3}{3} + 3x \Big|_1^4 \\ &= \frac{4^3}{3} + 3(4) - \left[\frac{1^3}{3} + 3(1)\right] = 30 \text{ square units} \end{aligned}$$

Area between Two Curves

Either of the curves $f(x)$ or $g(x)$ can, of course, be the x axis.

Suppose that we want the area A bounded by an upper curve $y = f(x)$ and a lower curve $y = g(x)$, between the limits a and b (Fig. 8-9). We draw a vertical element whose width is dx, whose height is $f(x) - g(x)$, and whose area dA is

$$dA = [f(x) - g(x)]\, dx$$

Integrating from a to b gives the total area

$$A = \int_a^b [f(x) - g(x)]\, dx$$

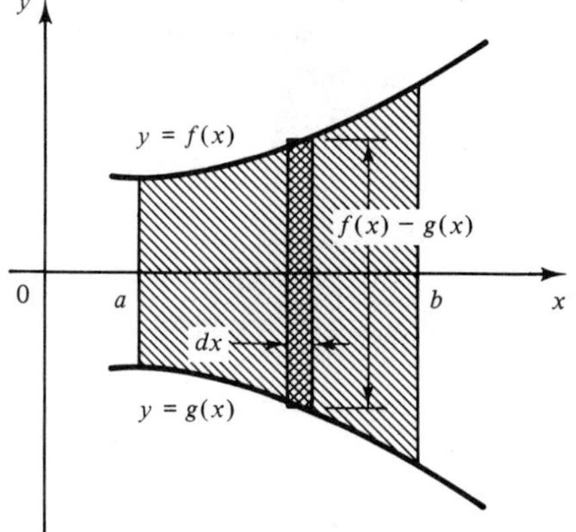

FIGURE 8-9. Area between two curves.

EXAMPLE 2: Find the area bounded by the curves $y = \sqrt{x}$ and $y = x - 3$, between $x = 1$ and $x = 4$ (Fig. 8-10).

Solution: Letting $f(x) = \sqrt{x}$ and $g(x) = x - 3$, we get

$$A = \int_a^b [f(x) - g(x)]\, dx = \int_1^4 [\sqrt{x} - x + 3]\, dx$$

$$= \left|\frac{2x^{3/2}}{3} - \frac{x^2}{2} + 3x\right|_1^4 = 6\tfrac{1}{6}$$

Limits Not Given

If we must find the area bounded by two curves and the limits a and b are not given, we must solve the given equations simultaneously to find their points of intersection.

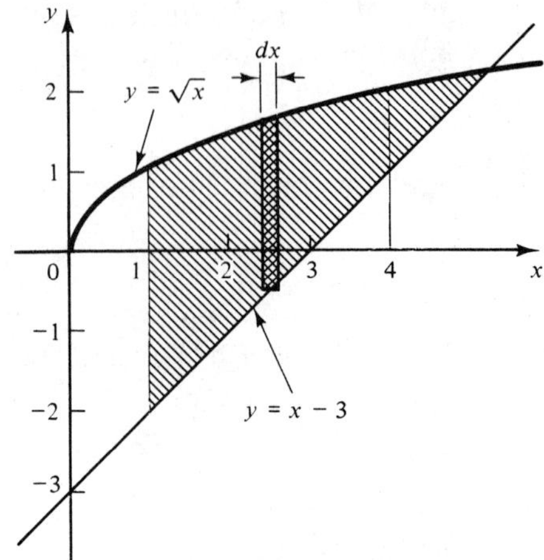

FIGURE 8-10

EXAMPLE 3: Find the area bounded by the parabola $y = 2 - x^2$ and the straight line $y = x$.

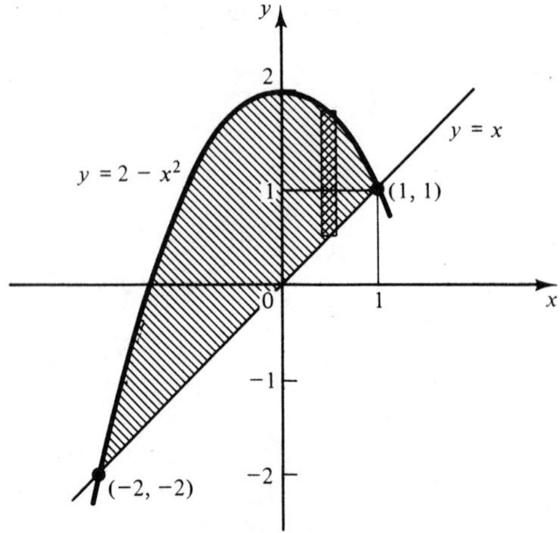

FIGURE 8-11

Solution: Our sketch (Fig. 8-11) shows two points of intersection. To find them, let us set one equation equal to the other,

$$2 - x^2 = x$$

or $x^2 + x - 2 = 0$. Factoring gives

$$(x + 2)(x - 1) = 0$$

so $x = -2$ and $x = 1$. Substituting back gives the points of intersection $(1, 1)$ and $(-2, -2)$.

We draw a vertical element of width dx and whose height is the upper curve minus the lower, or

$$2 - x^2 - x$$

The area dA of the strip is then

$$dA = (2 - x^2 - x)\, dx$$

We integrate, taking as limits the values of x (-2 and 1) found earlier by simultaneous solution of the given equations.

$$A = \int_{-2}^{1} (2 - x^2 - x)\, dx = 2x - \frac{x^3}{3} - \frac{x^2}{2}\bigg|_{-2}^{1} = 4\tfrac{1}{2}$$

Several Regions

Sometimes the given curves may cross in several places, or the limits may be specified in such a way as to define *more than one region,* as in Fig. 8-12. In such cases we find the area of each region separately, and then add them. Note that one of the given curves may be "upper" for one region and "lower" for another.

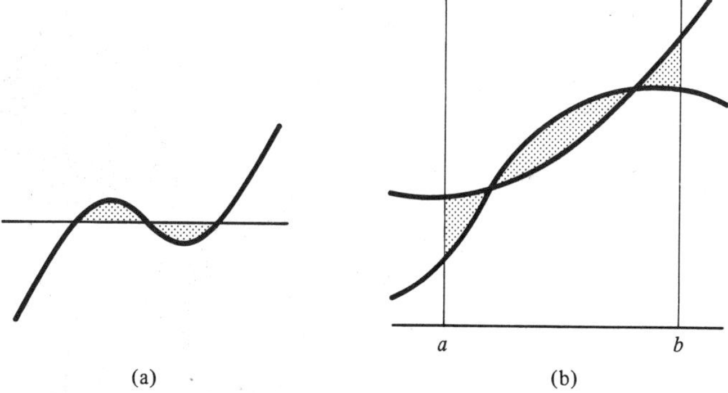

FIGURE 8-12. Intersecting curves may bound more than one region.

EXAMPLE 4: Find the area bounded by the curve $y = x^2 - 4$ and the x axis, between $x = 1$ and $x = 3$.

Solution: Our sketch (Fig. 8-13) shows two regions. For the first region the upper curve is $y = 0$ and the lower is $y = x^2 - 4$. The reverse is true for the second region. We set up two separate integrals,

$$A_1 = \int_{1}^{2} [0 - (x^2 - 4)]\, dx = 4x - \frac{x^3}{3}\bigg|_{1}^{2} = \frac{5}{3}$$

$$A_2 = \int_{2}^{3} [(x^2 - 4) - 0]\, dx = \frac{x^3}{3} - 4x\bigg|_{2}^{3} = \frac{7}{3}$$

Adding,

$$A = A_1 + A_2 = \frac{5}{3} + \frac{7}{3} = 4 \text{ square units}$$

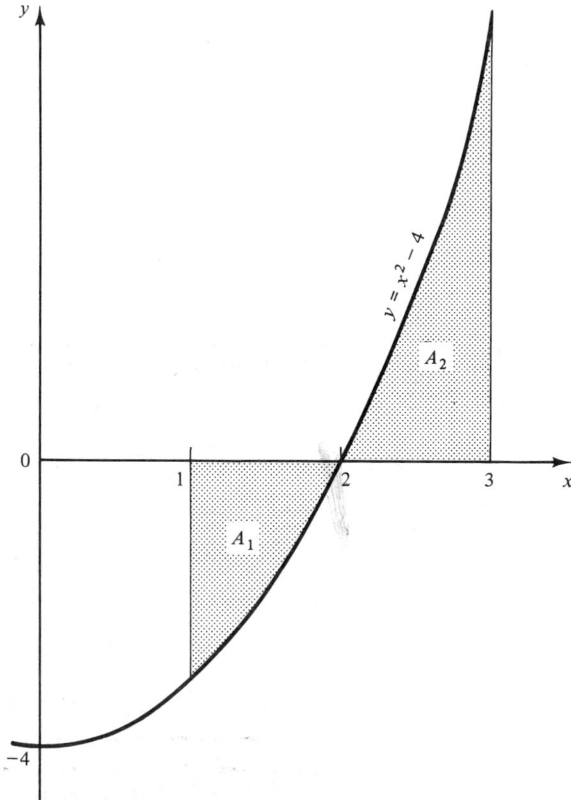

FIGURE 8-13

EXAMPLE 5: Find the area bounded by the curve $y = 1/x$ and the x axis (a) from $x = 1$ to 4 and (b) from $x = -1$ to 4.

Solution: Integrating,

$$A = \int_a^b \frac{dx}{x} = \ln |x| \Big|_a^b = \ln |b| - \ln |a|$$

(a) For the limits 1 to 4,

$$A = \ln 4 - \ln 1 = 1.386$$

(b) For the limits −1 to 4,

$$\begin{aligned} A &= \ln |4| - \ln |-1| \\ &= \ln 4 - \ln 1 = 1.386 \,(?) \end{aligned}$$

We appear to get the same area between the limits -1 and 4 as we did for the limits 1 and 4. However, a graph (Fig. 8-14) shows that the curve $y = 1/x$ *is discontinuous* at $x = 0$, so we cannot integrate over the interval -1 to 4.

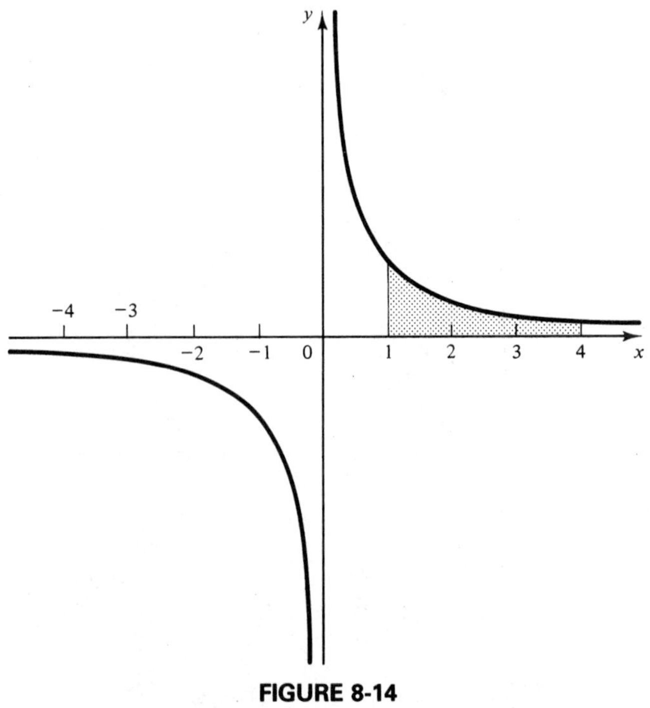

FIGURE 8-14

Common Error	Don't try to set up these area problems without making a sketch. Don't try to integrate across a discontinuity.

Horizontal Elements

So far we have used vertical elements of area to set up our problems. Often, however, the use of horizontal elements will make our work easier, as in the following example.

EXAMPLE 6: Find the first-quadrant area bounded by the curve $y = x^2 + 3$, the y axis, and the lines $y = 7$ and $y = 12$ (Fig. 8-15).

Solution: We locate a point (x, y) on the curve. If we were to draw a vertical element through (x, y), its height would be $12 - x$. Instead, we choose a *horizontal* element, whose length is simply x, and whose width is dy. So

$$dA = x\,dy = (y + 3)^{1/2}\,dy.$$

Integrating, we have

$$A = \int_{7}^{12} (y - 3)^{1/2}\,dy$$

$$= \left.\frac{2(y - 3)^{3/2}}{3}\right|_{7}^{12}$$

$$= \frac{2}{3}(9)^{3/2} - \frac{2}{3}(4)^{3/2} = \frac{38}{3} \text{ square units}$$

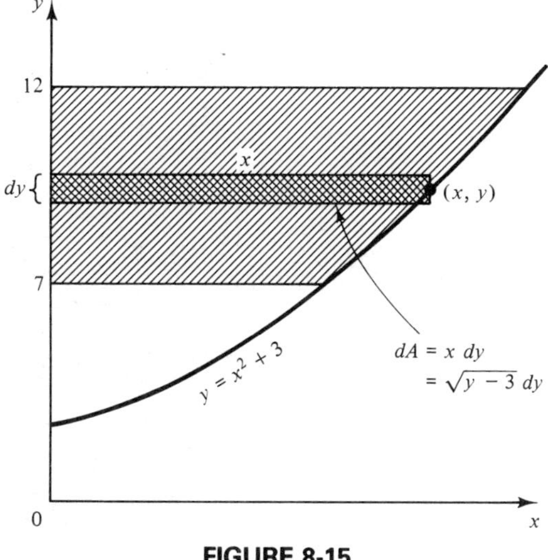

FIGURE 8-15

EXAMPLE 7: Find the area bounded by the curves $y^2 = 12x$ and $y^2 = 24x - 36$.

Solution: We first plot the two curves (Fig. 8-16) which we recognize from their equations to be parabolas opening to the right. We find their points of intersection by solving simultaneously,

$$24x - 36 = 12x$$

$$x = 3$$

$$y = \pm 6$$

Since we have symmetry about the x axis, let us solve for the first quadrant area, and later double it. We draw a horizontal strip of width dy and length $x_2 - x_1$. The area dA is then $dA = (x_2 - x_1)\,dy$. Integrating,

$$A = \int_{0}^{6} (x_2 - x_1)\,dy = \int_{0}^{6} \left(\frac{y^2}{24} + \frac{36}{24} - \frac{y^2}{12}\right) dy$$

$$= \int_0^6 \left(\frac{3}{2} - \frac{y^2}{24}\right) dy$$

$$= \frac{3y}{2} - \frac{y^3}{72}\bigg|_0^6$$

$$= \frac{3(6)}{2} - \frac{(6)^3}{72} = 6$$

By symmetry, the total area between the two curves is twice this, or 12 square units.

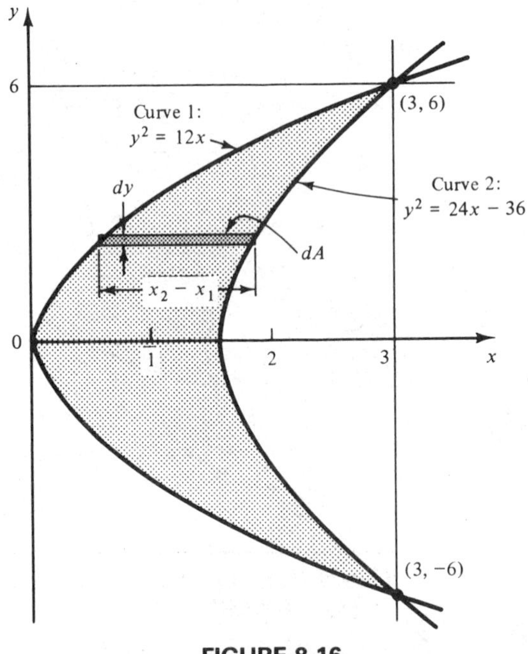

FIGURE 8-16

Common Error	Don't always assume that a vertical element is the best choice. Try setting up the integral in example 6 or 7 using a vertical element. What problems arise?

EXERCISE 3

Areas Bounded by the y Axis

Find the area bounded by the given curve, the y axis, and the given lines.

1. $y = x^2 + 2$ from $y = 3$ to 5

2. $8y^2 = x$ from $y = 0$ to 10

3. $y^3 = 4x$ from $y = 0$ to $y = 4$

4. $y = 4 - x^2$ in the first quadrant from $y = 0$ to $y = 3$

Areas Bounded by Both Axes

Find the first-quadrant area bounded by each curve and the coordinate axes.

5. $y^2 = 16 - x$

6. $y = x^3 - 8x^2 + 15x$

7. $x + y + y^2 = 2$

8. $\sqrt{x} + \sqrt{y} = 1$

Areas Bounded by Two Branches of a Curve

Find the area bounded by the given curve and given line.

9. $y^2 = x$ and $x = 4$

10. $y^2 = 2x$ and $x = 4$

11. $4y^2 = x^3$ and $x = 8$

12. $y^2 = x^2(x^2 - 1)$ and $x = 2$

Areas above and below the x Axis

13. Find the area bounded by the curve $10y = x^2 - 80$, the x axis, and the lines $x = 1$ and $x = 6$.

14. Find the area bounded by the curve $y = x^3$, the x axis, and the lines $x = -3$ and $x = 0$.

Find only the portion of the area below the x axis.

15. $y = x^3 - 4x^2 + 3x$

16. $y = x^2 - 4x + 3$

Periodic Functions

Find the area under one of each function.

17. $y = \sin x$

18. $v = 2 \cos x$

19. $y = 2 \sin \frac{1}{2}\pi x$

20. Find the area between the curve $y = \sin x$ and the x axis from $x = 1$ rad to $x = 3$ rad.

21. Find the area between the curve $y = \cos x$, the x axis, and the ordinates at $x = 0$ and $x = \frac{3}{2}\pi$.

Areas between Curves

22. Find the area bounded by the lines $y = 3x$, $y = 15 - 3x$, and the x axis.

23. Find the area bounded by the curves $y^2 = 4x$ and $2x - y = 4$.

24. Find the area bounded by the parabola $y = 6 + 4x - x^2$ and the chord joining $(-2, -6)$ and $(4, 6)$.

25. Find the area bounded by the curve $y^2 = x^3$ and the line $x = 4$.

26. Find the area bounded by the curve $y^3 = x^2$ and the chord joining $(-1, 1)$ and $(8, 4)$.

27. Find the area between the parabolas $y^2 = 4x$ and $x^2 = 4y$.

28. Find the area between the parabolas $y^2 = 2x$ and $x^2 = 2y$.

Areas of Geometric Figures

Hint: Integrate using rule 69 for Problem 33 and several of those to follow.

Use integration to verify the formula for the area of each figure.

29. square of side a

30. rectangle with sides a and b

31. triangle (Fig. 8-17a)

32. triangle (Fig. 8-17b)

33. circle of radius r

34. segment of circle (Fig. 8-17c)

35. ellipse (Fig. 8-17d)

36. parabola (Fig. 8-17e)

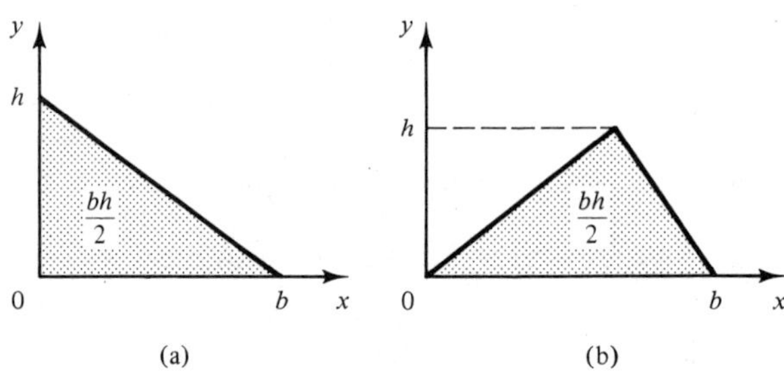

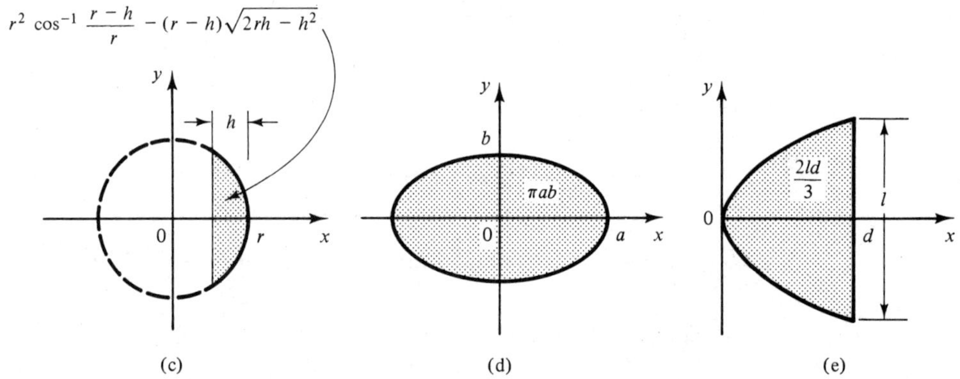

FIGURE 8-17. Areas of some geometric figures.

Applications

37. An elliptical culvert is partly full of water (Fig. 8-18). Find, by integration, the cross-sectional area of the water.

38. A mirror (Fig. 8-19) has a parabolic face. Find the volume of glass in the mirror.

39. Figure 8-20 shows a concrete column which has an elliptical cross section. Find the volume of concrete in the column.

9 ft
6 ft
18 ft

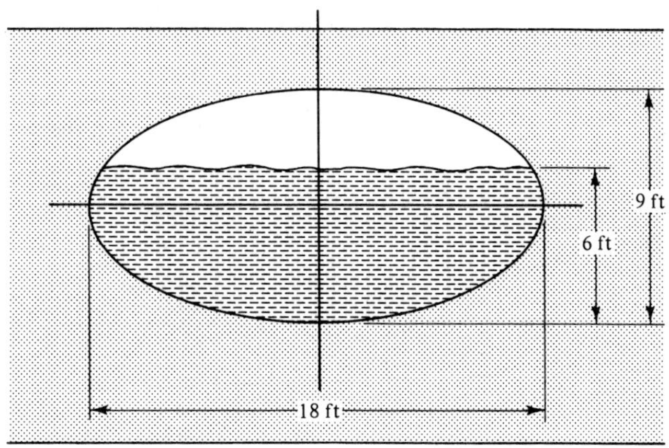

FIGURE 8-18. Elliptical culvert.

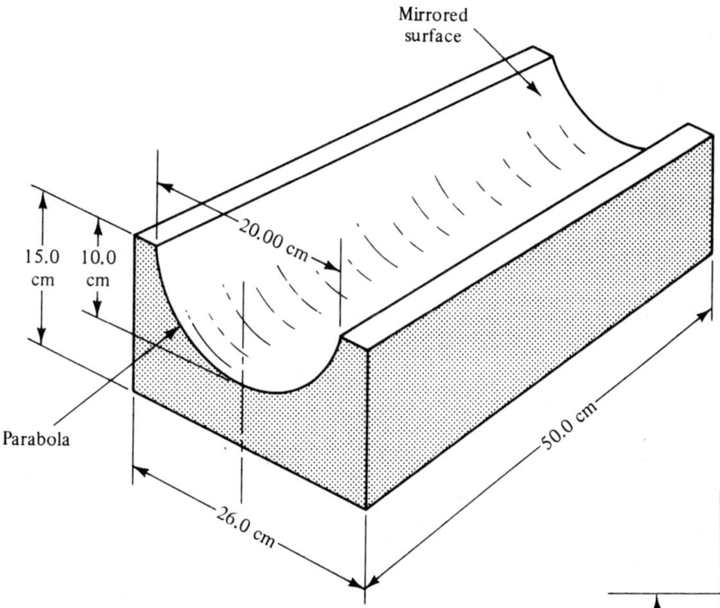

Mirrored
surface
20.00 cm
15.0
cm
10.0
cm
Parabola
50.0 cm
26.0 cm

FIGURE 8-19. Cylindrical mirror.

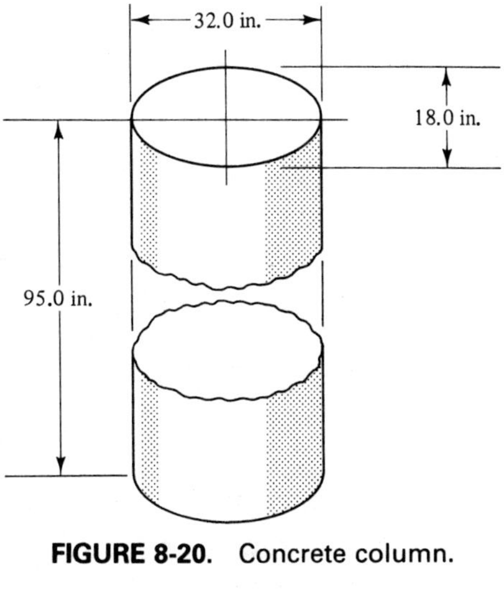

32.0 in.
18.0 in.
95.0 in.

FIGURE 8-20. Concrete column.

40. A concrete roof beam for an auditorium has a straight top edge and a parabolic lower edge (Fig. 8-21). Find the volume of concrete in the beam.

41. The deck of a certain ship has the shape of two intersecting parabolic curves (Fig. 8-22). Find the area of the deck.

42. A lens (Fig. 8-23) has a cross section formed by two intersecting circular arcs. Find by integration the cross-sectional area of the lens.

10.0 m
Parabola
0.50 m
1.25 m
0.50 m

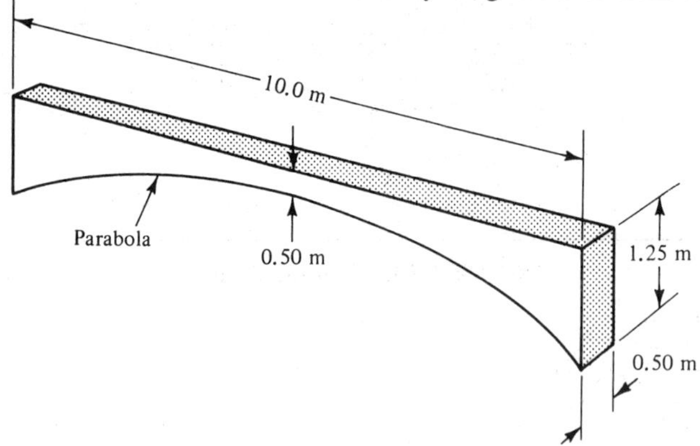

FIGURE 8-21. Roof beam.

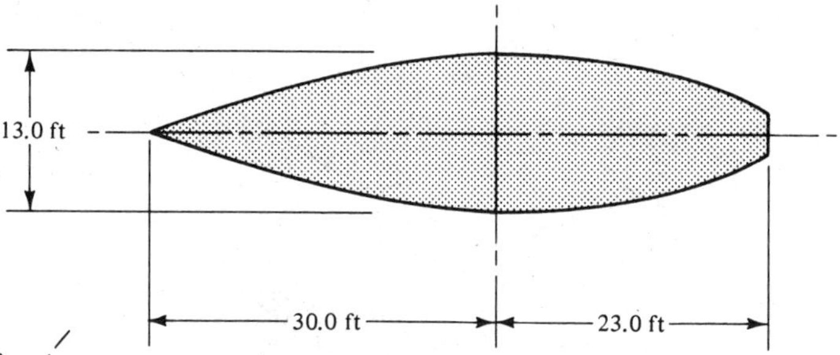

13.0 ft
30.0 ft
23.0 ft

FIGURE 8-22. Top view of a ship's deck.

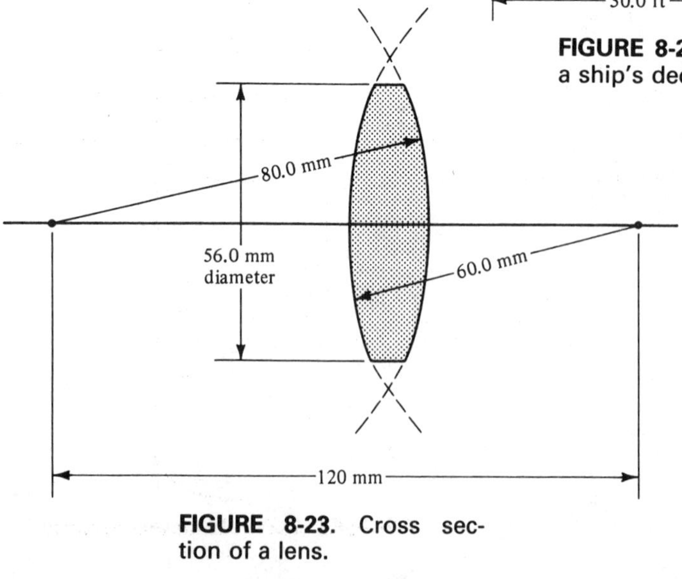

80.0 mm
56.0 mm
diameter
60.0 mm
120 mm

FIGURE 8-23. Cross section of a lens.

43. A window (Fig. 8-24) has the shape of a parabola above and a circular arc below. Find the area of the window.

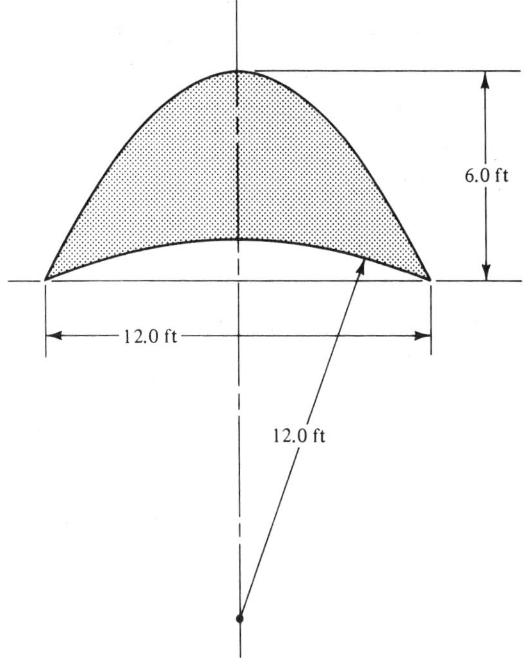

FIGURE 8-24. Window.

CHAPTER TEST

Evaluate each definite integral.

1. $\displaystyle\int_{-1}^{0} \frac{dx}{1-x}$

2. $\displaystyle\int_{0}^{2} \frac{x\,dx}{4+x^2}$

3. $\displaystyle\int_{0}^{\ln 3} e^{2x}\,dx$

4. $\displaystyle\int_{0}^{a} (\sqrt{a}-\sqrt{x})^2\,dx$

5. $\displaystyle\int_{2}^{7} (x^2-2x+3)\,dx$

6. $\displaystyle\int_{0}^{\pi} \cos 3x\,dx$

7. $\displaystyle\int_{-\pi}^{0} x \sin 2x^2\,dx$

8. $\displaystyle\int_{2}^{7} \sqrt{7-3x}\,dx$

9. Evaluate $\displaystyle\sum_{n=1}^{5} n^2(n-1)$.

10. Evaluate $\displaystyle\sum_{d=1}^{4} \frac{d^2}{d-1}$.

11. Find the approximate area under the curve $y = 5 + x^2$ from $x = 1$ to 9, by the midpoint method. Use panels 2 units wide.

12. Find the area bounded by the parabola $y^2 = 8x$, the x axis, and the line $x = 2$.

13. Find the entire area of the ellipse

$$\frac{x^2}{16} + \frac{y^2}{9} = 1$$

(use integral 69)

14. Find the area bounded by the coordinate axes and the curve $x^{1/2} + y^{1/2} = 2$.

15. Find the area between the curve $x^2 = 8y$ and $y = \dfrac{64}{x^2 + 16}$. (use integral 56)

16. Find the area bounded by the curves $y^2 = 8x$ and $x^2 = 8y$.

17. Find the area bounded by the curve $y = x^{3/2}$ and the line $x = 4$.

18. Find the area bounded by the curve $y = 1/x$ and the x axis, between the limits $x = 1$ and $x = 3$.

19. Find the area bounded by the curve $y^2 = x^3 - x^2$ and the line $x = 2$.

20. Find the area bounded by $xy = 6$, the lines $x = 1$ and $x = 6$, and the x axis.

Applications of the Definite Integral

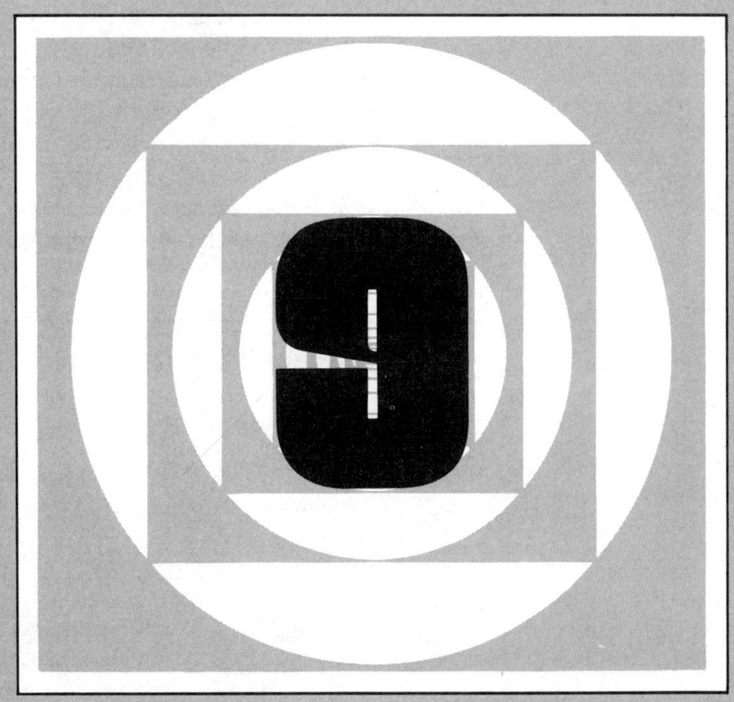

We have defined the definite integral in terms of the area under a curve, and in Chapter 8 used the definite integral to compute plane areas. From this you might get the idea that the definite integral is good only for finding areas, but nothing could be further from the truth. In this chapter and the next we show a wide range of applications other than areas.

In each case we set up our integral using the shortcut method of Sec. 8-3, that is, we define a small element of the quantity we want to compute, and then sum up all such elements by integration.

9-1. VOLUMES BY INTEGRATION

Solids of Revolution

When an area A (Fig. 9-1) is rotated about some axis L, it sweeps out a *solid of revolution*. It is clear from the figure that every cross section at right angles to the axis of rotation is a circle.

When the area A is rotated about an axis L located at some distance from the area, we get a solid of revolution with a cylindrical hole down its center (Fig. 9-2). The cross section of this solid, at right angles to the axis of rotation, consists of two concentric circles.

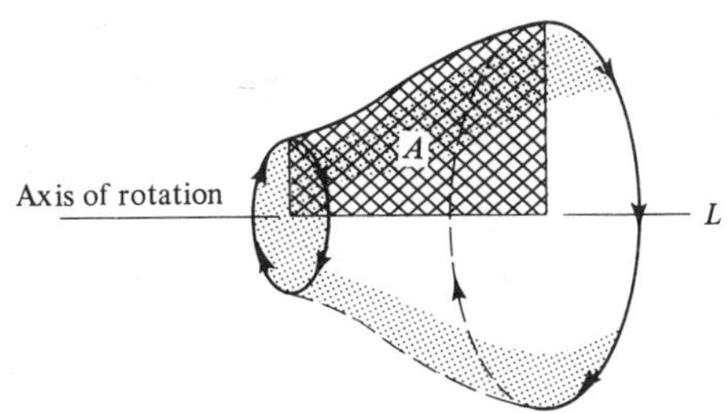

FIGURE 9-1. Solid of revolution.

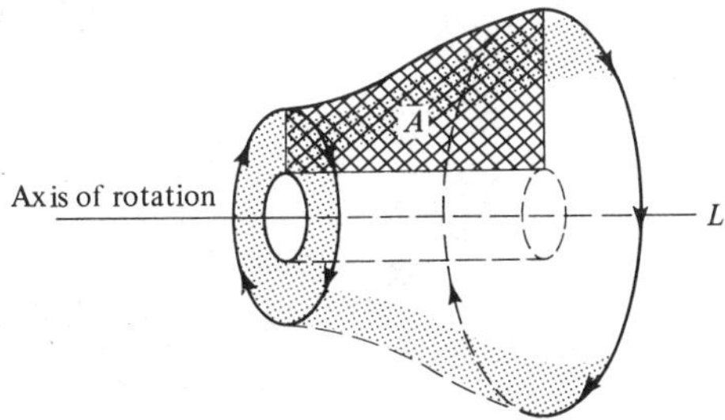

FIGURE 9-2. Solid of revolution with an axial hole.

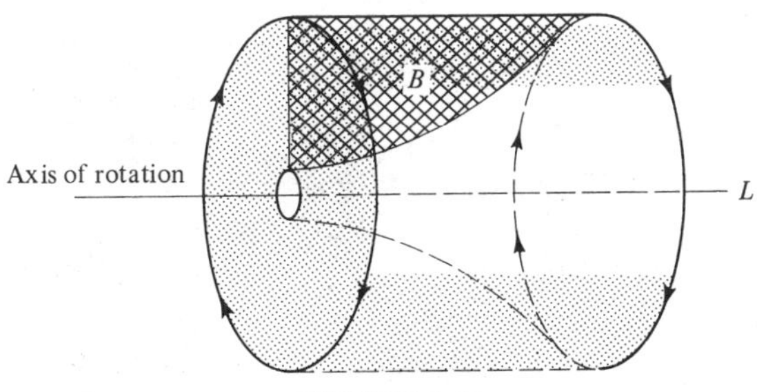

FIGURE 9-3

When the area B is rotated about axis L (Fig. 9-3), we get a solid of revolution with a hole of varying diameter down its center.

We first learn how to calculate the volume of a solid with no hole, and then cover "hollow" solids of revolution.

Volumes of Disks, Rings, and Shells

In the following sections we will have to compute the volumes of thin disks and rings and of thin-walled shells. For quick reference, the formula for the volume of each of these simple geometric figures are given in Fig. 9-4.

	Volume	Polar moment of inertia
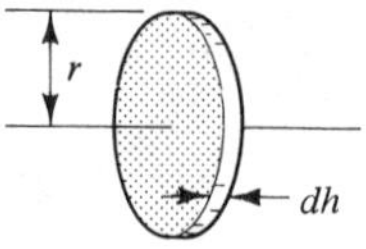 (a) Disk	$dV = \pi r^2\, dh$ (Eq. 351)	$dI = \frac{1}{2} m\pi r^4\, dh$ (Eq. 372)
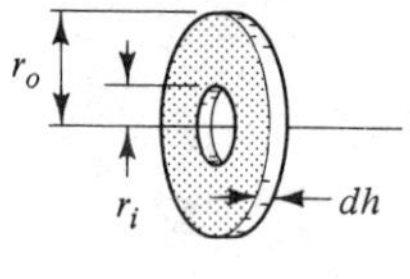 (b) Ring or washer	$dV = \pi(r_o^2 - r_i^2)\, dh$ (Eq. 353)	$dI = \frac{m\pi}{2}(r_o^4 - r_i^4)\, dh$ (Eq. 373)
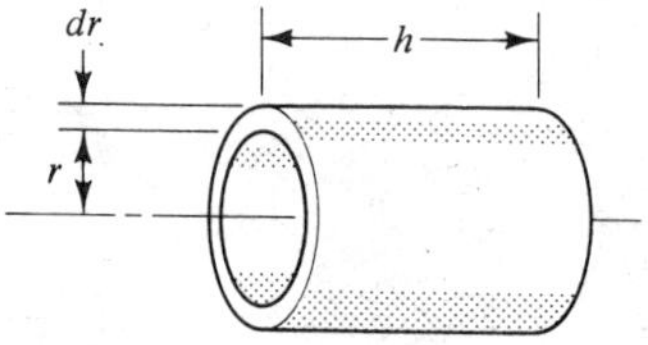 (c) Shell	$dV = 2\pi rh\, dr$ (Eq. 355)	$dI = 2\pi mr^3h\, dr$ (Eq. 374)

Don't worry about the polar moment of inertia formulas given in this table. We'll use them in Chapter 10.

FIGURE 9-4. Disks, rings, and shells.

Volumes by the Disk Method

We may think of a solid of revolution as being made up of a stack of thin disks, like a stack of coins of different sizes (Fig. 9-5). Each disk is called an *element* of the total volume. We let the radius of one such disk be r (which varies with the disk's position in the stack), and let the thickness be equal to dh. Since a disk (Fig. 9-4a) is a cylinder, we calculate its volume dV by Eq. 351,

$$dV = \pi r^2\, dh$$

Now using the shortcut method of Sec. 8-3 for setting up a definite integral, we "sum" the volumes of all such disk-shaped elements by integrating from one end of the solid to the other,

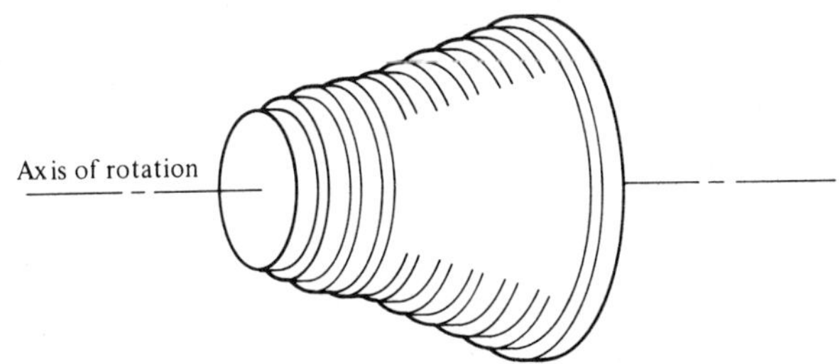

FIGURE 9-5. Solid of revolution approximated by a stack of thin disks.

Volumes by the Disk Method	$V = \pi \int_a^b r^2\, dh$	352

In an actual problem, we must express r and h in terms of x and y, as in the following example.

EXAMPLE 1: The area bounded by the curve $y = 8/x$, the x axis, and the lines $x = 1$ and $x = 8$, is rotated about the x axis. Find the volume generated.

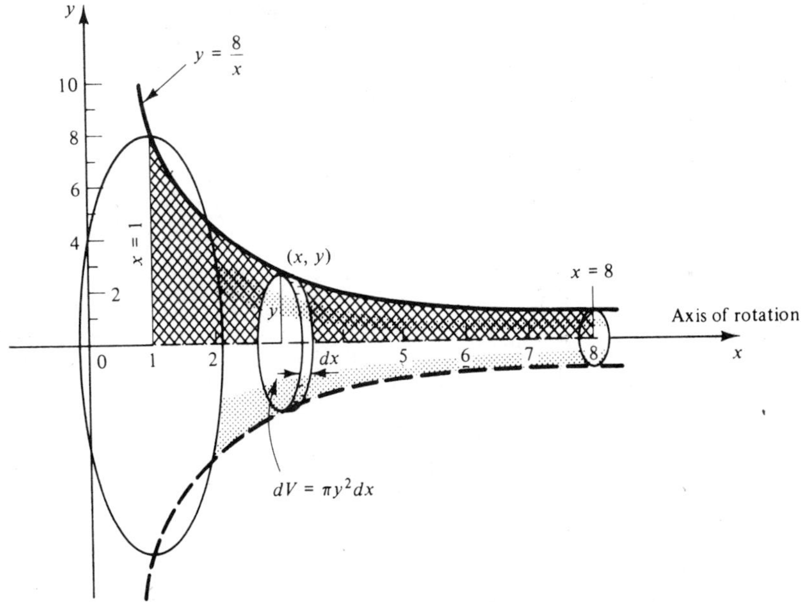

FIGURE 9-6. Volumes by the disk method.

Solution: We sketch the solid (Fig. 9-6) and a typical disk, touching the curve at some point (x, y). The radius r of the disk is equal to y, and the thickness dh of the disk is dx. So by Eq. 352,

$$V = \pi \int_a^b y^2\,dx$$

For y we substitute $8/x$, and for a and b, the limits 1 and 8,

$$V = \pi \int_1^8 \left(\frac{8}{x}\right)^2 dx = 64\pi \int_1^8 x^{-2}\,dx$$

Integrating,
$$= -64\pi x^{-1}\Big|_1^8$$

$$= -64\pi\left(\frac{1}{8} - 1\right) = 56\pi \text{ cubic units}$$

Common Error	Remember that all parts of the integral, including the limits, must be expressed in terms of the *same variable*.

Rotation about the y Axis

When our area is rotated about the y rather than the x axis, we take our element of area as a horizontal disk rather than a vertical one.

EXAMPLE 2: Find the volume generated when the area bounded by $y = x^2$, the y axis, and the lines $y = 1$ and $y = 4$, is rotated about the y axis.

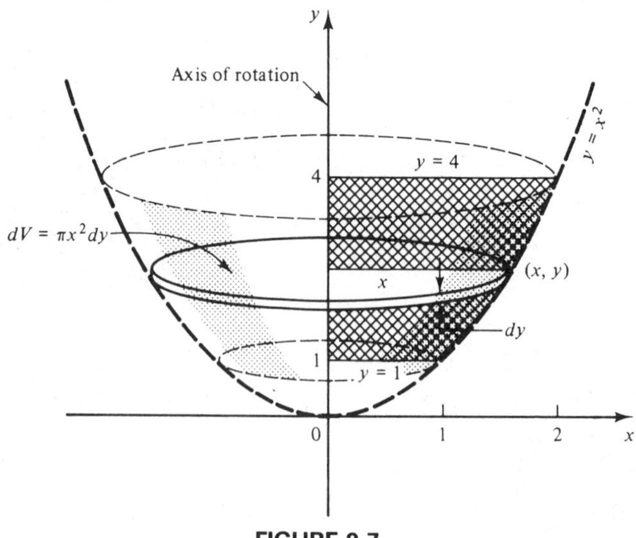

FIGURE 9-7

Solution: Our disk-shaped element of volume (Fig. 9-7) now has a radius x and a thickness dy. So by Eq. 352,

$$V = \pi \int_a^b x^2\, dy$$

Substituting y for x^2, and inserting the limits 1 and 4, we have

$$V = \pi \int_1^4 y\, dy = \pi \left[\frac{y^2}{2}\right]_1^4$$

$$= \frac{\pi}{2}(4^2 - 1^2) = \frac{15\pi}{2} \text{ cubic units}$$

Volumes by the Shell Method

Instead of using a thin disk for our element of volume, it is sometimes easier to use a thin-walled shell (Fig. 9-4c). The volume dV of a thin-walled cylindrical shell of radius r, height h, and wall thickness dr is

$$dV = \text{circumference} \times \text{height} \times \text{wall thickness}$$

$$= 2\pi rh\, dr$$

Integrating gives

Volumes by the Shell Method	$V = 2\pi \int_a^b rh\, dr$	**356**

As with the disk method, r and h must be expressed in terms of x and y in a particular problem.

EXAMPLE 3: The first-quadrant area bounded by the curve $y = x^2$, the y axis, and the line $y = 4$ (Fig. 9-8) is rotated about the y axis. Find the volume generated, by the shell method.

Solution: We sketch an element of volume in the shape of a shell, whose centerline is the y axis. The radius r of the shell is x, its thickness dr is dx, and its height h is the value of y on the upper curve minus that on the lower curve, $y_o - y_i$. The volume is then, by Eq. 356,

$$V = 2\pi \int x(y_o - y_i)\, dx$$

In this example the upper curve is the straight line $y = 4$ and the lower curve is the parabola. Substituting $y_o = 4$ and $y_i = x^2$ gives

$$V = 2\pi \int_0^2 x(4 - x^2)\, dx$$

$$= 2\pi \int_0^2 (4x - x^3)\, dx$$

$$= 2\pi \left[2x^2 - \frac{x^4}{4}\right]_0^2 = 2\pi \left[2(2)^2 - \frac{2^4}{4}\right]$$

$$= 8\pi \text{ cubic units}$$

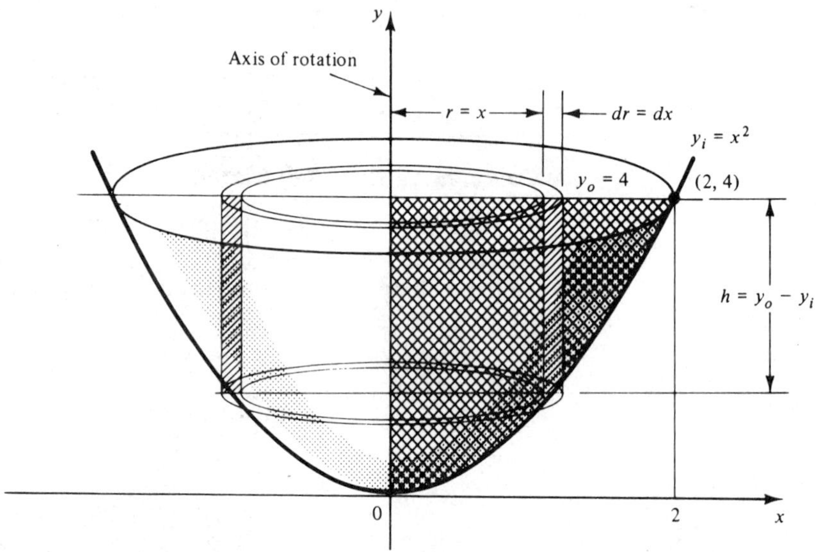

FIGURE 9-8. Volumes by the shell method.

Solid of Revolution with Hole

To find the volume of a solid of revolution with an axial hole, such as in Figs. 9-2 and 9-3, we can first find the volume of hole and solid separately, and then subtract. Or we can find it directly by either the ring or shell method, as in the following example.

EXAMPLE 4: The first-quadrant area bounded by the curve $y^2 = 4x$, the x axis, and the line $x = 4$ is rotated about the y axis. Find the volume generated (a) by the ring method and (b) by the shell method.

Solution: (a) *Ring method:* The volume dV of a thin ring (Fig. 9-4b) is given by

$$dV = \pi(r_o^2 - r_i^2)\, dh$$

The ring method is also called the *washer* method.

Integrating gives

Volume by the Ring Method	$V = \pi \int_a^b (r_o^2 - r_i^2)\, dh$	**354**

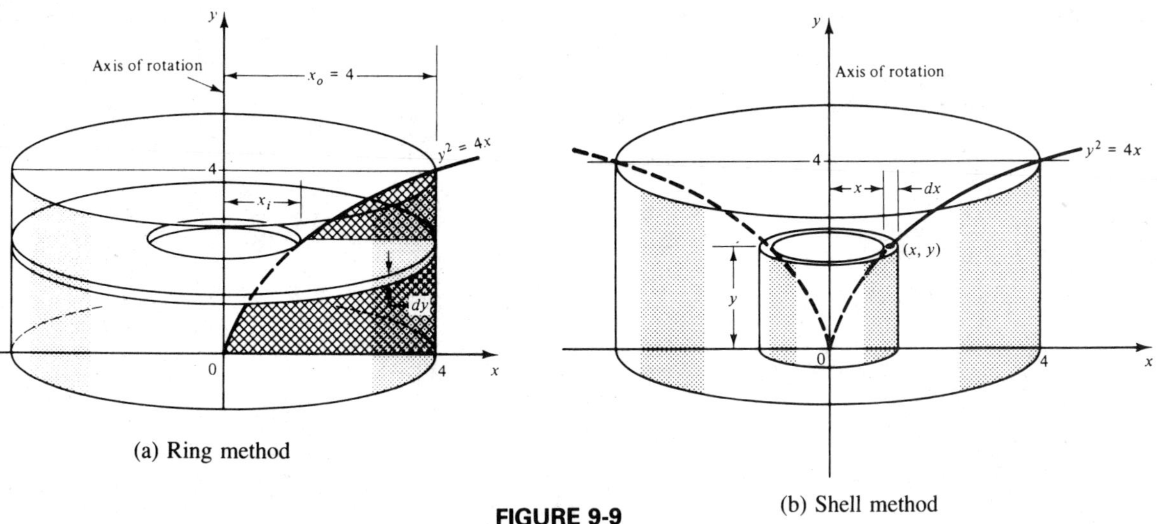

FIGURE 9-9

On our given solid (Fig. 9-9a) we show an element of volume in the shape of a ring, centered on the y axis. Its outer and inner radii are x_o and x_i, respectively, and its thickness dh is here dy. Then by Eq. 354,

$$V = \pi \int_a^b (x_o^2 - x_i^2)\, dy$$

In our problem, $x_o = 4$ and $x_i = y^2/4$. Substituting these values and placing the limits on y of 0 and 4 gives

$$V = \pi \int_0^4 \left(16 - \frac{y^4}{16}\right) dy$$

Integrating, we obtain

$$V = \pi \left[16y - \frac{y^5}{80}\right]_0^4 = \pi \left[16(4) - \frac{4^5}{80}\right] = \frac{256}{5}\pi \text{ cubic units}$$

(b) *Shell method:* On the given volume we indicate an element of volume in the shape of a shell (Fig. 9-9b). Its inner radius r is x, its thickness dr is dx, and its height h is y. Then by Eq. 355,

$$dV = 2\pi xy\, dx$$

So the total volume is

$$V = 2\pi \int xy\, dx$$

Replacing y with $2\sqrt{x}$ and placing limits on x we have

$$V = 2\pi \int_0^4 x(2x^{1/2})\,dx$$
$$= 4\pi \int_0^4 x^{3/2}\,dx$$
$$= 4\pi \left[\frac{x^{5/2}}{5/2}\right]_0^4$$
$$= \frac{8\pi}{5}(4)^{5/2} = \frac{256}{5}\pi \text{ cubic units}$$

as by the ring method.

Common Error	In Example 4 the radius x varies from 0 to 4. The limits of integration are therefore 0 to 4, not -4 to 4.

Rotation about a Noncoordinate Axis

We can, of course, get a volume of revolution by rotating a given area about some axis other than a coordinate axis. This often results in a solid with a hole in it. As with the preceding hollow figure, these can usually be set up using either shells, or with rings as in the following example.

EXAMPLE 5: The first-quadrant area bounded by the curve $y = x^2$, the y axis, and the line $y = 4$, is rotated about the line $x = 3$. Find the volume generated.

Solution: Through the point (x, y) on the curve (Fig. 9-10) we draw a ring-shaped element of volume, with outside radius of 3 units, inside radius of $(3 - x)$ units, and thickness dy. The volume dV of the element is then, by Eq. 353,

$$dV = \pi[3^2 - (3 - x)^2]\,dy$$
$$= \pi(9 - 9 + 6x - x^2)\,dy$$
$$= \pi(6x - x^2)\,dy$$

Substituting $\sqrt{y}$ for x gives

$$dV = \pi(6\sqrt{y} - y)\,dy$$

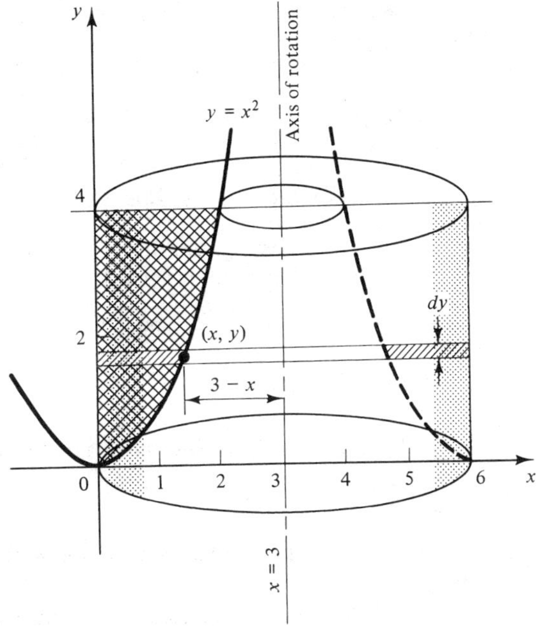

FIGURE 9-10

Integrating, we get

$$V = \int_0^4 (6y^{1/2} - y)\,dy = \pi\left[\frac{12y^{3/2}}{3} - \frac{y^2}{2}\right]_0^4$$

$$= \pi\left[4(4)^{3/2} - \frac{16}{2}\right] = 24\pi \text{ cubic units}$$

EXERCISE 1

Rotation about the x Axis

Find the volume generated by rotating the first-quadrant area bounded by each set of curves about the x axis. Use either the disk or the shell method.

1. $y = x^3$ and $x = 2$

2. $y = \dfrac{x^2}{4}$ and $x = 4$

3. $y = \dfrac{x^{3/2}}{2}$ and $x = 2$

4. $y^2 = x^3 - 3x^2 + 3x$ and $x = 1$

5. $y^2(2 - x) = x^3$ and $x = 1$

6. $\sqrt{x} + \sqrt{y} = 1$

7. $x^{2/3} + y^{2/3} = 1$ from $x = 0$ to $x = 1$

8. one arch of $y = \sin x$

9. the catenary $y = \frac{1}{2}(e^x + e^{-x})$ from $x = 0$ to $x = 1$.

10. $y^2 = x\dfrac{(x-3)}{(x-4)}$

Rotation about the y Axis

Find the volume generated by revolving about the y axis the first-quadrant area bounded by each set of curves.

11. $y = x^3$, the y axis, and $y = 8$

12. $2y^2 = x^3$, $x = 0$, and $y = 2$

13. $9x^2 + 16y^2 = 144$

14. $\left(\dfrac{x}{2}\right)^2 + \left(\dfrac{y}{3}\right)^{2/3} = 1$

15. one arch of the sine curve (use the shell method and integral 31)

16. $y^2 = 4x$, and $y = 4$

Rotation about a Noncoordinate Axis

Find the volume generated by rotating about the indicated axis the first-quadrant area bounded by each set of curves.

17. $x = 4$ and $y^2 = x^3$, about $x = 4$

18. $y = e^x$ and $x = 1$, about $x = 1$ (use integral 37)

19. $y = 3$ and $y = 4x - x^2$, about $y = 3$

20. $y^2 = x^3$ and $y = 8$, about $y = 9$

21. $y = -4$ and $y = 4 + 6x - 2x^2$, about $y = -4$

Applications

22. The nose cone of a certain rocket is a paraboloid of revolution (Fig. 9-11). Find its volume.

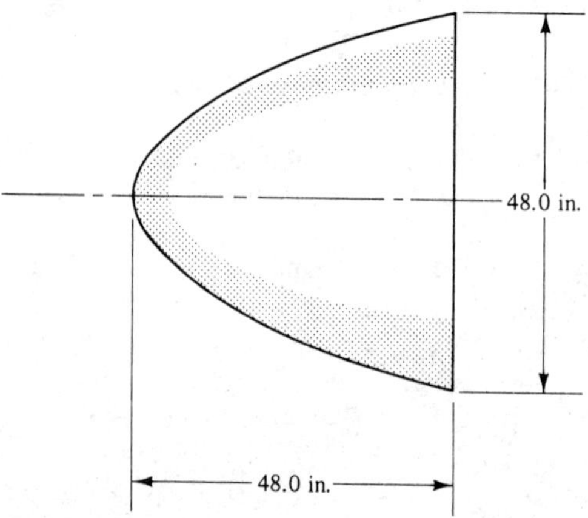

FIGURE 9-11. Rocket nose cone.

23. A wing tank for an airplane is a solid of revolution formed by rotating the curve $8y = 4x - x^2$, from $x = 0$ to $x = 3.5$, about the x axis (Fig. 9-12). Find the volume of the tank.

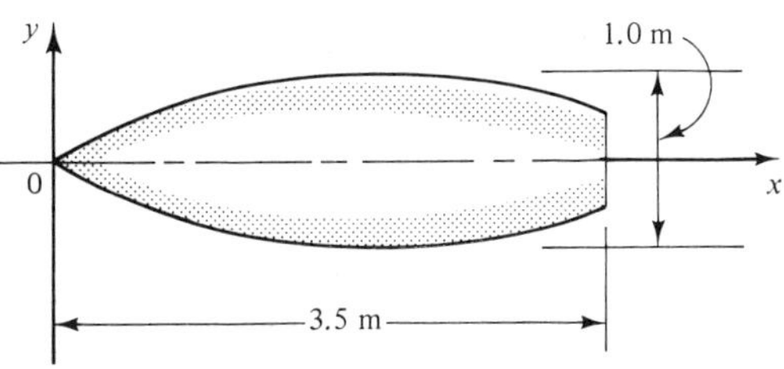

FIGURE 9-12. Airplane wing tank.

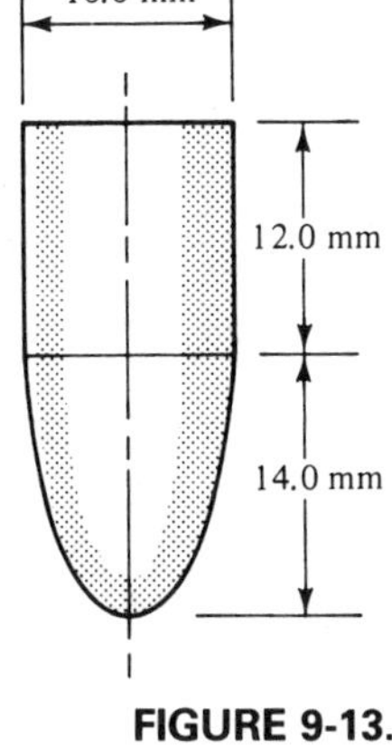

FIGURE 9-13. Bullet.

24. The bullet (Fig. 9-13) consists of a cylinder and a paraboloid of revolution, and is made of lead having a density of 11.3 g/cm^3. Find its weight.

25. The telescope mirror, shown in cross section in Fig. 9-14, is formed by rotating the area under the hyperbola $y^2/100 - x^2/1225 = 1$ about the y axis, and has a 20-cm-diameter hole at its center. Find the volume of glass in the mirror.

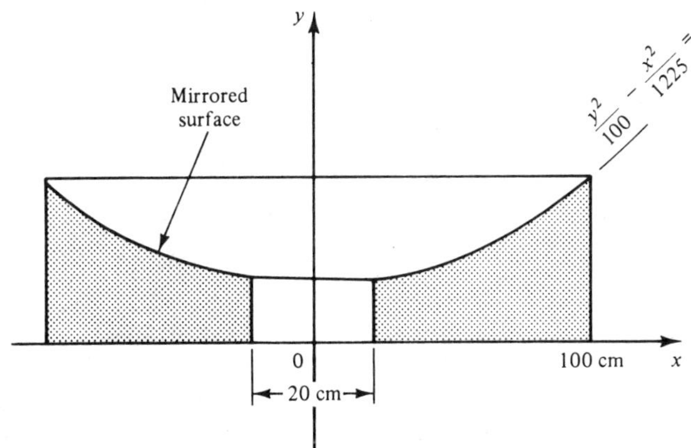

FIGURE 9-14. Telescope mirror.

9-2. LENGTH OF ARC

We are considering here (as usual) only smooth continuous curves.

In this section we develop a way to find the length of a curve between two endpoints, such as the distance PQ in Fig. 9-15. We will get the distance measured *along the curve,* as if the curve were stretched out straight, and then measured.

Again we use our intuitive method to set up the integral. We think of the curve as being made up of many short sections, each of length Δs. By the Pythagorean theorem,

$$(\Delta s)^2 \cong (\Delta x)^2 + (\Delta y)^2$$

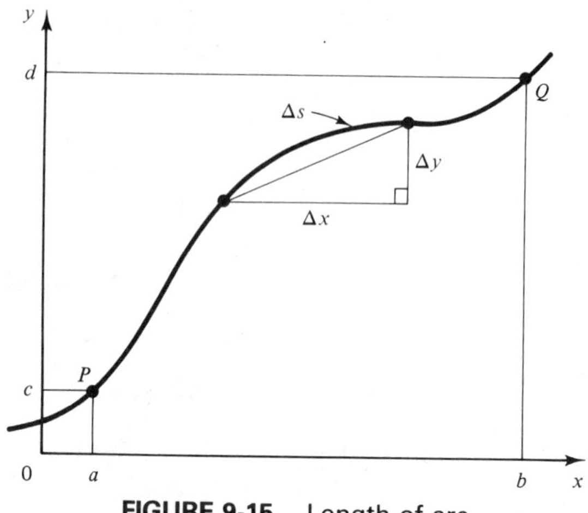

FIGURE 9-15. Length of arc.

Dividing by $(\Delta x)^2$ gives us

$$\frac{(\Delta s)^2}{(\Delta x)^2} \cong 1 + \frac{(\Delta y)^2}{(\Delta x)^2}$$

Taking the square root yields

$$\frac{\Delta s}{\Delta x} \cong \sqrt{1 + \left(\frac{\Delta y}{\Delta x}\right)^2}$$

We now let the number of these short sections of curve approach infinity as we let Δx approach zero.

$$\frac{ds}{dx} = \lim_{\Delta x \to 0} \frac{\Delta s}{\Delta x} = \sqrt{1 + \left(\frac{dy}{dx}\right)^2}$$

or

$$ds = \sqrt{1 + \left(\frac{dy}{dx}\right)^2}\, dx$$

Now thinking of integration as a summing process, we use it to add up all the small segments of length ds:

Length or Arc	$s = \int_a^b \sqrt{1 + \left(\frac{dy}{dx}\right)^2}\, dx$	**357**

Finding the length of arc is sometimes referred to as *rectifying* a curve.

EXAMPLE 1: Find the first quadrant length of the curve $y^2 = x^3$ from $x = 0$ to $x = 4$.

Solution: Let us first find $(dy/dx)^2$.

$$y = x^{3/2}$$

$$\frac{dy}{dx} = \frac{3}{2}x^{1/2}$$

$$\left(\frac{dy}{dx}\right)^2 = \frac{9x}{4}$$

Then by Eq. 357,

$$\begin{aligned} s &= \int_0^4 \sqrt{1 + \left(\frac{dy}{dx}\right)^2}\, dx = \int_0^4 \sqrt{1 + \frac{9x}{4}}\, dx \\ &= \frac{4}{9}\int_0^4 \left(1 + \frac{9x}{4}\right)^{1/2} \left(\frac{9}{4}\, dx\right) \\ &= \frac{8}{27}\left[\left(1 + \frac{9x}{4}\right)^{3/2}\right]_0^4 \\ &= \frac{8}{27}(10^{3/2} - 1) \cong 9.07 \end{aligned}$$

Another Form of the Arc Length Equation

Another form of the equation for arc length, which can be derived in a similar way to Eq. 357, is

Length of Arc	$s = \int_c^d \sqrt{1 + \left(\frac{dx}{dy}\right)^2}\, dy$	**358**

This equation is more useful when the equation of the curve is given in the form $x = f(y)$, instead of the more usual form $y = f(x)$.

EXAMPLE 2: Find the length of the curve $x = 4y^2$, between $y = 0$ and $y = 4$.

Solution: Taking the derivative,

$$\frac{dx}{dy} = 8y$$

so $(dx/dy)^2 = 64y^2$. Substituting into Eq. 358,

$$s = \int_0^4 \sqrt{1 + 64y^2}\,dy = \frac{1}{8}\int_0^4 \sqrt{1 + (8y)^2}\,(8\,dy)$$

When finding length of arc we usually wind up with messy integrals. Often they can be evaluated using rule 66.

Using integral 66, with $u = 8y$ and $a = 1$,

$$s = \frac{1}{8}\left[\frac{8y}{2}\sqrt{1 + 64y^2} + \frac{1}{2}\ln\left|8y + \sqrt{1 + 64y^2}\right|\right]_0^4$$

$$= \frac{1}{8}\left[4(4)\sqrt{1 + 64(16)} + \frac{1}{2}\ln\left|32 + \sqrt{1 + 64(16)}\right| - 0\right] = 64.3$$

EXERCISE 2

Many of these will require the use of rule 66.

Find the length of each curve.

1. $y^2 = x^3$ in the first quadrant from $x = 0$ to $x = \frac{5}{9}$

2. $6y = x^2$ from the origin to the point $(4, \frac{8}{3})$

3. $y = \ln \sec x$ from the origin to the point $(\pi/3, \ln 2)$

4. the circle $x^2 + y^2 = 36$ (use integral 61)

5. $x^{2/3} + y^{2/3} = 1$ from $x = 0$ to $x = 1$

6. $4y = x^2$ from $x = 0$ to $x = 4$

7. $2y^2 = x^3$ from the origin to $x = 10$

8. the arch of the parabola $y = 4x - x^2$ that lies above the x axis

9. the length in one quadrant of the curve $\left(\frac{x}{2}\right)^{2/3} + \left(\frac{y}{3}\right)^{2/3} = 1$

10. the catenary $y = \dfrac{a(e^{x/a} + e^{-x/a})}{2}$ from $x = 0$ to $x = 6$; use $a = 3$

Hint: Use Eq. 358 for Problems 11 and 12.

11. $y^3 = x^2$ between the points (0, 0) and (8, 4)

12. $y^2 = 8x$ from its vertex to one end of its latus rectum

Applications

13. Assuming the cable AB (Fig. 9-16) to be a parabola, find its length.

14. A roadway has a parabolic shape at the top of a hill (Fig. 9-17). If the road is 30 ft wide, find the cost of paving from P to Q, at the rate of \$35 per square foot.

15. The equation of the bridge arch in Fig. 9-18 is $y = 0.0625x^2 - 5x + 100$. Find its length.

16. Find the surface area of the curved portion of the mirror in Fig. 8-19.

17. Find the perimeter of the window in Fig. 8-24.

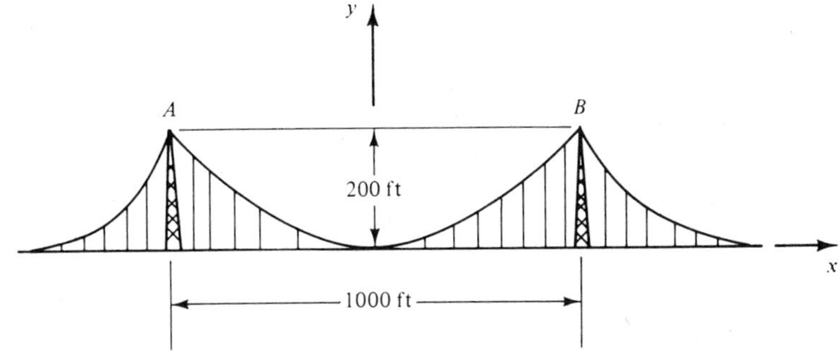

FIGURE 9-16. Suspension bridge.

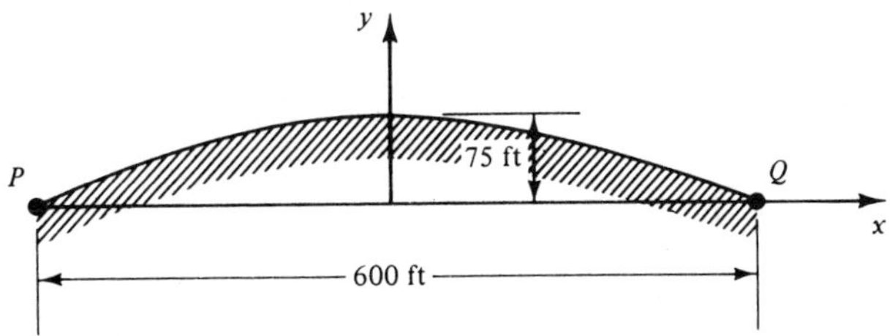

FIGURE 9-17. Road over a hill.

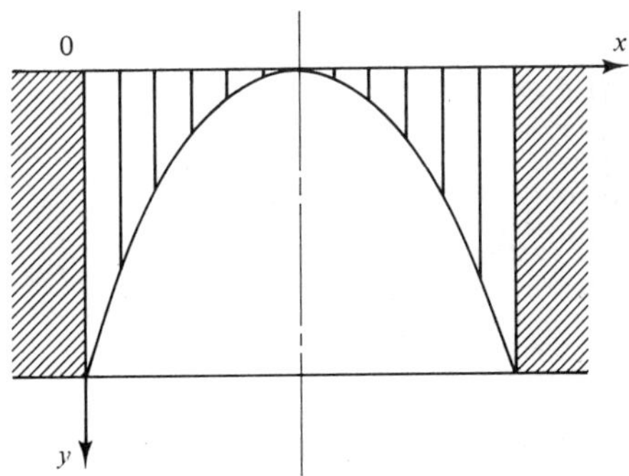

FIGURE 9-18. Parabolic bridge arch. Note that y axis is positive downward.

9-3. AREA OF SURFACE OF REVOLUTION

Surface of Revolution

In Sec. 9-1 we rotated an *area* about an axis and got a solid of revolution. We now find the *area* of the surface of such a solid. Alternatively, we can think of a surface of revolution as being generated when a curve rotates about some axis. Let us take the curve of Fig. 9-15 from Sec. 9-2, and rotate it about the x axis. The arc PQ sweeps out a surface of revolution (Fig. 9-19), while the small section ds sweeps out a hoop-shaped element of that surface.

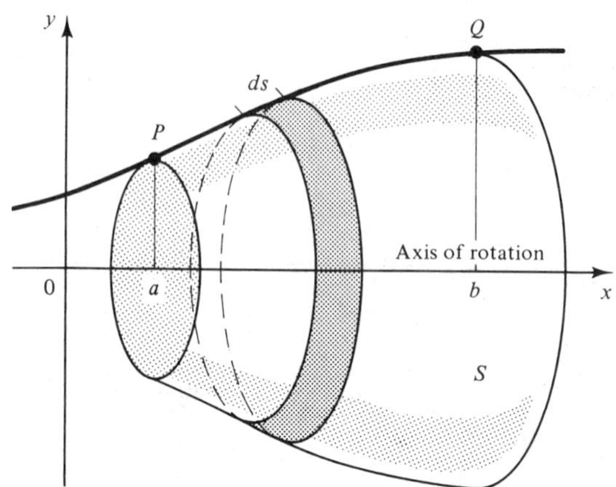

FIGURE 9-19. Surface of revolution.

The area of a hoop or circular band is equal to the length of the edge times the average circumference of the hoop. Our element ds is at a radius y from the x axis, so the area dS of the hoop is

$$dS = 2\pi y\ ds$$

We are using capital S for surface area and small s for arc length. Be careful not to confuse the two.

But from Sec. 9-2, we saw that

$$ds = \sqrt{1 + \left(\frac{dy}{dx}\right)^2}\ dx$$

So

$$dS = 2\pi y \sqrt{1 + \left(\frac{dy}{dx}\right)^2}\ dx$$

We then integrate from a to b, to sum up the areas of all such elements:

Area of Surface of Revolution about x Axis	$S = 2\pi \int_a^b y\sqrt{1 + \left(\frac{dy}{dx}\right)^2}\ dx$	**359**

EXAMPLE 1: The portion of the parabola $y^2 = 16x$ between $x = 0$ and $x = 16$ is rotated about the x axis. Find the area of the surface of revolution generated.

Solution: We first find the derivative. Solving for y gives us $y = 4x^{1/2}$. Then

$$\frac{dy}{dx} = 2x^{-1/2}$$

and

$$\left(\frac{dy}{dx}\right)^2 = 4x^{-1}$$

Surface area problems, like length of arc, usually result in difficult integrals.

Substituting into Eq. 359, we obtain

$$S = 2\pi \int_0^{16} 4x^{1/2}\sqrt{1 + 4x^{-1}}\, dx$$

$$= 8\pi \int_0^{16} (x + 4)^{1/2}\, dx$$

$$= \frac{16\pi}{3}(20^{3/2} - 4^{3/2}) \cong 1365 \text{ square units}$$

Rotation about the *y* Axis

The equation for the area of a surface of revolution whose axis of revolution is the *y* axis can be derived in a similar way. It is

Area of Surface of Revolution about y Axis	$S = 2\pi \int_a^b x\sqrt{1 + \left(\frac{dy}{dx}\right)^2}\, dx$	**360**

EXAMPLE 2: The portion of the curve $y = x^2$ lying between the points (0, 0) and (2, 4) is rotated about the y axis. Find the area of the surface generated.

Solution: Taking the derivative gives $dy/dx = 2x$, so

$$\left(\frac{dy}{dx}\right)^2 = 4x^2$$

Substituting into Eq. 360,

$$S = 2\pi \int_0^2 x\sqrt{1 + 4x^2}\, dx$$

$$= \frac{2\pi}{8} \int_0^2 (1 + 4x^2)^{1/2}(8x\, dx)$$

$$= \frac{\pi}{4} \cdot \frac{2(1 + 4x^2)^{3/2}}{3}\bigg|_0^2 = \frac{\pi}{6}(17^{3/2} - 1^{3/2}) = 36.2 \text{ square units}$$

EXERCISE 3

Find the area of the surface generated by rotating each curve about the x axis.

1. $y = \frac{x^3}{9}$ from $x = 0$ to $x = 3$

2. $y^2 = 2x$ from $x = 0$ to $x = 4$

3. $y^2 = 9x$ from $x = 0$ to $x = 4$

4. $y^2 = 4x$ from $x = 0$ to $x = 1$

5. $y^2 = 4 - x$ in the first quadrant

6. one arch of the curve $y = \sin x$

7. $y^2 = 24 - 4x$ from $x = 3$ to $x = 6$

8. $x^{2/3} + y^{2/3} = 1$ in the first quadrant

9. $y = \frac{x^3}{6} + \frac{1}{2x}$ from $x = 1$ to $x = 3$

10. $y = e^{-x}$ from $x = 0$ to $x = 100$

Find the area of the surface generated by rotating each curve about the y axis.

11. $y = 3x^2$ from $x = 0$ to 5

12. $y = 5x^2$ from $x = 2$ to 4

13. $y = 4 - x^2$ from $x = 0$ to 2

14. $y = 24 - x^2$ from $x = 2$ to 4

Geometric Figures

15. Find the surface area of a sphere by rotating the curve $x^2 + y^2 = r^2$ about a diameter.

16. Find the area of the curved surface of a cone by rotating about the x axis the line connecting the origin and the point (a, b).

Applications

17. Find the surface area of the nose cone shown in Fig. 9-11.

18. Find the cost of copper plating 10,000 bullets (Fig. 9-13) at the rate of \$15 per square meter.

19. Find the cost of grinding, polishing, and plating the curved surface of the mirror (Fig. 9-14) at 50 cents per square centimeter.

9-4. AVERAGE AND ROOT-MEAN-SQUARE VALUES

Average Value of a Function

The area A under the curve $y = f(x)$ (Fig. 9-20) between $x = a$ and $x = b$ is, by Eq. 339,

$$A = \int_a^b f(x)\, dx$$

The *average ordinate* of that function, within the same interval, is that value of y which will cause the rectangle *abcd* to have the same area as that under the curve, or

$$(b - a)y_{avg} = A = \int_a^b f(x)\, dx$$

so

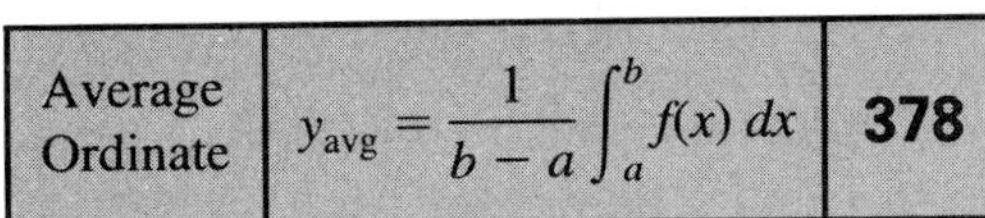

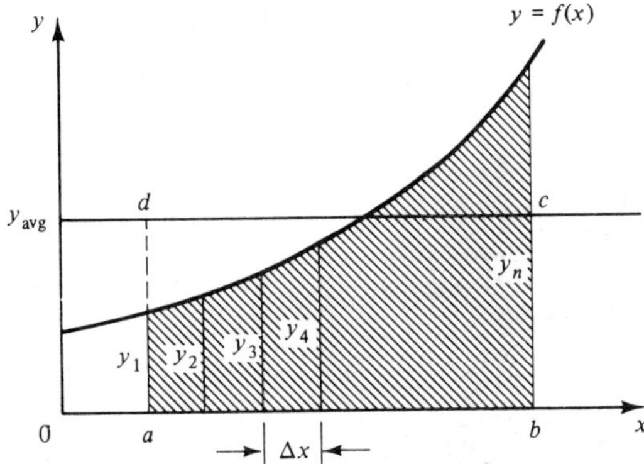

FIGURE 9-20. Average ordinate.

EXAMPLE 1: Find the average ordinate of a half-cycle of the sinusoidal voltage

$$v = V \sin \theta \qquad \text{volts}$$

Solution: By Eq. 378, with $a = 0$ and $b = \pi$

$$V_{avg} = \frac{V}{\pi - 0}\int_0^{\pi} \sin \theta\, d\theta$$

$$= \frac{V}{\pi}[-\cos \theta]_0^{\pi}$$

$$= \frac{V}{\pi}(-\cos \pi + \cos 0) = \frac{2}{\pi} V = 0.637\, V \qquad \text{volts}$$

Root-Mean-Square Value of a Function

The *root-mean-square* (rms) value of a function is the square root of the average of the squares of the ordinates. If, in Fig. 9-20 we take n values of y spaced apart by a distance Δx, the rms value is approximately

$$\text{rms} \cong \sqrt{\frac{y_1^2 + y_2^2 + y_3^2 + \cdots + y_n^2}{n}}$$

or, using summation notation,

$$\text{rms} \cong \sqrt{\frac{\sum_{i=1}^{n} y_i^2}{n}}$$

Multiplying numerator and denominator of the fraction under the radical by Δx,

$$\text{rms} \cong \sqrt{\frac{\sum_{i=1}^{n} y_i^2 \, \Delta x}{n \, \Delta x}}$$

But $n \, \Delta x$ is simply the width $(b - a)$ of the interval. If we now let n approach infinity, we get

$$\lim_{n \to \infty} \sum_{i=1}^{n} y_i^2 \, \Delta x = \int_a^b [f(x)]^2 \, dx$$

$$\text{rms} = \sqrt{\frac{1}{b - a} \int_a^b [f(x)]^2 \, dx} \qquad \textbf{379}$$

EXAMPLE 2: Find the rms value for the sinusoidal voltage of the preceding example.

Solution: We substitute into Eq. 379, with $a = 0$ and $b = \pi$

$$\text{rms} = \sqrt{\frac{1}{\pi - 0} \int_0^\pi V^2 \sin^2 \theta \, d\theta} = \sqrt{\frac{V^2}{\pi} \int_0^\pi \sin^2 \theta \, d\theta}$$

But by integral 16,

$$\int_0^\pi \sin^2 \theta \, d\theta = \frac{\theta}{2} - \frac{\sin 2\theta}{4} \bigg|_0^\pi$$

$$= \frac{\pi}{2} - \frac{\sin 2\pi}{4} = \frac{\pi}{2}$$

In electrical work, the rms value of an alternating current of voltage is also called the *effective* value. We see then that the effective value is 0.707 of the peak value.

So

$$\text{rms} = \sqrt{\frac{V^2}{\pi} \cdot \frac{\pi}{2}} = \frac{V}{\sqrt{2}} = 0.707\ V$$

EXERCISE 4

Find the average ordinate for each function in the given interval.

1. $y = x^2$ from 0 to 6

2. $y = x^3$ from -5 to 5

3. $y = \sqrt{1 + 2x}$ from 4 to 12

4. $y = \dfrac{x}{\sqrt{9 + x^2}}$ from 0 to 4

5. $y = \sin^2 x$ from 0 to $\pi/2$

6. $2y = \cos 2x + 1$ from 0 to π

Find the rms value for each function in the given interval.

7. $y = 2x + 1$ from 0 to 6

8. $y = \sin 2x$ from 0 to $\pi/2$

9. $y = x + 2x^2$ from 1 to 4

10. $y = 3 \tan x$ from 0 to $\pi/4$

11. $y = 2 \cos x$ from $\pi/6$ to $\pi/2$

12. $y = 5 \sin 2x$ from 0 to $\pi/6$

CHAPTER TEST

1. Find the volume generated when the area bounded by the parabolas $y^2 = 4x$ and $y^2 = 5 - x$ is rotated about the x axis.

2. Find the length of the curve $y = \frac{4}{5}x^2$ from the origin to $x = 4$.

3. The area bounded by the parabolas $y^2 = 4x$ and $y^2 = x + 3$ is rotated about the x axis. Find the surface area of the solid generated.

4. The area bounded by the parabola $y^2 = 4x$, from $x = 0$ to $x = 8$, is rotated about the x axis. Find the volume generated.

5. Find the volume of the solid generated by rotating the ellipse $x^2/16 + y^2/9 = 1$ about the x axis.

6. Find the length of the parabola $y^2 = 8x$ from the vertex to one end of the latus rectum.

7. The area bounded by the curves $x^2 = 4y$, $x - 2y + 4 = 0$, and the y axis is rotated about the y axis. Find the surface area of the volume generated.

8. Find the volume generated when the area bounded by the curve $y^2 = 16x$, from $x = 0$ to $x = 4$, is rotated about the x axis.

9. Find the volume generated when the area bounded by $y^2 = x^3$, the y axis, and $y = 8$ is rotated about the line $x = 4$.

10. The cables on a certain suspension bridge hang in the shape of a parabola. The towers are 100 m apart and the cables dip 10 m below the tops of the towers. Find the length of the cables.

11. The first-quadrant area bounded by the curves $y = x^3$ and $y = 4x$ is rotated about the x axis. Find the surface area of the solid generated.

12. Find the volume generated when the area bounded by $y^2 = x^3$, $x = 4$, and the x axis is rotated about the line $y = 8$.

13. Find the average ordinate for the function $y = \sin^2 x$, for $x = 0$ to 2π.

14. Find the rms value for the function $y = 2x + x^2$, for the interval $x = -1$ to $x = 3$.

Centroids and Moments

In this chapter we continue our applications of the definite integral. First we find moments and centroids of areas, starting with simple shapes and going on to those that require integration. Then we compute centroids of solids of revolution. This is followed by two applications involving centroids: finding the force due to fluid pressure, and computing the work required for various tasks.

Next we compute the second moment, or moment of inertia, for an area, and the polar moment of inertia of solids of revolution. Both of these quantities are useful when studying strength of materials or the motion of rigid bodies.

10-1. CENTROIDS

Center of Gravity and Centroid

The *center of gravity* (or *center of mass;* Fig. 10-1) of a body is the point where all the mass can be thought to be concentrated, without altering the affect that the earth's gravity has upon it. For simple shapes such as a sphere, cube, or cylinder, the center of gravity is exactly where you would expect it to be, at the center of the object.

Any object hung from a point will swing to where its center of gravity is directly below the point of suspension.

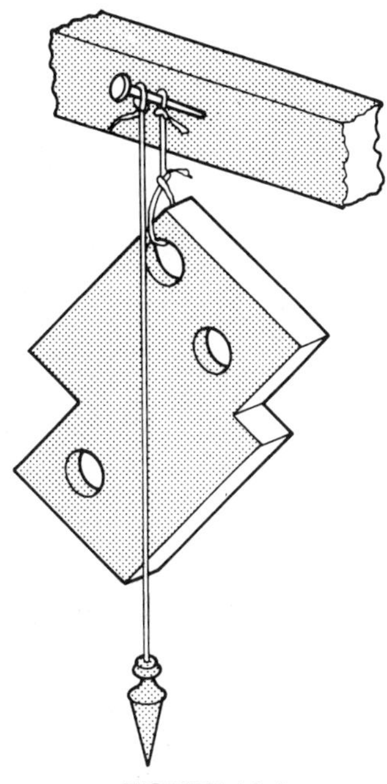

FIGURE 10-1

A plane area has no thickness, and hence has no weight or mass. Since it makes no sense to speak of center of mass or center of gravity for a weightless figure, we use instead the word *centroid*.

First Moment

We will learn about second moments (moments of inertia) in Sec. 10-4.

To calculate the position of the centroid in figures of various shapes, we need the idea of the *first moment*. It is now new to us. We know that the moment of a force about some point a is the product of the force F and the perpendicular distance d from the point to the line of action of the force. In a similar way, we speak of the *moment of an area* about some axis (Fig. 10-2a):

$$\text{moment of an area} = \text{area} \times \text{distance}$$

or *moment of a volume* (Figure 10-2b):

$$\text{moment of a volume} = \text{volume} \times \text{distance}$$

or *moment of a mass* (Fig. 10-2c):

$$\text{moment of a mass} = \text{mass} \times \text{distance}$$

In each case, the distance is that from the axis about which we take the moment, measured to some point on the area, volume, or mass. But to

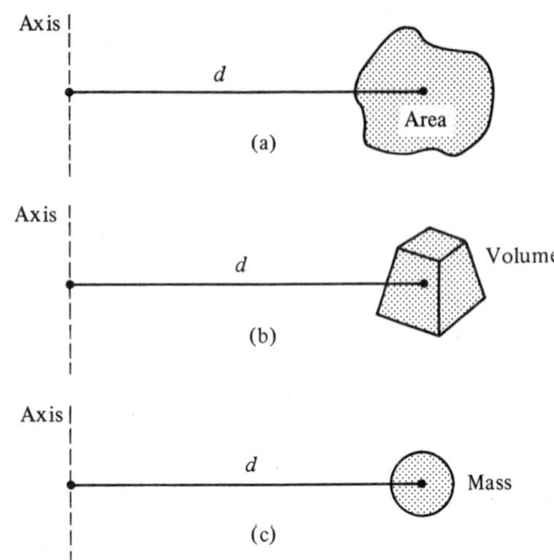

FIGURE 10-2. Moment of an area, a volume, and a mass.

which point on the figure shall we measure? To the *centroid* or *center of gravity*.

Centroids of Simple Shapes

If a shape has an axis of symmetry, the centroid wil be located on that axis. If there are two or more axes of symmetry, the centroid is found at the intersection of those axes.

To find the centroid of an area that can be subdivided into simple shapes whose centroid locations are known, we replace each shape by its centroid and proceed as in the following example.

EXAMPLE 1: Find the location of the centroid of the shape in Fig. 10-3a.

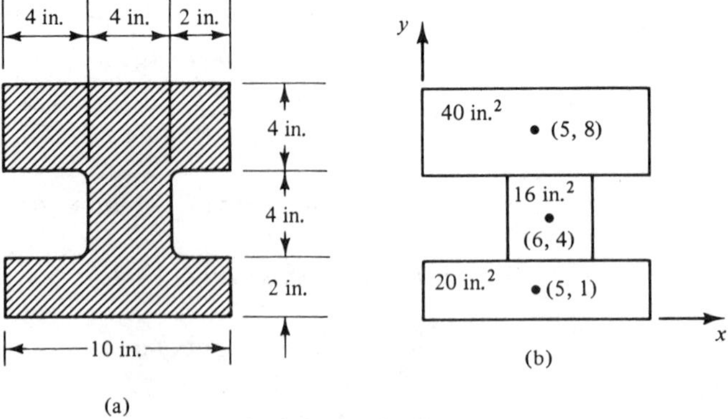

FIGURE 10-3

Solution: We subdivide the area into rectangles (Fig. 10-3b), locate the centroid, and compute the area of each. We also choose axes from which we will measure the coordinates of the centroid, $\bar{x}$ and $\bar{y}$. The total area is

$$40 + 16 + 20 = 76 \text{ in.}^2$$

The moment of the three areas about the y axis is

$$40(5) + 16(6) + 20(5) = 396 \text{ in.}^3$$

Then, since

$$\text{area} \times \text{distance to centroid} = \text{moment}$$

$$\bar{x} = \frac{\text{moment}}{\text{area}}$$

$$= \frac{396 \text{ in.}^3}{76 \text{ in.}^2} = 5.21 \text{ in.}$$

The moment about the x axis is

$$40(8) + 16(4) + 20(1) = 404 \text{ in.}^3$$

So

$$\bar{y} = \frac{404 \text{ in.}^3}{76 \text{ in.}^2} = 5.32 \text{ in.}$$

Centroids by Integration

If an area does not have axes of symmetry whose intersection gives us the location of the centroid, we can often find it by integration. We subdivide the area into thin strips, compute the first moment of each, sum these moments by integration, and then divide by the total area to get the distance to the centroid.

Consider the area bounded by the curves $y_1 = f_1(x)$ and $y_2 = f_2(x)$ and the lines $x = a$ and $x = b$ (Fig. 10-4). We draw a vertical element of area of width dx and height $(y_2 - y_1)$. Since the strip is narrow, all points on it may be considered to be the same distance x from the y axis. The moment dM_y of that strip about the y axis is then

$$dM_y = x\,dA = x(y_2 - y_1)\,dx$$

since $dA = (y_2 - y_1)\,dx$. We get the total moment M_y by integrating,

$$M_y = \int_a^b x(y_2 - y_1)\,dx$$

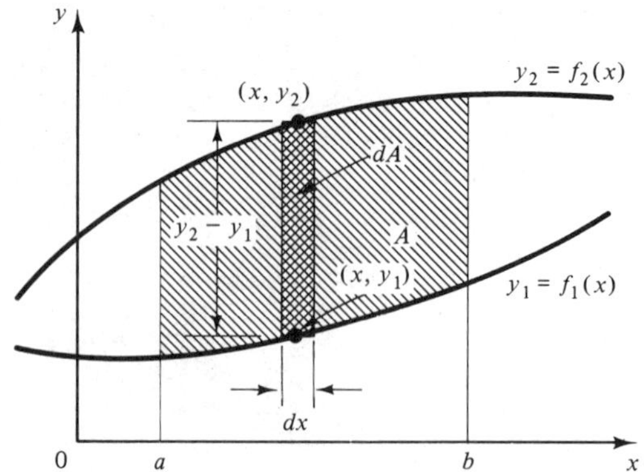

FIGURE 10-4. Centroid of irregular area found by integration.

But since the moment M_y is equal to the area A times the distance $\bar{x}$ to the centroid, we get $\bar{x}$ by dividing the moment by the area. So

Horizontal Distance to Centroid	$\bar{x} = \frac{1}{A}\int_a^b x(y_2 - y_1)\,dx$	**361**

To find $\bar{x}$ we must have the area A. It can be found by integration as in Chapter 8.

We now find the moment about the x axis. The centroid of the vertical element is at its midpoint, which is at a distance of $(y_1 + y_2)/2$ from the x axis. Thus the moment of the element about the x axis is

$$dM_x = \frac{y_1 + y_2}{2}(y_2 - y_1)\,dx$$

Equations 361 and 362 apply only for *vertical elements*. When using horizontal elements, interchange *x* and *y* in these equations.

Integrating and dividing by the area gives us $\bar{y}$.

Vertical Distance to Centroid	$\bar{y} = \frac{1}{2A}\int_a^b (y_1 + y_2)(y_2 - y_1)\,dx$	**362**

Our first example is for an area bounded by one curve and the x axis.

EXAMPLE 2: Find the centroid of the area bounded by the parabola $y^2 = 4x$, the x axis, and the line $x = 1$ (Fig. 10-5).

Solution: We need the area, so we find that first. We draw a vertical strip having an area $y\,dx$ and integrate.

$$A = \int y\,dx = 2\int_0^1 x^{1/2}\,dx = 2\left[\frac{x^{3/2}}{3/2}\right]_0^1 = \frac{4}{3}\text{ ft}^2$$

Then by Eq. 361,

$$\bar{x} = \frac{1}{A}\int_0^1 x(2x^{1/2} - 0)\,dx$$

$$= \frac{3}{4}\int_0^1 2x^{3/2}\,dx = \frac{3}{2}\frac{x^{5/2}}{(5/2)}\bigg|_0^1 = \frac{3}{5}\text{ ft}$$

and by Eq. 362,

$$\bar{y} = \frac{1}{2A}\int_0^1 (2x^{1/2} + 0)(2x^{1/2} - 0)\,dx$$

$$= \frac{3}{8}\int_0^1 4x\,dx = \frac{3}{8}\frac{4x^2}{2}\bigg|_0^1 = \frac{3}{4}\text{ ft}$$

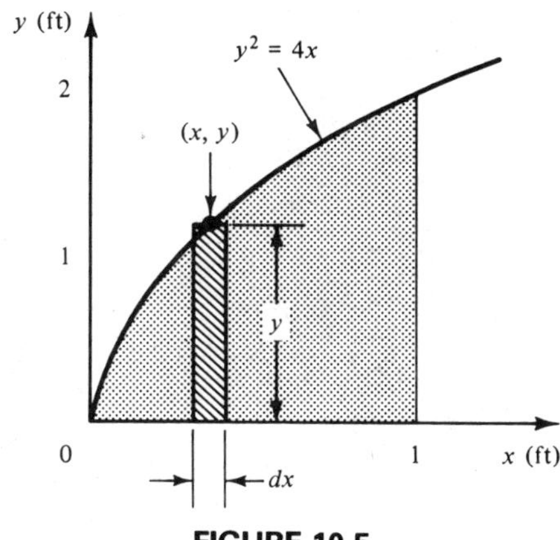

FIGURE 10-5

For areas that are not bounded by a curve and a coordinate axis, but are instead bounded by two curves, the work is only slightly more complicated, as shown in the following example.

EXAMPLE 3: Find the coordinates of the centroid of the area bounded by the curves $6y = x^2 - 4x + 4$ and $3y = 16 - x^2$.

Solution: We plot the curves (Fig. 10-6) and find their points of intersection by solving simultaneously. Multiplying the second equation by -2 and adding gives

$$3x^2 - 4x - 28 = 0$$

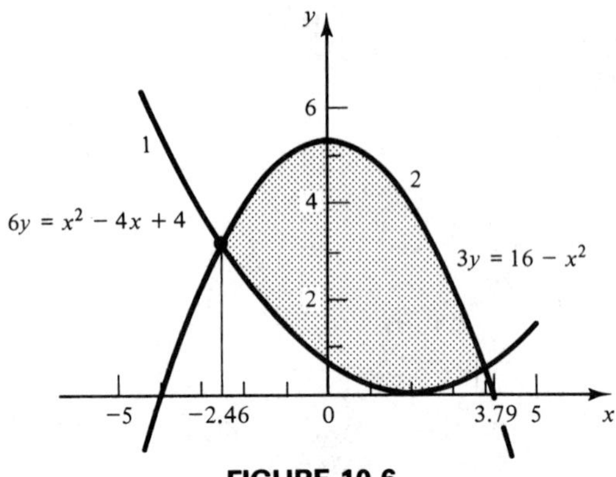

FIGURE 10-6

Solving by quadratic formula (work not shown) we find that the points of intersection are at $x = -2.46$ and $x = 3.79$. We take a vertical strip whose width is dx and whose height is $y_2 - y_1$, where

$$y_2 - y_1 = \frac{16}{3} - \frac{x^2}{3} - \frac{x^2}{6} + \frac{4x}{6} - \frac{4}{6} = \frac{28 + 4x - 3x^2}{6}$$

The area is then

$$\begin{aligned} A &= \int_{-2.46}^{3.79} (y_2 - y_1)\, dx \\ &= \frac{1}{6} \int_{-2.46}^{3.79} (28 + 4x - 3x^2)\, dx = 20.6 \end{aligned}$$

Then from Eq. 361,

$$\begin{aligned} AM_y &= \int_{-2.46}^{3.79} x(y_2 - y_1)\, dx \\ &= \frac{1}{6} \int_{-2.46}^{3.79} (28x + 4x^2 - 3x^3)\, dx = 13.6 \end{aligned}$$

Then

$$\bar{x} = \frac{M_y}{A} = \frac{13.6}{20.6} = 0.660$$

We now substitute into Eq. 362, with

$$y_1 + y_2 = \frac{36 - 4x - x^2}{6}$$

Thus

$$AM_x = \frac{1}{72}\int_{-2.46}^{3.79} (36 - 4x - x^2)(28 + 4x - 3x^2)\,dx$$

$$= \frac{1}{72}\int_{-2.46}^{3.79} (3x^4 + 8x^3 - 152x^2 + 32x + 1008)\,dx$$

$$= \frac{1}{72}\left|\frac{3x^5}{5} + \frac{8x^4}{4} - \frac{152x^3}{3} + \frac{32x^2}{3} + 1008x\right|_{-2.46}^{3.79} = 52.6$$

so

$$\bar{y} = \frac{M_x}{A} = \frac{52.6}{20.6} = 2.55$$

Centroids of Solids of Revolution by Integration

A solid of revolution is, of course, symmetrical about the axis of revolution, so the centroid must be on that axis. We only have to find the position of the centroid along that axis. The procedure is similar to that for an area. We think of the solid as being subdivided into many small elements of volume, find the sum of the moments for each element by integration, and set this equal to the moment of the entire solid (the product of its total volume and the distance to the centroid). We then divide by the volume to obtain the distance to the centroid.

EXAMPLE 4: Find the centroid of a hemisphere of radius r.

Solution: We place the hemisphere on coordinate axes (Fig. 10-7), and consider it as the solid obtained by rotating the first-quadrant portion of the curve $x^2 + y^2 = r^2$ about the x axis. Through the point (x, y) we draw an element of volume of radius y and thickness dx, at a distance x from the base of the hemisphere. Its volume is thus

$$dV = \pi y^2\,dx$$

and its moment about the base of the hemisphere is,

$$dM_y = \pi x y^2\,dx = \pi x(r^2 - x^2)\,dx$$

Of course, these methods work only for a solid that is homogeneous; that is, one whose density is the same throughout.

Integrating gives us the total moment,

$$M_y = \pi\int_0^r x(r^2 - x^2)\,dx = \pi\int_0^r (r^2x - x^3)\,dx$$

$$= \pi\left[\frac{r^2x^2}{2} - \frac{x^4}{4}\right]_0^r = \frac{\pi r^4}{4}$$

The total moment also equals the volume ($\frac{2}{3}\pi r^3$ for a hemisphere) times the distance $\bar{x}$ to the centroid, so

$$\bar{x} = \frac{\pi r^4/4}{2\pi r^3/3} = \frac{3r}{8}$$

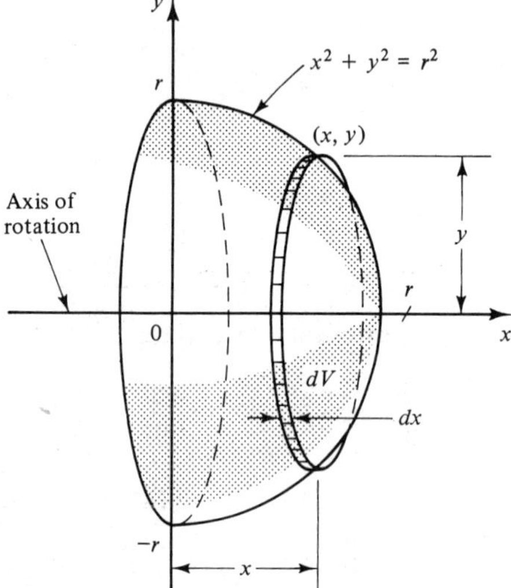

FIGURE 10-7. Finding the centroid of a hemisphere.

As with centroids of areas, we can write a formula for finding the centroid of a solid of revolution. For the volume V (Fig. 10-8) formed by rotating the curve $y = f(x)$ about the x axis, the distance to the centroid is:

Distance to the Centroid of Volume of Revolution about x Axis	$\bar{x} = \frac{\pi}{V}\int_a^b xy^2\,dx$	**363**

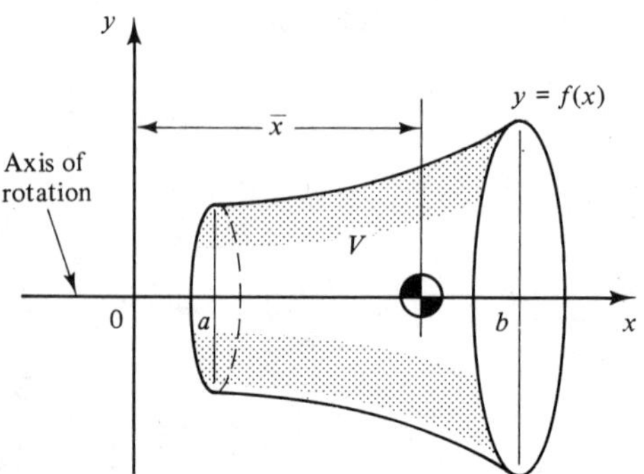

FIGURE 10-8. Centroid of a solid of revolution.

For volumes formed by rotation about the y axis, we simply interchange x and y in Eq. 363 and get

Distance to the Centroid of Volume of Revolution about y Axis	$\bar{y} = \frac{\pi}{V} \int_c^d yx^2\,dy$	**364**

These formulas work for solid figures only, not for those that have holes down their center.

EXERCISE 1

Without Integration

1. Find the centroid of four particles of equal mass located at (0, 0), (4, 2), (3, −5), and (−2, −3).

2. Find the centroid in Fig. 10-9a.

3. Find the centroid in Fig. 10-9b.

4. Find the centroid in Fig. 10-9c.

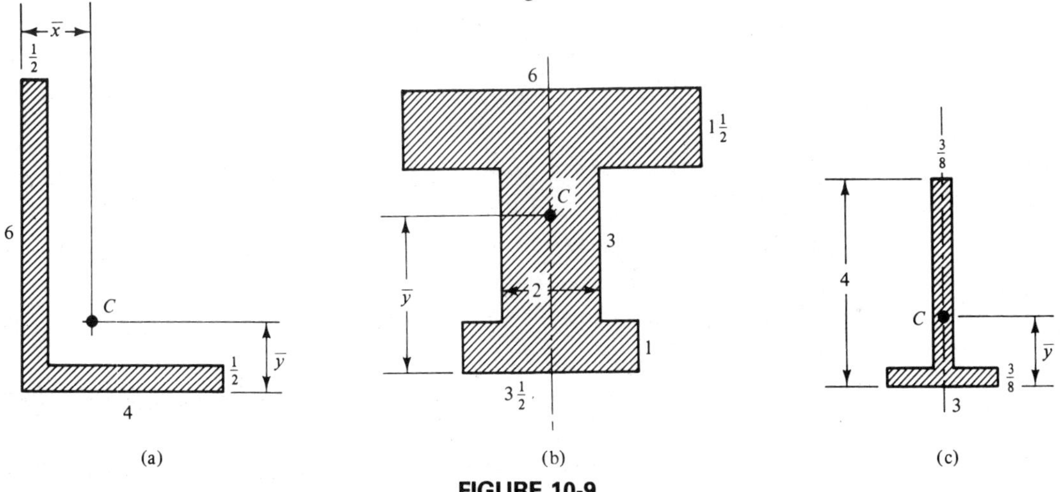

FIGURE 10-9

Centroids of Areas by Integration

Find the coordinates of the centroid of each area.

5. bounded by $y^2 = 4x$ and $x = 4$; find $\bar{x}$ and $\bar{y}$

6. bounded by $y^2 = 2x$ and $x = 5$; find $\bar{x}$ and $\bar{y}$

7. bounded by $y = \frac{1}{2}(e^x + e^{-x})$, the coordinate axes, and the line $x = 1$; find $\bar{x}$

8. bounded by $y = x^2$, the x axis, and $x = 3$; find $\bar{y}$

9. bounded by $y = e^x$, $x = 0$, and $x = 1$; find $\bar{y}$

10. bounded by $\sqrt{x} + \sqrt{y} = 1$ and the coordinate axes (area $= \frac{1}{6}$); find $\bar{x}$ and $\bar{y}$

Areas Bounded by Two Curves

11. bounded by $y = x^3$ and $y = 4x$, in the first quadrant; find $\bar{x}$

12. bounded by $x = 4y - y^2$ and $y = x$; find $\bar{y}$

13. bounded by $y^2 = x$ and $x^2 = y$; find $\bar{x}$ and $\bar{y}$

14. bounded by $y^2 = 4x$ and $y = 2x - 4$; find $\bar{y}$

15. bounded by $2y = x^2$ and $y = x^3$; find $\bar{x}$

16. bounded by $y = x^2 - 2x - 3$ and $y = 6x - x^2 - 3$ (area $= 21.33$); find $\bar{x}$ and $\bar{y}$

17. bounded by $y = x^2$ and $y = 2x + 3$, in the first quadrant; find $\bar{x}$

Centroids of Volumes of Revolution

Find the distance from the origin to the centroid of each volume.

18. formed by rotating the area bounded by $x^2 + y^2 = 4$, $x = 0$, $x = 1$, and the x axis about the x axis

19. formed by rotating the area bounded by $6y = x^2$, the line $x = 6$, and the x axis about the x axis

20. formed by rotating the first quadrant area under the curve $y^2 = 4x$, from $x = 0$ to $x = 1$, about the x axis

21. formed by rotating the area bounded by $y^2 = 4x$, $y = 6$, and the y axis about the y axis

22. formed by rotating the area bounded by $y = e^x$, $x = 0$ and $x = 1$ about the x axis

23. a paraboloid of revolution bounded by a plane through the focus perpendicular to the axis of symmetry

24. formed by rotating the first-quadrant portion of the ellipse $x^2/64 + y^2/36 = 1$ about the x axis

25. a right circular cone of height h, measured from its base

Applications

26. A certain airplane rudder (Fig. 10-10) consists of one quadrant $0AB$ of an ellipse, and a quadrant $0BC$ of a circle. Find the coordinates of the centroid.

27. The vane on a certain wind generator has the shape of a semicircle attached to a trapezoid (Fig. 10-11). Find the distance $\bar{x}$ to the centroid.

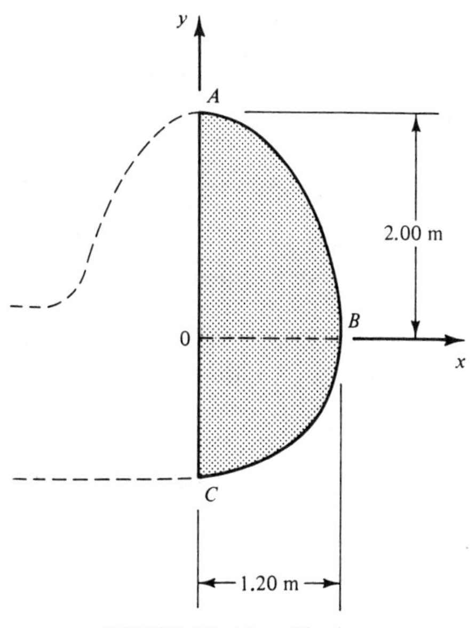

FIGURE 10-10. Airplane rudder.

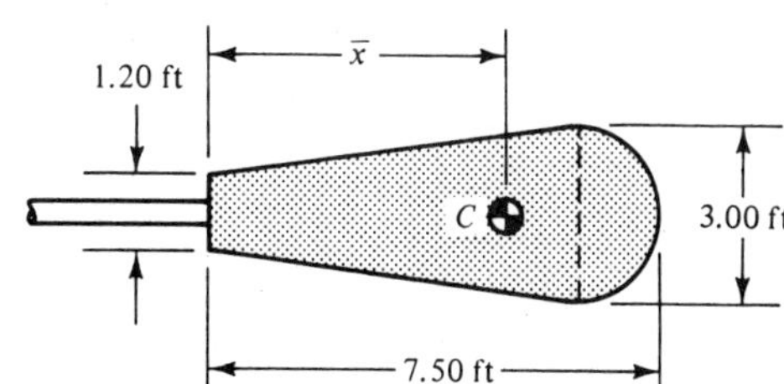

FIGURE 10-11. Wind vane.

28. A certain rocket (Fig. 10-12) consists of a cylinder attached to a paraboloid of revolution. Find the distance from the nose of the rocket to the centroid of the total volume.

29. An optical instrument contains a mirror in the shape of a paraboloid of revolution (Fig. 10-13) hollowed out of a cylindrical block of glass. Find the distance from the bottom of the mirror to the centroid of the mirror.

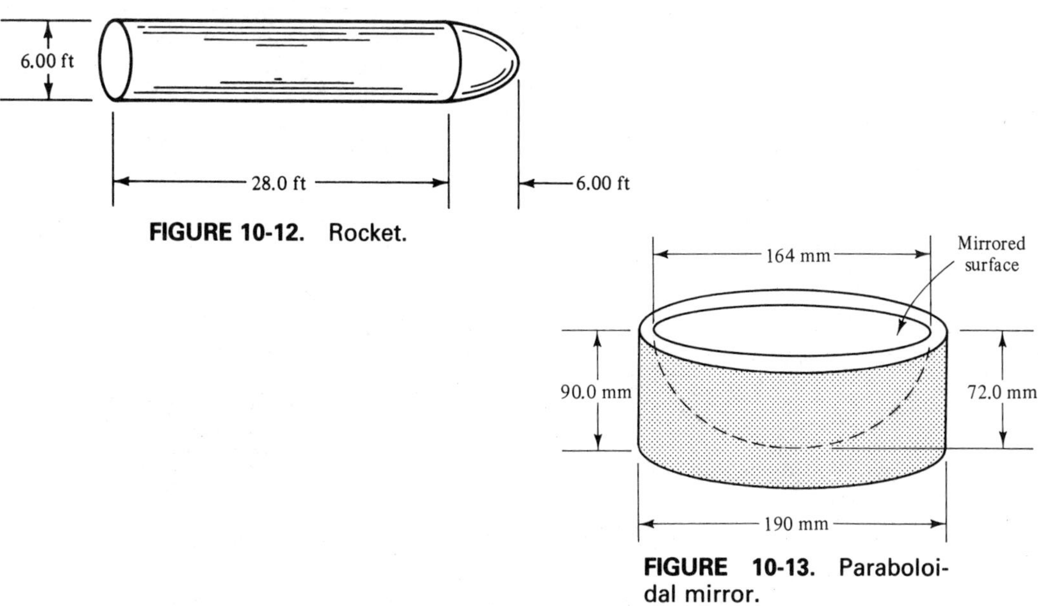

FIGURE 10-12. Rocket.

FIGURE 10-13. Paraboloidal mirror.

10-2. FLUID PRESSURE

The pressure at any point on a submerged surface varies directly with the depth of that point below the surface. Thus the pressure on a diver at a depth of 50 ft will be twice that at 25 ft.

The pressure on a submerged area is equal to the weight of the column of fluid above that area. Thus a square foot of area at a depth of 20 ft. supports a column of water having a volume of $20 \times 1^2 = 20$ ft^3. Since the density of water is 62.4 lb/ft^3, the weight of this column is $20 \times 62.4 = 1248$ lb, so the pressure is 1248 lb/ft^2 at a depth of 20 ft. Further, Pascal's law says that the pressure is the same in *all directions*, so the same 1248 lb/ft^2 will be felt by a surface that is horizontal, vertical, or at any angle.

The *force* exerted by the fluid can be found from

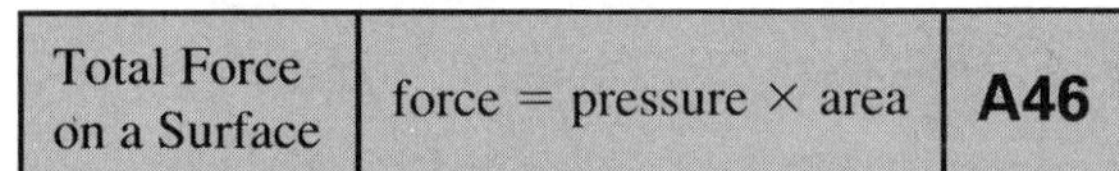

Total Force on a Surface	force = pressure × area	**A46**

The complication arises from an area that has points at various depths, and hence has different pressures over its surface. To compute the force on such a surface, we first compute the force on a narrow horizontal strip of area, assuming that the pressure is the same everywhere on that strip, and then add up the forces on all such strips by integration.

EXAMPLE 1: Find an expression for the force on the vertical area A submerged in a fluid of density δ (Fig. 10-14).

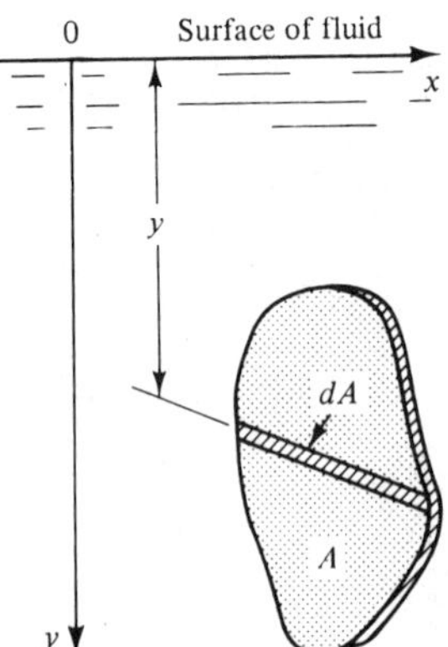

FIGURE 10-14. Force on a submerged surface.

Solution: Let us take our origin at the surface of the fluid, with the y axis downward. We draw a horizontal strip whose area is dA, located at a depth y below the surface. The pressure at depth y is $y\delta$, so the force dF on the strip is, by Eq. A46,

$$dF = y\delta \, dA = \delta(y \, dA)$$

Integrating, we get

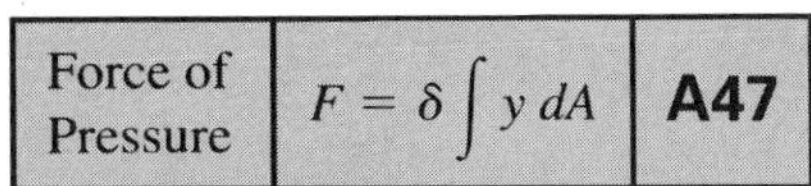

Force of Pressure	$F = \delta \int y \, dA$	**A47**

But the product $y \, dA$ is nothing but *the first moment of the area dA about the x axis*. Thus integration will give us the moment M_x of the entire area, about the x axis, multiplied by the density,

$$F = \delta \int y \, dA = \delta M_x$$

But the moment M_x is also equal to the area A times the distance $\bar{y}$ to the centroid, so

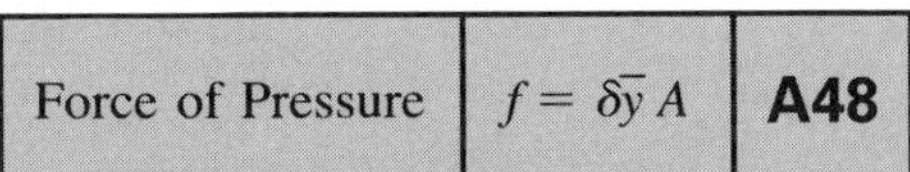

Force of Pressure	$f = \delta \bar{y} A$	**A48**

Thus *the force on a submerged, vertical surface is equal to the product of its area, the distance to its centroid, and the density of the fluid.*

We can compute the force on a submerged surface by using either Eq. A47 or Eq. A48. We give now an example of each method.

EXAMPLE 2: A vertical wall in a dam (Fig. 10-15) holds back water whose level is at the top of the wall. Find the total force on the wall by integration.

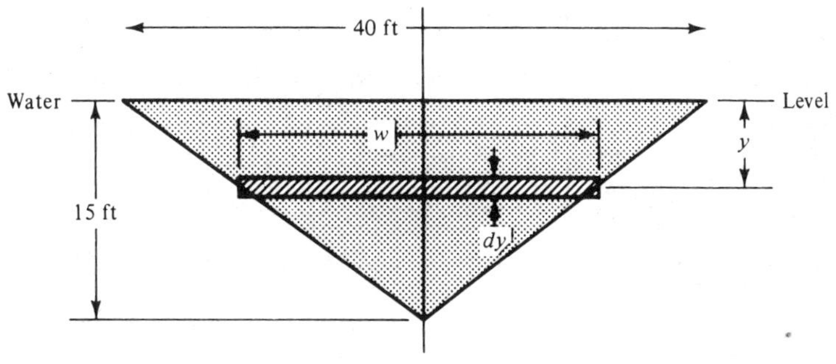

FIGURE 10-15

Solution: We sketch an element of area, with width w and height dy. So

$$dA = w\,dy \tag{1}$$

But by similar triangles

$$\frac{w}{40} = \frac{15 - y}{15}$$

$$w = \frac{40}{15}(15 - y)$$

Substituting into (1),

$$dA = \frac{40}{15}(15 - y)\,dy$$

Then by Eq. A47,

$$F = \delta \int y\,dA = \frac{40\delta}{15}\int_0^{15} y(15 - y)\,dy$$

$$= \frac{40\delta}{15}\int_0^{15} (15y - y^2)\,dy$$

$$= \frac{40\delta}{15}\left[\frac{15y^2}{2} - \frac{y^3}{3}\right]_0^{15}$$

$$= \frac{40(62.4)}{15}\left[\frac{15(15)^2}{2} - \frac{(15)^3}{3}\right] = 93{,}600 \text{ lb}$$

EXAMPLE 3: The area shown in Fig. 10-5 is submerged in water (density = 62.4 lb/ft^3) so that the origin is 10.0 ft below the surface. Find the force on the area using Eq. A48.

Solution: The area and the distance to the centroid have already been found in Example 2 of Sec. 10-1: area = $\frac{4}{3}$ ft^2 and $\bar{y} = \frac{3}{4}$ ft up from the origin, as shown in Fig. 10-16. The depth of the centroid below the surface is then

$$10.0 - \frac{3}{4} = 9\frac{1}{4} \text{ ft}$$

Thus, by Eq. A48,

$$\text{force} = \frac{62.4 \text{ lb}}{\text{ft}^3} \cdot 9\frac{1}{4} \text{ ft} \cdot \frac{4}{3} \text{ ft}^3 = 770 \text{ lb}$$

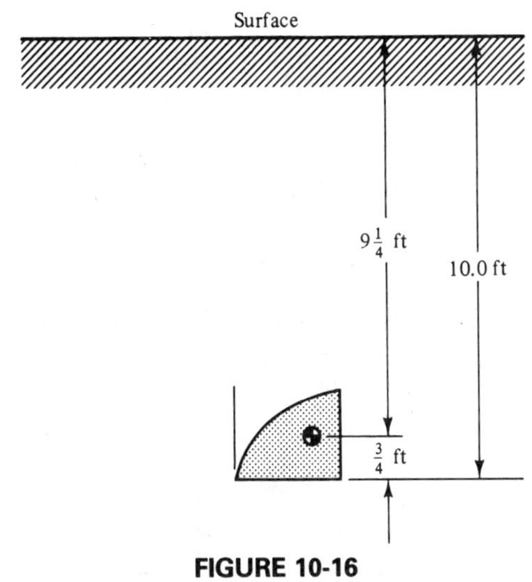

FIGURE 10-16

EXERCISE 2

1. A circular plate 6.0 ft in diameter is placed vertically in a dam with its center 50 ft below the surface of the water. Find the force on the plate.

2. A vertical rectangular plate in the wall of a reservoir has its upper edge 20.0 ft below the surface of the water. It is 6.0 ft wide and 4.0 ft high. Find the force on the plate.

3. A vertical rectangular gate in a dam is 10.0 ft wide and 6.0 ft high. Find the force on the gate when the water level is 8.0 ft above the top of the gate.

4. A vertical cylindrical tank has a diameter of 30.0 ft and a height of 50.0 ft. Find the total force on the curved surface when the tank is full of water.

5. A trough, whose cross section is an equilateral triangle with a vertex down, has sides 2.0 ft long. Find the total force on one end when the trough is full of water.

6. A horizontal cylindrical boiler 4.0 ft in diameter is half full of water. Find the force on one end.

7. A horizontal tank of oil (density = 60 lb/ft^3) has ends in the shape of an ellipse with horizontal axis 12.0 ft long and vertical axis 6.0 ft long (Fig. 10-17). Find the force on one end when the tank is half full.

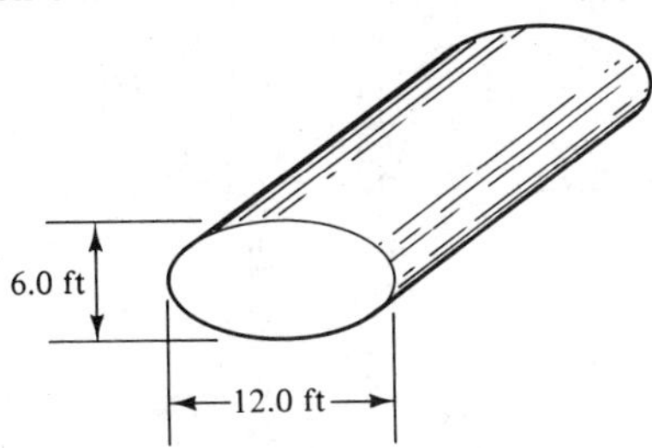

FIGURE 10-17. Oil tank.

8. The cross section of a certain trough is a parabola with vertex down. It is 2.0 ft deep and 4.0 ft wide at the top. Find the force on one end when the trough is full of water.

10-3. WORK

Definition

When a constant force acts on an object which moves in the direction of the force, the *work* done by the force is the product of the force and the distance moved by the object.

Work Done by a Constant Force	work = force × distance	**A6**

EXAMPLE 1: The work needed to lift a 100-lb weight a distance of 2 ft is 200 ft-lb.

Variable Force

Equation A6 applies when the force is *constant,* but this is not always the case. The force needed to stretch or compress a spring, for example, increases as the spring gets extended (Fig. 10-18). Another example of a variable force is that exerted by the expanding gases on the piston in an automobile engine.

If we let the variable force be represented by $F(x)$, acting in the x direction from $x = a$ to $x = b$, the work done by this force may be defined as

Work Done by a Variable Force	$W = \int_a^b F(x)\, dx$	**A7**

We first apply this formula to find the work done in stretching or compressing a spring.

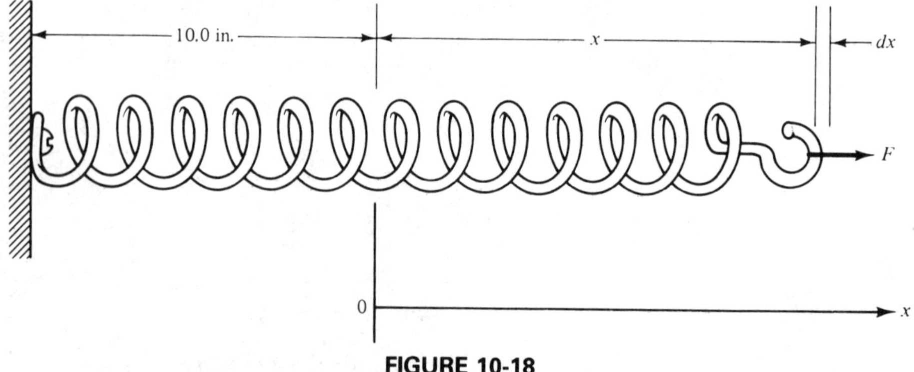

FIGURE 10-18

EXAMPLE 2: A certain spring (Fig. 10-18) has a free length (when no force is applied) of 10.0 in. The spring constant is 12.0 lb/in. Find the work needed to stretch the spring from a length of 12.0 in. to a length of 14.0 in.

Solution: We draw the spring partly stretched, as shown, taking our x axis in the direction of movement of the force, with zero at the free position of the end of the spring. The force needed to hold the spring in this position is equal to the spring constant k times the deflection x:

$$F = kx \tag{A12}$$

If we assume that the force does not change when stretching the spring an additional small amount dx, the work done is

$$dW = F\,dx = kx\,dx$$

We get the total work by integrating:

$$W = \int F\,dx = k\int_2^4 x\,dx = \frac{kx^2}{2}\bigg|_2^4$$

$$= \frac{12.0}{2}(4^2 - 2^2) = 72 \text{ in. lb}$$

Common Error	Be sure to measure spring deflections from the *free* end of the spring, not from the fixed end.

Another typical problem is that of finding the work needed to pump the fluid out of a tank. Such problems may be solved by noting that the work required is equal to the weight of the fluid times the distance which the centroid of the fluid (when still in the tank) must be raised.

EXAMPLE 3: A hemispherical tank (Fig. 10-19) is filled with water having a density of 62.4 lb/ft^3. Find the work needed to pump all the water to a height of 10.0 ft above the top of the tank.

Solution: The distance to the centroid of a hemisphere of radius r was found in Sec. 10-1 to be $3r/8$, so the centroid of our hemisphere is at a distance

$$\frac{3}{8}\cdot 8 = 3.0 \text{ ft}$$

as shown. It must therefore be raised a distance of 13.0 ft. We find the weight of the tankful of water by multiplying its volume times its density:

$$\text{weight} = \frac{2}{3}\pi(8^3)(62.4) = 66{,}900 \text{ lb}$$

The work done is then

$$\text{work} = 66{,}900 \text{ lb}(13.0 \text{ ft}) = 870{,}000 \text{ ft lb}$$

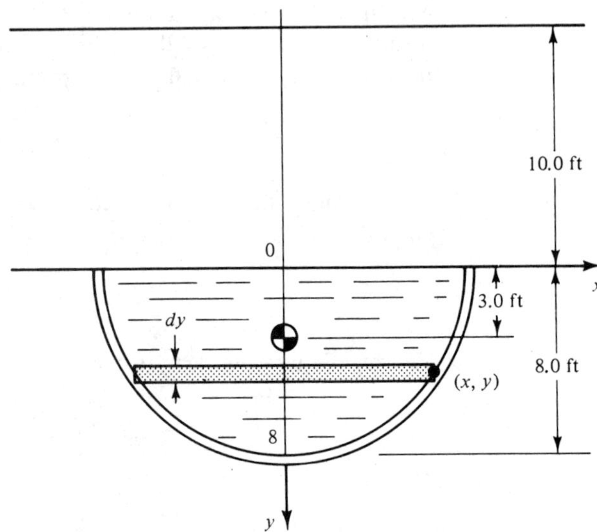

FIGURE 10-19. Hemispherical tank.

If we do not know the location of the centroid, we can find the work directly, by integration.

EXAMPLE 4: Repeat Example 3 by integration, assuming that the location of the centroid is not known.

Solution: We choose coordinate axes as shown in Fig. 10-19. Through some point (x, y) on the curve we draw an element of volume whose volume dV is

$$dV = \pi x^2\, dy$$

and whose weight is

$$62.4\, dV = 62.4\pi x^2\, dy$$

and since this element must be lifted a distance of $(10 + y)$ ft, the work required is

$$dW = (10 + y)(62.4\pi x^2\, dy)$$

Integrating,

$$W = 62.4\pi \int (10 + y)x^2\,dy$$

But, using the equation for a circle (Eq. 268),

$$x^2 = r^2 - y^2 = 64 - y^2$$

Substituting,

$$\begin{aligned} W &= 62.4\pi \int_0^8 (10 + y)(64 - y^2)\,dy \\ &= 62.4\pi \int_0^8 (640 + 64y - 10y^2 - y^3)\,dy \\ &= 62.4\pi \left[640y + 32y^2 - \frac{10y^3}{3} - \frac{y^4}{4}\right]_0^8 = 870{,}000 \text{ ft lb} \end{aligned}$$

as by the other method.

EXERCISE 3

Springs

1. A spring has a free length of 12.0 in., and a force of 50.0 lb will stretch it to a length of 13.0 in. How much work is needed to stretch the spring from a length of 14.0 in. to 16.0 in.?

2. A spring whose free length is 10.0 in. has a spring constant of 12.0 lb/in. Find the work needed to stretch this spring from 12.0 in. to 15.0 in.

3. A spring has a spring constant of 8.0 lb/in. and a free length of 5.0 in. Find the work required to stretch it from 6.0 in. to 8.0 in.

Tanks

4. Find the work required to pump all the water to the top of a vertical cylindrical tank, 16.0 ft in diameter and 20.0 ft deep, if it is completely filled at the start.

5. A hemispherical tank 12.0 ft in diameter is filled with water to a depth of 4.0 ft. How much work is needed to pump the water to the top of the tank?

6. A conical tank 20.0 ft deep and 20.0 ft across the top is full of water. Find the work needed to pump the water to a height of 15.0 ft above the top of the tank.

7. A tank has the shape of a frustum of a cone, with a top diameter of 8.0 ft, a bottom diameter of 12.0 ft, and a height of 10.0 ft. How much work is needed to pump the contents to a height of 10.0 ft above the tank, if filled with oil of density 50.0 lb/ft^3?

Gas Laws

8. Find the work needed to compress air initially at a pressure of 15.0 lb/in.^{2} from a volume of 200 ft^3 to 50.0 ft^3. (*Hint:* The work dW done in moving the piston (Fig. 10-20) is $dW = F\ dx$. Express both F and dx in terms of v by means of Eqs. 126 and A46, and integrate.)

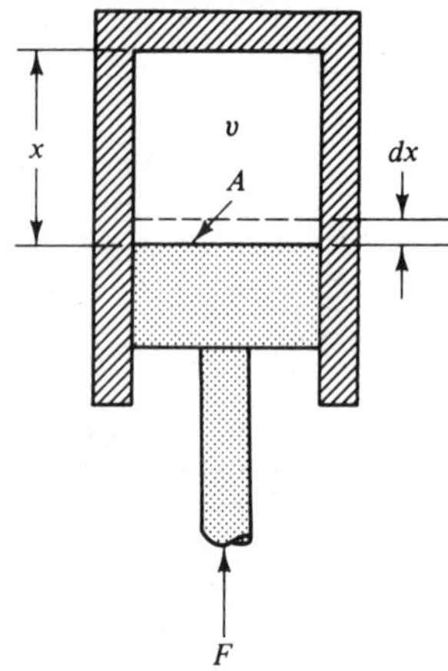

FIGURE 10-20. Piston and cylinder. Assume that the pressure and volume are related by the equation pv = constant.

9. Air is compressed from an initial pressure of 15.0 lb/in.^{2} and volume of 200 ft^3 to a pressure of 80.0 lb/in.^{2}. How much work was needed to compress the air?

10. If the pressure and volume of air are related by the equation $pv^{1.4} = k$, find the work needed to compress air initially at 14.5 lb/in.^{2} and 10.0 ft^3 to a pressure of 100 lb/in.^{2}.

Miscellaneous

11. The force of attraction (in pounds) between two masses separated by a distance d is equal to k/d^2, where k is a constant. If two masses are 50 ft apart, find the work needed to separate them another 50 ft. Leave your answer in terms of k.

12. Find the work needed to wind up a vertical cable 100 ft long, weighing 3.0 lb/ft.

13. A 500-ft-long cable weighs 1.0 lb/ft and is hanging from a tower with a 200-lb weight at its end. How much work is needed to raise the weight and the cable a distance of 20.0 ft?

10-4. MOMENT OF INERTIA

Moment of Inertia of an Area

Figure 10-21a and b each show a small area at a distance r from a line L in the same plane. In each case the dimensions of the area are such that we may consider all points on the area as being at the same distance r from the line L.

In Sec. 10-1 we defined the *first moment* of the area about L as being the product of the area times the distance to the line. We now define the *second moment*, or *moment of inertia I*, as the product of the area times the *square* of the distance to the line.

Moment of Inertia of an Element of Area	$i = Ar^2$	**365**

The distance r is called the *radius of gyration*.

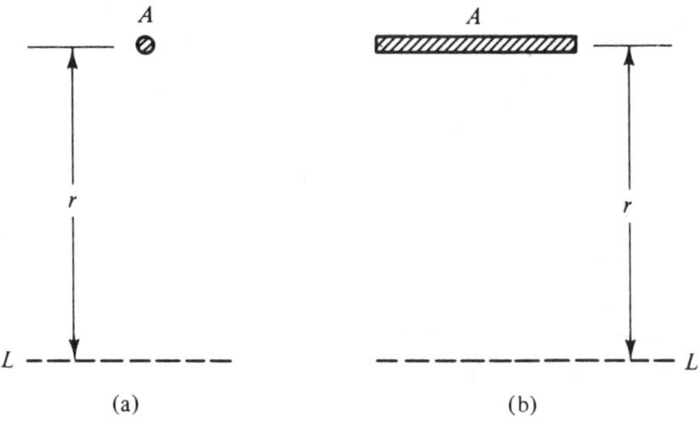

FIGURE 10-21

EXAMPLE 1: Find the moment of inertia of the thin strip in Fig. 10-21b if it has a length of 8.0 cm, a width of 0.20 cm, and is 7.0 cm from axis L.

Solution: The area of the strip is 8.0(0.20) = 1.6 cm^2, so the moment of inertia is, by Eq. 365,

$$I = 1.6(7.0)^2 = 78.4 \text{ cm}^3$$

Moment of Inertia of a Rectangle

In Example 1 our area was a thin strip parallel to the axis, with all points on the area at the same distance r from the axis. But what shall we use for r when dealing with an extended area, such as the rectangle

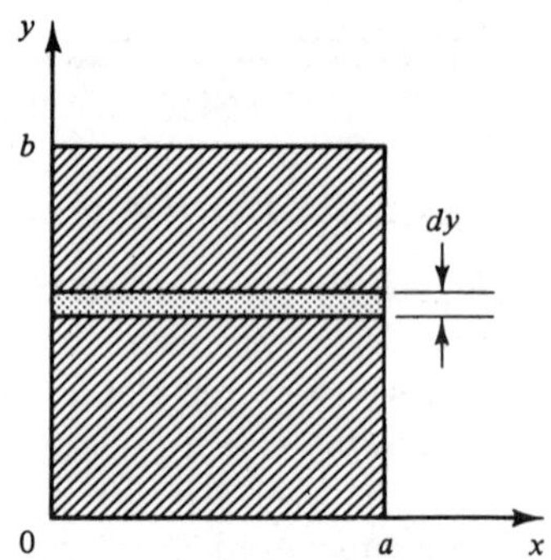

FIGURE 10-22. Moment of inertia of a rectangle.

in Fig. 10-22? Again calculus comes to our aid. Since we can easily compute the moment of inertia of a thin strip, we slice our area into many thin strips and add up their individual moments of inertia by integration.

EXAMPLE 2: Compute the moment of inertia of a rectangle about its base (Fig. 10-22).

Solution: We draw a single strip of area parallel to the axis about which we are taking the moment. This strip has a width dy and a length a. Its area is then

$$dA = a\,dy$$

All points on the strip are at a distance y from the x axis, so the moment of inertia, dI_x, is

$$dI_x = y^2(a\,dy)$$

We add up the moments of all such strips by integrating from $y = 0$ to $y = b$,

$$I_x = a\int_0^b y^2\,dy = \left.\frac{ay^3}{3}\right|_0^b = \frac{ab^3}{3}$$

Radius of Gyration

If all the area in a plane figure were squeezed into a single thin strip of equal area and placed parallel to the x axis at such a distance that it had the *same moment of inertia* as the original rectangle, it would be at a distance r that we call the *radius of gyration.* If I is the moment of inertia of some area A about some axis, then

$$Ar^2 = I$$

So

Radius of Gyration	$r = \sqrt{\frac{I}{A}}$	**369**

EXAMPLE 3: Find the radius of gyration for the rectangle in Example 2.

Solution: The area of the rectangle is *ab*, so by Eq. 369,

$$r = \sqrt{\frac{I}{A}} = \sqrt{\frac{ab^3}{3ab}} = \frac{b}{\sqrt{3}} \cong 0.577b$$

Note that the *centroid* is at a distance of $0.5b$ from the edge of the rectangle. This shows that *the radius of gyration is not equal to the distance to the centroid.*

Common Error	Do *not* use the distance to the *centroid* when computing moment of inertia.

Moment of Inertia of an Area by Integration

Being able to write the moment of inertia of a rectangular area now enables us to derive formulas for the moment of inertia of other areas, such as in Fig. 10-23. The area of the vertical strip shown in Fig. 10-23 is dA and its distance from the y axis is x, so its moment of inertia about the y axis is, by Eq. 365.

$$dI_y = x^2\, dA = x^2 y\, dx$$

since $dA = y\, dx$. The total moment is then found by integrating:

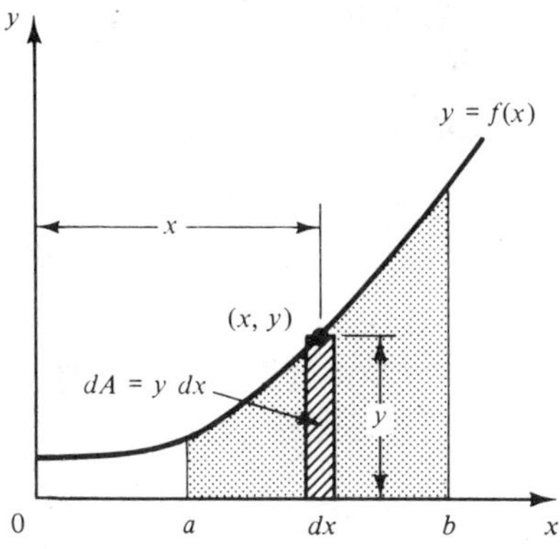

FIGURE 10-23

Moment of Inertia of an Area, about y Axis	$I_y = \int x^2 y\,dx$	**367**

To find the moment of inertia about the x axis, we use the result of an earlier example, that the moment of inertia of a rectangle about its base is $ab^3/3$. Thus the moment of inertia of the rectangle in Fig. 10-23 is

$$dI_x = \frac{1}{3} y^3\,dx$$

As before, the total moment of inertia is found by integrating:

Moment of Inertia of an Area about x Axis	$I_x = \frac{1}{3}\int y^3\,dx$	**366**

EXAMPLE 4: Find the moment of inertia about the x and y axes for the area under the curve $y = x^2$, from $x = 1$ to $x = 3$.

Solution: From Eq. 367,

$$I_y = \int_1^3 x^2(x^2)\,dx = \int_1^3 x^4\,dx$$

$$= \frac{x^5}{5}\bigg|_1^3 \cong 48.4$$

Now using Eq. 366, with $y^3 = (x^2)^3 = x^6$,

$$I_x = \frac{1}{3}\int_1^3 x^6\,dx = \frac{1}{3}\left[\frac{x^7}{7}\right]_1^3 = \frac{1}{21}(3^7 - 1^7) = \frac{2186}{21} \cong 104.1$$

EXAMPLE 5: Find the radius of gyration of the area in Example 4, about the x axis.

Solution: We first find the area of the figure. By Eq. 339,

$$A = \int_1^3 x^2\,dx = \frac{x^3}{3}\bigg|_1^3 = \frac{3^3}{3} - \frac{1^3}{3} \cong 8.667$$

From Example 4, $I_x = 104.1$. So by Eq. 369,

$$r_x = \sqrt{\frac{I_x}{A}} = \sqrt{\frac{104.1}{8.667}} = 3.466$$

Polar Moment of Inertia

Polar moment of inertia is needed when studying rotation of rigid bodies.

In the preceding sections we had found the moment of inertia of an *area* about some line in the plane of that area. Now we find moment of inertia of a *solid of revolution* about its axis of revolution. We call this the *polar moment of inertia*. We first find the polar moment of inertia for a thin-walled shell, then use that result to find the polar moment of inertia for any solid of revolution.

A thin-walled cylindrical shell (Fig. 9-4c) has a volume equal to the product of its circumference, wall thickness, and height,

$$dV = 2\pi rh\,dr \tag{355}$$

If we let m represent the mass per unit volume, the mass of the shell is then

$$dM = 2\pi mrh\,dr$$

Since we may consider all particles in this shell to be at a distance r from the axis of revolution, we obtain the moment of inertia of the shell by multiplying the mass by r^2,

Polar Moment of Inertia of a Shell	$dI = 2\pi mr^3h\,dr$	**374**

The polar amount of inertia of an entire solid of revolution is then found by integration.

Polar Moment of Inertia by the Shell Method	$I = 2\pi m \int r^3h\,dr$	**376**

EXAMPLE 6: Find the polar moment of inertia of the cylinder (Fig. 10-24) about its axis.

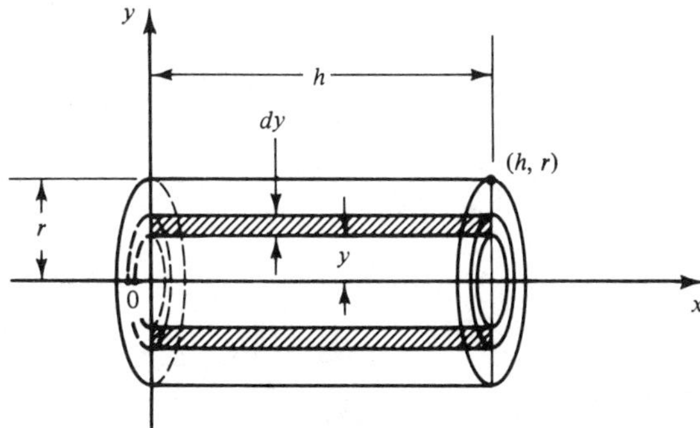

FIGURE 10-24. Polar moment of inertia of a solid cylinder.

Solution: We draw elements of volume in the form of concentric shells. The moment of inertia of each is

$$dI_x = 2\pi m y^3 h\, dy$$

Integrating yields

$$I_x = 2\pi mh \int_0^r y^3\, dy = 2\pi mh \left[\frac{y^4}{4}\right]_0^r$$

$$= \frac{\pi mhr^4}{2}$$

But since the volume of the cylinder is $\pi r^2 h$, we get

$$I_x = \frac{1}{2} mVr^2$$

In other words, *the polar moment of inertia of a cylinder is equal to half the product of its density, its volume, and the square of its radius.*

EXAMPLE 7: The first-quadrant area under the curve $y^2 = 8x$, from $x = 0$ to $x = 2$, is rotated about the x axis. Use the shell method to find the polar moment of inertia of the volume generated (Fig. 10-25).

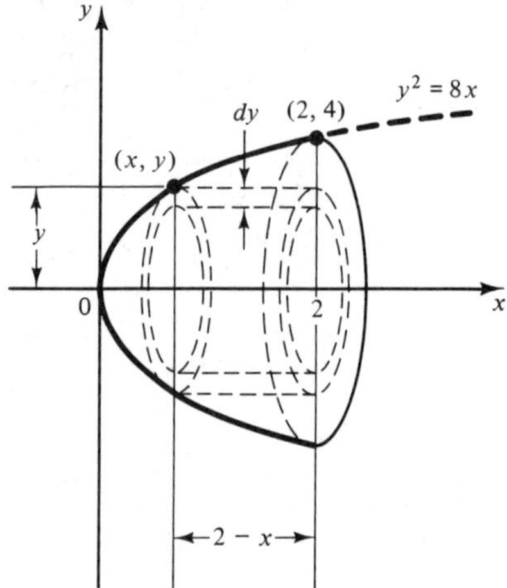

FIGURE 10-25. Polar moment of inertia by shell method.

Solution: Our shell element has a radius y, a length $2 - x$, and a thickness dy. By Eq. 374,

$$dI_x = 2\pi m y^3 (2 - x)\, dy$$

Replacing x by $y^2/8$ gives

$$dI_x = 2\pi m y^3\left(2 - \frac{y^2}{8}\right) dy$$

Integrating,

$$I_x = 2\pi m \int_0^4 \left(2y^3 - \frac{y^5}{8}\right) dy = 2\pi m \left|\frac{2y^4}{4} - \frac{y^6}{48}\right|_0^4 = 268m$$

Polar Moment of Inertia by the Disk Method

Sometimes the disk method will result in an integral that is easier to evaluate than that obtained by the shell method.

For the solid of revolution in Fig. 10-26, we choose a disk-shaped element of volume of radius r and thickness dh. Since it is a cylinder, we use the moment of inertia of a cylinder found in example 6,

$$dI = \frac{m\pi r^4\, dh}{2}$$

Integrating gives

Polar Moment of Inertia by the Disk Method	$I = \frac{m\pi}{2}\int_a^b r^4\, dh$	**375**

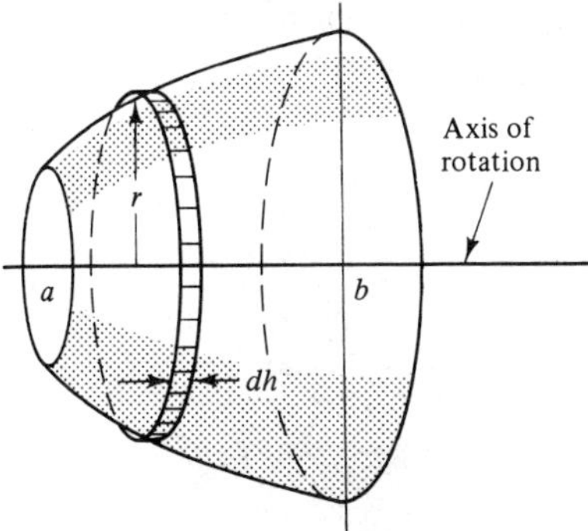

FIGURE 10-26

EXAMPLE 8: Repeat Example 7 by the disk method.

Solution: We use Eq. 375, with $r = y$ and $dh = dx$.

$$I_x = \frac{m\pi}{2}\int y^4\, dx$$

But $y^2 = 8x$, so $y^4 = 64x^2$. Substituting,

$$I_x = 32m\pi \int_0^2 x^2\,dx = 32m\pi \left[\frac{x^3}{3}\right]_0^2$$

$$= 32m\pi\left(\frac{2^3}{3}\right) \cong 268m$$

as before.

EXERCISE 4

Moment of Inertia of Plane Areas

1. Find the moment of inertia of the area bounded by $y = x$, $y = 0$, and $x = 1$ about the x axis.

2. Find the radius of gyration of the first-quadrant area bounded by $y^2 = 4x$, $x = 4$, and $y = 0$ about the x axis.

3. Find the radius of gyration of the area in Problem 2 about the y axis.

4. Find the moment of inertia of the area bounded by $x + y = 3$, $x = 0$, and $y = 0$ about the y axis.

5. Find the moment of inertia about the x axis of the first-quadrant area bounded by the curve $y = 4 - x^2$ and the coordinate axes.

6. Find the moment of inertia about the y axis of the area in Problem 5.

7. Find the moment of inertia about the x axis of the first-quadrant area bounded by the curve $y^3 = 1 - x^2$.

8. Find the moment of inertia about the y axis of the area in Problem 7.

9. Find the radius of gyration of the area under one arch of the sine curve $y = \sin x$ with respect to the x axis.

10. Find the moment of inertia of the area bounded by the curve $y = e^x$, the line $x = 1$, and the coordinate axes with respect to the x axis.

Polar Moment of Inertia

Find the polar moment of inertia of the volume formed when each first-quadrant area is rotated about the x axis.

11. bounded by $y = x$, $x = 2$, and the x axis

12. bounded by $y = x + 1$, from $x = 1$ to $x = 2$, and the x axis

13. bounded by the curve $y = x^2$, the line $x = 2$, and the x axis

14. bounded by $\sqrt{x} + \sqrt{y} = 2$ and the coordinate axes

Find the polar moment of inertia of each solid with respect to its axis in terms of the total mass M of the solid.

15. a right circular cone of height h and base radius r

16. a sphere of radius r

17. a paraboloid of revolution bounded by a plane through the focus perpendicular to the axis of symmetry

CHAPTER TEST

1. A conical tank is 8.0 ft in diameter at the top and 12.0 ft deep, and is filled with a liquid having a density of 80.0 lb/ft^3. How much work is required to pump all the liquid to the top of the tank?

2. Find the coordinates of the centroid of the area bounded by the x axis and a half-cycle of the sine curve, $y = \sin x$.

3. A cylindrical, horizontal tank is 6.0 ft in diameter and is half full of water. Find the force on one end of the tank.

4. Find the moment of inertia of a sphere of radius r and density m with respect to a diameter.

5. A spring has a free length of 12.0 in. and a spring constant of 5.45 lb/in. Find the work needed to compress it from a length of 11.0 in. to 9.0 in.

6. Find the coordinates of the centroid of the area bounded by the curve $y^2 = 2x$ and the line $y = x$.

7. A horizontal cylindrical tank 8.0 ft in diameter is half full of oil (60 lb/ft^3). Find the force on one end.

8. Find the moment of inertia of a right circular cone of base radius r, height h, and mass m with respect to its axis.

9. Find the centroid of a semicircle of radius 10.

10. A bucket that weighs 3.0 lb and has a volume of 2.0 ft^3 is filled with water. It is being raised at a rate of 5.0 ft/s while water leaks from the bucket at a rate of 0.01 ft^3/s. Find the work done in raising the bucket 100 ft.

11. Find the centroid of the area bounded by one quadrant of the ellipse $x^2/16 + y^2/9 = 1$.

12. Find the radius of gyration of the area bounded by the parabola $y^2 = 16x$ and its latus rectum, with respect to its latus rectum.

Methods of Integration

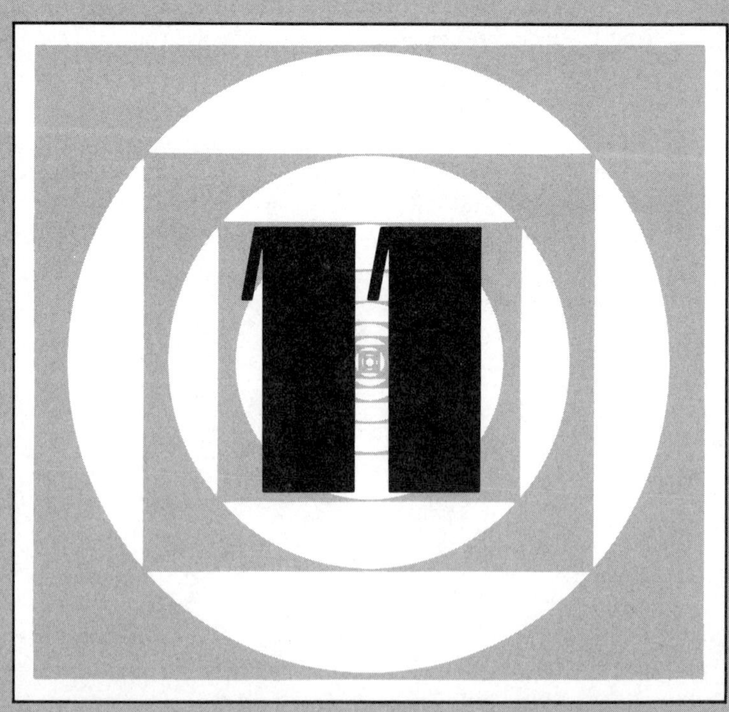

All practical methods of integration depend eventually on the use of a table of integrals. However, a given expression will seldom match any of the forms in the table, and will often look completely different. The methods given in this chapter will help us to manipulate a given expression into one of the listed forms.

But even with all our new methods there will always be expressions we cannot integrate, so we conclude this chapter with a few approximate methods of integration. These methods take on added importance because they can be used on the computer.

11-1. INTEGRATION BY PARTS

Derivation

Integration by parts is a useful method that enables us to split products into two parts for easier integration. We start with the rule for the derivative of a product (Eq. 310), in differential form,

$$d(uv) = u\,dv + v\,du$$

rearranging we get

$$u\,dv = d(uv) - v\,du$$

Integrating gives us

$$\int u\,dv = \int d(uv) - \int v\,du$$

or

$$6. \quad \int u\,dv = uv - \int v\,du$$

This rule is listed as number 6 in our table of integrals.

EXAMPLE 1: Integrate $\int x \cos 3x\,dx$.

Solution: We separate $x \cos 3x\,dx$ into two parts, one of which we call u and the other dv. But which part shall we call u and which dv?

A good rule of thumb is to take dv as *the most complicated part*, but one that you can still *easily integrate*. So we let

$$u = x \qquad \text{and} \qquad dv = \cos 3x\,dx$$

Now our rule 6 requires that we know du and v in addition to u and dv. We obtain du by differentiating u, and find v by integrating dv.

$$du = dx \qquad \text{and} \qquad v = \int \cos 3x\,dx$$

$$= \frac{1}{3}\int \cos 3x(3\,dx)$$

$$= \frac{1}{3}\sin 3x + C_1$$

Applying rule 6,

$$\underbrace{\int x \cos x\,dx}_{\int u\ dv} = \underbrace{x}_{u}\underbrace{\left(\frac{1}{3}\sin 3x + C_1\right)}_{v} - \int \underbrace{\left(\frac{1}{3}\sin 3x + C_1\right)}_{v}\underbrace{dx}_{du}$$

$$= \frac{x}{3}\sin 3x + C_1x - \frac{1}{9}\int \sin 3x(3\,dx) - \int C_1\,dx$$

$$= \frac{x}{3}\sin 3x + C_1x + \frac{1}{9}\cos 3x - C_1x + C$$

$$= \frac{x}{3}\sin 3x + \frac{1}{9}\cos 3x + C$$

Note that the constant C_1 obtained when integrating dv *does not appear in the final integral,* and in fact, the final result would be the same as if we had not introduced C_1 in the first place. So in the following examples we will follow the usual practice of dropping the constant C_1 when integrating dv.

EXAMPLE 2: Integrate $\int x \ln x\, dx$.

Solution: Let us choose

$$u = \ln x \qquad \text{and} \qquad dv = x\,dx$$

Then

$$du = \frac{1}{x}\,dx \qquad \text{and} \qquad v = \frac{x^2}{2} \text{ (plus a constant which we drop)}$$

By rule 6,

$$\int \underbrace{\ln x}_{u}\, \underbrace{(x\,dx)}_{dv} = \underbrace{(\ln x)}_{u} \underbrace{\left(\frac{x^2}{2}\right)}_{v} - \int \underbrace{\frac{x^2}{2}}_{v} \underbrace{\left(\frac{dx}{x}\right)}_{du}$$

$$= \frac{x^2}{2} \ln x - \frac{1}{2} \int x\,dx$$

$$= \frac{x^2}{2} \ln x - \frac{x^2}{4} + C$$

Sometimes our first choice of u and dv will result in an integral no easier to evaluate than our original.

EXAMPLE 3: Integrate $\int xe^x\, dx$.

Solution: *First try:* Let

$$u = e^x \qquad \text{and} \qquad dv = x\,dx$$

then

$$du = e^x\,dx \qquad \text{and} \qquad v = \frac{x^2}{2}$$

By rule 6,

$$\int xe^x\,dx = \frac{x^2e^x}{2} - \frac{1}{2}\int x^2e^x\,dx$$

we get an integral which is more difficult than the one we started with.

Second try: Let

$$u = x \qquad \text{and} \qquad dv = e^x\,dx$$

Then

$$du = dx \qquad \text{and} \qquad v = e^x$$

By rule 6,

$$\int xe^x\,dx = xe^x - \int e^x\,dx$$
$$= xe^x - e^x + C$$

Sometimes we may have to integrate by parts *twice* to get the final result.

EXAMPLE 4: Integrate $\int x^2 \sin x\,dx$.

Solution: Let

$$u = x^2 \qquad \text{and} \qquad dv = \sin x\,dx$$

Then

$$du = 2x\,dx \qquad \text{and} \qquad v = -\cos x$$

By rule 6,

$$\int x^2 \sin x\,dx = -x^2 \cos x + 2 \int x \cos x\,dx \tag{1}$$

Our new integral in (1) is similar to the original, but we note that the power of x has been reduced from 2 to 1. It might occur to us that by integrating by parts *again* we may reduce the exponent to zero. Let's try

$$u = x \qquad \text{and} \qquad dv = \cos x\,dx$$

Then

$$du = dx \qquad \text{and} \qquad v = \sin x$$

By rule 6,

$$\int x \cos x\,dx = x \sin x - \int \sin x\,dx$$
$$= x \sin x + \cos x + C_1$$

Substituting back into (1) gives us

$$\int x^2 \sin x\, dx = -x^2 \cos x + 2(x \sin x + \cos x + C_1)$$
$$= -x^2 \cos x + 2x \sin x + 2 \cos x + C$$

Once in a while, integrating by parts *twice* will result in an integral *which, except for its coefficient, is identical to the original one.* When this happens, we merely have to transpose these identical integrals to the same side of the equation, and combine them.

EXAMPLE 5: Integrate $\int e^x \cos x\, dx$.

Solution: We choose

$$u = e^x \qquad \text{and} \qquad dv = \cos x\, dx$$

Then

$$du = e^x\, dx \qquad \text{and} \qquad v = \sin x$$

So,

$$\int e^x \cos x\, dx = e^x \sin x - \int e^x \sin x\, dx \qquad (1)$$

The integral here is similar to the original, except that the cosine has changed to the sine. Integrating by parts again should switch it back to the cosine. We let

$$u = e^x \qquad \text{and} \qquad dv = \sin x\, dx$$

Then

$$du = e^x\, dx \qquad \text{and} \qquad v = -\cos x$$

and

$$\int e^x \sin x\, dx = -e^x \cos x + \int e^x \cos x\, dx$$

The integral here is now the same as the original. Substituting back into (1) yields

$$\int e^x \cos x\, dx = e^x \sin x - (-e^x \cos x + \int e^x \cos x\, dx)$$
$$= e^x \sin x + e^x \cos x - \int e^x \cos x\, dx$$

Two identical integrals can be added just like any other identical things: by combining coefficients.

Transposing the integral to the left side gives us

$$2 \int e^x \cos x\, dx = e^x \sin x + e^x \cos x$$

Finally, dividing by 2, we obtain

$$\int e^x \cos x\,dx = \frac{e^x}{2}(\sin x + \cos x) + C$$

EXERCISE 1

Integrate by parts, for the practice, even though you find a rule that fits.

Integrate by parts.

1. $\int x \sin x\,dx$ **2.** $\int x\sqrt{1-x}\,dx$ **3.** $\int x \sec^2 x\,dx$

4. $\int_0^2 x \sin\frac{x}{2}\,dx$ **5.** $\int x \cos x\,dx$ **6.** $\int x^2 \ln x\,dx$

7. $\int x \sin^2 3x\,dx$ **8.** $\int \cos^3 x\,dx$ **9.** $\int_1^3 \frac{\ln(x+1)dx}{\sqrt{x+1}}$

10. $\int \frac{x^3\,dx}{\sqrt{1-x^2}}$ **11.** $\int \frac{\ln(x+1)dx}{(x+1)^2}$ **12.** $\int x^3 \ln x\,dx$

13. $\int_0^4 xe^{2x}\,dx$ **14.** $\int x \tan^2 x\,dx$ **15.** $\int x^3\sqrt{1-x^2}\,dx$

16. $\int \frac{x^2\,dx}{(1+x^2)^2}$ **17.** $\int \frac{x^2\,dx}{(1-x^2)^{3/2}}$ **18.** $\int_{-2}^2 \frac{xe^x\,dx}{(1+x)^2}$

19. $\int \cos x \ln \sin x\,dx$

Integrate by parts twice.

20. $\int x^2 e^{2x}\,dx$ **21.** $\int x^2 e^{-x}\,dx$ **22.** $\int x^2 e^x\,dx$

23. $\int e^x \sin x\,dx$ **24.** $\int x^2 \cos x\,dx$ **25.** $\int_1^3 e^{-x} \sin 4x\,dx$

26. $\int e^x \cos x\,dx$ **27.** $\int e^{-x} \cos \pi x\,dx$

11-2. INTEGRATING RATIONAL FRACTIONS

Rational Algebraic Fractions

Recall that in a polynomial, all the powers of x are positive integers.

A rational algebraic fraction is one in which both numerator and denominator are polynomials.

EXAMPLE 1

(a) $\frac{x^2}{x^3-2}$ is a *proper* rational fraction because the numerator is of lower degree than the denominator.

(b) $\dfrac{x^3 - 2}{x^2}$ is an *improper* rational fraction.

(c) $\dfrac{\sqrt{x}}{x^3 - 2}$ is *not* a rational fraction.

Integrating Improper Rational Fractions

Perform the indicated division and integrate term by term.

EXAMPLE 2: Integrate $\displaystyle\int \frac{x^3 - 2x^2 - 5x - 2}{x + 1}\,dx$.

Solution: When we carry out the long division, we get a quotient of $x^2 - 3x - 2$, so

$$\int \frac{x^3 - 2x^2 - 5x - 2}{x + 1}\,dx = \int (x^2 - 3x - 2)dx$$

$$= \frac{x^3}{3} - \frac{3x^2}{2} - 2x + C$$

Common Error	It is easy to overlook the simple operation of dividing out the given expression. Be sure to consider it whenever you have to integrate an algebraic fraction.

Long division works fine if there is *no remainder,* as in Example 2. When there is a remainder, it will be a *proper* rational fraction. We now learn how to integrate a proper rational fraction.

Fractions with a Quadratic Denominator

We know that by *completing the square,* a quadratic trinomial $Ax^2 + Bx + C$ can be written in the form $u^2 \pm a^2$. This sometimes allows us to use rules 56 through 60, as in the following example.

EXAMPLE 3: Integrate $\displaystyle\int \frac{4\,dx}{x^2 + 3x - 1}$.

Solution: We start by writing the denominator in the form $u^2 \pm a^2$ by completing the square.

$$x^2 + 3x - 1 = \left(x^2 + 3x + \frac{9}{4}\right) - \frac{9}{4} - 1$$

$$= \left(x + \frac{3}{2}\right)^2 - \frac{13}{4}$$

$$= \left(x + \frac{3}{2}\right)^2 - \left(\frac{\sqrt{13}}{2}\right)^2$$

Our integral is then

$$\int \frac{4\,dx}{\left(x + \frac{3}{2}\right)^2 - \left(\frac{\sqrt{13}}{2}\right)^2}$$

Using rule 57,

$$\int \frac{du}{u^2 - a^2} = \frac{1}{2a} \ln \left| \frac{u - a}{u + a} \right| + C$$

with $u = x + 3/2$ and $a = \sqrt{13}/2$, we get

$$\int \frac{4\,dx}{x^2 + 3x - 1} = 4 \cdot \frac{1}{\frac{2\sqrt{13}}{2}} \ln \left| \frac{x + \frac{3}{2} - \frac{\sqrt{13}}{2}}{x + \frac{3}{2} + \frac{\sqrt{13}}{2}} \right| + C$$

or, switching to decimals,

$$= 1.109 \ln \left| \frac{x - 0.303}{x + 3.303} \right| + C$$

If the numerator in Example 3 had contained an x term, it would not have matched any rule in our table. Sometimes, however, the numerator can be separated into two parts, one of which is the derivative of the denominator.

EXAMPLE 4: Integrate

$$\int \frac{2x + 7}{x^2 + 3x - 1}\,dx$$

Solution: The derivative of the denominator is $2x + 3$. We can get this in the numerator by splitting the 7 into 3 and 4,

$$\int \frac{(2x + 3) + 4}{x^2 + 3x - 1}\,dx$$

We can now separate this integral into two,

$$\int \frac{2x + 3}{x^2 + 3x - 1}\, dx + \int \frac{4}{x^2 + 3x - 1}\, dx$$

The first of these two integrals is, by rule 7,

$$\int \frac{2x + 3}{x^2 + 3x - 1}\, dx = \ln |\, x^2 + 3x - 1 \,| + C_1$$

The second integral is the same as that in Example 3. Our complete integral is then

$$\int \frac{2x + 7}{x^2 + 3x - 1}\, dx = \ln |x^2 + 3x - 1| + 1.109 \ln \left| \frac{x - 0.303}{x + 3.303} \right| + C_2$$

Partial Fractions

When studying algebra, we learned how to combine several fractions into a single fraction by writing each fraction so that they all had a common denominator, and then added or subtracted the numerators as indicated. Now we do the *reverse*. Given a proper rational fraction, we *separate it* into several simpler fractions whose sum is the original fraction. These simpler fractions are called *partial fractions*.

We'll need partial fractions again when we study the Laplace transform.

EXAMPLE 5: The sum of

$$\frac{2}{x + 1} \quad \text{and} \quad \frac{3}{x - 2}$$

is

$$\frac{2(x - 2) + 3(x + 1)}{(x + 1)(x - 2)} = \frac{5x - 1}{x^2 - x - 2}$$

Therefore,

$$\frac{2}{x + 1} \quad \text{and} \quad \frac{3}{x - 2}$$

are the partial fractions of

$$\frac{5x - 1}{x^2 - x - 2}$$

The *first step* in finding partial fractions is to *factor the denominator* of the given fraction. The remaining steps then depend on the nature of those factors.

1. For each linear factor $ax + b$ in the denominator, there will be a partial fraction $A/(ax + b)$.
2. For the repeated linear factors $(ax + b)^n$, there will be n partial fractions,

The A's and B's are constants here.

$$\frac{A_1}{ax+b} + \frac{A_2}{(ax+b)^2} + \cdots + \frac{A_n}{(ax+b)^n}$$

3. For each quadratic factor $ax^2 + bx + c$, there will be a partial fraction $(Ax + B)/(ax^2 + bx + c)$.
4. For the repeated quadratic factors $(ax^2 + bx + c)^n$, there will be n partial fractions,

$$\frac{A_1x + B_1}{ax^2+bx+c} + \frac{A_2x + B_2}{(ax^2+bx+c)^2} + \cdots + \frac{A_nx + B_n}{(ax^2+bx+c)^n}$$

Denominator with Nonrepeated Linear Factors

This is the simplest case, and the method is best shown by an example.

EXAMPLE 6: Separate $\dfrac{x+2}{x^3 - x}$ into partial fractions.

Solution: We first factor the denominator into

$$x, \qquad x + 1, \qquad x - 1$$

Next we assume that each of these factors will be the denominator of a partial fraction, whose numerator is yet to be found, so let

$$\frac{x+2}{x^3 - x} = \frac{A}{x} + \frac{B}{x+1} + \frac{C}{x-1} \tag{1}$$

where A, B, and C are the unknown numerators. If our partial fractions are to be *proper,* the numerator of each must be of lower degree than the denominator. Since each denominator is of first degree, A, B, and C *must be constants.*

We now clear fractions by multiplying both sides of (1) by $x^3 - x$,

$$x + 2 = A(x^2 - 1) + Bx(x - 1) + Cx(x + 1)$$

or

$$x + 2 = (A + B + C)x^2 + (-B + C)x - A$$

For this equation to be true, the coefficients of like powers of x on both sides of the equation must be equal. The coefficient of x is 1, on the left side, and $(-B + C)$ on the right side. Therefore,

$$1 = -B + C \tag{2}$$

This is called the method of undetermined coefficients.

Similarly for the x^2 term,

$$0 = A + B + C \tag{3}$$

and for the constant term,

$$2 = -A \tag{4}$$

Solving (2), (3), and (4) simultaneously gives

$$A = -2, \qquad B = \frac{1}{2}, \qquad \text{and} \qquad C = \frac{3}{2}$$

You can check your work by recombining the partial fractions.

Substituting back into (1) gives us

$$\frac{x+2}{x^3 - x} = -\frac{2}{x} + \frac{1}{2(x+1)} + \frac{3}{2(x-1)}$$

EXAMPLE 7: Integrate $\int \frac{x+2}{x^3 - x}\, dx$.

Solution: From Example 6,

$$\int \frac{x+2}{x^3 - x}\, dx = \int \left[-\frac{2}{x} + \frac{1}{2(x+1)} + \frac{3}{2(x-1)} \right] dx$$

$$= -2 \ln |x| + \frac{1}{2} \ln |x+1| + \frac{3}{2} \ln |x-1| + C$$

Denominator with Repeated Linear Factors

We now cover fractions where a linear factor appears more than once in the denominator. Of course, there may also be one or more nonrepeated linear factors, as in the following example.

EXAMPLE 8: Separate the fraction $\dfrac{x^3+1}{x(x-1)^3}$ into partial fractions.

Solution: We assume that

$$\frac{x^3+1}{x(x-1)^3} = \frac{A}{x} + \frac{B}{(x-1)} + \frac{C}{(x-1)^2} + \frac{D}{(x-1)^3}$$

Multiplying both sides by $x(x-1)^3$ yields

$$\begin{aligned}x^3+1 &= A(x-1)^3 + Bx(x-1)^2 + Cx(x-1) + Dx\\ &= Ax^3 - 3Ax^2 + 3Ax - A + Bx^3 - 2Bx^2 + Bx + Cx^2 - Cx + Dx\\ &= (A+B)x^3 + (-3A-2B+C)x^2 + (3A+B-C+D)x - A\end{aligned}$$

Equating coefficients of like powers of x gives

$$\begin{aligned}A+B &= 1\\ -3A-2B+C &= 0\\ 3A+B-C+D &= 0\\ -A &= 1\end{aligned}$$

Solving this set of equations simultaneously gives

$$A=-1, \qquad B=2, \qquad C=1, \qquad D=2$$

so

$$\frac{x^3+1}{x(x-1)^3} = -\frac{1}{x} + \frac{2}{x-1} + \frac{1}{(x-1)^2} + \frac{2}{(x-1)^3}$$

Denominators with Quadratic Factors

We now deal with fractions that have one or more quadratic factors, in addition to any linear factors.

EXAMPLE 9: Separate the fraction $\dfrac{x+2}{x^3+3x^2-x}$ into partial fractions.

Solution: The factors of the denominator are x and x^2+3x-1, so we assume that

$$\frac{x+2}{(x^2+3x-1)x} = \frac{Ax+B}{x^2+3x-1} + \frac{C}{x}$$

Multiplying through by $(x^2 + 3x - 1)x$,

$$x + 2 = Ax^2 + Bx + Cx^2 + 3Cx - C$$
$$= (A + C)x^2 + (B + 3C)x - C$$

Equating the coefficients of x^2 gives

$$A + C = 0$$

and of x,

$$B + 3C = 1$$

and the constant terms,

$$-C = 2$$

Solving simultaneously gives

$$A = 2, \qquad B = 7, \qquad C = -2$$

So our partial fractions are

$$\frac{x + 2}{x^3 + 3x^2 - x} = \frac{2x + 7}{x^2 + 3x - 1} - \frac{2}{x}$$

EXAMPLE 10: Integrate $\int \frac{x + 2}{x^3 + 3x^2 - x}\,dx$.

Solution: Using the results of examples 4 and 9,

$$\int \frac{x + 2}{x^3 + 3x^2 - x} = \int \frac{2x + 7}{x^2 + 3x - 1}dx - \int \frac{2}{x}\,dx$$

$$= \ln |x^3 - 3x - 1| + 1.109 \ln \left|\frac{x - 0.303}{x + 3.303}\right| - 2 \ln |x| + C$$

EXERCISE 2

Integrate.

Nonrepeated Linear Factors

1. $\int \frac{(4x - 2)\,dx}{x^3 - x^2 - 2x}$
2. $\int_1^2 \frac{(3x - 1)\,dx}{x^3 - x}$
3. $\int_1^3 \frac{(2 - x^2)\,dx}{x^3 + 3x^2 + 2x}$
4. $\int \frac{x^2\,dx}{x^2 - 4x + 5}$
5. $\int \frac{x + 1}{x^3 + 2x^2 - 3x}\,dx$

Repeated Linear Factors

6. $\displaystyle\int_2^4 \frac{(x^3-2)\,dx}{x^3-x^2}$ **7.** $\displaystyle\int_0^1 \frac{x-5}{(x-3)^2}\,dx$ **8.** $\displaystyle\int_1^2 \frac{x^2\,dx}{(x-1)^3}$

9. $\displaystyle\int \frac{(2x-5)\,dx}{(x-2)^3}$ **10.** $\displaystyle\int \frac{dx}{x^3+x^2}$

11. $\displaystyle\int_2^3 \frac{dx}{(x^2+x)(x-1)^2}$ **12.** $\displaystyle\int_0^2 \frac{x-2}{(x+1)^3}\,dx$

Quadratic Factors

13. $\displaystyle\int_1^2 \frac{dx}{x^4+x^2}$ **14.** $\displaystyle\int_0^1 \frac{(x^2-3)\,dx}{(x+2)(x^2+1)}$ **15.** $\displaystyle\int_1^2 \frac{(4x^2+6)\,dx}{x^3+3x}$

16. $\displaystyle\int_1^4 \frac{(5x^2+4)\,dx}{x^3+4x}$ **17.** $\displaystyle\int \frac{x^2\,dx}{1-x^4}$ **18.** $\displaystyle\int_1^2 \frac{(x-3)\,dx}{x^3+x^2}$

19. $\displaystyle\int_0^1 \frac{5x\,dx}{(x+2)(x^2+1)}$ **20.** $\displaystyle\int_3^4 \frac{(5x^2-4x)\,dx}{x^4-16}$ **21.** $\displaystyle\int \frac{x^5\,dx}{(x^2+4)^2}$

This is also called *integration by rationalization*.

11-3. INTEGRATING BY MEANS OF ALGEBRAIC SUBSTITUTION

Expressions containing radicals are usually harder to integrate than those that do not. In this section we learn how to remove all radicals and fractional exponents from an expression by means of a suitable algebraic substitution.

Expressions containing Fractional Powers of *x*

If the expression to be integrated contains a factor of, say, $x^{1/3}$, we would substitute,

$$z = x^{1/3}$$

and the fractional exponents would vanish. If the same problem should also contain $x^{1/2}$, for example, we would substitute

$$z = x^{1/6}$$

because 6 is the lowest common denominator of the fractional exponents.

EXAMPLE 1: Integrate $\displaystyle\int \frac{x^{1/2}\,dx}{1+x^{3/4}}$.

Solution: Here x is raised to the $\frac{1}{2}$ power in one place, and to the $\frac{3}{4}$ power in another. The LCD of $\frac{1}{2}$ and $\frac{3}{4}$ is 4, so we substitute

$$z = x^{1/4}$$

So

$$x^{1/2} = (x^{1/4})^2 = z^2$$

and

$$x^{3/4} = (x^{1/4})^3 = z^3$$

and since

$$x = z^4$$
$$dx = 4z^3\,dz$$

Substituting into our given integral gives us

$$\int \frac{x^{1/2}\,dx}{1 + x^{3/4}} = \int \frac{z^2}{1 + z^3}(4z^3\,dz)$$
$$= 4\int \frac{z^5}{1 + z^3}\,dz$$

The integrand is now a rational fraction, such as those we studied in Sec. 11-2. As before, we divide numerator by denominator:

$$4\int \frac{z^5}{1 + z^3}\,dz = 4\int \left(z^2 - \frac{z^2}{1 + z^3}\right)dz$$
$$= \frac{4z^3}{3} - \frac{4}{3}\ln|1 + z^3| + C$$

Finally, we substitute back, $z = x^{1/4}$:

$$\int \frac{x^{1/2}\,dx}{1 + x^{3/4}} = \frac{4x^{3/4}}{3} - \frac{4}{3}\ln|1 + x^{3/4}| + C$$

Expressions containing Fractional Powers of a Binomial

The procedure here is very similar to the preceding case, where only x was raised to a fractional power.

EXAMPLE 2: Integrate $\int x\sqrt{3 + 2x}\, dx$.

Solution: Let $z = \sqrt{3 + 2x}$. Then

$$x = \frac{z^2 - 3}{2}$$

So

$$dx = \frac{1}{2}(2z)\, dz = z\, dz$$

Substituting, we get

$$\begin{aligned}\int x\sqrt{3 + 2x}\, dx &= \int \frac{z^2 - 3}{2} \cdot z \cdot z\, dz \\ &= \frac{1}{2}\int (z^4 - 3z^2)\, dz \\ &= \frac{z^5}{10} - \frac{3z^3}{6} + C\end{aligned}$$

Now substituting $\sqrt{3 + 2x}$ for z, we obtain

$$\int x\sqrt{3 + 2x}\, dz = \frac{1}{10}(3 + 2x)^{5/2} - \frac{1}{2}(3 + 2x)^{3/2} + C$$

Definite Integrals

When using the method of substitution with a definite integral, we may also substitute the limits of integration. This eliminates the need to substitute back to the original variable.

EXAMPLE 3: Evaluate $\int_0^3 \frac{x\, dx}{\sqrt{1 + x}}$.

Solution: We let $z = \sqrt{1 + x}$. So,

$$x = z^2 - 1$$

and

$$dx = 2z\, dz$$

We now compute the limits on z.

When $x = 0$,

$$z = \sqrt{1 + 0} = 1$$

and when $x = 3$,

$$z = \sqrt{1 + 3} = 2$$

Making the substitution yields

$$\int_0^3 \frac{x\,dx}{\sqrt{1+x}} = \int_1^2 \frac{z^2 - 1}{z} \cdot 2z\,dz$$

$$= 2\int_1^2 (z^2 - 1)dz = \left[\frac{2z^3}{3} - 2z\right]_1^2 = \frac{8}{3}$$

EXERCISE 3

Integrate.

1. $\displaystyle\int \frac{dx}{1 + \sqrt{x}}$

2. $\displaystyle\int \frac{dx}{x - \sqrt{x}}$

3. $\displaystyle\int \frac{dx}{x - \sqrt[3]{x}}$

4. $\displaystyle\int \frac{dx}{\sqrt{x} + \sqrt[4]{x^3}}$

5. $\displaystyle\int \frac{x\,dx}{\sqrt[3]{1 + x}}$

6. $\displaystyle\int \frac{x\,dx}{\sqrt{x - 1}}$

7. $\displaystyle\int \frac{x\,dx}{\sqrt{2 - 7x}}$

8. $\displaystyle\int \frac{\sqrt{x^2 - 1}\,dx}{x}$

9. $\displaystyle\int \frac{x^2\,dx}{(4x + 1)^{5/2}}$

10. $\displaystyle\int \frac{x\,dx}{(1 + x)^{3/2}}$

11. $\displaystyle\int \frac{(x + 5)\,dx}{(x + 4)\sqrt{x + 2}}$

12. $\displaystyle\int \frac{dx}{x^{5/8} + x^{3/4}}$

13. $\displaystyle\int \frac{dx}{x\sqrt{1 - x^2}}$

14. $\displaystyle\int \frac{(x^{3/2} - x^{1/3})\,dx}{6x^{1/4}}$

15. $\displaystyle\int x\sqrt[3]{1 + x}\,dx$

Evaluate.

16. $\displaystyle\int_0^3 \frac{dx}{(x + 2)\sqrt{x + 1}}$

17. $\displaystyle\int_0^1 \frac{x^{3/2}\,dx}{x + 1}$

18. $\displaystyle\int_0^4 \frac{dx}{1 + \sqrt{x}}$

19. $\displaystyle\int_0^{1/2} \frac{dx}{\sqrt{2x}(9 + \sqrt[3]{2x})}$

20. $\displaystyle\int_1^4 \frac{x\,dx}{\sqrt{4x + 2}}$

11-4. INTEGRATING BY TRIGONOMETRIC SUBSTITUTION

The two legs a and b of a right triangle are related to the hypotenuse c by the familiar Pythagorean theorem

$$a^2 + b^2 = c^2$$

Thus when we have an expression of the form $u^2 + a^2$, it might occur to us to represent it as the hypotenuse of a right triangle whose legs are u and a. Doing so enables us to integrate expressions involving the sum or difference of two squares, especially when they are under a radical sign. A similar substitution will work for other radical expressions, as given in the following table.

Expression	Substitution
$\sqrt{u^2 + a^2}$	let $u = a \tan \theta$
$\sqrt{a^2 - u^2}$	let $u = a \sin \theta$
$\sqrt{u^2 - a^2}$	let $u = a \sec \theta$

Integrals involving $u^2 + a^2$

EXAMPLE 1: Integrate $\int \frac{dx}{\sqrt{x^2 + 9}}$.

Solution: Our substitution, from the table above, is

$$x = 3 \tan \theta \tag{1}$$

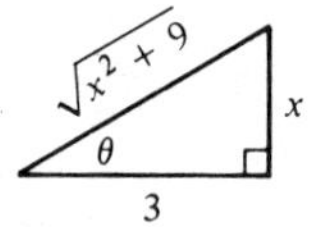

FIGURE 11-1

We sketch a right triangle (Fig. 11-1) with x as the leg opposite to θ and 3 as the leg adjacent to θ. The hypotenuse is then $\sqrt{x^2 + 9}$.

We now write our integral in terms of θ. Since $\cos \theta = 3/\sqrt{x^2 + 9}$,

$$\sqrt{x^2 + 9} = \frac{3}{\cos \theta} = 3 \sec \theta \tag{2}$$

Taking the derivative of (1) gives us

$$dx = 3 \sec^2 \theta \, d\theta \tag{3}$$

Substituting (2) and (3) into our original integral, we obtain

$$\int \frac{dx}{\sqrt{x^2 + 9}} = \int \frac{3 \sec^2 \theta \, d\theta}{3 \sec \theta} = \int \sec \theta \, d\theta$$

By rule 14,
$$= \ln |\sec \theta + \tan \theta| + C_1$$

We now return to our original variable, x. From Fig. 11-1

$$\sec \theta = \frac{\sqrt{x^2 + 9}}{3} \quad \text{and} \quad \tan \theta = \frac{x}{3}$$

So

$$\int \frac{dx}{\sqrt{x^2 + 9}} = \ln \left| \frac{\sqrt{x^2 + 9}}{3} + \frac{x}{3} \right| + C_1$$

$$= \ln|\sqrt{x^2+9}+x| - \ln 3 + C_1$$
$$= \ln|\sqrt{x^2+9}+x| + C$$

where C_1 and $-\ln 3$ have been combined into a single constant C.

Integrals involving $a^2 - u^2$

EXAMPLE 2: Integrate $\int \sqrt{9-4x^2}\,dx$.

Solution: Our radical is of the form $\sqrt{a^2-u^2}$, with $a = 3$ and $u = 2x$. Our substitution is then

$$2x = 3 \sin\theta$$

or

$$x = \frac{3}{2}\sin\theta$$

Then

$$dx = \frac{3}{2}\cos\theta\, d\theta$$

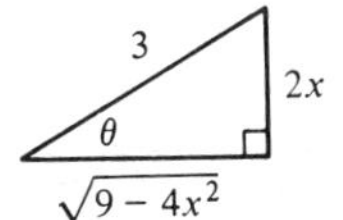

FIGURE 11-2

We sketch a right triangle with hypotenuse 3 and one leg $2x$, as in Fig. 11-2. Then

$$\cos\theta = \frac{\sqrt{9-4x^2}}{3}$$

so

$$\sqrt{9-4x^2} = 3\cos\theta$$

Substituting into our given integral gives us

$$\int \sqrt{9-4x^2}\,dx = \int (3\cos\theta)\left(\frac{3}{2}\cos\theta\, d\theta\right)$$
$$= \frac{9}{2}\int \cos^2\theta\, d\theta$$

By rule 17,
$$= \frac{9}{2}\left(\frac{\theta}{2} + \frac{\sin 2\theta}{4}\right) + C$$

We now use Eq. 170 to replace $\sin 2\theta$ with $2 \sin\theta \cos\theta$,

$$\int \sqrt{9 - 4x^2}\, dx = \frac{9}{4}(\theta + \sin\theta\cos\theta) + C$$

We now substitute back to return to the variable x. From Fig. 11-2 we see that

$$\sin\theta = \frac{2x}{3}$$

and that

$$\theta = \sin^{-1}\frac{2x}{3}$$

and

$$\cos\theta = \frac{\sqrt{9 - 4x^2}}{3}$$

So

$$\int \sqrt{9 - 4x^2}\, dx = \frac{9}{4}\left(\sin^{-1}\frac{2x}{3} + \frac{2x}{3}\cdot\frac{\sqrt{9 - 4x^2}}{3}\right) + C$$

$$= \frac{9}{4}\left(\sin^{-1}\frac{2x}{3} + \frac{2x\sqrt{9 - 4x^2}}{9}\right) + C$$

Definite Integrals

When evaluating a *definite* integral, we may substitute the limits on x after substituting back into x, or we may *change the limits* when we change variables, as shown in the following example.

EXAMPLE 3: Integrate $\int_2^4 \sqrt{x^2 - 4}\, dx$.

Solution: We substitute $x = 2\sec\theta$, so

$$dx = 2\sec\theta\tan\theta\, d\theta$$

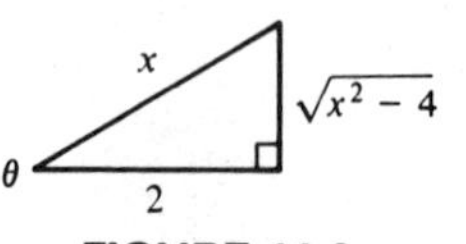

FIGURE 11-3

We sketch the triangle (Fig. 11-3) from which

$$\sqrt{x^2 - 4} = 2\tan\theta$$

Our integral is, then,

$$\int_2^4 \sqrt{x^2-4}\,dx = \int (2\tan\theta)(2\sec\theta\tan\theta\,d\theta)$$
$$= 4\int \tan^2\theta\sec\theta\,d\theta$$
$$= 4\int (\sec^2\theta - 1)\sec\theta\,d\theta$$
$$= 4\int (\sec^3\theta - \sec\theta)\,d\theta$$

which we can integrate by rules 14 and 26. But what about limits? What happens to θ when x goes from 2 to 4? From Fig. 11-3 we see that

$$\cos\theta = \frac{2}{x}$$

Note that θ is always acute, so when taking the inverse function we need only consider angles less than 90°.

when $x = 2$, $\cos\theta = 1$	when $x = 4$, $\cos\theta = \frac{1}{2}$
$\theta = 0°$	$\theta = 60°$

So our integral is then

$$\int_{x=2}^{x=4} \sqrt{x^2-4}\,dx = 4\int_{\theta=0°}^{\theta=60°} (\sec^3\theta - \sec\theta)\,d\theta$$

By rules 14 and 26,

$$= 4\left[\frac{1}{2}\sec\theta\tan\theta + \frac{1}{2}|\ln\sec\theta + \tan\theta| - \ln|\sec\theta + \tan\theta|\right]_{0°}^{60°}$$
$$= 2\left[\sec\theta\tan\theta - \ln|\sec\theta + \tan\theta|\right]_{0°}^{60°}$$
$$= 2[2(1.732) - \ln|2 + 1.732|] - 2[0 - \ln|1 + 0|] = 4.294$$

EXERCISE 4

Integrate.

1. $\int \frac{5\,dx}{(5-x^2)^{3/2}}$
2. $\int \frac{\sqrt{9-x^2}\,dx}{x^4}$
3. $\int \frac{dx}{x\sqrt{x^2+4}}$
4. $\int_3^4 \frac{dx}{x^3\sqrt{x^2-9}}$
5. $\int \frac{dx}{x^2\sqrt{4-x^2}}$
6. $\int \frac{dx}{(x^2+2)^{3/2}}$
7. $\int \frac{x^2\,dx}{\sqrt{4-x^2}}$
8. $\int_1^5 \frac{dx}{x\sqrt{25-x^2}}$
9. $\int \frac{\sqrt{16-x^2}\,dx}{x^2}$

10. $\displaystyle\int \frac{dx}{\sqrt{x^2 + 2x}}$ **11.** $\displaystyle\int_3^6 \frac{x^2\,dx}{\sqrt{x^2 - 6}}$ **12.** $\displaystyle\int_3^5 \frac{x^2\,dx}{(x^2 + 8)^{3/2}}$

13. $\displaystyle\int \frac{dx}{x^2\sqrt{x^2 - 7}}$

11-5. IMPROPER INTEGRALS

Up to now, we have assumed that the limits in the definite integral

$$\int_a^b f(x)\,dx \tag{1}$$

were finite, and that the integrand $f(x)$ was continuous for all values of x in the interval $a \leq x \leq b$. If either or both of these conditions is not met, the integral is called an *improper integral.*

Thus an integral is called improper if

1. one or both limits is infinite, or
2. the integrand is discontinuous at some x within the interval $a \leq x \leq b$.

EXAMPLE 1
(a) The integral

$$\int_2^\infty x^2\,dx$$

is improper because one limit is infinite.
(b) The integral

$$\int_0^5 \frac{x^2}{x - 3}\,dx$$

is improper because the integrand, $f(x) = x^2/(x - 3)$, is discontinuous at $x = 3$.

We now treat each type separately.

Infinite Limit

The value of a definite integral with one or both limits infinite is defined by the following relations:

$$\int_a^\infty f(x)\,dx = \lim_{b\to\infty} \int_a^b f(x)\,dx \tag{2}$$

$$\int_{-\infty}^b f(x)\,dx = \lim_{a\to-\infty} \int_a^b f(x)\,dx \tag{3}$$

$$\int_{-\infty}^\infty f(x)\,dx = \lim_{\substack{b\to\infty\\ a\to-\infty}} \int_a^b f(x)\,dx \tag{4}$$

When the limit has a finite value, we say that the integral *converges,* or *exists. When the limit is infinite, we say that the integral diverges, or does not exist.*

EXAMPLE 2: Does the improper integral

$$\int_1^\infty \frac{dx}{x}$$

converge or diverge?

Solution: From (2),

$$\int_1^\infty \frac{dx}{x} = \lim_{b\to\infty} \int_1^b \frac{dx}{x} = \lim_{b\to\infty} \Big[\ln x\Big]_1^b$$

$$= \lim_{b\to\infty} [\ln b] = \infty$$

Since the limit is infinite, we say that the given integral diverges, or does not exist.

EXAMPLE 3

$$\int_{-\infty}^0 e^x\,dx = \lim_{a\to\infty} \Big[e^x\Big]_a^0 = \lim_{a\to\infty} [1 - e^a] = 1$$

Thus the given limit converges and has a value of 1.

EXAMPLE 4: Find the area under the curve $y = 1/x^2$ (Fig. 11-4) from $x = 2$ to infinity.

Solution: $\displaystyle\int_2^\infty \frac{dx}{x^2} = \lim_{b\to\infty} \int_2^b \frac{dx}{x^2} = \lim_{b\to\infty} \left| -\frac{1}{x} \right|_2^b$

$$= \lim_{b\to\infty} \left| -\frac{1}{b} + \frac{1}{2} \right| = \frac{1}{2}$$

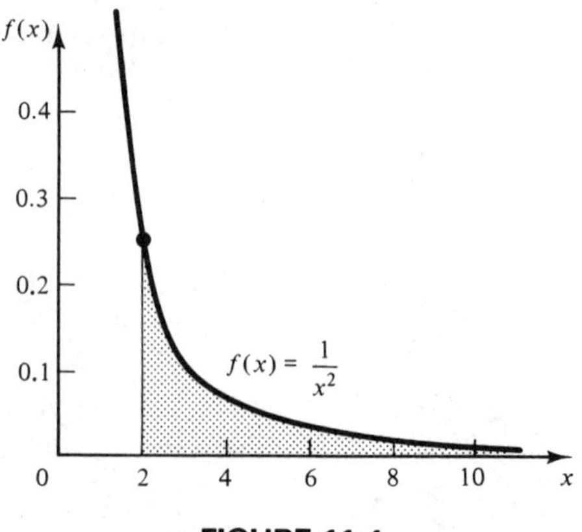

FIGURE 11-4

Thus the area under the curve is not infinite, as might be expected, but has a finite value of 1/2.

In many cases we need not carry out the limiting process as in Example 4, but can treat the infinite limit as if it were a number. Thus, repeating Example 4,

$$\int_2^\infty \frac{dx}{x^2} = \left| -\frac{1}{x} \right|_2^\infty = \left| -\frac{1}{\infty} + \frac{1}{2} \right| = \frac{1}{2}$$

Discontinuous Integrand

We now cover the case where the integrand is discontinuous at or between the limits.

If the integrand $f(x)$ in (1) is continuous for all values of x in the interval $a \le x \le b$ except at b, then the integral (1) is defined by

$$\int_a^b f(x)\, dx = \lim_{x \to b^-} \int_a^x f(x)\, dx \tag{5}$$

where the notation $x \to b^-$ means that x approaches b from values that are less than b. Similarly, if the integrand is discontinuous only at the lower limit a, we have

$$\int_a^b f(x)\, dx = \lim_{x \to a^+} \int_x^b f(x)\, dx \tag{6}$$

where $x \to a^+$ means that x approaches a from values greater than a.

EXAMPLE 5: Evaluate the improper integral $\int_0^4 \frac{dx}{\sqrt{x}}$.

Solution: Here our function $f(x) = 1/\sqrt{x}$ is discontinuous at the lower limit, 0. Thus from (6),

$$\int_0^4 \frac{dx}{\sqrt{x}} = \lim_{x\to 0^+} \int_x^4 \frac{dx}{\sqrt{x}} = \lim_{x\to 0^+} \left[2\sqrt{x}\right]_x^4$$

$$= 4 - \lim_{x\to 0^+} 2\sqrt{x} = 4$$

EXERCISE 5

Evaluate each improper integral if it exists.

Infinite Limit

1. $\int_1^{\infty} \frac{dx}{x^3}$
2. $\int_3^{\infty} \frac{dx}{(x-2)^3}$
3. $\int_{-\infty}^{1} e^x\,dx$
4. $\int_5^{\infty} \frac{dx}{\sqrt{x-1}}$
5. $\int_{-\infty}^{\infty} \frac{dx}{1+x^2}$
6. $\int_0^{\infty} \frac{dx}{(x^2+4)^{3/2}}$

Discontinuous Integrand

7. $\int_0^1 \frac{dx}{\sqrt{1-x}}$
8. $\int_0^1 \frac{dx}{x^3}$
9. $\int_0^{\pi/2} \tan x\,dx$
10. $\int_1^2 \frac{dx}{x\sqrt{x^2-1}}$
11. $\int_{-1}^{1} \frac{dx}{\sqrt{1-x^2}}$
12. $\int_0^2 \frac{dx}{(x-1)^{2/3}}$

11-6 APPROXIMATE INTEGRATION

The Need for Approximate Methods

We have studied many rules and methods for evaluating a definite integral. But unfortunately many functions cannot be integrated by these methods. Further, all our methods so far require that the function to be integrated be known in the form of an equation. But in practice, our function may simply be a graph, or perhaps a table of point pairs. In these cases we need an approximate method.

All our approximate methods are based on the fact that the definite integral can be interpreted as the area bounded by the function and the x axis, between the given limits. Thus, even though the problem we are solving may have nothing to do with area (we could be finding work, or a volume, for example), we attack it by finding the area under a curve.

There are many approximate methods for finding the area under a curve, and even mechanical devices such as the planimeter for doing the same. We have already described the *midpoint method* in Sec. 8-2 and here we cover the *average ordinate method,* the *trapezoid rule,* and *Simpson's rule.*

Average Ordinate Method

In Sec. 9-4 we showed how to compute the average ordinate, by dividing the area under a curve by the total interval $b - a$.

$$y_{avg} = \frac{1}{b-a}\int_a^b f(x)\,dx \tag{378}$$

Here we reverse the procedure. We compute the average ordinate (simply by averaging the ordinates of points on the curve), and use it to compute the area. Thus

Average Ordinate Method	$A \cong \int_a^b f(x)\,dx = y_{avg}(b-a)$	**346**

Notice that the data points do not have to be equally spaced, although accuracy is usually better if they are.

EXAMPLE 1: A curve passes through the points

y	1	2	4	5	7	7.5	9
y	3.51	4.23	6.29	7.81	7.96	8.91	9.25

Find the area under the curve.

Solution: The average of the seven ordinates is

$$y_{avg} \cong \frac{3.51 + 4.23 + 6.29 + 7.81 + 7.96 + 8.91 + 9.25}{7} = 6.85$$

So $A \cong 6.85(9 - 1) = 54.8$ square units.

Trapezoid Rule

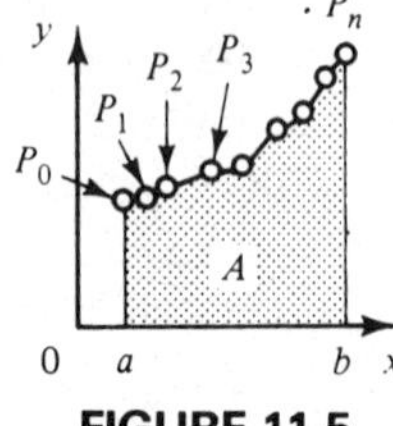

FIGURE 11-5

Let us suppose that our function is in the form of a table of point pairs, and that we plot each of these points as in Fig. 11-5. Joining these points with a smooth curve would give us a graph of the function, and it is the area under this curve that we seek.

We subdivide this area into a number of vertical *panels* by connecting the points with straight lines, and by dropping a perpendicular from each point to the x axis (Fig. 11-6). Each panel then has the shape of a *trapezoid*—hence the name of the rule. If our first data point has the coordinates (x_0, y_0) and the second point is (x_1, y_1), the area of the first panel is, by Eq. 111,

$$\frac{1}{2}(x_1 - x_0)(y_1 + y_0)$$

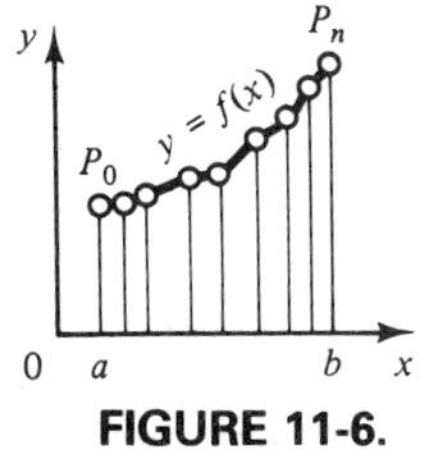

FIGURE 11-6. Trapezoid rule.

and the sum of all the panels gives us the approximate area under the curve:

Trapezoid Rule	$A \cong \frac{1}{2}[(x_1 - x_0)(y_1 + y_0) + (x_2 - x_1)(y_2 + y_1) + \cdots + (x_n - x_{n-1})(y_n + y_{n-1})]$	**347**

If the points are equally spaced in the horizontal direction at a distance of

$$h = x_1 - x_0 = x_2 - x_1 = \cdots = x_n - x_{n-1}$$

Equation 347 reduces to:

Trapezoid Rule, Equal Spacing	$A \cong h\left[\frac{1}{2}(y_0 + y_n) + y_1 + y_2 + \cdots + y_{n-1}\right]$	**348**

EXAMPLE 2: Find the area under the graph of the function given by the following table of ordered pairs:

x	0	1	2	3	4	5	6	7	8	9	10
y	4	3	4	7	12	19	28	39	52	67	84

Solution: By Eq. 348 with $h = 1$,

$$A \cong 1\left[\frac{1}{2}(4 + 84) + 3 + 4 + 7 + 12 + 19 + 28 + 39 + 52 + 67\right]$$

$$= 275$$

As a check, we note that the points given are actually points on the curve $y = x^2 - 2x + 4$. Integrating gives us

$$\int_0^{10} (x^2 - 2x + 4)\, dx = \frac{x^3}{3} - x^2 + 4x \bigg|_0^{10} = 273\frac{1}{3}$$

Our approximate area thus agrees with the exact value within about 0.6%.

When the function is given in the form of an equation or a graph, we are free to select as many points as we want, for use in Eq. 348. The more points selected, the greater will be the accuracy, and the greater will be the labor.

EXAMPLE 3: Evaluate the integral

$$\int_0^{10} \left(\frac{x^2}{10} + 2\right) dx$$

taking points spaced 1 unit apart.

Solution: Making a table of point pairs for the function

$$y = \frac{x^2}{10} + 2$$

x	0	1	2	3	4	5	6	7	8	9	10
y	2	2.1	2.4	2.9	3.6	4.5	5.6	6.9	8.4	11.1	12

sum = 47.5

Then Eq. 348,

$$\int_0^{10} \left(\frac{x^2}{10} + 2\right) dx \cong 1\left[\frac{1}{2}(2 + 12) + 47.5\right] \cong 54.5$$

Prismoidal Formula

Let us find the area under a parabola (Fig. 11-7a). Since the area does not depend on the location of the origin, let us place the y axis halfway between a and b (Fig. 11-7b) so that it divides our area into two panels of width h. Let y_0, y_1, and y_2 be the heights of the curve at $x = -h$, 0, and h, respectively. Then

$$\text{Area} = \int_{-h}^{h} y\, dx = \int_{-h}^{h} (Ax^2 + Bx + C)\, dx = \frac{Ax^3}{3} + \frac{Bx^2}{2} + Cx \bigg|_{-h}^{h}$$

$$= \frac{2}{3}Ah^3 + 2Ch = \frac{h}{3}(2Ah^2 + 6C) \tag{1}$$

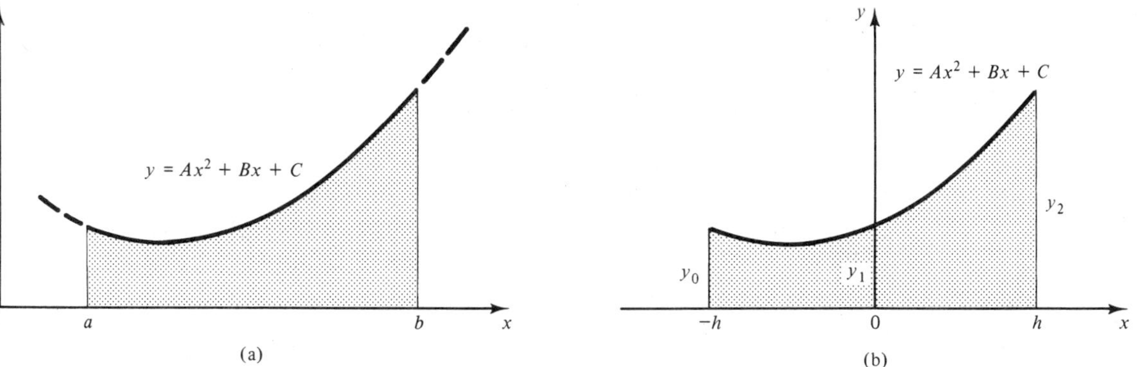

FIGURE 11-7. Areas by prismoidal formula.

We now get A and C in terms of y_0, y_1, and y_2 by substituting the three points on the curve, one by one, into the equation of the parabola, $y = Ax^2 + Bx + C$:

$$y(-h) = y_0 = A(-h)^2 + B(-h) + C = Ah^2 - Bh + C$$

$$y(0) = y_1 = A(0) + B(0) + C = C$$

$$y(h) = y_2 = Ah^2 + Bh + C$$

Adding y_0 and y_2 gives

$$y_0 + y_2 = 2Ah^2 + 2C = 2Ah^2 + 2y_1$$

from which $A = (y_0 - 2y_1 + y_2)/2h^2$. Substituting into (1),

$$\text{Area} = \frac{h}{3}(2Ah^2 + 6C) = \frac{h}{3}\left(2h^2 \frac{y_0 - 2y_1 + y_2}{2h^2} + 6y_1\right)$$

or

Our main use for the Prismoidal Formula will be in the derivation of Simpson's rule.

Prismoidal Formula	$\text{Area} = \frac{h}{3}(y_0 + 4y_1 + y_2)$	**349**

EXAMPLE 4: A curve passes through the points (1, 2.05), (3, 5.83), and (5, 17.9). Assuming the curve to be a parabola, find the area bounded by the curve and the x axis, from $x = 1$ to 5.

Solution: Substituting into the prismoidal formula with $h = 2$, $y_0 = 2.05$, $y_1 = 5.83$, and $y_1 = 17.9$,

$$A \cong \frac{2}{3}[2.05 + 4(5.83) + 17.9] = 28.8 \qquad \text{square units}$$

Named for the English mathematician Thomas Simpson (1710–1761).

Simpson's Rule

The prismoidal formula gives us the area under a curve that is a parabola, or *the approximate area under a curve that is assumed to be a parabola,* if we know (or can find) the coordinates of three equally spaced points on the curve. When there are more than three points on the curve, as in Fig. 11-5, we apply the prismoidal formula to each group of three, in turn.

Thus the area of each pair of panels is

Panels 1 and 2 $h(y_0 + 4y_1 + y_2)/3.$

Panels 3 and 4 $h(y_2 + 4y_3 + y_4)/3.$

⋮

Panels $n - 1$ and n $h(y_{n-2} + 4y_{n-1} + y_n)/3$

Adding these areas gives the approximate area under the entire curve, assuming it to be made up of parabolic segments.

Simpson's Rule	$A = \int_a^b f(x)\,dx = \frac{h}{3}(y_0 + 4y_1 + 2y_2 + 4y_3 + \cdots + 4y_{n-1} + y_n)$	350

If we had an equation rather than a table of data, we would start by choosing an even number of intervals and then make a table of point pairs.

EXAMPLE 5: A curve passes through the following data points.

x	0	1	2	3	4	5	6	7	8	9	10
y	4.02	5.23	5.66	6.05	5.81	5.62	5.53	5.71	6.32	7.55	8.91

Find the approximate area under the curve using Simpson's rule.

Solution: Substituting, with $h = 1$,

$$A \cong \frac{4.02 + 4(5.23) - 2(5.66) + 4(6.05) + 2(5.81) + 4(5.62) + 2(5.53) + 4(5.71) + 2(6.32) + 4(7.55) + 8.91}{3}$$

$$= 60.1 \text{ square units}$$

EXERCISE 6

Evaluate each integral using either the average ordinate method, the trapezoid rule, or Simpson's rule. Choose 10 panels of equal width. Check your answer by integrating.

1. $\int_1^3 \frac{1 + 2x}{x + x^2}\,dx$

2. $\int_1^{10} \frac{dx}{x}$

3. $\int_1^8 (4x^{1/3} + 3)\,dx$

4. $\int_0^{\pi} \sin x\,dx$

Find the area bounded by each curve and the x axis, by any method. Use 10 panels.

5. $y = \frac{1}{\sqrt{x}}$ from $x = 4$ to 9

6. $y = \frac{2x}{1 + x^2}$ from $x = 2$ to 3

7. $y = x\sqrt{1 - x^2}$ from $x = 0$ to 1

8. $y = \frac{x + 6}{\sqrt{x + 4}}$ from $x = 0$ to 5

Find the area bounded by the given data points and the x axis. Use any method.

9.

x	0	1	2	3	4	5	6	7	8	9	10
y	415	593	577	615	511	552	559	541	612	745	893

10.

x	0	1	2	3	4	5	6	7	8	9	10
y	3.02	4.63	4.76	5.08	6.31	6.60	6.23	6.48	7.27	8.93	9.11

11.

x	0	1	2	3	4	5	6	7	8	9	10
y	24.0	25.2	25.6	26.0	25.8	25.6	25.5	25.7	26.3	27.5	28.9

12.

x	0	1	2	3	4	5	6	7	8	9	10
y	462	533	576	625	591	522	563	511	602	745	821

Applications

13. A ship's deck is measured every 30.0 ft, and has the following widths, in feet:

5.5 34.0 53.5 57.2 61.0 59.4 59.0 56.8 45.2 12.0

Find the area of the deck.

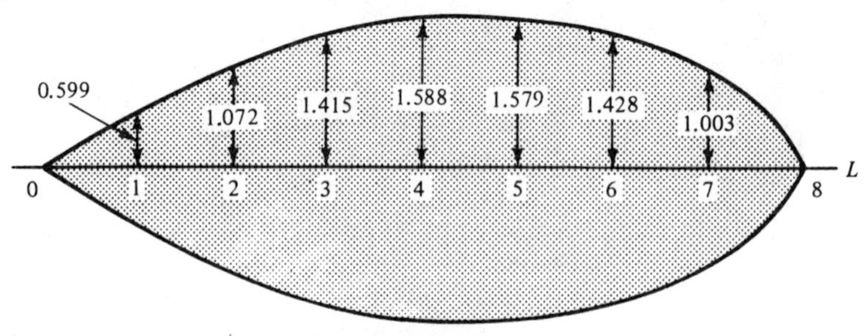

FIGURE 11-8

14. The dimensions, in inches, of a streamlined strut are shown in Fig. 11-8. Find the cross-sectional area.

15. The cross-sectional area A of a certain ship at various depths D below the waterline are:

D(ft)	0	3	6	9	12	15	18
A(ft^2)	7500	7150	6640	5680	4225	2430	260

Find the volume of the ship's hull below the waterline.

16. The pressure and volume of an expanding gas in a cylinder are measured and found to be

Volume (in.^{3})	20	40	60	80	100
Pressure (lb/in.^{2})	68.7	31.4	19.8	14.3	11.3

Find the work done by the gas by finding the area under the p–v curve.

One nanometer (nm) equals 10^{-9} meter.

17. The light output of a certain lamp (W/nm) as a function of the wavelength of the light (nm) is given as

Wavelength	460	480	500	520	540	560	580	600	620
Output	0	21.5	347	1260	1360	939	338	64.6	0

Find the total output wattage of the lamp by finding the area under the output curve.

18. A certain tank has the following cross-sectional areas at a distance x from one end:

x(ft)	0	0.3	0.6	0.9	1.2	1.5	1.8
Area (ft^2)	28.3	27.3	26.4	24.2	21.6	17.3	13.9

Find the volume of the tank.

Computer

19. Write a program or a spreadsheet that will accept as input the coordinates of any number of data points, and then compute and print the area below those points, using either the average ordinate method, the trapezoid rule, or Simpson's rule.

CHAPTER TEST

Integrate. Try some of the definite integrals by an approximate method.

1. $\int \cot^2 x \csc^4 x\, dx$

2. $\int \sqrt{5 - 3x^2}\, dx$

3. $\int \frac{dx}{9x^2 - 4}$

4. $\int \sin x \sec^2 x\, dx$

5. $\int \frac{(3x - 1)\, dx}{x^2 + 9}$

6. $\int \frac{x^2\, dx}{(9 - x^2)^{3/2}}$

7. $\int \frac{dx}{x\sqrt{1 + x^2}}$

8. $\int \frac{(5x^2 - 3)\, dx}{x^3 - x}$

9. $\int \frac{dx}{2x^2 - 2x + 1}$

10. $\int \frac{dx}{\sqrt{1 + x^2}}$

11. $\int \frac{x\, dx}{\cos^2 x}$

12. $\int \sqrt{1 - 4x^2}\, dx$

13. $\int \frac{(x + 2)\, dx}{x^2 + x + 1}$

14. $\int \frac{\sqrt{x^2 - 1}\, dx}{x^3}$

15. $\int_0^5 \frac{(x^2 - 3)\, dx}{(x + 2)(x + 1)^2}$

16. $\int \frac{dx}{x^2\sqrt{5 - x^2}}$

17. $\int_0^4 \frac{9x^2\, dx}{(2x + 1)(x + 2)^2}$

18. $\int \frac{dx}{\sqrt{4 - (x + 3)^2}}$

19. $\int \frac{dx}{9x^2 - 1}$

20. $\int \frac{(3x - 7)\, dx}{(x - 2)(x - 3)}$

21. $\int \frac{x\, dx}{(1 - x)^5}$

22. $\int \frac{dx}{1 - \sqrt{x}}$

23. $\int \cot^3 x \sin x\, dx$

24. $\int \frac{3x\, dx}{\sqrt[3]{x + 1}}$

25. $\int_0^1 \frac{(3x^2 + 7x)\, dx}{(x + 1)(x + 2)(x + 3)}$

26. $\int \frac{(5x + 9)\, dx}{(x - 9)^{3/2}}$

27. $\int_0^1 \frac{(2x^2 + x + 3)\, dx}{(x + 1)(x^2 + 1)}$

28. $\int_2^\infty \frac{dx}{x\sqrt{x^2 - 4}}$

29. $\int_0^2 \frac{dx}{(x - 2)^{2/3}}$

30. $\int_0^\infty e^{-x} \sin x\, dx$

31. $\int_0^1 \frac{dx}{x^2\sqrt{x^2 + 1}}$

First-Order Differential Equations

A *differential equation* is one that contains one or more derivatives. We have already solved some simple differential equations in Chapter 7 and here we go on to solve more difficult types. We do first-order differential equations in this chapter, and second order in Chapter 13. Our main applications will be in exponential growth and decay, motion, and electrical circuits.

12-1. DIFFERENTIAL EQUATIONS

A *differential equation* is one that contains one or more derivatives.

EXAMPLE 1: Some differential equations, using different notation for the derivative, are

(a) $\frac{dy}{dx} + 5 = 2xy$

(b) $y'' - 4y' + xy = 0$

(c) $Dy + 5 = 2xy$

(d) $D^2y - 4D + xy = 0$

Differential Form

A differential equation containing the derivative dy/dx is put into *differential form* simply by multiplying through by dx.

EXAMPLE 2: Example 1(a), in differential form, is

$$dy + 5\,dx = 2xy\,dx$$

Ordinary versus Partial Differential Equations

We get an *ordinary* differential equation when our differential equation contains only two variables, and "ordinary" derivatives, as in Examples 1 and 2. When the equation contains *partial* derivatives, due to the presence of three or more variables, we have a *partial differential equation.*

EXAMPLE 3: $\dfrac{\partial x}{\partial t} = 5\,\dfrac{\partial y}{\partial t}$ is a partial differential equation.

We cover only *ordinary* differential equations in this book.

Order

The *order* of a differential equation is the order of the highest-order derivative in the equation.

EXAMPLE 4

(a) $\dfrac{dy}{dx} - x = 2y$ is of *first* order.

(b) $\dfrac{d^2y}{dx^2} - \dfrac{dy}{dx} = 3x$ is of *second* order.

(c) $5y''' - 3y'' = xy$ is of *third* order.

Degree

It's important to recognize type here for the same reason as for other equations. It lets us pick the right method of solution.

The *degree of a derivative* is the power to which that derivative is raised.

EXAMPLE 5: $(y')^2$ is of second degree.

The *degree of a differential equation* is the degree of the highest-order derivative in the equation. The equation must be rationalized and cleared of fractions before its degree can be determined.

EXAMPLE 6
(a) $(y'')^3 - 5(y')^4 = 7$ is a second-order equation of *third* degree.
(b) To find the degree of the differential equation

$$\frac{x}{\sqrt{y' - 2}} = 1$$

we clear fractions and square both sides, getting

$$\sqrt{y' - 2} = x$$

or

$$y' - 2 = x^2$$

which is of first degree.

Common Error	Don't confuse order and degree. The symbols $\dfrac{d^2y}{dx^2}$ and $\left(\dfrac{dy}{dx}\right)^2$ have different meanings.

Solving a Simple Differential Equation

We have already solved some simple differential equations by multiplying both sides of the equation by dx and integrating.

EXAMPLE 7: Solve the differential equation $dy/dx = x^2 - 3$.

Solution: We first put our equation into differential form, getting $dy = (x^2 - 3)\,dx$. Integrating we get

$$y = \int (x^2 - 3)\,dx = \frac{x^3}{3} - 3x + C$$

Checking a Solution

Any *function* which satisfies a differential equation is called a *solution* of that equation. Thus, to check a solution, we substitute it and its derivatives into the original equation and see if an identity is obtained.

EXAMPLE 8: Is the function $y = e^{2x}$ a solution of the differential equation $y'' - 3y' + 2y = 0$?

Solution: Taking the first and second derivatives of the function gives

$$y' = 2e^{2x} \quad \text{and} \quad y'' = 4e^{2x}$$

Substituting the function and its derivatives into the differential equation gives

$$4e^{2x} - 6e^{2x} + 2e^{2x} = 0$$

which is an identity. Thus $y = e^{2x}$ is a solution to the differential equation. We will see shortly that it is one of many solutions to the given equation.

General and Particular Solutions

We have just seen that the function $y = e^{2x}$ is a solution of the differential equation $y'' - 3y' + 2y = 0$. But is it the *only* solution? Try the following functions yourself, and you will see that they are also solutions to the given equation.

$$y = 4e^{2x} \tag{1}$$

$$y = Ce^{2x} \tag{2}$$

$$y = C_1e^{2x} + C_2e^{x} \tag{3}$$

where C, C_1, and C_2 are *arbitrary constants.*

The solution of an nth-order differential equation can have at most n arbitrary constants. A solution having the maximum number of constants is called the *general solution.*

The differential equation in our example is of second order, so the solution can have up to two arbitrary constants. Thus (3) is the general solution, while (1) and (2) are called *particular solutions.* When we later solve a differential equation, we will first obtain the general solution. Then, by using other given information, we will evaluate the arbitrary constants to obtain a particular solution. The "other given information" is referred to as *boundary conditions* or *initial conditions.*

EXERCISE 1

Give the order and degree of each equation, and state whether it is ordinary or partial.

1. $\dfrac{dy}{dx} + 3xy = 5$

2. $y'' + 3y' = 5x$

3. $D^3y - 4D = 2xy$

4. $\dfrac{\partial^2 y}{\partial x^2} + 4y = 7$

5. $3(y'')^4 - 5y' = 3y$

6. $4\dfrac{dy}{dx} - 3\left(\dfrac{d^2y}{dx^2}\right)^3 = x^2y$

Solve each differential equation.

7. $\dfrac{dy}{dx} = 7x$

8. $2y' = x^2$

9. $4x - 3y' = 5$

10. $3Dy = 5x + 2$

11. $dy = x^2\, dx$

12. $dy - 4x\, dx = 0$

Show that each function is a solution to the differential equation.

13. $y' = \dfrac{2y}{x}, \qquad y = Cx^2$

14. $\dfrac{dy}{dx} = \dfrac{x^2}{y^3}, \qquad 4x^3 - 3y^4 = C$

15. $Dy = \dfrac{2y}{x}, \qquad y = Cx^2$

16. $y' \cot x + 3 + y = 0, \qquad y = C \cos x - 3$

12-2. VARIABLES SEPARABLE

Given a differential equation of first order, $dy/dx = f(x, y)$, it is sometimes possible to *separate* the variables x and y so that when we multiply both sides by dx we can put the given equation into the form

Variables Separable	$f(y)\, dy = g(x)\, dx$	**380**

If this is possible, we can obtain a solution simply by integrating term by term.

EXAMPLE 1: Solve the differential equation $y' = x^2/y$.

Solution: 1. Rewrite the equation in differential form. We do this by replacing y' with dy/dx and multiplying by dx.

$$dy = \frac{x^2\, dx}{y}$$

2. Separate the variables. We do this here by multiplying both sides by y, and get $y\, dy = x^2\, dx$. The variables are now separated, and our equation is in the form of Eq. 380.

3. Integrate

$$\int y\, dy = \int x^2\, dx$$

$$\frac{y^2}{2} = \frac{x^3}{3} + C_1$$

where C_1 is an arbitrary constant.

4. Simplify the answer. We can do this by multiplying by the LCD 6.

We'll often leave our answer in implicit form, as we have done here.

$$3y^2 = 2x^3 + C$$

where we have replaced $6C_1$ by C.

Simplifying the Solution

Often the simplification of a solution to a differential equation will involve several steps.

EXAMPLE 2: Solve the differential equation $dy/dx = 4xy$.

Solution: Multiplying both sides by dx/y and integrating,

$$\frac{dy}{y} = 4x\,dx$$

$$\ln |y| = 2x^2 + C$$

The solution to a DE can take on different forms, depending on how you simplify it. Don't be discouraged if your solution does not at first match the one in the answer key.

We can leave our solution in this form or solve for y.

$$|y| = e^{2x^2+C} = e^{2x^2}e^c$$

Let us replace e^c by another constant k. Since k can be positive or negative, we can remove the absolute value symbols from y, getting

$$y = ke^{2x^2}$$

We will often interchange one arbitrary constant with another of a different form, such as replacing C_1 by $8C$ or by $\ln C$ or by $\sin C$ or by e^c, or any other form that will help us to simplify an expression.

The laws of exponents or of logarithms will often be helpful in simplifying an answer, as in the following example.

EXAMPLE 3: Solve the differential equation $dy/dx = y/(5 - x)$.

Solution: 1. Going to differential form, $dy = y\,dx/(5 - x)$.
2. Separating the variables,

$$\frac{dy}{y} = \frac{dx}{5 - x}$$

3. Integrating gives $\ln y = -\ln (5 - x) + C_1$.

4. Simplifying, $\ln y + \ln (5 - x) = C_1$. Using our laws of logarithms,

$$\ln y(5 - x) = C_1$$

or $y(5 - x) = C$, where $C = e^{c_1}$. So

$$y = \frac{C}{5 - x}$$

where $x \neq 5$.

Logarithmic, Exponential, or Trigonometric Equations

In order to separate the variables in certain equations, it may be necessary to use our laws of exponents or logarithms, or the trigonometric identities.

EXAMPLE 4: Solve the equation $4y'e^{4y} \cos x = e^{2y} \sin x$.

Solution: We replace y' with dy/dx and multiply through by dx, getting $4e^{4y} \cos x \, dy = e^{2y} \sin x \, dx$. We can eliminate x on the left side by dividing through by $\cos x$. Similarly, we can eliminate y from the right by dividing by e^{2y}.

$$\frac{4e^{4y}}{e^2y} dy = \frac{\sin x}{\cos x} dx$$

or $4e^{2y} = \tan x \, dx$. Integrating gives the solution

$$2\, e^{2y} = -\ln |\cos x| + C$$

Particular Solution

We can evaluate the constant in our solution to a differential equation when given suitable *boundary conditions*, as in the following example.

EXAMPLE 5: Solve the equation $2y(1 + x^2)y' + x(1 + y^2) = 0$, subject to the condition that $y = 2$ when $x = 0$.

Solution: Separating variables and integrating,

$$\frac{2y\,dy}{1 + y^2} + \frac{x\,dx}{1 + x^2} = 0$$

$$\int \frac{2y\,dy}{1 + y^2} + \frac{1}{2}\int \frac{2\,x\,dx}{1 + x^2} = 0$$

$$\ln |1 + y^2| + \frac{1}{2} \ln |1 + x^2| = C$$

Simplifying, we drop the absolute value signs since $(1 + x^2)$ and $(1 + y^2)$ cannot be negative. Then we multiply by 2 and apply the laws of logarithms,

$$2 \ln (1 + y^2) + \ln (1 + x^2) = 2C$$

$$\ln (1 + x^2)(1 + y^2)^2 = 2C$$

$$(1 + x^2)(1 + y^2)^2 = e^{2c}$$

Applying the boundary conditions that $y = 2$ when $x = 0$ gives

$$e^{2c} = (1 + 0)(1 + 2^2)^2 = 25$$

Our complete equation is then $(1 + x^2)(1 + y^2)^2 = 25$

EXERCISE 2

General Solution

We'll have applications later in this chapter.

Find the general solution to each differential equation.

1. $y' = \dfrac{x}{y}$

2. $\dfrac{dy}{dx} = \dfrac{2y}{x}$

3. $dy = x^2 y\, dx$

4. $y' = xy^3$

5. $y' = \dfrac{x^2}{y^3}$

6. $y' = \dfrac{x^2 + x}{y - y^2}$

7. $xy\, dx - (x^2 + 1)\, dy = 0$

8. $y' = x^3 y^5$

9. $(1 + x^2)\, dy + (y^2 + 1)\, dx = 0$

10. $y' = x^2 e^{-3y}$

11. $\sqrt{1 + x^2}\, dy + xy\, dx = 0$

12. $y^2\, dx = (1 - x)\, dy$

13. $(y^2 + 1)\, dx = (x^2 + 1)\, dy$

14. $y^3\, dx = x^3\, dy$

15. $(2 + y)\, dx + (x - 2)\, dy = 0$

16. $y' = \dfrac{e^{x-y}}{e^x + 1}$

17. $(x - xy^2)\, dx = -(x^2 y + y)\, dy$

With Exponential Functions

18. $dy = e^{-x}\, dx$

19. $ye^{2x} = (1 + e^{2x})y'$

20. $e^y(y' + 1) = 1$

21. $e^{x-y}\, dx + e^{y-x}\, dy = 0$

With Trigonometric Functions

22. $(3 + y)\, dx + \cot x\, dy = 0$

23. $\tan y\, dx + (1 + x)\, dy = 0$

24. $\tan y\, dx + \tan x\, dy = 0$

25. $\cos x \sin y \, dy + \sin x \cos y \, dx = 0$

26. $\sin x \cos^2 y \, dx + \cos^2 x \, dy = 0$

27. $4 \sin x \sec y \, dx = \sec x \, dy$

Particular Solution

Find the particular solution to each differential equation.

28. $x \, dx = 4y \, dy$, $\quad x = 5$ when $y = 2$

29. $y^2 y' = x^2$, $\quad x = 0$ when $y = 1$

30. $\sqrt{x^2 + 1} \, y' + 3xy^2 = 0$, $\quad x = 1$ when $y = 1$

31. $y' \sin y = \cos x$, $\quad x = \pi/2$ when $y = 0$

32. $x(y + 1)y' = y(1 + x)$, $\quad x = 1$ when $y = 1$

12-3. EXACT DIFFERENTIAL EQUATIONS

Even if we cannot separate the variables, a differential equation might still be solved by integrating a *combination* of terms.

EXAMPLE 1: Solve $y \, dx + x \, dy = x \, dx$.

Solution: The variables are not now separated, and we see that no amount of manipulation will separate them. However, the combination of terms $y \, dx + x \, dy$ on the left side may ring a bell. In fact, it is the derivative of the product of x and y (Eq. 310),

$$\frac{d(xy)}{dx} = x\frac{dy}{dx} + y\frac{dx}{dx}$$

or

$$d(xy) = x \, dy + y \, dx$$

This, then, is the left side of our given equation. The right side contains only x's, so we integrate,

$$\int (y \, dx + x \, dy) = \int x \, dx$$

$$\int d(xy) = \int x \, dx$$

$$xy = \frac{x^2}{2} + C$$

or

$$y = \frac{x}{2} + \frac{C}{x}$$

When the left side of a differential equation is the exact differential of some function (as in Example 1) we call that equation an *exact differential equation*.

Integrable Combinations

The expression $y\,dx + x\,dy$ from Example 1 is called an *integrable combination*. Some of the most frequently used combinations are

$x\,dy + y\,dx = d(xy)$	**381**
$\dfrac{x\,dy - y\,dx}{x^2} = d\left(\dfrac{y}{x}\right)$	**382**
$\dfrac{y\,dx - x\,dy}{y^2} = d\left(\dfrac{x}{y}\right)$	**383**
$\dfrac{x\,dy - y\,dx}{x^2 + y^2} = d\left(\tan^{-1}\dfrac{y}{x}\right)$	**384**

EXAMPLE 2: Solve $dy/dx = y(1 - xy)/x$.

Solution: We go to differential form and clear denominators by multiplying through by $x\,dx$, getting $x\,dy = y(1 - xy)\,dx$. Removing parentheses,

$$x\,dy = y\,dx - xy^2\,dx$$

On the lookout for an integrable combination, we move the $y\,dx$ term to the left side.

$$x\,dy - y\,dx = -xy^2\,dx$$

Any expression that we multiply by (such as $-1/y^2$ here) to make our equation exact, is called an *integrating factor*.

We see now that the left side will be the differential of x/y if we multiply by $-1/y^2$.

$$\frac{y\,dx - x\,dy}{y^2} = x\,dx$$

Integrating gives $x/y = x^2/2 + C_1$, or

$$y = \frac{2x}{x^2 + C}$$

Particular Solution

As before, we substitute boundary conditions to evaluate the constant of integration.

EXAMPLE 3: Solve $2xy\,dy - 4x\,dx + y^2\,dx = 0$, such that $x = 1$ when $y = 2$.

Solution: It looks as though the first and third terms might be an integrable combination, so we transpose the $-4x\,dx$.

$$2xy\,dy + y^2\,dx = 4x\,dx$$

The left side is the derivative of the product xy^2,

$$d(xy^2) = 4x\,dx$$

Integrating gives $xy^2 = 2x^2 + C$. Substituting the boundary conditions, we get $C = xy^2 - 2x^2 = 1(2)^2 - 2(1)^2 = 2$, so our complete equation is

$$xy^2 = 2x^2 + 2$$

EXERCISE 3

Integrable Combinations

Find the general solution of each differential equation.

1. $y\,dx + x\,dy = 7\,dx$

2. $x\,dy = (4 - y)\,dx$

3. $x\dfrac{dy}{dx} = 3 - y$

4. $y + xy' = 9$

5. $2xy' = x - 2y$

6. $x\dfrac{dy}{dx} = 2x - y$

7. $x\,dy = (3x^2 + y)\,dx$

8. $(x + y)\,dx + x\,dy = 0$

9. $3x^2 + 2y + 2xy' = 0$

10. $(1 - 2x^2y)\dfrac{dy}{dx} = 2xy^2$

11. $(2x - y)y' = x - 2y$

12. $\dfrac{y\,dx - x\,dy}{y^2} = x\,dx$

13. $y\,dx - x\,dy = 2y^2\,dx$

14. $(x - 2x^2y)\,dy = y\,dx$

15. $(4y^3 + x)\dfrac{dy}{dx} = y$

16. $(y - x)y' + 2xy^2 + y = 0$

17. $3x - 2y^2 - 4xyy' = 0$

18. $\dfrac{x\,dy - y\,dx}{x^2 + y^2} = 5\,dy$

Find the particular solution to each differential equation.

19. $4x = y + xy'$, $x = 3$ when $y = 1$

20. $y\,dx = (x - 2x^2y)\,dy$, $x = 1$ when $y = 2$

21. $y = (3y^3 + x)\dfrac{dy}{dx}$, $x = 1$ when $y = 1$

22. $4x^2 = -2y - 2xy'$, $x = 5$ when $y = 2$

23. $3x - 2y = (2x - 3y)\dfrac{dy}{dx}$, $x = 2$ when $y = 2$

24. $5x = 2y^2 + 4xyy'$, $x = 1$ when $y = 4$

12-4. HOMOGENEOUS FIRST-ORDER DIFFERENTIAL EQUATIONS

If each variable in a function is replaced by t times the variable, and a power of t can be factored out, we say that the function is a *homogeneous function*.

EXAMPLE 1: Is $\sqrt{x^4 + xy^3}$ a homogeneous function?

Solution: We replace x with tx and y with ty and get

$$\sqrt{(tx)^4 + (tx)(ty)^3} = \sqrt{t^4x^4 + t^4xy^3} = t^2\sqrt{x^4 + xy^3}$$

Since we were able to factor out a t^2, we say that our function is homogeneous to the second degree.

A *homogeneous polynomial* is one in which every term is of the same degree.

EXAMPLE 2: $x^2 + xy - y^2$ is a homogeneous polynomial of degree 2.

A *homogeneous differential equation* is one of the form

First-Order Homogeneous	$M\,dx + N\,dy = 0$	385

where M and N are functions of x and y, and are homogeneous of the same degree.

EXAMPLE 3: $(x^2 + y^2)\,dx + xy\,dy = 0$ is homogeneous.

Solving a Homogeneous Linear Differential Equation

Sometimes we can transform a homogeneous differential equation whose variables cannot be separated into one whose variables can be separated by making the substitution

$$y = vx$$

EXAMPLE 4: Solve $x\dfrac{dy}{dx} - y = \sqrt{x^2 - y^2}$.

Solution: We see that this is homogeneous and of first degree. We make the substitution $y = vx$ to transform the given equation into one whose variables can be separated. However, when we substitute for y we must also substitute for dy/dx. Taking the derivative of $y = vx$, we get

$$\frac{dy}{dx} = v + x\frac{dv}{dx}$$

Substituting

$$x\left(v + x\frac{dv}{dx}\right) - vx = \sqrt{x^2 + v^2x^2}$$

which reduces to

$$x\frac{dv}{dx} = \sqrt{1 + v^2}$$

Separating variables,

$$\frac{dv}{\sqrt{1 + v^2}} = \frac{dx}{x}$$

Integrating by rule 62,

$$\ln|v + \sqrt{1 + v^2}| = \ln|x| + C_1$$

$$\ln\left|\frac{v + \sqrt{1 + v^2}}{x}\right| = C_1$$

$$\frac{v + \sqrt{1 + v^2}}{x} = e^{C_1}$$

$$v + \sqrt{1 + v^2} = Cx$$

where $C = e^{C_1}$. Now subtracting v from both sides, squaring, and simplifying gives

$$C^2x^2 - 2Cvx = 1$$

Finally, substituting back, $v = y/x$, we get

$$C^2x^2 - 2Cy = 1$$

EXERCISE 4

First-Order Homogeneous

Find the general solution to each differential equation.

1. $(x - y)\,dx - 2x\,dy = 0$

2. $(3y - x)\,dx = (x + y)\,dy$

These are not easy.

3. $(x^2 - xy)y' + y^2 = 0$

4. $(x^2 - xy)\,dy + (x^2 - xy + y^2)\,dx = 0$

5. $xy^2\,dy - (x^3 + y^3)\,dx = 0$

6. $2x^3y' + y^3 - x^2y = 0$

Find the particular solution.

7. $x - y = 2xy'$, $\quad x = 1, y = 1$

8. $3xy^2\,dy = (3y^3 - x^3)\,dx$, $\quad x = 1, y = 2$

9. $(2x + y)\,dx = y\,dy$, $\quad x = 2, y = 1$

10. $(x^3 + y^3)\,dx - xy^2\,dy = 0$, $\quad x = 1, y = 0$

12-5. FIRST-ORDER LINEAR DIFFERENTIAL EQUATIONS

When describing the degree of a term, we usually add the degrees of each variable in the term. Thus x^2y^3 is of fifth degree.

Sometimes, however, we want to describe the degree of a term with regard to just one of the variables. Thus we say that x^2y^3 is of second degree in x, and of third degree in y.

In determining the degree of a term, we must also consider any derivatives in that term. Thus the term $xy\,dy/dx$ is of first degree in x and of second degree in y.

A first-order differential equation is called *linear* if each term is of first degree or less *in the dependent variable y*.

EXAMPLE 1:
(a) $y' + x^2y = e^x$ is linear.
(b) $y' + xy^2 = e^x$ is not linear because y is squared in the second term.
(c) $y\, dy/dx - xy = 5$ is not linear because we must add the exponents of y and dy/dx, making the first term of second degree.

A first-order linear differential equation can always be written in the standard form

First-Order Linear	$\dfrac{dy}{dx} + Py = Q$	**386**

where P and Q are functions of x only.

EXAMPLE 2: Write the equation $xy' - e^x + y = xy$ in the form of Eq. 386.

Solution: Transposing gives $xy' + y - xy = e^x$. Factoring,

$$xy' + (1 - x)y = e^x$$

Dividing by x gives us the standard form,

$$y' + \frac{1 - x}{x}y = \frac{e^x}{x}$$

where $P = (1 - x)/x$ and $Q = e^x/x$.

Integrating Factor

The left side of a first-order linear differential equation can always be made into an integrable combination by multiplying by an *integrating factor R*. We now find such a factor.

Multiplying Eq. 386 by R, the left side becomes

$$R\frac{dy}{dx} + yRP \tag{1}$$

Let us try to make the left side *the exact derivative of the product Ry of y and the integrating factor R*. The derivative of Ry is

$$R\frac{dy}{dx} + y\frac{dR}{dx} \tag{2}$$

But (1) and (2) will be equal if $dR/dx = RP$. Since P is a function of x only, we can separate variables,

$$\frac{dR}{R} = P\,dx$$

We omit the constant of integration because we seek only one integrating factor.

Integrating, $\ln R = \int P\,dx$, or

Integrating Factor	$R = e^{\int P dx}$	**387**

Thus, *multiplying a given linear equation by the integrating factor* $R = e^{\int P dx}$ *will make the left side of the equation the exact derivative of* Ry.

EXAMPLE 3: Solve $\dfrac{dy}{dx} + \dfrac{4y}{x} = 3$.

Solution: Our equation is in standard form with $P = 4/x$ and $Q = 3$. Then

$$\int P\,dx = 4\int \frac{dx}{x} = 4 \ln x = \ln x^4$$

Can you show why $e^{\ln x^4} = x^4$?

Our integrating factor R is then

$$R = e^{\int P dx} = e^{\ln x^4} = x^4$$

Multiplying our given equation by x^4 and going to differential form,

$$x^4\,dy + 4x^3y\,dx = 3x^4\,dx$$

Notice that the left side is now the derivative of y times the integrating factor, $d(x^4y)$. Integrating,

$$x^4y = \frac{3x^5}{5} + C$$

or

$$y = \frac{3x}{5} + \frac{C}{x^4}$$

In summary, to solve the first-order linear equation

$$y' + Py = Q$$

multiply by an integrating factor $e^{\int P dx}$, and the solution to the equation becomes

Solution to First-Order Linear DE	$ye^{\int P dx} = \int Qe^{\int P dx}\,dx$	**388**

Next we try an equation having trigonometric functions.

EXAMPLE 4: Solve $y' + y \cot x = \csc x$.

Solution: This is in standard form, with $P = \cot x$ and $Q = \csc x$. Thus $\int P\,dx = \int \cot x\,dx = \ln|\sin x|$. The integrating factor is e raised to the $\ln|\sin x|$, or simply $\sin x$. Then, by Eq. 388,

$$y \sin x = \int \csc x \sin x\,dx$$
$$= \int dx$$

so

$$y \sin x = x + C$$

Particular Solution

As before, we find the general solution and then substitute the boundary conditions to evaluate C.

EXAMPLE 5: Solve $y' + \left(\frac{1-2x}{x^2}\right) y = 1$ given that $y = 2$ when $x = 1$.

Solution: We are in standard form with $P = (1 - 2x)/x^2 = x^{-2} - 2x^{-1}$. Integrating gives $\int P\,dx = -1/x - 2 \ln x = -1/x - \ln x^2$. Our integrating factor is then

$$R = e^{\int P dx} = e^{-1/x - \ln x^2} = \frac{e^{-1/x}}{e^{\ln x^2}} = \frac{1}{x^2 e^{1/x}}$$

Substituting into Eq. 388,

$$\frac{y}{x^2 e^{1/x}} = \int \frac{dx}{x^2 e^{1/x}}$$
$$= \int e^{-1/x} x^{-2}\,dx = e^{-1/x} + C$$

When $x = 1$ and $y = 2$, $C = 2/e - 1/e = 1/e$, so

$$\frac{y}{x^2 e^{1/x}} = e^{-1/x} + \frac{1}{e}$$

or

$$y = x^2\left(1 + \frac{e^{1/x}}{e}\right)$$

Equations Reducible to Linear Form

One type of nonlinear equation easily reducible to linear form is

After James Bernoulli (1654–1705), the most famous of this family of Swiss mathematicians.

Bernoulli's Equation	$\frac{dy}{dx} + Py = Qy^n$	**396**

where P and Q are functions of x, as before. We solve such an equation by making the substitution

$$z = y^{1-n}$$

EXAMPLE 6: Solve the equation $y' + y/x = x^2y^6$.

Solution: This is a Bernoulli equation with $n = 6$. We first divide through by y^6 and get

$$\frac{1}{y^6}\frac{dy}{dx} + \frac{y^{-5}}{x} = x^2$$

We substitute $z = y^{1-6} = y^{-5}$, and

$$\frac{dz}{dx} = -\frac{5}{y^6}\frac{dy}{dx}$$

Substituting,

$$-\frac{1}{5}\frac{dz}{dx} + \frac{1}{x}z = x^2$$

or

$$\frac{dz}{dx} - \frac{5}{x}z = -5x^2$$

This equation is linear in z, with $P = -5/x$ and $Q = -5x^2$. We now find the integrating factor. The integral of P is $\int (-5/x)\,dx = -5 \ln x = \ln x^{-5}$, so the integrating factor is

$$R = e^{\int P dx} = x^{-5} = \frac{1}{x^5}$$

From Eq. 388,

$$\frac{z}{x^5} = -5 \int x^{-3}\, dx = \frac{-5x^{-2}}{-2} + C$$

We solve for z and then substitute back $z = 1/y^5$,

$$z = \frac{5}{2}x^3 + Cx^5 = \frac{1}{y^5}$$

Our final equation is then

$$y^5 = \frac{1}{(5x^3/2 + Cx^5}$$

or

$$y^5 = \frac{2}{5x^3 + C_1x^5}$$

EXERCISE 5

First-Order Linear

Find the general solution to each differential equation.

1. $\dfrac{dy}{dx} + \dfrac{y}{x} = 4$
2. $y' + \dfrac{y}{x} = 3x$
3. $xy' = 4x^3 - y$
4. $\dfrac{dy}{dx} + xy = 2x$
5. $y' = x^2 - x^2y$
6. $y' - \dfrac{y}{x} = \dfrac{-1}{x^2}$
7. $y' = \dfrac{3 - xy}{2x^2}$
8. $y' = x + \dfrac{2y}{x}$
9. $xy' = 2y - x$
10. $(x + 1)y' - 2y = (x + 1)^4$
11. $y' = \dfrac{2 - 4x^2y}{x + x^3}$
12. $(x + 1)y' = 2(x + y + 1)$
13. $xy' + x^2y + y = 0$
14. $(1 + x^3)\, dy = (1 - 3x^2y)\, dx$

With Exponential and Logarithmic Expressions

15. $y' + y = e^x$
16. $y' = e^{2x} + y$
17. $y' = 2y + 4e^{2x}$
18. $xy' - e^x + y + xy = 0$
19. $y' = \dfrac{4 \ln x - 2x^2y}{x^3}$

With Trigonometric Expressions

20. $\dfrac{dy}{dx} + y \sin x = 3 \sin x$ **21.** $y' + y = \sin x$ **22.** $y' + 2xy = 2x \cos x^2$

23. $y' = 2 \cos x - y$ **24.** $\dfrac{dx}{y'} = \sec x - y \cot x$

Bernoulli's Equation

25. $y' + \dfrac{y}{x} = 3x^2y^2$ **26.** $xy' + x^2y^2 + y = 0$

27. $y' = y - xy^2(x + 2)$ **28.** $y' + 2xy = xe^{-x^2}y^3$

Particular Solution

Find the particular solution to each differential equation.

29. $xy' + y = 4x, \quad x = 1$ when $y = 5$

30. $\dfrac{dy}{dx} + 5x = x - xy, \quad x = 2$ when $y = 1$

31. $y' + \dfrac{y}{x} = 5, \quad x = 1$ when $y = 2$

32. $y' = 2 + \dfrac{3y}{x}, \quad x = 2$ when $y = 6$

33. $y' = \tan^2 x + y \cot x, \quad x = \dfrac{\pi}{4}$ when $y = 2$

34. $\dfrac{dy}{dx} + 5y = 3e^x, \quad x = 1$ when $y = 1$

12-6. GEOMETRIC APPLICATIONS OF DIFFERENTIAL EQUATIONS

Now that we are able to solve some simple differential equations of first degree, we turn to applications. Here we must not only solve the equation, but must first *set up* the differential equation. The geometric problems we do first will help to prepare us for the physical applications that follow.

Setting Up a Differential Equation

When reading the problem statement, look for the words "slope" or "rate of change." Each of these is represented by the first derivative.

EXAMPLE 1

(a) The statement "*the slope of a curve at every point is equal to twice the ordinate*" is represented by the differential equation

$$\frac{dy}{dx} = 2y$$

(b) The statement "*the ratio of abscissa to ordinate at each point on a curve is proportional to the rate of change at that point*" can be written

$$\frac{x}{y} = k\frac{dy}{dx}$$

(c) The statement "*the slope of a curve at every point is inversely proportional to the square of the ordinate at that point*" can be described by the equation

$$\frac{dy}{dx} = \frac{k}{y^2}$$

Finding an Equation Whose Slope Is Specified

Once the equation is written, it is solved by the methods of the preceding sections.

EXAMPLE 2: The slope of a curve at each point is one-tenth the product of the ordinate and the square of the abscissa, and the curve passes through the point (2, 3). Find the equation of the curve.

Solution: The differential equation is

$$\frac{dy}{dx} = \frac{x^2y}{10}$$

In solving a differential equation, we first see if the variables can be separated. In this case they can be.

$$\frac{10\,dy}{y} = x^2\,dx$$

Integrating,

$$10 \ln y = \frac{x^3}{3} + C$$

At (2, 3)

$$C = 10 \ln 3 - \frac{8}{3} = 8.32$$

Our curve thus has the equation $\ln y = x^3/30 + 0.832$.

Tangents and Normals to Curves

In setting up problems involving tangents and normals to curves, recall that

$$\text{slope of the tangent} = \frac{dy}{dx} = y'$$

$$\text{slope of the normal} = -\frac{1}{y'}$$

EXAMPLE 3: A curve (Fig. 12-1) passes through the point (4, 2). If from any point P on the curve, the line OP and the tangent PT is drawn, the triangle OPT is isosceles. Find the equation of the curve.

Solution: The slope of the tangent is dy/dx, and the slope of OP is y/x. These must be equal but of opposite signs.

$$\frac{dy}{dx} = -\frac{y}{x}$$

We can solve this equation by separation of variables or as the separable combination $x\,dy + y\,dx = 0$. Either way gives the hyperbola

$$xy = C$$

(see Eq. 299). At the point (4, 2), $C = 4(2) = 8$. So our equation is $xy = 8$.

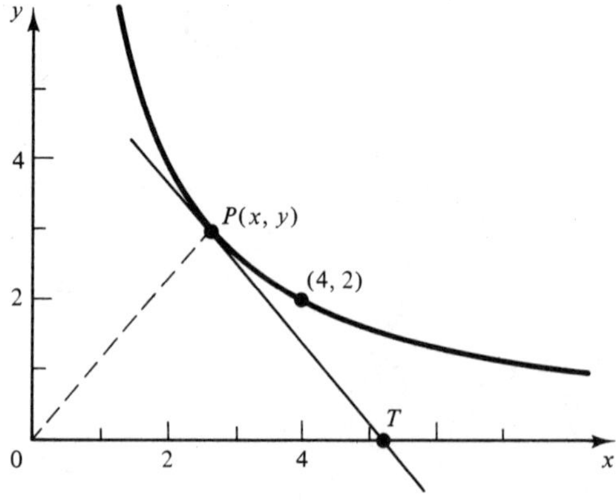

FIGURE 12-1

Orthogonal Trajectories

In Sec. 7-3 we graphed a relation that had an arbitrary constant and got a *family of curves*. For example, the relation $x^2 + y^2 = C^2$ represents a family of circles of radius C, whose center is at the origin. Another curve that cuts each curve of the family at right angles is called an *orthogonal trajectory* to that family.

To find the orthogonal trajectory to a family of curves,

1. Differentiate the equation of the family to get the slope.
2. Eliminate the constant contained in the original equation.
3. Take the negative reciprocal of that slope to get the slope of the orthogonal trajectory.
4. Solve the resulting differential equation to get the equation of the orthogonal trajectory.

EXAMPLE 4: Find the equation of the orthogonal trajectories to the parabolas $y^2 = px$.

Solution: 1. The derivative is $2yy' = p$, so

$$y' = \frac{p}{2y}$$

2. The constant p, from the original equation, is y^2/x, so $y' = y/2x$.

Common Error	Be sure to eliminate the constant (p in this example) before continuing.

3. The slope of the orthogonal trajectory is, by Eq. 259, the negative reciprocal of the slope of the given family,

$$y' = \frac{-2x}{y}$$

4. Separating variables,

$$y\,dy = -2x\,dx$$

Integrating,

$$\frac{y^2}{2} = \frac{-2x^2}{2} + C_1$$

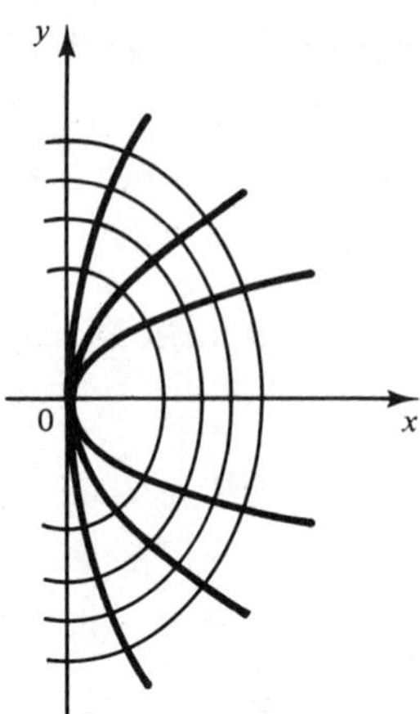

FIGURE 12-2. Orthogonal ellipses and parabolas. Each intersection is at 90°.

or the family of ellipses, $2x^2 + y^2 = C$ (Fig. 12-2).

EXERCISE 6

Slopes of Curves

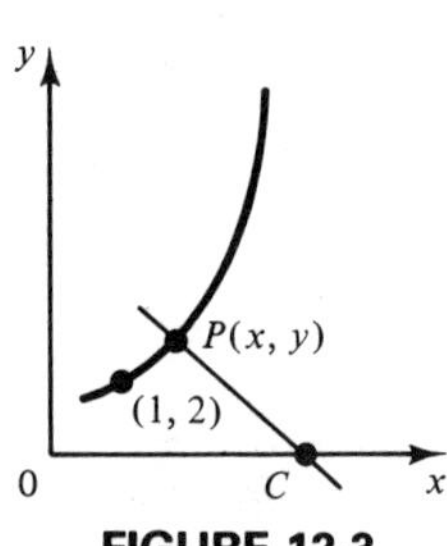

FIGURE 12-3

FIGURE 12-4

1. Find the equation of the curve that passes through the point (2, 9) and whose slope is $y' = x + 1/x + y/x$.

2. The slope of a certain curve at any point is equal to the reciprocal of the ordinate at that point. Write the equation of the curve if it passes through the point (1, 3).

3. Find the equation of a curve whose slope at any point is equal to the abscissa of that point divided by the ordinate, and which passes through the point (3, 4).

4. Find the equation of a curve that passes through (1, 1) and whose slope at any point is equal to the product of the ordinate and abscissa.

5. A curve passes through the point (2, 3) and has a slope equal to the sum of the abscissa and ordinate at each point. Find its equation.

Tangents and Normals

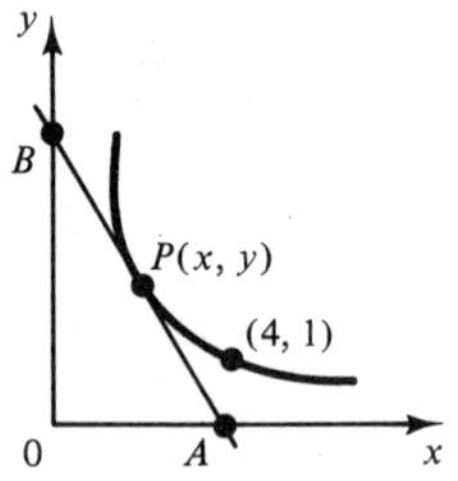

6. The x intercept OC of the normal (Fig. 12-3) is given by $OC = x + y'y$. Write the equation of the curve passing through (1, 2) for which OC is equal to three times the abscissa of P.

The equations for the various distances associated with tangents and normals were given in Chapter 2, Exercise 2.

7. A certain first-quadrant curve (Fig. 12-4) passes through the point (4, 1). If a tangent is drawn through any point P, the portion AB of the tangent that lies between the coordinate axes is bisected by P. Find the curve, given that

$$AP = -\left(\frac{y}{y'}\right)\sqrt{1 + (y')^2} \qquad \text{and} \qquad BP = x\sqrt{1 + (y')^2}$$

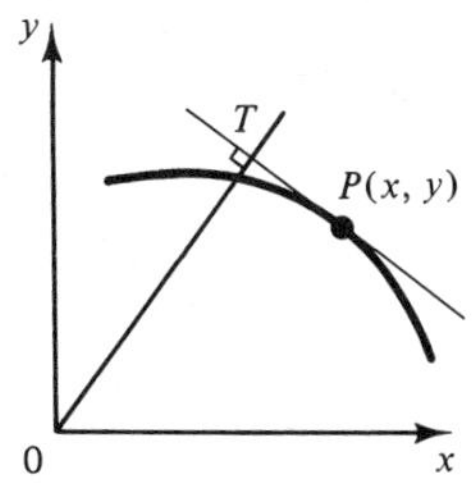

FIGURE 12-5

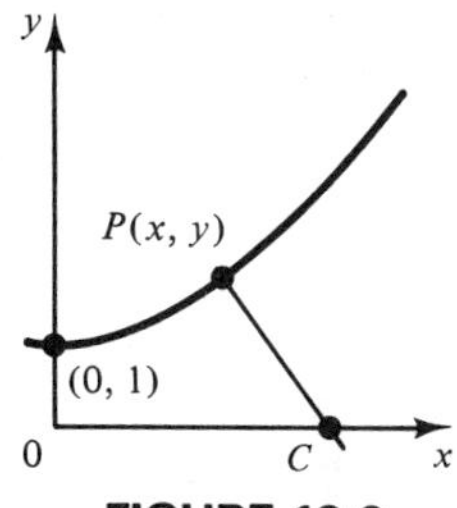

FIGURE 12-6

8. A tangent PT is drawn to a curve at a point P (Fig. 12-5). The distance OT from the origin to a tangent through P is given by

$$OT = \frac{xy' - y}{\sqrt{1 + (y')^2}}$$

Find the equation of the curve passing through the point (2, 4) so that OT is equal to the abscissa of P.

9. Find the equation of the curve passing through (0, 1) for which the length PC of a normal through P equals the square of the ordinate of P (Fig. 12-6), where

$$PC = y\sqrt{1 + (y')^2}$$

10. Find the curve passing through (4, 4) such that the distance OT (Fig. 12-5) is equal to the ordinate of P.

Orthogonal Trajectories

Write the equation of the orthogonal trajectories to each family of curves.

11. the circles, $x^2 + y^2 = r^2$

12. the parabolas, $x^2 = ay$

13. the hyperbolas $x^2 - y^2 = Cy$

12-7. EXPONENTIAL GROWTH AND DECAY

Equations 197, 199, and 200 for exponential growth and decay can be derived by means of compound interest formulas. Here we derive the equation for exponential growth to an upper limit by solving the differential equation that describes such growth. The derivation of equations for exponential growth and decay are left as an exercise.

EXAMPLE 1: A quantity starts from zero and grows with time such that its rate of growth is proportional to the difference between the final amount a and the present amount y. Find an equation for y as a function of time.

Solution: The amount present at time t is y, and the rate of growth of y we write as dy/dt. Since the rate of growth is proportional to $(a - y)$, we write the differential equation

$$\frac{dy}{dt} = n(a - y)$$

where n is a constant of proportionality. Separating variables,

$$\frac{dy}{a - y} = n\,dt$$

Integrating,

$$-\ln(a - y) = nt + C$$

Going to exponential form and simplifying,

$$a - y = e^{-nt-C} = e^{-nt}e^{-C} = C_1e^{-nt}$$

where $C_1 = e^{-C}$. Applying the initial conditions that $y = 0$ when $t = 0$ gives $C_1 = a$, so our equation becomes $a - y = ae^{-nt}$ or

Exponential Growth to an Upper Limit	$y = a(1 - e^{-nt})$	**200**

Motion in a Resisting Fluid

Here we continue our study of motion. In Chapter 5 we showed that the instantaneous velocity is given by the derivative of the displacement, and that the instantaneous acceleration is given by the derivative of the velocity (or by the second derivative of the displacement). In Chapter 7 we solved simple differential equations to find displacement given the velocity or acceleration. Here we do a type of problem that leads to equations that we were not able to solve in Chapter 7.

In this type of problem, an object falls through a fluid (usually air or water) which exerts a *resisting force* which is proportional to the velocity of the object and in the opposite direction. We set up these problems using Newton's second law, $F = ma$, and we'll see that the motion follows the law for exponential growth described by Eq. 200.

EXAMPLE 2: A crate falls from rest from an airplane. The air resistance is proportional to the crate's velocity, and the crate reaches a limiting speed of 218 ft/s. (a) Write an equation for the crate's velocity. (b) Find the crate's velocity after 0.75 s.

Solution: (a) By Newton's second law,

$$F = ma = \frac{W}{g}\frac{dv}{dt}$$

where W is the weight of the crate, dv/dt is the acceleration, and $g = 32.2$ ft/s². Taking the downward direction as positive, the resultant force F is equal to $W - kv$, where k is a constant of proportionality. So

$$W - kv = \frac{W}{32.2}\frac{dv}{dt}$$

We can find k by noting that the acceleration must be zero when the limiting speed (218 ft/s) is reached. Thus

$$W - 218k = 0$$

Notice that the weight W has dropped out.

So $k = W/218$. Our differential equation, after multiplying by $218/W$, is then $218 - v = 6.77\ dv/dt$. Separating variables and integrating,

$$\frac{dv}{218 - v} = 0.148\ dt$$

$$\ln|218 - v| = -0.148t + C$$

$$218 - v = e^{-0.148t+C} = e^{-0.148t}e^{C}$$

$$= C_1 e^{-0.148t}$$

Our equation for v is of the same form as Eq. 200, for exponential growth to an upper limit.

Since $v = 0$ when $t = 0$, $C_1 = 218 - 0 = 218$. Then

$$v = 218(1 - e^{-0.148t})$$

(b) When $t = 0.75$ s,

$$v = 218(1 - e^{-0.111}) = 22.9 \text{ ft/s}$$

EXERCISE 7

Exponential Growth

1. A quantity grows with time such that its rate of growth dy/dt is proportional to the present amount y. Use this statement to derive the equation for exponential growth, $y = ae^{nt}$.

We'll have more problems involving exponential growth and decay in the following section on electric circuits.

2. A biomedical company finds that a certain bacteria used for crop insect control will grow exponentially at the rate of 12% per hour. Starting with 1000 bacteria, how many will they have after 10 h?

3. If the U.S. energy consumption in 1979 was 37 million barrels (bbl) per day oil equivalent and is growing exponentially at a rate of 6.9% per year, estimate the daily oil consumption in the year 2000.

Exponential Decay

4. A quantity decreases with time such that its rate of decrease dy/dt is proportional to the present amount y. Use this statement to derive the equation for exponential decay, $y = ae^{-nt}$.

5. An iron ingot is 1850°F above room temperature. If it cools exponentially at 3.5% per minute, find its temperature (above room temperature) after 2.50 h.

6. A certain pulley in a tape drive is rotating at 2550 r/min. After the power is shut off, its speed decreases exponentially at a rate of 12.5% per second. Find the pulley's speed after 5.00 s.

Exponential Growth to an Upper Limit

7. A forging, initially at 0°F, is placed in a furnace at 1550°F, where its temperature rises exponentially at the rate of 6.5% per minute. Find its temperature after 25.0 min.

8. If we assume that the compressive strength of concrete increases exponentially with time to an upper limit of 4000 lb/in.^{2}, and that the rate of increase is 52.5% per week, find the strength after 2 weeks.

Motion in a Resisting Medium

9. A 45.5-lb carton is initially at rest. It is then pulled horizontally by a 19.4-lb force in the direction of motion, and resisted by a frictional force which is equal (in pounds) to four times the carton's velocity (in ft/s). Show that the differential equation of motion is $dv/dt = 13.7 - 2.83v$.

10. For the carton in Problem 9, find the velocity after 1.25 s.

11. An instrument package is dropped from an airplane. It falls from rest through air whose resisting force is proportional to the speed of the package. The terminal speed is 155 ft/s. Show that the acceleration is given by the differential equation $a = dv/dt = g - gv/155$.

12. Find the speed of the instrument package in Problem 11 after 0.50 s.

13. Find the displacement of the instrument package in Problem 11 after 1.00 s.

14. A 157-lb stone falls from rest from a cliff. If the air resistance is proportional to the square of the stone's speed, and the limiting speed of the stone is 125 ft/s, show that the differential equation of motion is $dv/dt = g - gv^2/15{,}625$.

15. Find the time for the velocity of the stone in Problem 14 to be 60 ft/s.

16. A 15.0-lb ball is thrown downward from an airplane with a speed of 21.0 ft/s. If we assume the air resistance to be proportional to the ball's speed, and the limiting speed is 135 ft/s, show that the velocity of the ball is given by $v = 135 - 114e^{-t/4.19}$ ft/s.

17. Find the time at which the ball in Problem 16 is going at a speed of 70.0 ft/s.

18. A box falls from rest and encounters air resistance proportional to the cube of the speed. The limiting speed is 12.5 ft/s. Show that the acceleration is given by the differential equation $60.7\ dv/dt = 1953 - v^3$.

12-8. SERIES *RL* AND *RC* CIRCUITS

Series *RL* Circuit

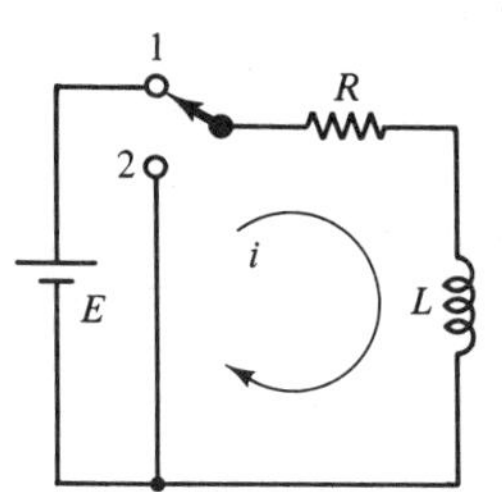

FIGURE 12-7. RL circuit.

Figure 12-7 shows a resistance of R ohms in series with an inductance of L henrys. The switch can either connect these elements to a battery of voltage E (position 1, charge) or to a short circuit (position 2, discharge). In either case our object is to find the current i in the circuit. We will see that it is composed of two parts; a *steady-state* current that flows long after the switch has been thrown, and a *transient* current that dies down shortly after the switch is thrown.

EXAMPLE 1: After being in position 2 for a long time, the switch (Fig. 12-7) is thrown into position 1 at $t = 0$. Write an expression for (a) the current i and (b) the voltage across the inductor.

Solution: (a) The voltage V_L across an inductance L is given by Eq. A86, $V_L = L\,di/dt$. Using Kirchhoff's voltage law (Eq. A62) gives

$$Ri + L\frac{di}{dt} = E \tag{1}$$

or

$$\frac{di}{dt} + \frac{R}{L}i = \frac{E}{L} \tag{2}$$

We recognize this as a first-order linear differential equation. Our integrating factor is

$$e^{\int R/L\,dt} = e^{Rt/L}$$

In electrical problems, we will use *k* for the constant of integration, saving *C* for capacitance.

Multiplying (2) by the integrating factor and integrating,

$$ie^{Rt/L} = \frac{E}{L}\int e^{Rt/L}\,dt = \frac{E}{R}e^{Rt/L} + k$$

Dividing by $e^{Rt/L}$,

$$i = \frac{E}{R} + ke^{-Rt/L} \tag{3}$$

We now evaluate k by noting that $i = 0$ at $t = 0$. So $ke^0 = k = -E/R$. Substituting into (3),

Current in a Charging Inductor	$i = \frac{E}{R}(1 - e^{-Rt/L})$	**A87**

The first term in this expression (E/R) is the steady-state current, and the second term $(E/R)\,(e^{-Rt/L})$ is the transient current.

(b) From (1), the voltage across the inductor is

Note that the equation for the voltage across the inductor is of the same form as for exponential decay (Eq. 199), and that the equation for the current is the same as that for exponential growth to an upper limit (Eq. 200).

$$V_L = L\frac{di}{dt} = E - Ri = E - E + Ee^{-Rt/L}$$

So,

Voltage across a Charging Inductor	$V_L = Ee^{-Rt/L}$	**A88**

Series *RC* Circuit

We now analyze the *RC* circuit as we did the *RL* circuit.

EXAMPLE 2: A fully charged capacitor (Fig. 12-8) is discharged by throwing the switch from 1 to 2 at $t = 0$. Write an expression for (a) the voltage across the capacitor and (b) the current i.

Solution: (a) By Kirchhoff's law, the voltage v_R across the resistor at any instant must be equal to the voltage v across the capacitor, but of opposite sign. Further, the current through the resistor, v_R/R, or $-v/R$, must be equal to the current $C\,dv/dt$ in the capacitor.

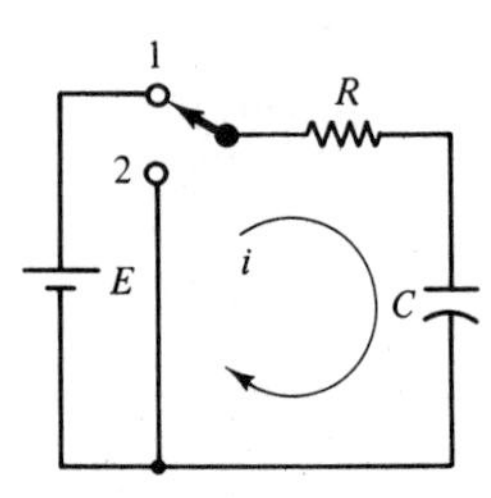

FIGURE 12-8. RC circuit.

$$-\frac{v}{R} = C\frac{dv}{dt}$$

Separating variables and integrating,

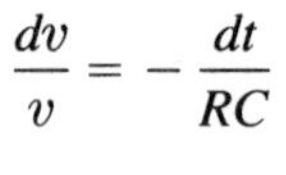

$$\frac{dv}{v} = -\frac{dt}{RC}$$

$$\ln v = -\frac{t}{RC} + k$$

But at $t = 0$ the voltage across the capacitor is the battery voltage E, so $k = \ln E$. Substituting,

$$\ln v - \ln E = \ln\frac{v}{E} = -\frac{t}{RC}$$

Or, in exponential form, $v/E = e^{-t/RC}$, or

Voltage across a Discharging Capacitor	$v = Ee^{-t/RC}$	**A84**

Equations A83 and A84 are both for exponential decay.

(b) We get the current through the resistor (and the capacitor) by dividing the voltage v by R,

Current in a Discharging Capacitor	$i = \frac{E}{R} e^{-t/RC}$	**A83**

EXAMPLE 3: For the circuit of Fig. 12-8, $R = 1540\ \Omega$, $C = 125\ \mu\text{F}$, and $E = 115$ V. If the switch is thrown from position 2 to position 1 at $t = 0$, find the current and the voltage across the capacitor at $t = 60$ ms.

Solution: We first compute $1/RC$.

$$\frac{1}{RC} = \frac{1}{1540 \times 125 \times 10^{-6}} = 5.19$$

Then from Eqs. A83 and A84,

$$i = \frac{E}{R} e^{-t/RC} = \frac{115}{1540} e^{-5.19t}$$

and

$$v = Ee^{-t/RC} = 115\, e^{-5.19t}$$

At $t = 0.060$ s, $e^{-5.19t} = 0.732$, so

$$i = \frac{115}{1540}(0.732) = 0.0547 \text{ A} = 54.7 \text{ mA}$$

and

$$v = 115(0.732) = 84.2 \text{ V}$$

Alternating Source

We now consider the case where the *RL* circuit or *RC* circuit is connected to an alternating rather than a direct source of voltage.

EXAMPLE 4: A switch (Fig. 12-9) is closed at $t = 0$, thus applying an alternating voltage of amplitude E to a resistor and inductor in series. Write an expression for the current.

Solution: By Kirchhoff's voltage law, $Ri + L\, di/dt = E \sin \omega t$, or

$$\frac{di}{dt} + \frac{\mathrm{R}}{\mathrm{L}} i = \frac{E}{L} \sin \omega t$$

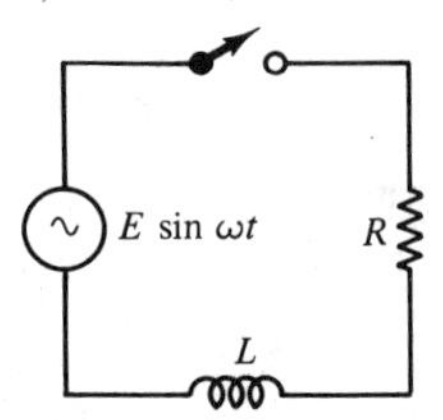

FIGURE 12-9. RL circuit with ac source.

This is a first-order linear differential equation. Our integrating factor is $e^{\int R/L\,dt} = e^{Rt/L}$. Thus

$$ie^{Rt/L} = \frac{E}{L} \int e^{Rt/L} \sin \omega t \, dt$$

$$= \frac{E}{L} \frac{e^{Rt/L}}{(R^2/L^2 + \omega^2)} \left(\frac{R}{L} \sin \omega t - \omega \cos \omega t \right) + k$$

by rule 42. We now divide through by $e^{Rt/L}$, and after some manipulation get

$$i = E \frac{R \sin \omega t - \omega L \cos \omega t}{R^2 + \omega^2 L^2} + ke^{-Rt/L}$$

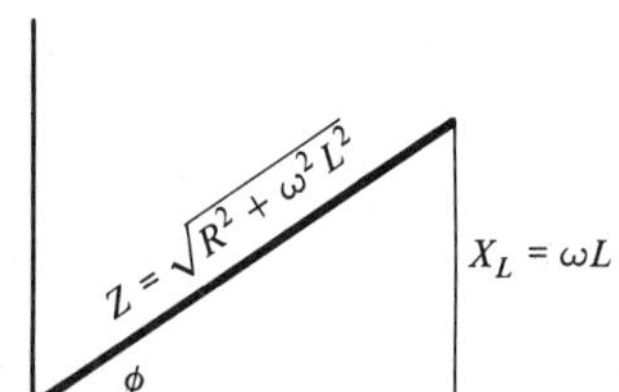

FIGURE 12-10. Impedance triangle.

From the impedance triangle (Fig. 12-10) we see that $R^2 + \omega^2 L^2 = Z^2$, the square of the impedance. Further, by Ohm's law for AC, $E/Z = I$, the amplitude of the current wave. Thus

$$i = \frac{E}{Z} \left(\frac{R}{Z} \sin \omega t - \frac{\omega L}{Z} \cos \omega t \right) + ke^{-Rt/L}$$

$$= I \left(\frac{R}{Z} \sin \omega t - \frac{\omega L}{Z} \cos \omega t \right) + ke^{-Rt/L}$$

Again from the impedance triangle, $R/Z = \cos \phi$ and $\omega L/Z = \sin \phi$. Substituting,

$$i = I(\sin \omega t \cos \phi - \cos \omega t \sin \phi) + ke^{-Rt/L}$$

$$= I \sin (\omega t - \phi) + ke^{-Rt/L}$$

which we get by means of the trigonometric identity for the sine of the difference of two angles (Eq. 167). Evaluating k, we note that $i = 0$ when $t = 0$, so

$$k = -I \sin (-\phi) = I \sin \phi = \frac{IX_L}{Z}$$

where, from the impedance triangle, $\sin \phi = X_L/Z$. Substituting,

$$i = \underbrace{I \sin (\omega t - \phi)}_{\text{steady-state}} + \underbrace{\frac{IX_L}{Z} e^{-Rt/L}}_{\text{transient}}$$

Our current thus has two parts; (1) a steady-state alternating current of magnitude I, out of phase with the applied voltage by an angle ϕ, and (2) a transient current with an initial value of IX_L/Z, which decays exponentially.

EXERCISE 8

Series RL Circuit

1. If the inductor in Fig. 12-7 is discharged by throwing the switch from position 1 to 2, show that the current decays exponentially according to the function $i = (E/R)e^{-Rt/L}$.

2. The voltage across an inductor is equal to $L\ di/dt$. Show that the magnitude of the voltage across the inductance in Problem 1 decays exponentially according to the function $v = Ee^{-Rt/L}$.

3. When the switch in Fig. 12-7 is thrown from 2 to 1, (charging) we saw that the current grows exponentially to an upper limit and is given by $i = (E/R)(1 - e^{-Rt/L})$. Show that the voltage across the inductance ($L\ di/dt$) decays exponentially and is given by $v = Ee^{-Rt/L}$.

4. For the circuit of Fig. 12-7, $R = 382\ \Omega$, $L = 4.75$ H, and $E = 125$ V. If the switch is thrown from 2 to 1 (charging), find the current and the voltage across the inductance at $t = 2.00$ ms.

Series RC Circuit

5. Show that the current to the capacitor (Fig. 12-8) during charging is given by $i = (E/R)e^{-t/RC}$.

6. Using the facts that $i = dq/dt$ and $v = q/C$, show that the voltage across the capacitor in Problem 5 is given by $v = E(1 - e^{-t/RC})$.

7. For the circuit of Fig. 12-8, $R = 538\ \Omega$, $C = 525\ \mu F$, and $E = 125$ V. If the switch is thrown from 2 to 1 (charging), find the current and the voltage across the capacitor at $t = 2.00$ ms.

Circuits in Which R, L, and C Are Not Constant

8. For the circuit of Fig. 12-7, $L = 2.00\ H$, $E = 60.0$ V, and the resistance decreases with time according to the expression $R = 4.00/(t + 1)$. Show that the current i is given by $i = 10(t + 1) - 10(t + 1)^{-2}$.

9. For the circuit in Problem 8, find the current at $t = 1.55$ ms.

10. For the circuit of Fig. 12-7, $R = 10.0\ \Omega$, $E = 100$ V, and the inductance varies with time according to the expression $L = 5.00t + 2.00$. Show that the current i is given by the expression $i = 10.0 - 40/(5t + 2)^2$.

11. For the circuit in Problem 10, find the current at $t = 4.82$ ms.

12. For the circuit of Fig. 12-7, $E = 300$ V, the resistance varies with time according to the expression $R = 4.00t$, and the inductance varies with time according to the expression $L = t^2 + 4.00$. Show that the current i as a function of time is given by $i = 100t(t^2 + 12)/(t^2 + 4)^2$.

13. For the circuit in Problem 12, find the current at $t = 1.85$ ms.

14. For the circuit of Fig. 12-8, $C = 2.55\ \mu F$, $E = 625$ V, and the resistance varies with current according to the expression $R = 493 + 372i$. Show that the differential equation for current is $di/dt + 1.51i\ di/dt + 795i = 0$.

Series RL or RC Circuit with Alternating Current

15. For the circuit of Fig. 12-9, $R = 233\ \Omega$, $L = 5.82$ H, and $e = 58.0 \sin 377t$ volts. If the switch is thrown from position 2 to position 1 when e is zero and increasing, show that the current is given by $i = 26.3 \sin(377t - 83.9°) - 26.1e + 40t$ mA.

16. For the circuit in Problem 15, find the current at $t = 2.00$ ms.

17. For the circuit in Fig. 12-8, the applied voltage is alternating and is given by $e = E \sin \omega t$. If the switch is thrown from 2 to 1 (charging) when e is zero and increasing, show that the current is given by $i = (E/Z)[\sin(\omega t + \phi) - e^{-t/RC} \sin \phi]$. In your derivation, follow the steps used for the RL circuit with an ac source.

18. For the circuit in Problem 17, $R = 837\ \Omega$, $C = 2.96\ \mu F$, and $e = 58.0 \sin 377t$. Find the current at $t = 1.00$ ms.

CHAPTER TEST

Find the general solution to each differential equation.

1. $xy + y + xy' = e^x$

2. $y + xy' = 4x^3$

3. $y' + y - 2 \cos x = 0$

4. $y + x^2y^2 + xy' = 0$

5. $2y + 3x^2 + 2xy' = 0$

6. $y' - 3x^2y^2 + \frac{y}{x} = 0$

7. $y^2 + (x^2 - xy)y' = 0$

8. $y' = e^{-y} - 1$

9. $(1 - x)\frac{dy}{dx} = y^2$

10. $y' \tan x + \tan y = 0$

11. $y + 2xy^2 + (y - x)y' = 0$

Find the particular solution to each differential equation.

12. $x\,dx = 2y\,dy$, $\quad x = 3$ when $y = 1$

13. $y' \sin y = \cos x$, $\quad x = \frac{\pi}{4}$ when $y = 0$

14. $xy' + y = 4x$, $\quad x = 2$ when $y = 1$

15. $y\,dx = (x - 2x^2y)\,dy$, $\quad x = 2$ when $y = 1$

16. $3xy^2\,dy = (3y^3 - x^3)\,dx$, $\quad x = 3$ when $y = 1$

17. A gear is rotating at 1550 r/min. Its speed decreases exponentially at a rate of 9.5% per second, after the power is shut off. Find the gear's speed after 6.00 s.

18. For the circuit of Fig. 12-7, $R = 1350\ \Omega$, $L = 7.25$ H, and $E = 225$ V. If the switch is thrown from position 2 to position 1, find the current and the voltage across the inductance at $t = 3.00$ ms.

19. Write the equation of the orthagonal trajectories to each family of parabolas, $x^2 = 4y$.

20. For the circuit of Fig. 12-8, $R = 2550\ \Omega$, $C = 145\ \mu$F, and $E = 95$ V. If the switch is thrown from position 2 to position 1, find the current and the voltage across the capacitor at $t = 5.00$ ms.

21. Find the equation of the curve that passes through the point (1, 2) and whose slope is $y' = 2 + y/x$.

22. An object is dropped and falls from rest through air whose resisting force is proportional to the speed of the package. The terminal speed is 275 ft/s. Show that the acceleration is given by the differential equation $a = dv/dt = g - gv/275$.

23. A certain yeast is found to grow exponentially at the rate of 15% per hour. Starting with 500 g of yeast, how many grams will there be after 15 h?

Second-Order Differential Equations

We continue our study of differential equations with the second-order equation. These, you recall, will have second derivatives. They may, of course, also have first derivatives, but no third or higher derivatives.

Here we solve types of second-order equations that are fairly simple but of great practical importance just the same. This will be borne out by the applications to mechanical vibrations and electrical circuits presented in this chapter.

13-1. VARIABLES SEPARABLE

The General Second-Order Equation

A linear differential equation of second order can be written in the form

$$Py'' + Qy' + Ry = S$$

where P, Q, R, and S are constants or functions of x.

A second-order linear differential equation *with constant coefficients is one where* P, Q, and R are *constants,* although S can be a function of x, such as

Second-Order Right Side Not Zero	$ay'' + by' + cy = f(x)$	**394**

where a, b, and c are constants. This is the type of equation we will solve in this chapter.

Equation 394 is said to be *homogeneous* if $f(x) = 0$, and is called *nonhomogeneous* if $f(x)$ is not zero. We will, instead, usually refer to these equations as "right-hand side zero" and "right-hand side not zero."

Operator Notation

We'll usually use the more familiar y' notation rather than the D operator.

Differential equations are often written using the D operator we first introduced in Chap. 3, where

$$Dy = y', \qquad D^2y = y'', \qquad D^3y = y''', \qquad \text{etc.}$$

Thus Eq. 394 can be written

$$aD^2y + bDy + c = f(x)$$

Variables Separable

We develop methods for solving the general second-order equation in Sec. 13-2. However, simple differential equations of second-order which are lacking a first derivative term can be solved by separation of variables, as in the following example.

EXAMPLE 1: Solve the equation $y'' = 3 \cos x$ (where x is in radians) if $y' = 1$ at the point (2, 1).

Solution: Replacing y'' by $d(y')/dx$ and multiplying both sides by dx to separate variables,

$$d(y') = 3 \cos x \, dx$$

Integrating,

$$y' = 3 \sin x + C_1$$

Since $y' = 1$ when $x = 2$ rad, $C_1 = 1 - 3 \sin 2 = -1.73$, so

$$y' = 3 \sin x - 1.73$$

or $dy = 3 \sin x \, dx - 1.73 \, dx$. Integrating again,

$$y = -3 \cos x - 1.73x + C_2$$

At the point (2, 1), $C_2 = 1 + 3 \cos 2 + 1.73(2) = 3.21$. Our solution is then

$$y = -3 \cos x - 1.73x + 3.21$$

EXERCISE 1

Solve each equation for y.

1. $y'' = 5$

2. $y'' = x$

3. $y'' = 3e^x$

4. $y'' = \sin 2x$

5. $y'' - x^2 = 0$, where $y' = 1$ at the point (0, 0)

6. $xy'' = 1$, where $y' = 2$ at the point (1, 1)

13-2. SECOND-ORDER EQUATIONS WITH CONSTANT COEFFICIENTS AND RIGHT-HAND SIDE ZERO

Solving a Second-Order Equation with Right Side Equal to Zero

If the right side, $f(x)$, in Eq. 394 is zero, we have the (homogeneous) equation

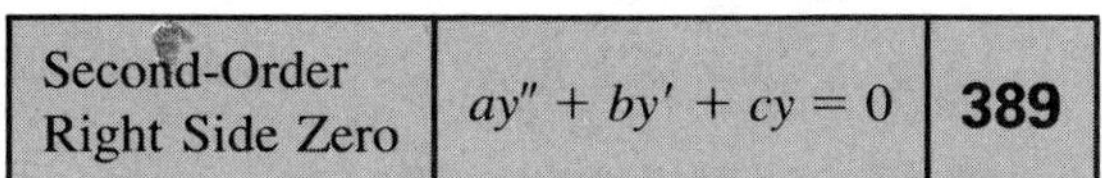

Second-Order Right Side Zero	$ay'' + by' + cy = 0$	**389**

To solve this equation, we note that the sum of the three terms on the left side must equal zero. So a solution must be a value of y that will make these terms alike, so they may be combined to give a sum of zero. Also note that each term on the left contains y or its first or second derivative. Thus *a possible solution is a function such that it and its derivatives are like terms.* Recall from Chapter 6 that one such function was the exponential function

$$y = e^{mx}$$

It has derivatives $y' = me^{mx}$ and $y'' = m^2e^{mx}$, which are all like terms. We thus try this for our solution. Substituting $y = e^{mx}$ and its derivatives into Eq. 389,

$$am^2e^{mx} + bme^{mx} + ce^{mx} = 0$$

Factoring,

$$e^{mx}(am^2 + bm + c) = 0$$

Since e^{mx} can never be zero, this equation is satisfied only when $am^2 + bm + c = 0$. This is called the *auxiliary* or *characteristic equation.*

Note that the coefficients in the auxiliary equation are the same as those in the original DE. We can thus get the auxiliary equation by inspection of the DE.

Auxiliary Equation	$am^2 + bm + c = 0$	**390**

The auxiliary equation is a quadratic, and has two roots, which we call m_1 and m_2. Thus we get *two solutions* to Eq. 389, $y = e^{m_1x}$ and $y = e^{m_2x}$.

General Solution

We now show that if y_1 and y_2 are each solutions to $ay'' + by' + cy = 0$, then $y = c_1y_1 + c_2y_2$ is also a solution. If $y = c_1y_1 + c_2y_2$ then

$$y' = c_1y_1' + c_2y_2' \quad \text{and} \quad y'' = c_1y_1'' + c_2y_2''$$

Substituting into the differential equation (389),

$$a(c_1y_1'' + c_2y_2'') + b(c_1y_1' + c_2y_2') + c(c_1y_1 + c_2y_2) = 0$$

This simplifies to

$$c_1(ay_1'' + by_1' + cy_1) + c_2(ay_2'' + by_2' + cy_2) = 0$$

But $ay_1'' + by_1' + cy_1 = 0$ and $ay_2'' + by_2' + cy_2 = 0$, so

$$c_1(0) + c_2(0) = 0$$

showing that $y = c_1y_1 + c_2y_2$ is a solution to the differential equation. Since e^{m_1x} and $y = e^{m_2x}$ are solutions to Eq. 389, the complete solution is then

General Solution	$y = c_1e^{m_1x} + c_2e^{m_2x}$	**391**

The roots of a quadratic can be real and unequal, real and equal, or nonreal. We now give examples of each case.

Roots Real and Unequal

EXAMPLE 1: Solve the equation $y'' - 3y' + 2y = 0$.

Solution: We get the auxiliary equation by inspection.

$$m^2 - 3m + 2 = 0$$

It factors into $(m - 1)$ and $(m - 2)$. Setting each factor equal to zero gives $m = 1$ and $m = 2$. Our solution, by Eq. 391, is then

$$y = c_1e^x + c_2e^{2x}$$

Sometimes one root of the auxiliary equation will be zero, as in the next example.

EXAMPLE 2: Solve the equation $y'' - 5y' = 0$.

Solution: The auxiliary equation is $m^2 - 5m = 0$, which factors into $m(m - 5) = 0$. Setting each factor equal to zero gives

$$m = 0 \qquad \text{and} \qquad m = 5$$

Our solution is then

$$y = c_1 + c_2e^{5x}$$

If the auxiliary equation cannot be factored, we use the quadratic formula to find its roots.

EXAMPLE 3: Solve $4.82y'' + 5.85y' - 7.26y = 0$.

Solution: The auxiliary equation is $4.82m^2 + 5.85m - 7.26 = 0$. By the quadratic formula,

$$m = \frac{-5.85 \pm \sqrt{(5.85)^2 - 4(4.82)(-7.26)}}{2(4.82)} = 0.762, \ -1.98$$

Our solution is then

$$y = c_1e^{0.762x} + c_2e^{-1.98x}$$

Roots Real and Equal

If $b^2 - 4ac$ is zero, the auxiliary equation has the double root

$$m = \frac{-b \pm \sqrt{b^2 - 4ac}}{2a} = -\frac{b}{2a}$$

Our solution $y = c_1e^{mx} + c_2e^{mx}$ seems to contain two arbitrary constants, but actually does not. Factoring gives

$$y = (c_1 + c_2)e^{mx} = c_3e^{mx}$$

where $c_3 = c_1 + c_2$. A solution with one constant cannot be a complete solution for a second-order equation.

Let us *assume a second solution* $y = ue^{mx}$, where u is some function of x which we are free to choose. Differentiating,

$$y' = mue^{mx} + u'e^{mx}$$

and

$$y'' = m^2ue^{mx} + mu'e^{mx} + mu'e^{mx} + u''e^{mx}$$

Substituting into $ay'' + by' + cy = 0$ gives

$$a(m^2ue^{mx} + 2mu'e^{mx} + u''e^{mx}) + b(mue^{mx} + u'e^{mx}) + cue^{mx} = 0$$

which reduces to

$$e^{mx}[u(am^2 + bm + c) + u'(2am + b) + au''] = 0$$

But $am^2 + bm + c = 0$. Also, $m = -b/2a$, so $2am + b = 0$. Our equation then becomes $e^{mx}(u'') = 0$. Since e^{mx} cannot be zero, we have $u'' = 0$. Thus any u that has a zero second derivative will make ue^{mx} a solution to the differential equation. The simplest u (not a constant) for which $u'' = 0$ is $u = x$. Thus xe^{mx} is a solution to the differential equation, and the complete solution to Eq. 389 is

Equal Roots	$y = c_1e^{mx} + c_2xe^{mx}$	**392**

EXAMPLE 4: Solve $y'' - 6y' + 9y = 0$.

Solution: The auxiliary equation $m^2 - 6m + 9 = 0$ has the double root $m = 3$. Our solution is then

$$y = c_1e^{3x} + c_2xe^{3x}$$

Euler's Formula

When the roots of the auxiliary equation are nonreal, our solution will contain expressions of the form e^{jbx}. In the following section we will want to simplify such expressions using *Euler's formula,* which we derive here.

Let $z = \cos\theta + j\sin\theta$, where $j = \sqrt{-1}$. Then

$$\frac{dz}{d\theta} = -\sin\theta + j\cos\theta$$

Multiplying by j (and recalling that $j^2 = -1$) gives

$$j\frac{dz}{d\theta} = -j\sin\theta - \cos\theta = -z$$

Multiplying by $-j$ we get $dz/d\theta = jz$. We now separate variables and integrate

$$\frac{dz}{z} = j\,d\theta$$

so

$$\ln z = j\theta + c$$

When $\theta = 0$, $z = \cos 0 + j \sin 0 = 1$. So $c = \ln z - j\theta = \ln 1 - 0 = 0$. Thus $\ln z = j\theta$, or, in exponential form, $z = e^{j\theta}$. But $z = \cos \theta + j \sin \theta$, so

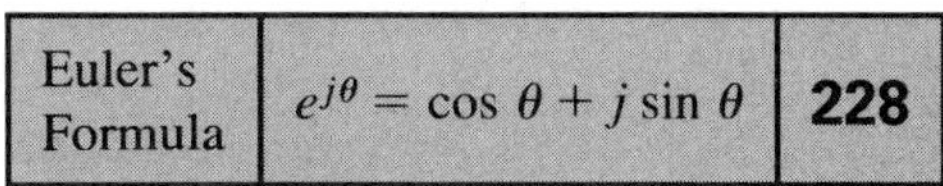

Euler's Formula	$e^{j\theta} = \cos \theta + j \sin \theta$	**228**

We can get another useful form of Eq. 228 by setting $\theta = bx$. Thus

$$e^{jbx} = \cos bx + j \sin bx$$

Further,

$$\begin{aligned} e^{-jbx} &= \cos(-bx) + j \sin(-bx) \\ &= \cos bx - j \sin bx \end{aligned}$$

since $\cos(-A) = \cos A$, and $\sin(-A) = -\sin A$.

Roots Not Real

If the auxiliary equation has the nonreal roots $a + jb$ and $a - jb$, our solution becomes

$$\begin{aligned} y &= c_1 e^{(a+jb)x} + c_2 e^{(a-jb)x} \\ &= c_1 e^{ax} e^{jbx} + c_2 e^{ax} e^{-jbx} = e^{ax}(c_1 e^{jbx} + c_2 e^{-jbx}) \\ &= e^{ax}[c_1(\cos bx + j \sin bx) + c_2(\cos bx - j \sin bx)] \\ &= e^{ax}[(c_1 + c_2) \cos bx + j(c_1 - c_2) \sin bx] \end{aligned}$$

Replacing $c_1 + c_2$ by C_1, and $j(c_1 - c_2)$ by C_2 gives

Nonreal Roots	$y = e^{ax}(C_1 \cos bx + C_2 \sin bx)$	**393a**

A more compact form of the solution may be obtained by using the trigonometric identity

$$\sin(bx + \phi) = \cos bx \sin \phi + \sin bx \cos \phi$$

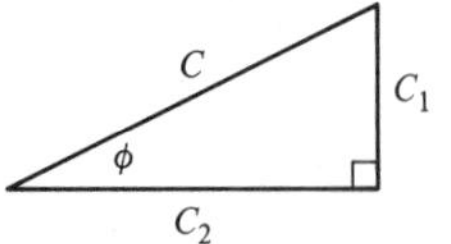

FIGURE 13-1

If ϕ is one angle of a right triangle with sides C_1 and C_2 and hypotenuse C (Fig. 13-1), then

$$\sin(bx + \phi) = \frac{C_1}{C}\cos bx + \frac{C_2}{C}\sin bx$$

or

$$C\sin(bx + \phi) = C_1 \cos bx + C_2 \sin bx$$

Thus the solution to the differential equation can take the alternative form

We'll find this form handy for applications.

Nonreal Roots	$y = Ce^{ax}\sin(bx + \phi)$	**393b**

EXAMPLE 5: Solve $y'' - 4y' + 13y = 0$.

Solution: The auxiliary equation $m^2 - 4m + 13 = 0$ has the roots $m = 2 \pm j3$. Substituting into Eq. 393a with $a = 2$ and $b = 3$ gives

$$y = e^{2x}(C_1 \cos 3x + C_2 \sin 3x)$$

or, using the alternative form, Eq. 393b,

$$y = Ce^{2x}\sin(3x + \phi)$$

Summary

The solutions to a second-order differential equation with right side zero and with constant coefficients are summarized here.

Solution to $ay'' + by' + cy = 0$

Roots of Auxiliary Equation $am^2 + bm + c = 0$	Solution	
Real and Unequal	$y = c_1e^{m_1x} + c_2e^{m_2x}$	**391**
Real and Equal	$y = c_1e^{mx} + c_2xe^{mx}$	**392**
Nonreal	(a) $y = e^{ax}(C_1 \cos bx + C_2 \sin bx)$ or (b) $y = Ce^{ax}\sin(bx + \phi)$	**393**

Particular Solution

As before, we use the boundary conditions to find the two constants in the solution.

EXAMPLE 6: Solve $y'' - 4y' + 3y = 0$ if $y' = 5$ at (1, 2).

Solution: The auxiliary equation $m^2 - 4m + 3 = 0$ has roots $m = 1$ and $m = 3$. Our solution is then

$$y = c_1e^x + c_2e^{3x}$$

At (1, 2) we get

$$2 = c_1e + c_2e^3 \qquad (1)$$

Here we have one equation and two unknowns. We get a second equation by taking the derivative of y.

$$y' = c_1e^x + 3c_2e^{3x}$$

Substituting $y' = 5$ when $x = 1$ gives

$$5 = c_1e + 3c_2e^3 \qquad (2)$$

Subtracting (1) from (2) gives $2c_2e^3 = 3$, or

$$c_2 = \frac{3}{2e^3} = 0.0747$$

Then from (1),

$$c_1 = \frac{2 - (0.0747)e^3}{e} = 0.184$$

Our final equation is then

$$y = 0.184e^x + 0.0747e^{3x}$$

Third-Order Differential Equations

The method of the preceding sections can be extended to simple third-order equations that can be easily factored.

EXAMPLE 7: Solve $y''' - 4y'' - 11y' + 30y = 0$.

Solution: We write the auxiliary equation by inspection,

$$m^3 - 4m^2 - 11m + 30 = 0$$

which factors, by trial and error, into

$$(m - 2)(m - 5)(m + 3) = 0$$

giving roots of 2, 5, and -3. The solution to the given equation is then

$$y = C_1e^{2x} + C_2e^{5x} + C_3e^{-3x}$$

EXERCISE 2

Find the general solution to each differential equation.

Second Order, Roots of Auxiliary Equation Real and Unequal

1. $y'' - 6y' + 5y = 0$

2. $2y'' - 5y' - 3y = 0$

3. $y'' - 3y' + 2y = 0$

4. $y'' + 4y' - 5y = 0$

5. $y'' - y' - 6y = 0$

6. $y'' + 5y' + 6y = 0$

7. $5y'' - 2y' = 0$

8. $y'' + 4y' + 3y = 0$

9. $6y'' + 5y' - 6y = 0$

10. $y'' - 4y' + y = 0$

Second Order, Roots of Auxiliary Equation Real and Equal

11. $y'' - 4y' + 4y = 0$

12. $y'' - 6y' + 9y = 0$

13. $y'' - 2y' + y = 0$

14. $9y'' - 6y' + y = 0$

15. $y'' + 4y' + 4y = 0$

16. $4y'' + 4y' + y = 0$

17. $y'' + 2y' + y = 0$

18. $y'' - 10y' + 25y = 0$

Second Order, Roots of Auxiliary Equation Not Real

19. $y'' + 4y' + 13y = 0$

20. $y'' - 2y' + 2y = 0$

21. $y'' - 6y' + 25y = 0$

22. $y'' + 2y' + 2y = 0$

23. $y'' + 4y = 0$

24. $y'' + 2y = 0$

25. $y'' - 4y' + 5y = 0$

26. $y'' + 10y' + 425y = 0$

Particular Solution

Solve each differential equation. Use the given boundary conditions to find the constants of integration.

27. $y'' + 6y' + 9y = 0$, $\quad y = 0$ and $y' = 3$ when $x = 0$

28. $y'' + 3y' + 2y = 0$, $\quad y = 0$ and $y' = 1$ when $x = 0$

29. $y'' - 2y' + y = 0, \quad y = 5$ and $y' = -9$ when $x = 0$

30. $y'' + 3y' - 4y = 0, \quad y = 4$ and $y' = -2$ when $x = 0$

31. $y'' - 2y' = 0, \quad y = 1 + e^2$ and $y' = 2e^2$ when $x = 1$

32. $y''+ 2y' + y = 0, \quad y = 1$ and $y' = -1$ when $x = 0$

33. $y'' - 4y = 0, \quad y = 1$ and $y' = -1$ when $x = 0$

34. $y'' + 9y = 0, \quad y = 2$ and $y' = 0$ when $x = \pi/6$

35. $y'' + 2y' + 2y = 0, \quad y = 0$ and $y' = 1$ when $x = 0$

36. $y'' + 4y' + 13y = 0, \quad y = 0$ and $y' = 12$ when $x = 0$

Third Order

37. $y''' - 2y'' - y' + 2y = 0$

38. $y''' - y' = 0$

39. $y''' - 6y'' + 11y' - 6y = 0$

40. $y''' + y'' - 4y' - 4y = 0$

41. $y''' - 3y'' - y' + 3y = 0$

42. $y''' - 7y' + 6y = 0$

43. $4y''' - 3y' + y = 0$

44. $y''' - y'' = 0$

13-3. SECOND-ORDER EQUATIONS WITH RIGHT SIDE NOT ZERO

Particular Integral and Complementary Function

We'll now see that the solution to a differential equation is made up of *two parts,* the *complementary function* and the *particular integral.* We show this now for a first-order equation, and later, for a second-order equation.

The solution to a first-order differential equation, say

$$y' + \frac{y}{x} = 4 \tag{1}$$

can be found by the methods of Chapter 12. The solution to (1) is $y = c/x + 2x$. Note that the solution has two parts. Let us label one part y_p and the other y_c. Thus $y = y_c + y_p$, where $y_c = c/x$ and $y_p = 2x$.

Let us substitute only $y_c = c/x$ into the left side of (1).

$$-\frac{c}{x^2} + \frac{c}{x^2} = 0$$

We get zero instead of the required 4. Thus y_c *does not satisfy (1).* It does however, satisfy the reduced equation, obtained by setting the right side equal to zero. It is called the *complementary function.*

We now substitute only $y_p = 2x$ into the left side of (1).

$$2 + \frac{2x}{x} = 4$$

We see that y_p *does* satisfy (1), and is hence a solution. But it cannot be a complete solution because it has no arbitrary constant. It is called a *particular integral*. The quantity y_c had the required constant but did not, by itself, satisfy (1).

However, *the sum of y_c and y_p satisfies (1) and has the required number of constants, and is hence the complete solution.*

Complete Solution	$y = y_c + y_p = \begin{pmatrix}\text{complementary}\\ \text{function}\end{pmatrix} + \begin{pmatrix}\text{particular}\\ \text{integral}\end{pmatrix}$	**395**

Second-Order Equation

We have seen that the solution to a first-order equation is made up of a complementary function and a particular integral. But is the same true of the second-order equation

$$ay'' + by' + cy = f(x)? \tag{394}$$

If a particular integral y_p is a solution to Eq. 394, we get, on substituting,

$$ay_p'' + by_p' + cy_p = f(x) \tag{2}$$

If the complementary function y_c is a complete solution to the reduced equation

$$ay'' + by' + cy = 0 \tag{389}$$

we get

$$ay_c'' + by_c' + cy_c = 0 \tag{3}$$

Adding (2) and (3),

$$a(y_p'' + y_c'') + b(y_p' + y_c') + c(y_p + y_c) = f(x)$$

Since the sum of the two derivatives is the derivative of the sum, we get

$$a(y_p + y_c)'' + b(y_p + y_c)' + c(y_p + y_c) = f(x)$$

This shows that $y_p + y_c$ is a solution to (2).

EXAMPLE 1: Given that the solution to $y'' - 5y' + 6y = 3x$ is

$$y = \underbrace{c_1e^{3x} + c_2e^{2x}}_{\text{complementary function}} + \underbrace{\frac{x}{2} + \frac{5}{12}}_{\text{particular integral}}$$

prove to yourself, by substitution, that the complementary function will make the left side of the given equation equal to zero, and that the particular integral will make the left side equal to $3x$.

Finding the Particular Integral

We already know how to find the complementary function y_c. We set the right side of the given equation to zero and then solve that (reduced) equation just as we did in the preceding section. Now we will see how to find the particular integral y_p.

Starting with Eq. 394, let us isolate y. We get

$$y = \frac{1}{c}[f(x) - ay'' - by']$$

For this equation to balance, y must contain terms similar to those in $f(x)$. Further, y must contain terms similar to those in its own first and second derivatives. Thus it seems reasonable to try a solution consisting of the sum of $f(x)$, $f'(x)$, and $f''(x)$, each with an (as yet) undetermined coefficient.

EXAMPLE 2: Find y_p for the equation $y'' - 5y' + 6y = 3x$.

This is sometimes called the *trial function*.

Solution: Here $f(x) = 3x$, $f'(x) = 3$, and $f''(x) = 0$. Then

$$y_p = Ax + B$$

where A and B are constants yet to be found.

We used the method of undetermined coefficients before when finding partial fractions.

Method of Undetermined Coefficients

To find the constants in y_p, we substitute y_p and its first and second derivatives into the differential equation. We then equate coefficients of like terms and solve for the constants.

EXAMPLE 3: Determine the constants in y_p in Example 2, and find the complete solution to the given equation given the complementary function $y_c = c_1e^{3x} + c_2e^{2x}$.

Solution: We had $y_p = Ax + B$. Taking derivatives,

$$y_p' = A \quad \text{and} \quad y_p'' = 0$$

Substituting into the original equation,

$$0 - 5A + 6(Ax + B) = 3x$$

$$6Ax + (6B - 5A) = 3x$$

This equation is satisfied when $6A = 3$, or $A = \frac{1}{2}$, and $6B - 5A = 0$. Thus $B = 5A/6 = \frac{5}{12}$. Our particular integral is then $y_p = x/2 + \frac{5}{12}$, and the complete solution is

$$y = y_c + y_p = c_1e^{3x} + c_2e^{2x} + \frac{x}{2} + \frac{5}{12}$$

General Procedure

Thus to solve a second-order linear differential equation with constant coefficients,

$$ay'' + by' + cy = f(x) \tag{394}$$

1. Find the *complementary function* y_c.
2. Write the *particular integral* y_p. It should contain each term from the right side $f(x)$ (less coefficients) as well as the first and second derivatives of each term of $f(x)$ (less coefficients). Discard any duplicates.
3. If a term in y_p is a duplicate of one in y_c, multiply that term in y_p by x^n, using the lowest n that will *eliminate the duplication.*
4. *Write* y_p, each term with an undetermined coefficient.
5. *Substitute* y_p and its first and second derivatives into the differential equation.
6. *Evaluate the coefficients* by the method of undetermined coefficients.
7. *Combine* y_c and y_p to obtain the complete solution.

Remember that when we say "duplicate," we mean terms that are alike, regardless of numerical coefficient.

We illustrate these steps in the following example.

EXAMPLE 4: Solve $y'' - y' - 6y = 36x + 50 \sin x$.

Solution: 1. The auxiliary equation $m^2 - m - 6 = 0$ has the roots $m = 3$ and $m = -2$, so the complementary function is $y_c = c_1e^{3x} + c_2e^{-2x}$.

2. The terms in $f(x)$ and their derivatives (less coefficients) are

Term	x	$\sin x$
First derivative	Constant	$\cos x$
Second derivative	0	$\sin x$

We discard the duplicate sin x term.

3. No term in y_p is a duplicate of one in y_c.
4. Our particular integral is then

$$y_p = A + Bx + C\sin x + D\cos x$$

5. The derivatives of y_p are

$$y_p' = B + C\cos x - D\sin x$$

and

$$y_p'' = -C\sin x - D\cos x$$

Substituting into the differential equation gives

$$(-C\sin x - D\cos x) - (B + C\cos x - D\sin x) - 6(A + Bx + C\sin x + D\cos x) = 36x + 50\sin x$$

6. Collecting terms and equation coefficients of like terms gives the equations

$$-6B = 36, \qquad -B - 6A = 0, \qquad D - 7C = 50, \qquad -7D - C = 0$$

from which $A = 1$, $B = -6$, $C = -7$, and $D = 1$. Our particular integral is, then, $y_p = 1 - 6x - 7\sin x + \cos x$.

7. The complete solution is thus $y_c + y_p$, or

$$y = c_1e^{3x} + c_2e^{-2x} + 1 - 6x - 7\sin x + \cos x$$

Duplicate Terms in the Solution

The terms in the particular integral y_p must be *independent*. If y_p has duplicate terms, only one need be kept. However, if a term in y_p is a duplicate of one in the complementary function y_c (except for the coefficient), *multiply that term in y_p by x^n, using the lowest n that will eliminate the duplication.*

EXAMPLE 5: Solve the equation $y'' - 4y' + 4y = e^{2x}$.

Solution: 1. The complementary function (work not shown) is

$$y_c = c_1e^{2x} + c_2xe^{2x}.$$

2. Our particular integral should contain e^{2x} and its derivatives, which are also of the form e^{2x}. But since these are duplicates, we need e^{2x} only once.

3. But e^{2x} is a duplicate of c_1e^{2x} in y_c. If we multiply by x, we see that xe^{2x} is a duplicate of the *second* term in y_c. We thus need x^2e^{2x}.

4. Our particular integral is thus $y_p = Ax^2e^{2x}$.

5. Taking derivatives

$$y_p' = 2Ax^2e^{2x} + 2Axe^{2x}$$

and

$$y_p'' = 4Ax^2e^{2x} + 8Axe^{2x} + 2Ae^{2x}$$

Substituting into the differential equation gives

$$4Ax^2e^{2x} + 8Axe^{2x} + 2Ae^{2x} - 4(2Ax^2e^{2x} + 2Axe^{2x}) + 4(Ax^2e^{2x}) = e^{2x}$$

6. Collecting terms and solving for A gives $A = \frac{1}{2}$.

7. Our complete solution is then

$$y = c_1e^{2x} + c_2xe^{2x} + \frac{1}{2}x^2e^{2x}$$

EXERCISE 3

Solve each differential equation.

With Algebraic Expressions

1. $y'' - 4y = 12$

2. $y'' + y' - 2y = 3 - 6x$

3. $y'' - y' - 2y = 4x$

4. $y'' + y' = x + 2$

5. $y'' - 4y = x^3 + x$

With Exponential Expressions

6. $y'' + 2y' - 3y = 42e^{4x}$

7. $y'' - y' - 2y = 6e^x$

8. $y'' + y' = 6e^x + 3$

9. $y'' - 4y = 4x - 3e^x$

10. $y'' - y = e^x + 2e^{2x}$

11. $y'' + 4y' + 4y = 8e^{2x} + x$

12. $y'' - y = x^3e^x$

With Trigonometric Expressions

13. $y'' + 4y = \sin 2x$

14. $y'' + y' = 6 \sin 2x$

15. $y'' + 2y' + y = \cos x$

16. $y'' + 4y' + 4y = \cos x$

17. $y'' + y = 2 \cos x - 3 \cos 2x$

18. $y'' + y = \sin x + 1$

With Exponential and Trigonometric Expressions

19. $y'' + y = e^x \sin x$

20. $y'' + y = 10e^x \sin x$

21. $y'' - 4y' + 5y = e^{2x} \sin x$

22. $y'' + 2y' + 5y = 3e^{-x} \sin x - 10$

Particular Solution

23. $y'' - 4y' = 8, \quad y = y' = 0$ when $x = 0$

24. $y'' + 2y' - 3y = 6, \quad y = 0$ and $y' = 2$ when $x = 0$

25. $y'' + 4y = 2, \quad y = 0$ when $x = 0$ and $y = \frac{1}{2}$ when $x = \pi/4$

26. $y'' + 4y' + 3y = 4e^{-x}, \quad y = 0$ and $y' = 2$ when $x = 0$

27. $y'' - 2y' + y = 2e^x, \quad y' = 2e$ at $(1, 0)$

28. $y'' - 9y = 18 \cos 3x + 9, \quad y = -1$ and $y' = 3$ when $x = 0$

29. $y'' + y = -2 \sin x, \quad y = 0$ at $x = 0$ and $x = \pi/2$

13-4. MECHANICAL VIBRATIONS

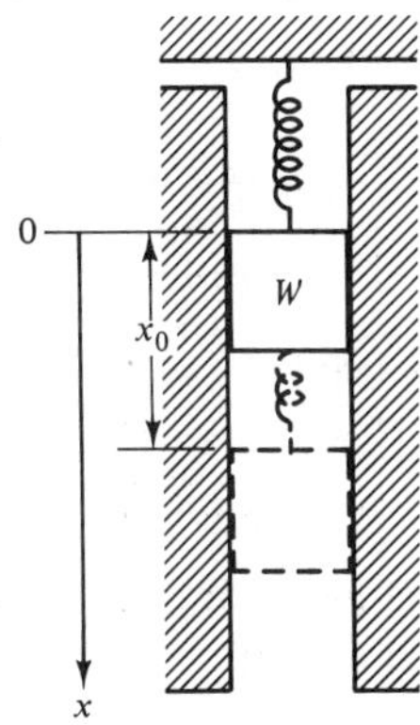

FIGURE 13-2

Free Vibrations

An important use for second-order differential equations is the analysis of mechanical vibrations. We first consider *free vibrations,* such as a vibrating spring, and later study *forced vibrations,* such as those caused by an unbalanced motor.

A block of weight W hangs from a spring with spring constant k (Fig. 13-2). The block is pulled down a distance x_0 from its rest position and released, its motion retarded by a frictional force proportional to the velocity dv/dt of the block. By Newton's second law of motion, the force F on the body equals the product of the mass m and its acceleration a.

$$ma = F$$

But $m = W/g$ (where g is the acceleration due to gravity) and a is the second derivative of the displacement x. Further, the force on the block is equal to the spring force kx acting upward, plus the frictional force $c\, dx/dt$, where c is called the coefficient of friction. So if the block is moving in the positive (downward) direction, we have

$$\frac{W}{g}\frac{d^2x}{dt^2} = -kx - c\frac{dx}{dt}$$

Rearranging gives

$$\frac{d^2x}{dt^2} + \frac{cg}{W}\frac{dx}{dt} + \frac{kg}{W}x = 0 \tag{1}$$

Making the substitutions

$$2a = \frac{cg}{W} \qquad \text{and} \qquad \omega_n^2 = \frac{kg}{W}$$

our equation becomes

$$x'' + 2ax' + \omega_n^2 x = 0 \tag{2}$$

This is a second-order linear differential equation with a right side of zero, which we solve as before. The auxiliary equation $m^2 + 2am + \omega_n^2 = 0$ has the roots

$$m = \frac{-2a + \sqrt{4a^2 - 4\omega_n^2}}{2} = -a \pm \sqrt{a^2 - \omega_n^2} \tag{3}$$

We saw that the solution to a second-order differential equation depends on the nature of the roots of the auxiliary equation. Now we will see that each of these cases corresponds to a particular type of motion. These are

Roots	*Type of motion*	
Nonreal	Underdamped	$a < \omega_n$
	No damping (simple harmonic motion)	$a = 0$
Real, equal	Critically damped	$a = \omega_n$
Real, unequal	Overdamped	$a > \omega_n$

We consider the underdamped and overdamped motions with free vibrations, and then the underdamped case with forced vibrations.

Underdamped Free Vibrations ($a < \omega_n$)

When a is less than ω_n, the auxiliary equation (3) gives the nonreal roots

$$\begin{aligned} m &= -a \pm j\sqrt{\omega_n^2 - a^2} \\ &= -a \pm j\omega_d \end{aligned}$$

if we make the substitution $\omega_d^2 = \omega_n^2 - a^2$. The solution to (2) is then (using the alternative form)

$$x = Ce^{-at} \sin(\omega_d t + \phi)$$

a damped sine wave of maximum amplitude C. From the physical problem, we know that the maximum amplitude is the initial displacement x_0, so

$$x = x_0 e^{-at} \sin(\omega_d t + \phi)$$

Also, at $t = 0$,

$$\frac{x}{x_0} = 1 = \sin \phi$$

from which $\phi = \pi/2$. Then $\sin(\omega_d t + \pi/2) = \cos \omega_d t$, so

Underdamped Free Vibrations	$x = x_0 e^{-at} \cos \omega_d t$	**A39**

The motion is thus a cosine wave whose amplitude decreases exponentially. The angular velocity ω_d is called the *damped angular velocity.* From our earlier substitution,

Damped Angular Velocity	$\omega_d = \sqrt{\omega_n^2 - a^2}$	**A40**

EXAMPLE 1: The block in Fig. 13-2 weighs 25.9 lb, the spring constant is 110 lb/in., and $c = 1.26$ lb/in./s. It is pulled down 0.625 in. from the rest position and released with zero velocity at $t = 0$. Find (a) the damped angular velocity, (b) the frequency, and (c) the period. (d) Write an equation for the displacement. Take $g = 386$ in./s^2.

Solution: (a) We first find a and ω_n.

$$a = \frac{cg}{2W} = \frac{(1.26 \text{ lb/in./s})(386 \text{ in./s}^2)}{2(25.9 \text{ lb})} = 9.39 \text{ rad/s}$$

and

$$\omega_n = \sqrt{\frac{kg}{W}} = \sqrt{\frac{110 \text{ lb/in. } (386 \text{ in./s}^2)}{25.9 \text{ lb}}} = 40.5 \text{ rad/s}$$

Since $a < \omega_n$, we have underdamping. Then by Eq. A40,

$$\omega_d = \sqrt{(40.5)^2 - (9.39)^2} = 39.4 \text{ rad/s}$$

(b) The frequency f is then

$$f = \frac{\omega_d}{2\pi} = \frac{39.4}{2\pi} = 6.27 \text{ Hz (cycles/s)}$$

(c) The period P is the reciprocal of the frequency, so

$$P = \frac{1}{f} = \frac{1}{6.27} = 0.159 \text{ s}$$

(d) By Eq. A39,

$$x = x_0 e^{-at} \cos \omega_d t = 0.625 e^{-9.39t} \cos 39.4t \qquad \text{in.}$$

The displacement is graphed in Fig. 13-3, showing a cosine wave enclosed within an "envelope" which decreases exponentially.

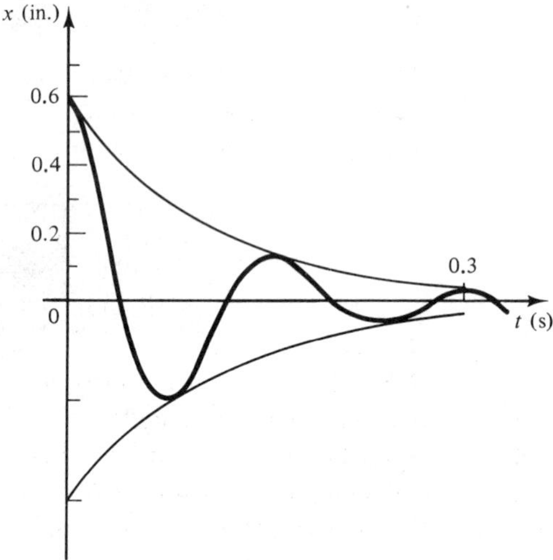

FIGURE 13-3. Damped vibrations.

Simple Harmonic Motion ($a = 0$)

When there is no damping, we have a special case of underdamped motion called *simple harmonic motion.* The damping coefficient c is 0 and hence a is zero, and Eq. A39 reduces to

Simple Harmonic Motion	$x = x_0 \cos \omega_n t$	**A36**

We see that the displacement is a cosine function of amplitude x_0 (it would be a sine function if we had chosen $x = 0$, rather that $x = x_0$, at $t = 0$). The quantity ω_n is called the *undamped angular velocity*. From before,

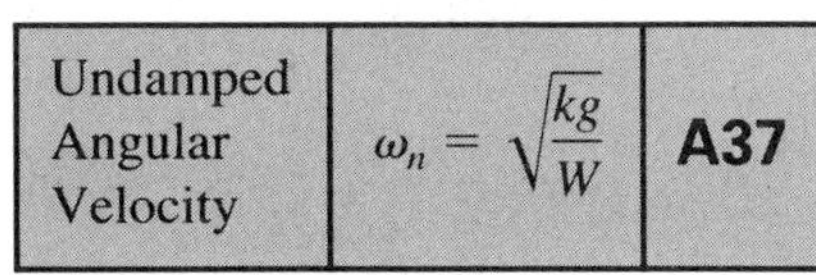

Undamped Angular Velocity	$\omega_n = \sqrt{\dfrac{kg}{W}}$	**A37**

The frequency obtained by dividing ω_n by 2π is called the *natural frequency* f_n.

Natural Frequency	$f_n = \dfrac{\omega_n}{2\pi}$	**A38**

EXAMPLE 2: If the damping coefficient c in Example 1 is zero, find (a) the undamped angular velocity, (b) the natural frequency, and (c) the period. Write equations for (d) the displacement and (e) the velocity.

Solution: (a) The undamped angular velocity ω_n is, from Example 1, 40.5 rad/s.
(b) By Eq. A38,

$$f_n = \frac{\omega_n}{2\pi} = \frac{40.5}{2\pi} = 6.44 \text{ Hz}$$

(c) The period is the reciprocal of the frequency, so

$$P = \frac{1}{f_n} = \frac{1}{6.44} = 0.155 \text{ s}$$

(d) By Eq. A36,

$$x = x_0 \cos \omega_n t = 0.625 \cos 40.5t \qquad \text{in.}$$

This is graphed in Fig. 13-4.

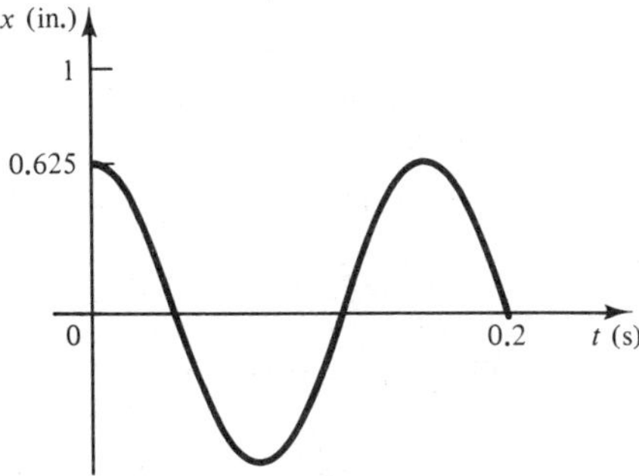

FIGURE 13-4. Undamped vibrations.

(e) Taking the derivative,

$$v = \frac{dx}{dt} = -\omega_n x_0 \sin \omega_n t = -(40.5)(0.625) \sin 40.5t$$

$$= -25.3 \sin 40.5t \qquad \text{in./s}$$

Overdamped Free Vibrations ($a > \omega_n$)

When a is greater than ω_n, the auxiliary equation (3) has the real and unequal roots

$$m = -a \pm \sqrt{a^2 - \omega_n^2}$$

The solution to the differential equation (2) is then

Overdamped Free Vibrations	$x = C_1 e^{m_1 t} + C_2 e^{m_2 t}$	**A41**

We evaluate C_1 and C_2 by substituting the initial values, as shown in the following example.

EXAMPLE 3: If $c = 7.46$ lb/in./s for the block in examples 1 and 2, write an equation for the displacement and the velocity.

Solution: We first find a.

$$a = \frac{cg}{2W} = \frac{(7.46 \text{ lb/in./s})(386 \text{ in./s}^2)}{2(25.9 \text{ lb})} = 55.6 \text{ rad/s}$$

Since $a > \omega_n$ (40.5 from before), we have overdamping. Then

$$\sqrt{a^2 - \omega_n^2} = \sqrt{(55.6)^2 - (40.5)^2} = 38.1 \text{ rad/s}$$

So

$$m_1 = -55.6 - 38.1 = -93.7 \qquad \text{and} \qquad m_2 = -55.6 + 38.1 = -17.5$$

By Eq. A41,

$$x = C_1 e^{-93.7t} + C_2 e^{-17.5t}$$

Taking the derivative,

$$v = -93.7C_1 e^{-93.7t} - 17.5C_2 e^{-17.5t}$$

Substituting $x = 0.625$ and $v = 0$ at $t = 0$ in the equations for x and v gives

$$C_1 + C_2 = 0.625$$

and

$$93.7C_1 + 17.5C_2 = 0$$

Solving simultaneously gives $C_1 = -0.143$ and $C_2 = 0.768$. Substituting back, we get

$$x = -0.143e^{-93.7t} + 0.768e^{-17.5t} \quad \text{in.}$$

and

$$v = 13.4e^{-93.7t} - 13.4e^{-17.5t} \quad \text{in./s}$$

The displacement is graphed in Fig. 13-5.

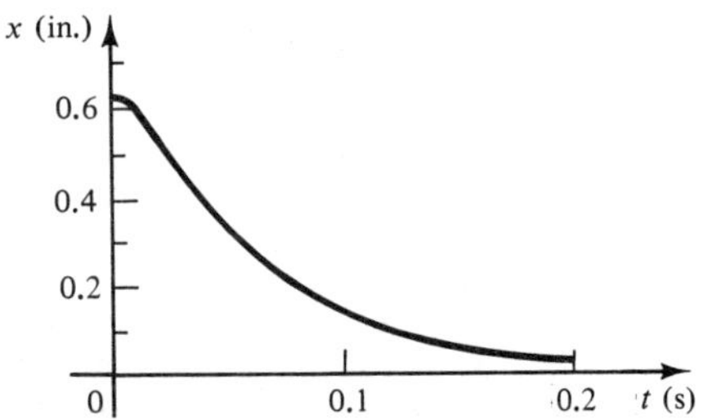

FIGURE 13-5. Overdamped vibrations.

Underdamped Forced Vibrations

To the forces acting on the block in Fig. 13-2 we now add an alternating force $P \sin \omega t$ of amplitude P and angular velocity ω. Our equation of motion is then

$$\frac{W}{g}\frac{d^2x}{dt^2} = -kx - c\frac{dx}{dt} + P \sin \omega t$$

Rearranging gives

$$\frac{d^2x}{dt^2} + \frac{cg}{W}\frac{dx}{dt} + \frac{kg}{W}x = \frac{Pg}{W}\sin \omega t \tag{1}$$

or, replacing cg/W with $2a$ and kg/W with ω_n^2, as before,

$$x'' + 2ax' + \omega_n^2 x = \frac{Pg}{W} \sin \omega t \tag{2}$$

This is a second-order linear differential equation with the right side not zero. As with free vibrations, the motion under forced vibrations can be either underdamped, critically damped, or overdamped, depending on the roots of the auxiliary equation. We cover only underdamping, which has greater practical importance than the others.

The complementary function is the same as found for free vibrations,

$$x_c = x_0 e^{-at} \cos \omega_d t$$

where ω_d is the damped angular velocity given by Eq. A40.

The particular integral x_p is $A \sin \omega t + B \cos \omega t$, or, using the alternative form,

$$x_p = x_m \sin (\omega t + \phi) \tag{3}$$

which is a sine wave of amplitude x_m. The complete solution is then $x_c + x_p$, or

$$x = \underbrace{x_0 e^{-at} \cos \omega_d t}_{\text{transient}} + \underbrace{x_m \sin (\omega t + \phi)}_{\text{steady-state}} \tag{4}$$

Thus the motion consists of two parts, called the *transient* motion, which dies out with time, and the *steady-state* motion, which continues as long as the force $P \sin \omega t$ is applied. We usually care little about the transient motion, since it dies quickly, but the magnitude x_m of the steady-state motion is important since this will determine, for example, whether a machine will destroy itself. We now solve for x_m.

The derivatives of (3) are

$$x_p' = \omega x_m \cos (\omega t + \phi)$$

$$x_p'' = -\omega^2 x_m \sin (\omega t + \phi)$$

Substituting into (2) gives

$$-\omega^2 x_m \sin (\omega t + \phi) + 2a\omega x_m \cos (\omega t + \phi) + \omega_n^2 x_m \sin (\omega t + \phi) = \frac{Pg}{W} \sin \omega t$$

When $\omega t + \phi = 0$, $\sin (\omega t + \phi) = 0$, $\cos (\omega t + \phi) = 1$, and $\sin \omega t = \sin (-\phi) = -\sin \phi$. Substituting,

$$2a\omega x_m = -\frac{Pg}{W} \sin \phi \tag{5}$$

When $\omega t + \phi = \pi/2$, $\sin(\omega t + \phi) = 1$, $\cos(\omega t + \phi) = 0$. Substituting gives

$$-\omega^2 x^m + \omega_n^2 x_m = \frac{Pg}{W} \sin\left(\frac{\pi}{2} - \phi\right) \tag{6}$$

Squaring both sides of (5) and (6) and adding gives

$$4a^2\omega^2 x_m^2 + x_m^2(\omega_n^2 - \omega^2)^2 = \left(\frac{PG}{W}\right)^2 (\sin^2\phi + \cos^2\phi)$$

or, since $\sin^2\phi + \cos^2\phi = 1$,

$$x_m^2[4a^2\omega^2 + (\omega_n^2 - \omega^2)^2] = \left(\frac{Pg}{W}\right)^2$$

so

Maximum Deflection	$x_m = \dfrac{Pg}{W\sqrt{4a^2\omega^2 + (\omega_n^2 - \omega^2)^2}}$	**A42**

EXAMPLE 4: Again using the block in Fig. 13-2 with a damping coefficient of 1.26 lb/in./s, we apply an alternating force 8.25 sin 25.0*t*. Find the maximum displacement x_m.

Solution: From preceding examples we have $\omega_n = 40.5$ rad/s, $a = 9.39$ rad/s, $g = 386$ in./s^2, and $W = 25.9$ lb. From Eq. A42, with $P = 8.25$ lb and $\omega = 25.0$ rad/s,

$$x_m = \frac{Pg}{W\sqrt{4a^2\omega^2 + (\omega_n^2 - \omega^2)^2}}$$

$$= \frac{8.25(386)}{25.9\sqrt{4(9.39)^2(25.0)^2 + [(40.5)^2 - (25.0)^2]^2}} = 0.112 \text{ in.}$$

The deflection we would get if the 8.25 lb was simply hung on the end of the spring, called the *static deflection,* is

$$\frac{8.25 \text{ lb}}{110 \text{ lb/in.}} = 0.075 \text{ in.}$$

The deflection we get is about 50% greater.

Resonance

Note that the denominator in Eq. A42 will be a minimum when $\omega = \omega_n$. This condition, called *resonance,* is greatly feared in mechanical systems and structures because the displacement x_m can get destructively large, especially if there is little damping.

EXAMPLE 5: The displacement x_m computed for values of ω from 0 to 50 rad/s is given in the following table and plotted in Fig. 13-6.

ω	x_m
0	0.075
5	0.076
10	0.080
15	0.086
20	0.098
25	0.118
30	0.155
35	0.232
40	0.326
45	0.215
50	0.126

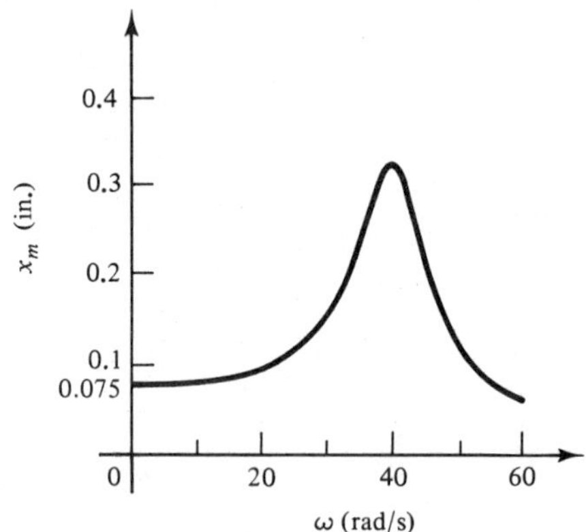

FIGURE 13-6. Resonance.

Note the peak displacement when $\omega = \omega_n$ (40.5 rad/s) and the static deflection, 0.075 in., when $\omega = 0$.

EXERCISE 4

Simple Harmonic Motion

1. The displacement x of an object at t seconds is given by $x = 3.75 \cos 182t$ in. Find (a) the period and (b) the amplitude of this motion.

2. What is the earliest time at which the displacement of the object in Problem 1 is -3 in.?

3. The equation of motion of a certain wood block bobbing in water is $x'' + 225x = 0$, where x is in centimeters and t is in seconds. The initial conditions are $x = 0$ and $v = 15.0$ cm/s at $t = 0$. Write an equation for x as a function of time.

4. In Problem 3, find (a) the maximum displacement and (b) the earliest time at which it occurs.

5. A 2.00-lb weight hangs motionless from a spring and is seen to stretch the spring 8.00 in. from its free length. It is pulled down an additional 1.00 in. and released. Write the equation of motion of the weight, taking zero at the original (motionless) position.

6. The maximum velocity in simple harmonic motion occurs when an object passes through its zero position. Find the maximum velocity and its time of occurrence for the weight in Problem 5.

7. The motion of a pendulum hanging from a long string and swinging through small angles approximates simple harmonic motion. If the initial conditions are $x_0 = 1.50$ cm and $a_0 = -3.00$ cm/s^2, write an equation for the displacement as a function of time.

8. Find the period for the motion of the pendulum in Problem 7.

Damped Vibrations

9. The equation of motion of a certain car shock absorber is given by $x = 2.50e^{-3t} \cos 55t$ in., where t is in seconds. Make a graph of x versus t.

10. In Problem 9, is the motion underdamped, overdamped, or critically damped?

11. For the shock absorber in Problem 9, find x when t is 0.30 s.

12. For the shock absorber in Problem 9, use Newton's method to find the earliest time at which x is half its initial value.

13. An object has a differential equation of motion given by $x'' + 5x' + 4x = 0$, with the initial conditions $x_0 = 1.50$ cm and $v_0 = 15.3$ cm/s. Write an equation for x as a function of time.

14. For the weight of Problem 5, assume a resisting force numerically equal to 3.00 times the velocity, in ft/s. Write an equation for the displacement.

15. In Problem 14, (a) what type of damping do we have and (b) what weight W will produce critical damping?

Forced Vibrations

16. A 52.8-lb weight hangs from a spring ($k = 287$ lb/in.). The coefficient of damping is 3.12 lb/in./s. Find the maximum deflection if an alternating force of $22.1 \sin 31.6t$ lb is applied to the weight.

17. What must be the frequency of the applied force to produce resonance for the weight of Problem 16?

18. Find the displacement of the weight in Problem 16 at resonance.

19. A 4.82-lb weight hangs from a spring ($k = 1.95$ lb/in.) The coefficient of damping is 0.113 lb/in./s. Find the maximum deflection if an alternating force of $3.44 \sin 28.5t$ lb is applied to the weight.

20. What must be the frequency of the applied force to produce resonance for the weight of Problem 19?

21. Find the displacement of the weight in Problem 19 at resonance.

13-5. *RLC* CIRCUITS

In Chapter 12 we studied the *RL* circuit and the *RC* circuit. Each gave rise to a first-order differential equation. We'll now see that the *RLC* circuit will result in a second-order differential equation.

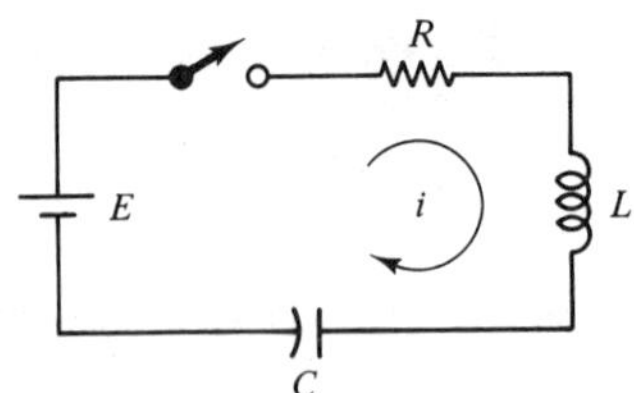

FIGURE 13-7. *RLC* circuit with dc source.

A switch (Fig. 13-7) is closed at $t = 0$. The sum of the voltage drops must equal the applied voltage, so

$$Ri + L\frac{di}{dt} + \frac{q}{C} = E \tag{1}$$

Replacing q by $\int i\,dt$ and differentiating gives

$$R\frac{di}{dt} + L\frac{d^2i}{dt^2} + \frac{i}{C} = 0$$

or

$$Li'' + Ri' + \left(\frac{1}{C}\right)i = 0 \tag{2}$$

Notice the similarity between the derivation of these circuit equations and those for mechanical vibrations.

This is a second-order linear differential equation, which we now solve as we did before. The auxiliary equation $Lm^2 + Rm + 1/C = 0$ has the roots

$$\begin{aligned} m &= \frac{-R \pm \sqrt{R^2 - 4L/C}}{2L} \\ &= -\frac{R}{2L} \pm \sqrt{\frac{R^2}{4L^2} - \frac{1}{LC}} \end{aligned}$$

We now let

$$a^2 = \frac{R^2}{4L^2} \quad \text{and} \quad \omega_n^2 = \frac{1}{LC}$$

so that

$$\omega_n = \sqrt{\frac{1}{LC}} \qquad \textbf{A89}$$

and our roots become

$$m = -a \pm \sqrt{a^2 - \omega_n^2}$$

or

$$m = -a \pm j\omega_d$$

where $\omega_d^2 = \omega_n^2 - a^2$. As with mechanical vibrations, we have three possible cases.

Roots	*Type of solution*	
Nonreal	Underdamped	$a < \omega_n$
	No damping (series *LC* circuit)	$a = 0$
Real, equal	Critically damped	$a = \omega_n$
Real, unequal	Overdamped	$a > \omega_n$

We consider the underdamped and overdamped cases with a dc source, and the underdamped case with an ac source.

Underdamped, $a < \omega_n$

We first consider the case where R is not zero but is low enough so that $a < \omega_n$. The roots of the auxiliary equation are then nonreal and the current is

$$i = e^{-at}(k_1 \sin \omega_d t + k_2 \cos \omega_d t) \tag{8}$$

Since i is zero at $t = 0$ we get $k_2 = 0$. The current is then

$$i = k_1 e^{-at} \sin \omega_d t$$

Taking the derivative,

$$\frac{di}{dt} = k_1 \omega_d e^{-at} \cos \omega_d t - a k_1 e^{-at} \sin \omega_d t$$

At $t = 0$ the capacitor behaves as a short circuit, and there is also no voltage drop across the resistor (since $i = 0$). The entire voltage E then appears across the inductor. Since $E = L\, di/dt$, then $di/dt = E/L$, so

$$k_1 = \frac{E}{\omega_d L}$$

The current is then

$$i = \frac{E}{\omega_d L} e^{-at} \sin \omega_d t \qquad \textbf{A91}$$

where, from our previous substitution,

$$\omega_d = \sqrt{\omega_n^2 - a^2} = \sqrt{\omega_n^2 - \frac{R^2}{4L^2}} \qquad \textbf{A92}$$

We get a damped sine wave whose amplitude decreases exponentially with time.

EXAMPLE 1: A switch (Fig. 13-7) is closed at $t = 0$. If $R = 225$ Ω, $L = 1.50$ H, $C = 4.75$ μF, and $E = 75.4$ V, write an expression for the instantaneous current.

Solution: We first compute LC,

$$LC = 1.50(4.75 \times 10^{-6}) = 7.13 \times 10^{-6}$$

Then, by Eq. A89,

$$\omega_n = \sqrt{\frac{1}{LC}} = \sqrt{\frac{10^6}{7.13}} = 375 \text{ rad/s}$$

and

$$a = \frac{R}{2L} = \frac{225}{2(1.50)} = 75.0 \text{ rad/s}$$

The angular velocity is then, by Eq. A92,

$$\omega_d = \sqrt{\omega_n^2 - a^2} = \sqrt{(375)^2 - (75.0)^2} = 367 \text{ rad/s}$$

The instantaneous current is then

$$i = \frac{E}{\omega_d L} e^{-at} \sin \omega_d t = \frac{75.4}{367(1.50)} e^{-75t} \sin 367t \qquad \text{A}$$

$$= 137\, e^{-75t} \sin 367t \qquad \text{mA}$$

This curve is plotted in Fig. 13-8, showing the damped sine wave.

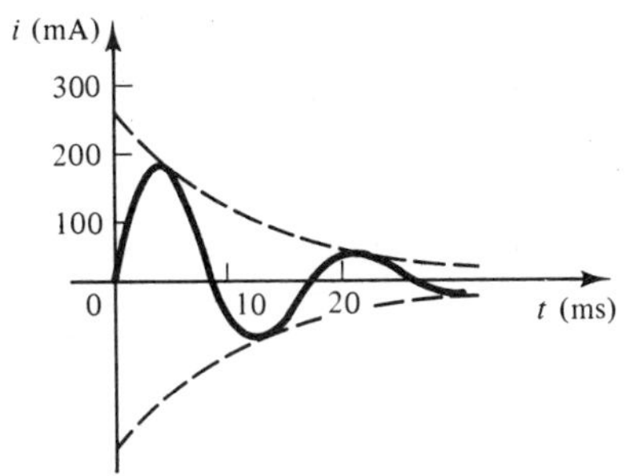

FIGURE 13-8

No Resistance: The Series *LC* Circuit

This, of course, is a theoretical case, for a real circuit always has some resistance.

When the resistance R is zero, $a = 0$ and $\omega_d = \omega_n$. From Eq. A91 we get

$$i = \frac{E}{\omega_n L} \sin \omega_n t \qquad \textbf{A90}$$

which represents a sine wave with amplitude $E/\omega_n L$.

EXAMPLE 2: Repeat Example 1 with $R = 0$.

Solution: The value of ω_n, from before, is 375 Hz. The amplitude of the current wave is

$$\frac{E}{\omega_n L} = \frac{75.4}{375(1.50)} = 0.134 \qquad \text{A}$$

so the instantaneous current is

$$i = 134 \sin 375t \text{ mA}$$

Overdamped, $a > \omega_n$

If the resistance is relatively large, so that $a > \omega_n$, the auxiliary equation has the real and unequal roots

$$m_1 = -a + j\omega_d \qquad \text{and} \qquad m_2 = -a - j\omega_d$$

The current is then

$$i = k_1 e^{m_1 t} + k_2 e^{m_2 t} \tag{1}$$

Since $i(0) = 0$ we have

$$k_1 + k_2 = 0 \tag{2}$$

Taking the derivative of (1),

$$\frac{di}{dt} = m_1 k_1 e^{m_1 t} + m_2 k_2 e^{m_2 t}$$

Since $di/dt = E/L$ at $t = 0$,

$$\frac{E}{L} = m_1 k_1 + m_2 k_2 \tag{3}$$

Solving (2) and (3) simultaneously gives

$$k_1 = \frac{E}{(m_1 - m_2)L} \quad \text{and} \quad k_2 = -\frac{E}{(m_1 - m_2)L}$$

where $m_1 - m_2 = -a + j\omega_d + a + j\omega_d = 2j\omega_d$. The current is then

$i = \dfrac{E}{2j\omega_d L}[e^{(-a+j\omega_d)t} - e^{(-a-j\omega_d)t}]$	**A93**

EXAMPLE 3: For the circuit of Examples 1 and 2, let $R = 2550\ \Omega$ and compute the instantaneous current.

Solution: $a = \dfrac{R}{2L} = \dfrac{2550}{2(1.50)} = 850$ rad/s. Since $\omega_n = 375$ rad/s, we have

$$\omega_d = \sqrt{(375)^2 - (850)^2} = j763 \text{ rad/s}$$

Then $-a - j\omega_d = -87.0$ and $-a + j\omega_d = -1613$. From Eq. A93,

$$i = \frac{75.4}{2(-763)(1.50)}(e^{-1613t} - e^{-87.0t})$$

$$= 32.9(e^{-87.0t} - e^{-1613t}) \qquad \text{mA}$$

This equation is graphed in Fig. 13-9.

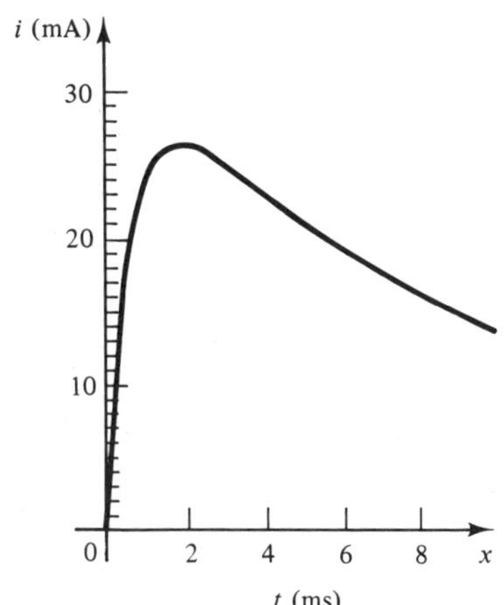

FIGURE 13-9. Current in an overdamped circuit.

Low Resistance; AC Source

Up to now we have considered only a dc source. We now repeat the low resistance case with an alternating voltage $E \sin \omega t$.

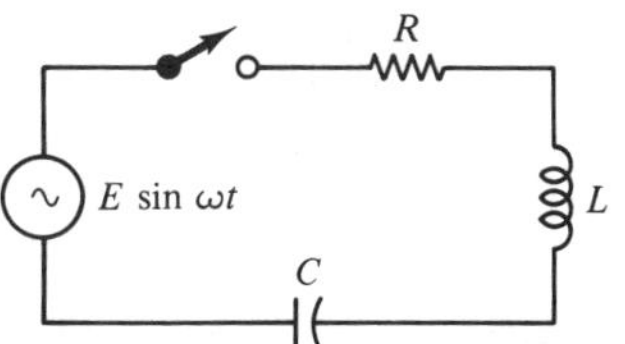

FIGURE 13-10. *RLC* circuit with ac source.

A switch (Fig. 13-10) is closed at $t = 0$. The sum of the voltage drops must equal the applied voltage, so

$$Ri + L\frac{di}{dt} + \frac{q}{C} = E \sin \omega t \tag{1}$$

Replacing q by $\int i\, dt$ and differentiating gives

$$R\frac{di}{dt} + L\frac{d^2i}{dt^2} + \frac{i}{C} = \omega E \cos \omega t$$

or

$$Li'' + Ri' + \left(\frac{1}{C}\right)i = \omega E \cos \omega t \tag{2}$$

The complementary function is the same as was calculated for the dc case,

$$i_c = e^{-at}(k_1 \sin \omega_d t + k_2 \cos \omega_d t)$$

The particular integral i_p will have a sine term and a cosine term,

$$i_p = A \sin \omega t + B \cos \omega t$$

In Chapter 14 we'll do this type of problem by the Laplace transform.

Taking first and second derivatives,

$$i' = \omega A \cos \omega t - \omega B \sin \omega t$$

$$i'' = -\omega^2 A \sin \omega t - \omega^2 B \cos \omega t$$

Substituting into (2), we get

$$-L\omega^2 A \sin \omega t - L\omega^2 B \cos \omega t - R\omega B \sin \omega t + R\omega A \cos \omega t + \frac{A}{C} \sin \omega t$$

$$+ \frac{B}{C} \cos \omega t = \omega E \cos \omega t$$

Equating the coefficients of the sine terms gives

$$-L\omega^2 A - R\omega B + \frac{A}{C} = 0$$

from which

$$-RB = A\left(\omega L - \frac{1}{\omega C}\right) = AX \tag{3}$$

where X is the reactance of the circuit. Equating the coefficients of the cosine terms gives

$$-L\omega^2 B + R\omega A + \frac{B}{C} = \omega E$$

or

$$RA - E = B\left(\omega L - \frac{1}{\omega C}\right) = BX$$

Solving (3) and (4) simultaneously gives

$$A = -\frac{RE}{R_2 + X^2} = \frac{RE}{Z^2}$$

where Z is the impedance of the circuit, and

$$B = -\frac{EX}{R^2 + X^2} = -\frac{EX}{Z^2}$$

Our particular integral thus becomes

$$y_p = -\frac{RE}{Z^2}\sin\omega t - \frac{EX}{Z^2}\cos\omega t$$

$$= -\frac{E}{Z^2}(R\sin\omega t - X\cos\omega t)$$

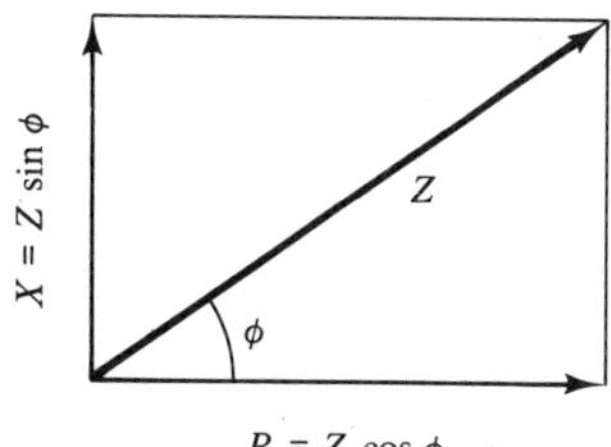

FIGURE 13-11. Impedance triangle.

From the impedance triangle (Fig. 13-11)

$$R = Z\cos\phi \qquad \text{and} \qquad X = Z\sin\phi$$

where ϕ is the phase angle. Thus

$$R\sin\omega t - X\cos\omega t = Z\sin\omega t\cos\phi - Z\cos\omega t\sin\phi$$

$$= Z\sin(\omega t - \phi)$$

by the trigonometric identity (Eq. 167). Thus i_p becomes $E/Z\sin(\omega t - \phi)$, or

$$i_p = I_{\max}\sin(\omega t - \phi)$$

since E/Z gives the maximum current $I_{\max}$. The total current is the sum of the complementary function i_c and the particular integral i_p,

$$i = \underbrace{e^{-at}(k_1\sin\omega_d t + k_2\cos\omega_d t)}_{\text{transient}} + \underbrace{I_{\max}\sin(\omega t - \phi)}_{\text{steady-state}}$$

Thus the current is made up of two parts; a *transient* part that dies quickly with time, and a *steady-state* part that continues as long as the ac source is connected. We are usually interested only in the steady-state current.

EXAMPLE 4: Find the steady-state current for an *RLC* circuit if $R = 345\ \Omega$, $L = 0.726$ H, $C = 41.4\ \mu$F, and $e = 155 \sin 285t$.

Solution: By Eqs. A94 and A95.

$$X_L = \omega L = 285(0.726) = 207\ \Omega$$

and

$$X_c = \frac{1}{\omega C} = \frac{1}{285(41.4 \times 10^{-6})} = 84.8\ \Omega$$

The total reactance is, by Eq. A96,

$$X = X_L - X_c = 207 - 84.8 = 122\ \Omega$$

The impedance Z is found from Eq. A97:

$$Z = \sqrt{R^2 + X^2} = \sqrt{(345)^2 + (122)^2} = 366\ \Omega$$

The phase angle, from Eq. A98, is

$$\phi = \tan^{-1}\frac{X}{R} = \tan^{-1}\frac{122}{345} = 19.5°$$

The steady-state current is then

$$i_{ss} = \frac{155}{366}\sin(285t - 19.5°) = 0.423 \sin(285t - 19.5°) \qquad \text{A}$$

Figure 13-12 shows the applied voltage and the steady-state current, with a phase difference of 19.5°, or 0.239 ms.

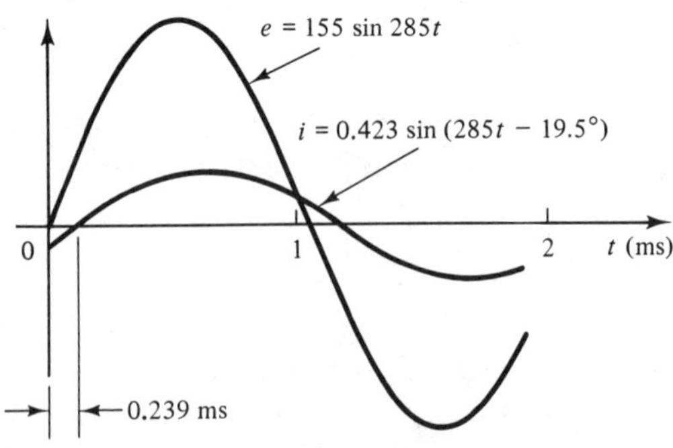

FIGURE 13-12

Resonance

The current in a series *RLC* circuit will be a maximum when the impedance Z is zero. This will occur when the reactance X is zero, so

$$X = \omega L - \frac{1}{\omega C} = 0$$

Solving for ω, we get $\omega^2 = 1/LC$ or ω_n^2. Thus

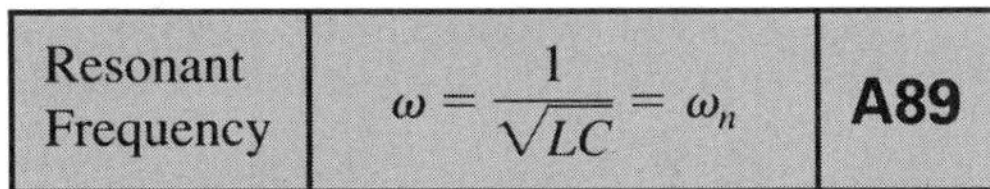

Resonant Frequency	$\omega = \dfrac{1}{\sqrt{LC}} = \omega_n$	**A89**

EXAMPLE 5: Find the resonant frequency for the circuit of Example 4, and write an expression for the steady-state current at that frequency.

Solution: The resonant frequency is

$$\omega_n = \frac{1}{\sqrt{LC}} = \frac{1}{\sqrt{0.726(41.4 \times 10^{-6})}} = 182 \text{ rad/s}$$

Since $X_L = X_c$, then $X = 0$, $Z = R$, and $\phi = 0$. Thus I_{max} is 155/345 = 0.448 A and the steady-state current is then

$$i_{ss} = 448 \sin 182t \qquad \text{mA}$$

EXERCISE 5

Series LC Circuit

1. In an *LC* circuit, $C = 1.00\ \mu$F, $L = 1.00$ H, and $E = 100$ V. At $t = 0$ the charge and the current are both zero. Show that $i = 0.1 \sin 1000t$ A.

2. For Problem 1, take the integral of i to show that $q = 10^{-4}$ (1 − cos 1000t) coulomb (C).

3. Repeat Problem 2 with $E = 0$ and an initial charge of 255×10^{-6} C (i_0 is still zero).

4. Find an expression for the current as a function of time for the circuit of Problem 3.

Series RLC Circuit with DC Source

5. In an *RLC* circuit, $R = 1.55\ \Omega$, $C = 250\ \mu$F, $L = 0.125$ H, and $E = 100$ V. The current and charge are 0 when $t = 0$. Show that $i = 4.47e^{-6.2t} \sin 179t$ A.

6. Integrate i in Problem 5 to show that $q = -e^{-6.2t}(0.866 \sin 179t + 25.0 \cos 179t) + 25.0$ mC.

7. In an *RLC* circuit, $R = 1.75\ \Omega$, $C = 4.25$ F, $L = 1.50$ H, and $E = 100$ V. The current and charge are 0 when $t = 0$. Show that $i = -77.6e^{-1.01t} + 77.8e^{-0.155t}$ A.

8. For the circuit in Problem 7, show that $q = 76.8e^{-1.01t} - 502e^{-0.155t} + 425$ C.

Series RLC Circuit with AC Source

9. For an *RLC* circuit, $R = 10.5\ \Omega$, $L = 0.125$ H, $C = 225\ \mu$F, and $e = 100 \sin 175t$. Show that the steady-state current is $i = 9.03 \sin(175t + 18.4°)$.

10. Find the resonant frequency for Problem 9.

11. Find the steady-state current for an *RLC* circuit if $R = 1550\ \Omega$, $L = 0.350$ H, $C = 20.0\ \mu$F, and $e = 250 \sin 377t$.

12. If $R = 10.5\ \Omega$, $L = 0.175$ H, $C = 1.50 \times 10^{-3}$ F, and $e = 175 \sin 55t$, find the amplitude of the steady-state current.

13. Find the resonant frequency for the circuit of Problem 12.

14. Find the amplitude of the steady-state current at resonance for the circuit in Problem 12.

15. For the circuit of Fig. 13-10, $R = 110\ \Omega$, $L = 5.60 \times 10^{-6}$ H, and $\omega = 6.00 \times 10^5$ rad/s. What value of C will produce resonance?

16. If an *RLC* circuit has the values $R = 10\ \Omega$, $L = 0.200$ H, $C = 500\ \mu$F, and $e = 100 \sin \omega t$, find the resonant frequency.

17. For the circuit in Problem 16, find the amplitude of the steady-state current at resonance.

CHAPTER TEST

1. $y'' = \sin 3x$

2. $4y'' - 3y' = 0$

3. $y'' + 5y = 0$

4. $y'' + 2y' - 3y = 7$, $y = 1$ and $y' = 2$ when $x = 0$

5. $y'' - y' - 2y = 5x$

6. $y'' - 2y = 2x^3 + 3x$

7. $y'' - 4y' - 5y = e^{4x}$

8. $y''' - 2y' = 0$

9. $y'' - 5y' = 6$, $y = y' = 1$ when $x = 0$

10. $y'' + 2y' - 3y = 0$

11. $y'' - 7y' - 18y = 0$

12. $y'' + 6y' + 9y = 0$, $y = 0$ and $y' = 2$ when $x = 0$

13. $y''' - 6y'' + 11y' - 6y = 0$

14. $y'' - 2y = 3x^3e^x$

15. $y'' + 2y = 3 \sin 2x$

16. $y'' + y' = 5 \sin 3x$

17. $y'' + 3y' + 2y = 0$, $y = 1$ and $y' = 2$ when $x = 0$

18. $y'' + 5y = 0$

19. $y'' + 3y' - 4y = 0$

20. $y'' - 2y' + y = 0$, $y = 0$ and $y' = 2$ when $x = 0$

21. $y'' + 5y' - y = 0$

22. $y'' - 4y' + 4y = 0$

23. $y'' + 3y' - 4y = 0$, $y = 0$ and $y' = 2$ when $x = 0$

24. $y'' + 4y' + 4y = 0$

25. $y'' + 6y' + 9y = 0$

26. $y'' - 2y' + y = 0$

27. $9y'' - 6y' + y = 0$

28. $y'' - x^2 = 0$, where $y' = 4$ at the point $(1, 2)$

29. A 5.00-lb weight hangs motionless from a spring, which is seen to stretch 7.00 in. from its free length. The weight is pulled down another 1.50 in. and released. Write the equation of motion of the weight, taking zero at the original (motionless) position.

30. In Problem 29, assume a resisting force numerically equal to 2.50 times the velocity, in ft/s. Write an equation for the displacement.

31. For an LC circuit, $C = 1.75\ \mu\text{F}$, $L = 2.20$ H, and $E = 110$ V. At $t = 0$ the charge and the current are both zero. Find i as a function of time.

32. In Problem 31, find the charge as a function of time.

33. For an RLC circuit, $R = 3.75\ \Omega$, $C = 150\ \mu\text{F}$, $L = 0.100$ H, and $E = 120$ V. The current and charge are 0 when $t = 0$. Find i_{ss} as a function of time.

34. In Problem 33, find an expression for the steady-state charge.

The Laplace Transform

The methods we learned in Chapters 12 and 13 are often called *classical* methods. In this chapter we learn another method, called the *Laplace transform*.

The Laplace transform enables us to transform a differential equation into an algebraic one. We can then solve the algebraic equation and apply an inverse transform to obtain the solution to the differential equation.

The Laplace transform is good for solving *initial value* problems, that is, when the value of the function is known at $t = 0$. Thus we will deal here with functions of time rather than with functions of x.

We will use the Laplace transform to solve first- and second-order differential equations with constant coefficients. We'll then do some applications, and compare our results with those obtained by classical methods in Chapters 12 and 13.

14-1. THE LAPLACE TRANSFORM OF A FUNCTION

Laplace Transform of Some Simple Functions

If y is some function of time, so that $y = f(t)$, the Laplace transform of that function is defined by the improper integral

After the French analyst, probabilist, astronomer, and physicist Pierre Laplace, (1749–1827).

Laplace Transform	$\mathcal{L}[f(t)] = \int_0^\infty f(t)e^{-st}\,dt$	**397**

Note that the variable of integration is s, so the transformed expression *will be a function of s*, which we call $F(s)$. Thus

$$\mathcal{L}[f(t)] = F(s)$$

When can transform an expression by direct integration, using Eq. 397.

Glance back at Sec. 11-5 if you have forgotten how to evaluate an inproper integral.

EXAMPLE 1: If $y = f(t) = 1$, then

$$\mathcal{L}[f(t)] = \int_0^\infty (1)e^{-st}\,dt$$

$$= -\frac{1}{s}\int_0^\infty e^{-st}\,(-s\,dt)$$

$$= -\frac{1}{s}e^{-st}\Big|_0^\infty = -\frac{1}{s}(0-1) = \frac{1}{s}$$

Thus $\mathcal{L}[1] = 1/s$.

Rule 2 from our table of integrals says that $\int af(x)\,dx = a\int f(x)\,dx$. It follows then that

$$\mathcal{L}[af(t)] = a\mathcal{L}[f(t)]$$

The Laplace transform of a constant times a function is equal to the constant times the transform of the function.

EXAMPLE 2: If $\mathcal{L}[1] = 1/s$, as above, then

$$\mathcal{L}[5] = 5\mathcal{L}[1] = \frac{5}{s}$$

EXAMPLE 3: If $y = f(t) = t$, then

$$\mathcal{L}[t] = \int_0^\infty te^{-st}\,dt$$

Using integral 37 with $u = t$ and $a = -s$,

$$\mathscr{L}[t] = \frac{e^{-st}}{s^2}(-st - 1)\Big|_0^{\infty} = 0 - \left(-\frac{1}{s^2}\right) = \frac{1}{s^2}$$

Note that in each case the transformed expression is a function of s only.

EXAMPLE 4: Find $\mathscr{L}[\sin at]$.

Solution: Using integral 41,

$$\mathscr{L}[\sin at] = \int_0^{\infty} e^{-st} \sin at\, dt$$

$$= \frac{e^{-st}}{s^2 + a^2}(-s \sin at - a \cos at)\Big|_0^{\infty}$$

$$= 0 - \frac{1}{s^2 + a^2}(-a) = \frac{a}{s^2 + a^2}$$

Transform of a Sum

If we have the sum of several terms,

$$f(t) = a\, g(t) + b\, h(t) \cdots$$

then

$$\mathscr{L}[f(t)] = \int_0^{\infty} [a\, g(t) + b\, h(t) + \cdots]e^{-st}\, dt$$

$$= a\int_0^{\infty} g(t)e^{-st}\, dt + b\int_0^{\infty} h(t)e^{-st}\, dt + \cdots$$

so

$$4.\quad \mathscr{L}[a\, g(t) + b\, h(t) \cdots] = a\mathscr{L}[g(t)] + b\mathscr{L}[h(t)] + \cdots$$

The Laplace transform of a sum of several terms is the sum of the transforms of each term.

EXAMPLE 5: If $y = 4 + 5t - 2 \sin 3t$, then

$$\mathscr{L}[y] = \frac{4}{s} + \frac{5}{s^2} - 2\frac{3}{s^2 + 9}$$

Transform of a Derivative

To solve differential equations, we must be able to take the transform of a derivative. Let $f'(t)$ be the derivative of some function $f(t)$. Then

$$\mathscr{L}[f'(t)] = \int_0^{\infty} f'(t)e^{-st}\, dt$$

Integrating by parts, we let $u = e^{-st}$ and $dv = f'(t)\,dt$. So $du = -se^{-st}\,dt$ and $v = \int f'(t)\,dt = f(t)$. Then

$$\begin{aligned}\mathcal{L}[f'(t)] &= f(t)e^{-st}\Big|_0^\infty + s\int_0^\infty f(t)e^{-st}\,dt \\ &= f(t)e^{-st}\Big|_0^\infty + s\mathcal{L}[f(t)] \\ &= 0 - f(0) + s\mathcal{L}[f(t)]\end{aligned}$$

so

Transform of a Derivative	2. $\mathcal{L}[f'(t)] = s\mathcal{L}[f(t)] - f(0)$

Thus if we have a function $y = f(t)$, then

$$\mathcal{L}[y'] = s\mathcal{L}[y] - y(0)$$

Note that the transform of a derivative contains $f(0)$ or $y(0)$, the value of the function at $t = 0$. Thus *to evaluate the Laplace transform of a derivative, we must know the initial conditions.*

EXAMPLE 6: Take the Laplace transform of both sides of the differential equation $y' = 6t$, with the initial condition that $y(0) = 5$.

Solution: We get

$$\mathcal{L}[y'] = \mathcal{L}[6t]$$

$$s\mathcal{L}[y] - y(0) = \frac{6}{s^2}$$

or

$$s\mathcal{L}[y] - 5 = \frac{6}{s^2}$$

since $y(0) = 5$. Later we will solve such an equation for $\mathcal{L}[y]$ and then do an *inverse transform* to obtain y.

Similarly, we can find the transform of the second derivative,

Transform of the Second Derivative	3. $\mathcal{L}[f''(t)] = s^2\mathcal{L}[f(t)] - sf(0) - f'(0)$

For a function $y = f(t)$,

$$\mathcal{L}[y''] = s^2\mathcal{L}[y] - sy(0) - y'(0)$$

EXAMPLE 7: Transform both sides of the second-order differential equation $y'' - 3y' + 4y = 5t$ if $y(0) = 6$ and $y'(0) = 7$.

Solution: We get

$$\underbrace{\{s^2\mathscr{L}[y] - sy(0) - y'(0)\}}_{\mathscr{L}[y''']} - =\underbrace{\{s\mathscr{L}[y] - y(0)\}}_{\mathscr{L}[y']} + \div\mathscr{L}[y] = \frac{5}{s^2}$$

Substituting $y(0) = 6$ and $y'(0) = 7$,

$$s^2\mathscr{L}[y] - 6s - 7 - 3s\mathscr{L}[y] - 3(6) + 4\mathscr{L}[y] = \frac{5}{s^2}$$

or

$$s^2\mathscr{L}[y] - 3s\mathscr{L}[y] + 4\mathscr{L}[y] - 6s - 25 = \frac{5}{s^2}$$

Table of Laplace Transforms

The transforms of some common functions are given in Table 14-1. Instead of using Eq. 397 to transform a function, we simply look it up in the table.

Here n is a positive integer.

TABLE 14-1. Short Table of Laplace Transforms

	$f(t)$	$\mathscr{L}[f(t)] = F(s)$
1	$f(t)$	$F(s) = \int_0^\infty e^{-st}f(t)\,dt$
2	$f'(t)$	$s\mathscr{L}[f(t)] - f(0)$
3	$f''(t)$	$s^2\mathscr{L}[f(t)] - sf(0) - f'(0)$
4	$a\,g(t) + b\,h(t) + \cdots$	$a\mathscr{L}[g(t)] + b\mathscr{L}[h(t)] + \cdots$
5	1	$\frac{1}{s}$
6	t	$\frac{1}{s^2}$
7	t^n	$\frac{n!}{s^{n+1}}$
8	$\frac{t^{n-1}}{(n-1)!}$	$\frac{1}{s^n}$
9	e^{at}	$\frac{1}{s-a}$
10	$1 - e^{-at}$	$\frac{a}{s(s+a)}$
11	te^{at}	$\frac{1}{(s-a)^2}$

	$f(t)$	$\mathcal{L}[f(t)] = F(s)$
12	$e^{at}(1 + at)$	$\dfrac{s}{(s-a)^2}$
13	$t^n e^{at}$	$\dfrac{n!}{(s-a)^{n+1}}$
14	$t^{n-1}e^{-at}$	$\dfrac{(n-1)!}{(s+a)^n}$
15	$e^{-at} - e^{-bt}$	$\dfrac{b-a}{(s+a)(s+b)}$
16	$ae^{-at} - be^{-bt}$	$\dfrac{s(a-b)}{(s+a)(s+b)}$
17	$\sin at$	$\dfrac{a}{s^2 + a^2}$
18	$\cos at$	$\dfrac{s}{s^2 + a^2}$
19	$t \sin at$	$\dfrac{2as}{(s^2 + a^2)^2}$
20	$t \cos at$	$\dfrac{s^2 - a^2}{(s^2 + a^2)^2}$
21	$1 - \cos at$	$\dfrac{a^2}{s(s^2 + a^2)}$
22	$at - \sin at$	$\dfrac{a^3}{a^2(s^2 + a^2)}$
23	$e^{-at} \sin bt$	$\dfrac{b}{(s+a)^2 + b^2}$
24	$e^{-at} \cos bt$	$\dfrac{s+a}{(s+a)^2 + b^2}$
25	$\sin at - at \cos at$	$\dfrac{2a^3}{(s^2 + a^2)^2}$
26	$\sin at + at \cos t$	$\dfrac{2as^2}{(s^2 + a^2)^2}$
27	$\cos at - \frac{1}{2} at \sin at$	$\dfrac{s^3}{(s^2 + a^2)^2}$
28	$\dfrac{b}{a^2}(e^{-at} + at - 1)$	$\dfrac{b}{s^2(s+a)}$
29	$\int_0^t f(t)\,dt$	$\dfrac{\mathcal{L}[f(t)]}{s}$

EXAMPLE 8: Find the transform of t^3e^{2t} by table.

Solution: Our function matches transform 13, with $n = 3$ and $a = 2$, so

$$\mathcal{L}[t^3e^{2t}] = \frac{3!}{(s-2)^{3+1}} = \frac{6}{(s-2)^4}$$

EXERCISE 1

Transforms by Direct Integration

Find the Laplace transform of each function by direct integration.

1. $f(t) = 6$

2. $f(t) = t$

3. $f(t) = t^2$

4. $f(t) = 2t^2$

5. $f(t) = \cos 5t$

6. $f(t) = e^t \sin t$

Transforms by Table

Use Table 14-1 to find the Laplace transform of each function.

7. $f(t) = t^2 + 4$

8. $f(t) = t^3 - 2t^2 + 3t - 4$

9. $f(t) = 3e^t + 2e^{-t}$

10. $f(t) = 5te^{3t}$

11. $f(t) = \sin 2t + \cos 3t$

12. $f(t) = 3 + e^{4t}$

13. $f(t) = 5e^{3t} \cos 5t$

14. $f(t) = 2e^{-4t} - t^2$

15. $f(t) = 5e^t - t \sin 3t$

16. $f(t) = t^3 + 4t^2 - 3e^t$

Transforms of Derivatives

Find the Laplace transform of each expression and substitute the given initial conditions.

17. $y' + 2y, \quad y(0) = 1$

18. $3y' + 2y, \quad y(0) = 3$

19. $y' - 4y, \quad y(0) = 0$

20. $5y' - 3y, \quad y(0) = 2$

21. $y'' + 3y' - y, \quad y(0) = 1, y'(0) = 3$

22. $y'' - y' + 2y, \quad y(0) = 1, y'(0) = 0$

23. $2y'' + 3y' + y, \quad y(0) = 2, y'(0) = 3$

24. $3y'' - y' + 2y, \quad y(0) = 2, y'(0) = 1$

14-2. INVERSE TRANSFORMS

Before we can use the Laplace transform to solve differential equations, we must be able to transform a function of s back to a function of t. The *inverse Laplace transform* is denoted by $\mathcal{L}^{-1}$. Thus, if $\mathcal{L}[f(t)] = F(s)$, then

We'll see that finding the inverse transform is harder than finding the transform.

Inverse Laplace Transform	$\mathcal{L}^{-1}[F(s)] = f(t)$	**398**

We use the table of Laplace transforms to do inverse transforms.

EXAMPLE 1: If $F(s) = 4/(s^2 + 16)$, find $f(t)$.

Solution: We search Table 14-1 for an expression of similar form and find transform 17,

$$\mathcal{L}[\sin at] = \frac{a}{s^2 + a^2}$$

which matches our expression if $a = 4$. Thus

$$f(t) = \sin 4t$$

EXAMPLE 2: Find y if $\mathcal{L}[y] = 5/(s - 7)^4$.

Solution: From the table we find transform 13,

$$\mathcal{L}[t^n e^{at}] = \frac{n!}{(s-a)^{n+1}}$$

In order to match our function, we must have $a = 7$ and $n = 3$. The numerator then must be $3! = 3(2) = 6$. So we rewrite our function as

$$\mathcal{L}[y] = \frac{5}{6}\frac{6}{(s-7)^4}$$

Its inverse transform is then

$$y = \frac{5}{6}t^3e^{7t}$$

Completing the Square

Sometimes we must complete the square to make our function match one in Table 14.1.

EXAMPLE 3: Find y if $\mathscr{L}[y] = \dfrac{s-1}{s^2+4s+20}$.

This is no different from the method we used in Sec. 11-2, when we completed the square on the denominator of rational fractions.

Solution: This does not now match any functions in our table, but it will if we complete the square on the denominator.

$$\begin{aligned} s^2+4s+20 &= (s^2+4s \qquad) + 20 \\ &= (s^2+4s+4)+20-4 \\ &= (s+2)^2+4^2 \end{aligned}$$

The denominator is now of the same form as in transform 24. However, to use transform 24, the numerator must be $s+2$, not $s-1$. So let us add 3 and subtract 3 from the numerator, so that $s-1=(s+2)-3$. Then

$$\begin{aligned} \mathscr{L}[y] &= \frac{s-1}{s^2+4s+20} = \frac{(s+2)-3}{(s+2)^2+4^2} \\ &= \frac{s+2}{(s+2)^2+4^2} - \frac{3}{(s+2)^2+4^2} \\ &= \frac{s+2}{(s+2)^2+4^2} - \frac{3}{4}\,\frac{4}{(s+2)^2+4^2} \end{aligned}$$

Now the first expression matches transform 24 and the second matches transform 23. Making the transform,

$$\begin{aligned} y &= e^{-2t}\cos 4t - \frac{3}{4}e^{-2t}\sin 4t \\ &= e^{-2t}\left(\cos 4t - \frac{3}{4}\sin 4t\right) \end{aligned}$$

Partial Fractions

We used partial fractions in Sec. 11-2 to make a given expression match one listed in the table of integrals. Now we use partial fractions to make an expression match one listed in our table of Laplace transforms.

We can also complete the square here, but would find that the resulting expression does not match a table entry.

EXAMPLE 4: Find y if $\mathscr{L}[y] = 12/(s^2 - 2s - 8)$.

Solution: We separate $\mathscr{L}[y]$ into partial fractions,

$$\mathscr{L}[y] = \frac{12}{s^2 - 2s - 8} = \frac{A}{s - 4} + \frac{B}{s + 2}$$

so

$$\begin{aligned} 12 &= A(s + 2) + B(s - 4) \\ &= (A + B)s + (2A + 4B) \end{aligned}$$

from which $A + B = 0$ and $2A + 4B = 12$. Solving simultaneously gives $A = 2$ and $B = -2$, so

$$\mathscr{L}[y] = \frac{12}{s^2 - 2s - 8} = \frac{2}{s - 4} - \frac{2}{s + 2}$$

Using transform 9, we get

$$y = 2e^{4t} - 2e^{-2t}$$

EXERCISE 2

Inverse Transforms

Find the inverse transform of each function.

1. $\dfrac{1}{s}$
2. $\dfrac{3}{s^2}$
3. $\dfrac{2}{s^3}$
4. $\dfrac{1}{s^2 + 3s}$
5. $\dfrac{4}{4 + s^2}$
6. $\dfrac{s}{(s - 6)^2}$
7. $\dfrac{3s}{s^2 + 2}$
8. $\dfrac{4}{s^3 + 9s}$
9. $\dfrac{s + 4}{(s - 9)^2}$
10. $\dfrac{3s}{(s^2 + 4)^2}$
11. $\dfrac{5}{(s + 2)^2 + 9}$
12. $\dfrac{s + 2}{s^2 - 6s + 8}$
13. $\dfrac{2s^2 + 1}{s(s^2 + 1)}$
14. $\dfrac{s^2}{s^2 + 2s + 1}$

15. $\dfrac{5s + 2}{s^2(s - 1)(s + 2)}$

16. $\dfrac{1}{(s + 1)(s + 2)^2}$

17. $\dfrac{2}{s^2 + s - 2}$

18. $\dfrac{s + 1}{s^2 + 2s}$

19. $\dfrac{2s}{s^2 + 5s + 6}$

20. $\dfrac{3s}{(s^2 + 4)(s^2 + 1)}$

14-3. SOLVING DIFFERENTIAL EQUATIONS BY THE LAPLACE TRANSFORM

To solve a differential equation with the Laplace transform, (1) take the transform of each side of the equation, (2) solve for $\mathscr{L}[y] = F(s)$, (3) manipulate $F(s)$ until it matches one or more table entries, and (4) take the inverse transform to find $y = f(t)$. We start with a very simple example.

EXAMPLE 1: Solve the first-order differential equation $y' + y = 2$ if $y(0) = 0$.

Solution: 1. We transform both sides and get

$$\mathscr{L}[y'] + \mathscr{L}[y] = \mathscr{L}[2]$$

$$s\mathscr{L}[y] - y(0) + \mathscr{L}[y] = \frac{2}{s}$$

2. We substitute 0 for $y(0)$ and solve for $\mathscr{L}[y]$.

$$(s + 1)\mathscr{L}[y] = \frac{2}{s}$$

$$\mathscr{L}[y] = \frac{2}{s(s + 1)}$$

3. Our function will match transform 10 if we write it as

$$\mathscr{L}[y] = 2\frac{1}{s(s + 1)}$$

4. Taking the inverse transform,

$$y = 2(1 - e^{-t})$$

EXAMPLE 2: Solve the first-order differential equation $y' - 3y + 4 = 9t$ if $y(0) = 2$.

Solution: 1. We transform both sides and get

$$\mathcal{L}[y'] - \mathcal{L}[3y] + \mathcal{L}[4] = \mathcal{L}[9t]$$

$$s\mathcal{L}[y] - y(0) - 3\mathcal{L}[y] + \frac{4}{s} = \frac{9}{s^2}$$

2. We substitute 2 for $y(0)$ and solve for $\mathcal{L}[y]$.

$$(s - 3)\mathcal{L}[y] = \frac{9}{s^2} - \frac{4}{s} + 2$$

$$\mathcal{L}[y] = \frac{9}{s^2(s - 3)} - \frac{4}{s(s - 3)} + \frac{2}{s - 3}$$

3. We match our expressions with transforms 28, 10, and 9,

$$\mathcal{L}[y] = \frac{9}{s^2(s - 3)} + \frac{4}{3}\frac{-3}{s(s - 3)} + 2\frac{1}{s - 3}$$

4. Taking the inverse transform,

$$y = e^{3t} - 3t - 1 + \frac{4}{3}(1 - e^{3t}) + 2e^{3t}$$

$$= \frac{5}{3}e^{3t} - 3t + \frac{1}{3}$$

EXAMPLE 3: Solve the second-order differential equation $y'' + 4y - 3 = 0$, where y is a function of t, if y and y' are both zero at $t = 0$.

Solution: 1. Taking the Laplace transform of both sides,

$$\mathcal{L}[y''] + \mathcal{L}[4y] - \mathcal{L}[3] = 0$$

$$s^2\mathcal{L}[y] - sy(0) - y'(0) + 4\mathcal{L}[y] - \frac{3}{s} = 0$$

Substituting $y(0) = 0$ and $y'(0) = 0$ gives

$$s^2\mathcal{L}[y] - 0 - 0 + 4\mathcal{L}[y] - \frac{3}{s} = 0$$

2. Solving for $\mathcal{L}[y]$,

$$(s^2 + 4)\mathcal{L}[y] = \frac{3}{s}$$

$$\mathcal{L}[y] = \frac{3}{s(s^2 + 4)}$$

3. We match it to transform 21,

$$\mathcal{L}[y] = \frac{3}{s(s^2 + 4)} = \frac{3}{4}\frac{4}{s(s^2 + 2^2)}$$

4. Taking the inverse transform,

$$y = \frac{3}{4}(1 - \cos 2t)$$

EXERCISE 3

Solve each differential equation by the Laplace transform.

First-Order Equations

1. $y' - 3y = 0$, $y(0) = 1$

2. $2y' + y = 1$, $y(0) = 3$

3. $4y' - 2y = t$, $y(0) = 0$

4. $y' + 5y = e^{2t}$, $y(0) = 2$

5. $3y' - 2y = t^2$, $y(0) = 3$

6. $y' - 3y = \sin t$, $y(0) = 0$

7. $y' + 2y = \cos 2t$, $y(0) = 0$

8. $4y' - y = 3t^3$, $y(0) = 0$

Second-Order Equations

9. $y'' + 2y' - 3y = 0$, $y(0) = 0$, $y'(0) = 2$

10. $y'' + y' + y = 1$, $y(0) = 0$, $y'(0) = 0$

11. $y'' + 3y' = 3$, $y(0) = 1$, $y'(0) = 2$

12. $y'' + 2y = 2$, $y(0) = 0$, $y'(0) = 3$

13. $2y'' + y = 4t$, $y(0) = 3$, $y'(0) = 0$

14. $y'' + y' + 5y = t$, $y(0) = 1$, $y'(0) = 2$

15. $y'' + 4y' + 3y = t$, $y(0) = 2$, $y'(0) = 2$

16. $y'' + 4y' = 2t^3$, $y(0) = 0$, $y'(0) = 0$

17. $3y'' + y' = \sin t$, $y(0) = 2$, $y'(0) = 3$

18. $2y'' + y' + 2y = 3$, $y(0) = 2$, $y'(0) = 1$

19. $y'' - 2y' + y = e^t$, $y(0) = 0$, $y'(0) = 0$

20. $2y'' + 32y = \cos 2t$, $y(0) = 0$, $y'(0) = 1$

21. $y'' + 2y' + 3y = te^t, \quad y(0) = 0, \quad y'(0) = 0$

22. $3y'' + 2y' - y = \sin 3t, y(0) = 0, \quad y'(0) = 0$

Motion in a Resisting Medium

23. A 15.0 kg object is dropped from rest through air whose resisting force is equal to 1.85 times the object's speed, in m/s. Find its speed after 1.00s.

Exponential Growth and Decay

24. A certain steel ingot is 1900°F and cools at a rate (in °F/min) equal to 1.25 times its present temperature (°F). Find its temperature after 5.00 min.

25. The rate of growth (bacteria/h) of a colony of bacteria is equal to 2.5 times the number present at any instant. How many bacteria are there after 24 h if there are 5000 at first?

Mechanical Vibrations

26. A weight that hangs from a spring is pulled down 1.00 in. at $t = 0$ and released from rest. The differential equation of motion is $x'' + 6.25x' + 25.8x = 0$. Write the equation for x as a function of time.

27. An alternating force is applied to a weight such that the equation of motion is $x'' + 6.25x' = 45.3 \cos 2.25t$. If v and x are zero at $t = 0$, write an equation for x as a function of time.

14-4. ELECTRICAL APPLICATIONS

Many of the differential equations for motion and electric circuits covered in Chapters 12 and 13 are nicely handled by the Laplace transform. However, since the Laplace transform is used mainly for electric circuits, we emphasise that application here. The method will be illustrated by examples of several types of circuits, with dc and ac sources.

Series *RC* Circuit with DC Source

EXAMPLE 1: A capacitor (Fig. 14-1) is discharged through a resistor by throwing the switch from position 1 to position 2 at $t = 0$. Find the current i.

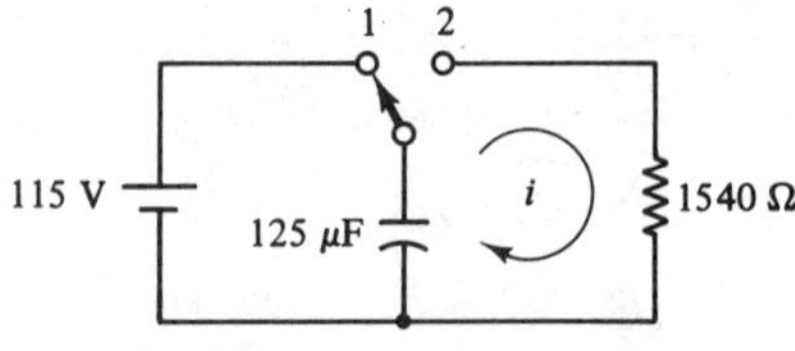

FIGURE 14-1. RC circuit.

Solution: Summing the voltages around the loop gives

$$Ri + \frac{1}{C}\int_0^t i\,dt - v_c = 0$$

Substituting values and rearranging,

$$1540i + \frac{1}{125 \times 10^{-6}}\int_0^t i\,dt = 115$$

Taking the transform of each term,

$$1540\mathscr{L}[i] + 8000\frac{\mathscr{L}[i]}{s} = \frac{115}{s}$$

We solve for $\mathscr{L}[i]$ and rewrite our expression in the form of a table entry.

$$\mathscr{L}[i]\left(1540 + \frac{8000}{s}\right) = \frac{115}{s}$$

$$\mathscr{L}[i]\left(\frac{1540s + 8000}{s}\right) = \frac{115}{s}$$

$$\mathscr{L}[i] = \frac{115}{1540s + 8000} = \frac{0.0747}{s + 5.19}$$

Taking the inverse transform using transform 9,

$$i = 0.0747\,e^{-5.19t}$$

This is the same result obtained by classical methods in Sec. 12-8, Example 3.

EXAMPLE 2: A capacitor (Fig. 14-2) has an initial charge of 35 V, with the polarity as marked. Find the current if the switch is closed at $t = 0$.

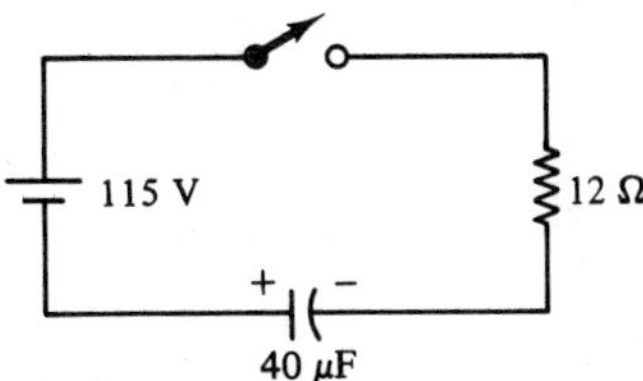

FIGURE 14-2. RC circuit.

Solution: Summing the voltages around the loops gives

$$Ri + \frac{1}{C}\int_0^t i\,dt - v_c - E = 0$$

Substituting values and rearranging,

$$12i + \frac{1}{40 \times 10^{-6}} \int_0^t i\,dt = 150$$

Taking the transform of each term,

$$12\mathscr{L}[i] + 25{,}000 \frac{\mathscr{L}[i]}{s} = \frac{150}{s}$$

Solving for $\mathscr{L}[i]$,

$$\mathscr{L}[i] \left(12 + \frac{25{,}000}{s}\right) = \frac{150}{s}$$

$$\mathscr{L}[i] \left(\frac{12s + 25{,}000}{s}\right) = \frac{150}{s}$$

$$\mathscr{L}[i] = \frac{150}{12s + 25{,}000} = \frac{12.5}{s + 2080}$$

Taking the inverse transform using transform 9,

$$i = 12.5e^{-2080t}$$

Series *RL* Circuit with DC Source

EXAMPLE 3: The initial current in an *RL* circuit (Fig. 14-3) is 1 A in the direction shown. The switch is closed at $t = 0$. Find the current.

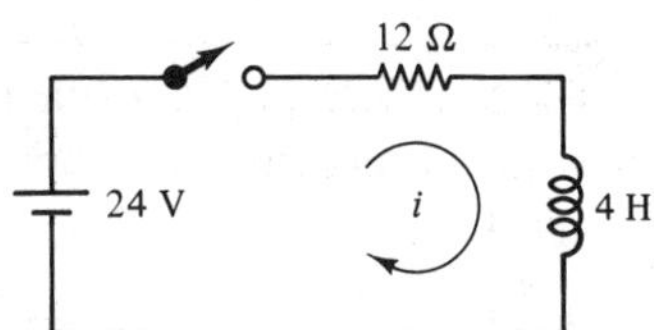

FIGURE 14-3. RL circuit.

Solution: Summing the voltages around the loop gives

$$E = Ri + L\frac{di}{dt}$$

Substituting values and rearranging,

$$24 = 12i + 4i'$$

or

$$6 = 3i + i'$$

Taking the transform of each term,

$$\frac{6}{s} = 3\mathscr{L}[i] + s\mathscr{L}[i] - i(0)$$

Substituting 1 A for $i(0)$ and solving for $\mathscr{L}[i]$,

$$\frac{6}{s} = \mathscr{L}[i](s + 3) - 1$$

$$\mathscr{L}[i] = \frac{6}{s(s + 3)} + \frac{1}{s + 3}$$

Taking the inverse transform using transforms 9 and 10 gives

$$i = 2(1 - e^{-3t}) + e^{-3t}$$

or

$$i = 2 - e^{-3t}$$

This current is graphed in Fig. 14-4.

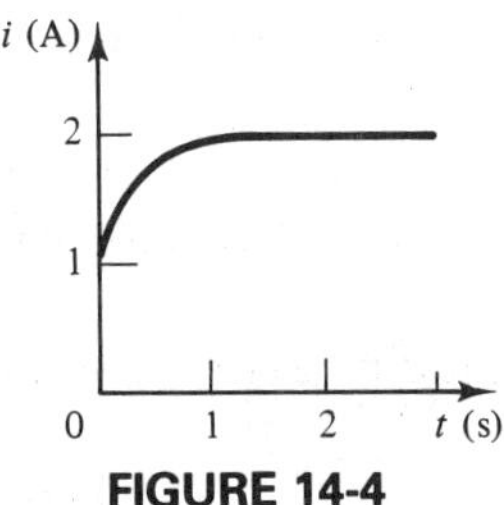

FIGURE 14-4

Series *RL* Circuit with AC Source

EXAMPLE 4: A switch (Fig. 14-5) is closed at $t = 0$ when the applied voltage is zero and increasing. Find the current.

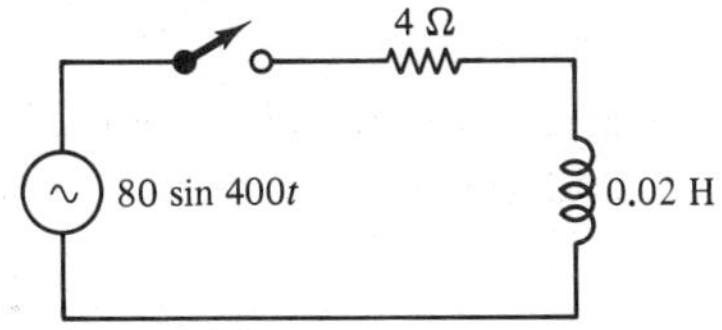

FIGURE 14-5. RL circuit with ac source.

Solution: Summing the voltages around the loop gives

$$Ri + L\frac{di}{dt} = v$$

Substituting values and rearranging,

$$4i + 0.02\frac{di}{dt} = 80 \sin 400t$$

Taking the transform of each term,

$$4\mathscr{L}[i] + 0.02[s\mathscr{L}[i] - i(0)] = \frac{80(400)}{s^2 + (400)^2}$$

Substituting 0 for $i(0)$ and solving for $\mathscr{L}[i]$,

$$\mathscr{L}[i](0.02s + 4) = \frac{32{,}000}{s^2 + (400)^2}$$

$$\mathscr{L}[i] = \frac{32{,}000}{[s^2 + (400)^2](0.02s + 4)} = \frac{1{,}600{,}000}{[s^2 + (400)^2](s + 200)}$$

$$= \frac{1{,}600{,}000}{(s - j400)(s + j400)(s + 200)}$$

Splitting this expression into partial fractions,

$$\frac{1{,}600{,}000}{(s - j400)(s + j400)(s + 200)} = \frac{A}{s - j400} + \frac{B}{s + j400} + \frac{C}{s + 200}$$

Multiplying both sides by $(s - j400)$ gives

$$\frac{1{,}600{,}000}{(s + j400)(s + 200)} = A + \frac{B(s - j400)}{s + j400} + \frac{C(s - j400)}{s + 200}$$

Letting $s = j400$ causes the partial fractions containing B and C to vanish, and leaves

$$A = \frac{1{,}600{,}000}{(j400 + j400)(j400 + 200)} = 2(-2 - j)$$

after simplification (not shown). We find B and C by the same method, and get $B = 2(-2 + j)$, and $C = 8$. Substituting back,

$$\mathscr{L}[y] = \frac{2(-2 - j)}{s - j400} + \frac{2(-2 + j)}{s + j400} + \frac{8}{s + 200}$$

We multiply the numerator and denominator of the first two fractions each by the conjugate of its denominator,

$$\mathcal{L}[i] = \frac{2(-2-j)(s+j400)}{(s-j400)(s+j400)} + \frac{2(-2+j)(s-j400)}{(s+j400)(s-j400)} + \frac{8}{s+200}$$

We combine the first two terms over a common denominator and collect terms, getting

$$\mathcal{L}[i] = \frac{1600-8s}{s^2+(400)^2} + \frac{8}{s+200}$$

$$= 4\frac{400}{s^2+(400)^2} - 8\frac{s}{s^2+(400)^2} + 8\frac{1}{s+200}$$

Taking the inverse transform using transforms 9, 17, and 18,

$$i = 4\sin 400t - 8\cos 400t + 8e^{-200t}$$

We now combine the first two terms in the form $I \sin(400t + \theta)$, where

$$I = \sqrt{4^2+8^2} = 8.94 \quad \text{and} \quad \theta = \tan^{-1}\left(-\frac{8}{4}\right) = -63.4° = -1.107 \text{ rad}$$

so

$$i = 8.94 \sin(400t - 1.107) + 8e^{-200t}$$

A graph of this wave, as well as the applied voltage wave, is shown in Fig. 14-6.

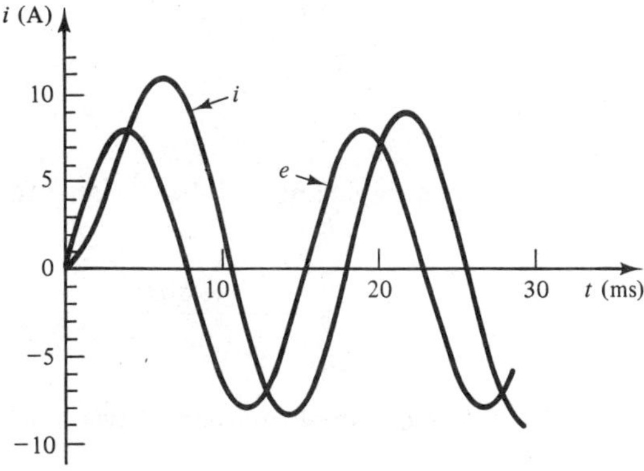

FIGURE 14-6

Series *RLC* Circuit with DC Source

EXAMPLE 5: A switch (Fig. 14-7) is closed at $t = 0$, and there is no initial charge on the capacitor. Write an expression for the current.

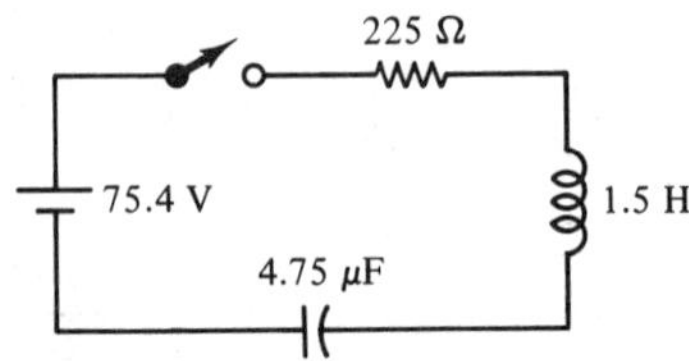

FIGURE 14-7. RLC circuit.

Solution: 1. The differential equation for this circuit is

$$Ri + L\frac{di}{dt} + \frac{1}{C}\int_0^t i\,dt = E$$

2. Transforming each term and substituting values gives

$$225\mathscr{L}[i] + 1.5\{s\,\mathscr{L}[i] - i(0)\} + \frac{10^6\mathscr{L}[i]}{4.75s} = \frac{75.4}{s}$$

3. Setting $i(0)$ to 0 and solving for $\mathscr{L}[i]$,

$$\mathscr{L}[i] = \frac{50.3}{s^2 + 150s + 140{,}000}$$

4. We now try to match this expression with those in the table. Let us factor the denominator.

$$\begin{aligned} s^2 + 150s + 140{,}000 &= (s^2 + 150s + 5625) + 140{,}000 - 5625 \\ &= (s + 75)^2 - (-134{,}000) \\ &= (s + 75)^2 - j^2(367)^2 \\ &= (s + 75 + j367)(s + 75 - j367) \end{aligned}$$

after replacing -1 by j^2, and factoring the difference of two squares. So

$$\mathscr{L}[i] = \frac{50.3}{s^2 + 150s + 140{,}000} = \frac{50.3}{(s + 75 + j367)(s + 75 - j367)}$$

We can now find partial fractions,

$$\frac{50.3}{(s + 75 + j367)(s + 75 - j367)} = \frac{A}{s + 75 + j367} + \frac{B}{s + 75 - j367}$$

Multiplying through by $s + 75 + j367$ gives

$$\frac{50.3}{s + 75 - j367} = A + \frac{B(s + 75 + j367)}{s + 75 - j367}$$

Setting $s = -75 - j367$ we get

$$A = \frac{50.3}{-75 - j367 + 75 - j367} = \frac{50.3}{-j734} = j0.0685$$

We find B by multiplying through by $s + 75 - j367$ and then setting $s = j367 - 75$, and we get $B = -j0.0625$ (work not shown). Substituting,

$$\mathscr{L}[i] = \frac{j0.0685}{s + 75 + j367} - \frac{j0.0685}{s + 75 - j367}$$

This still doesn't match any table entry. Let us now combine the two fractions over a common denominator.

$$\begin{aligned}\mathscr{L}[i] &= \frac{j0.0685}{s + 75 + j367} - \frac{j0.0685}{s + 75 - j367} \\ &= \frac{j0.0685(s + 75 - j367) - j0.0685(s + 75 + j367)}{(s + 75 + j367)(s + 75 - j367)} \\ &= \frac{-j^2 50.4}{s^2 + 150s + 140{,}000} = 0.137\frac{367}{(s + 75)^2 + (367)^2}\end{aligned}$$

5. We finally have a match, with transform 23. Taking the inverse transform gives

$$i = 137e^{-75t}\sin 367t \qquad \text{mA}$$

which agrees with the result found in Sec. 13-5, Example 1.

EXERCISE 4

1. The current in a certain RC circuit satisfies the equation $172i' + 2750i = 115$. If i and i' are zero at $t = 0$, write an equation for i.

2. In an RL circuit, $R = 3750\ \Omega$ and $L = 0.150$ H. It is connected to a dc source of 250 V at $t = 0$. If i and i' are zero at $t = 0$, write an equation for the instantaneous current.

3. The current in an RLC circuit satisfies the equation $0.55i'' + 482i' + 7350i = 120$. If i and i' are zero at $t = 0$, write an equation for i.

4. In an RLC circuit, $R = 3750\ \Omega$, $L = 0.150$ H, and $C = 1.25\ \mu$F. It is connected to a dc source of 150 V at $t = 0$. If i and i' are zero at $t = 0$, write an equation for the instantaneous current.

5. The current in a certain RLC circuit satisfies the equation $3.15i + 0.0223i' + 1680 \int_0^t i\,dt = 0$. If q and i are zero at $t = 0$, write an equation for the instantaneous current.

6. In an RL circuit, $R = 4150\ \Omega$ and $L = 0.127$ H. It is connected to a dc source of 28.4 V at $t = 0$. If i and i' are zero at $t = 0$, write an equation for the instantaneous current.

7. In an RL circuit, $R = 8370\ \Omega$, and $L = 0.250$ H. It is connected to a dc source of 50.5 V at $t = 0$. If i and i' are zero at $t = 0$, write an equation for the instantaneous current.

CHAPTER TEST

Find the Laplace transform by direct integration.

1. $f(t) = 2t$
2. $f(t) = 3t^2$
3. $f(t) = \cos 3t$

Find the Laplace transform by table.

4. $f(t) = 2t^3 + 3t$
5. $f(t) = 3te^{2t}$
6. $f(t) = 2t + 3e^{2t}$
7. $f(t) = 3e^{-t} - 2t^2$
8. $f(t) = 3t^3 + 5t^2 + 4e^t$
9. $y' + 3y, \quad y(0) = 1$
10. $y' - 6y, \quad y(0) = 1$
11. $3y' + 2y, \quad y(0) = 0$
12. $2y'' + y' - 3y, \quad y(0) = 1, y'(0) = 3$
13. $y'' + 3y' + 4y, \quad y(0) = 1, y'(0) = 3$

Find the inverse transform of each function.

14. $F(s) = \dfrac{6}{4 + s^2}$
15. $F(s) = \dfrac{s}{(s - 3)^2}$
16. $F(s) = \dfrac{s + 6}{(s - 2)^2}$
17. $F(s) = \dfrac{2s}{(s^2 + 5)^2}$
18. $F(s) = \dfrac{3}{(s + 4)^2 + 5}$
19. $F(s) = \dfrac{s + 1}{s^2 - 6s + 8}$
20. $F(s) = \dfrac{4s^2 + 2}{s(s^2 + 3)^2}$
21. $F(s) = \dfrac{3s}{s^2 + 2s + 1}$
22. $F(s) = \dfrac{3}{s^2 + s - 4}$
23. $F(s) = \dfrac{4s}{s^2 + 4s + 4}$

Solve by the Laplace transform.

24. $y' + 2y = 0, \quad y(0) = 1$
25. $y' - 3y = t^2, \quad y(0) = 2$
26. $3y' - y = 2t^5, \quad y(0) = 1$
27. $y'' + 2y' + y = 2, \quad y(0) = 0, \quad y'(0) = 0$
28. $y'' + 3y' + 2y = t, \quad y(0) = 1, \quad y'(0) = 2$
29. $y'' + 2y = 4t, \quad y(0) = 3, \quad y'(0) = 0$
30. $y'' + 3y' + 2y = e^{2t}, \quad y(0) = 2, \quad y'(0) = 1$

31. A 21.5 lb ball is dropped from rest through air whose resisting force is equal to 2.75 times the ball's speed, in ft/s. Write an expression for the speed of the ball.

32. A weight that hangs from a spring is pulled down 0.50 cm at $t = 0$ and released from rest. The differential equation of motion is $x'' + 3.22x' + 18.5 = 0$. Write the equation for x as a function of time.

33. The current in a certain *RC* circuit satisfies the equation $8.24i' + 149i = 100$. If i is zero at $t = 0$, write an equation for i.

34. The current in an *RLC* circuit satisfies the equation $5.45i'' + 2820i' + 9730i = 0$. If i and i' are zero at $t = 0$, write an equation for i.

35. In an *RLC* circuit, $R = 3750\ \Omega$, $L = 0.150$ H, and $C = 1.25\ \mu$F. It is connected to a dc source of 150 V at $t = 0$. If i and i' are zero at $t = 0$, write an equation for the instantaneous current.

36. In an *RL* circuit, $R = 3750\ \Omega$ and $L = 0.150$ H. It is connected to an ac source at $28.4 \sin 84t$ V at $t = 0$. If i and i' are zero at $t = 0$, write an equation for the instantaneous current.

Numerical Solution of Differential Equations

There are many differential equations that cannot be solved by the methods of Chapters 12 through 14. Further, even if an equation can be solved analytically, we may want to do a computer solution, for which analytical methods are not useful. In both cases we can use a numerical method.

We start with Euler's method, the earliest and simplest numerical method for solving differential equations, but not the most accurate. It does, however, serve as a good introduction to the other methods, which include predictor–corrector methods and the Runge–Kutta method. We apply these methods to both first- and second-order equations.

15-1. FIRST-ORDER DIFFERENTIAL EQUATIONS

Euler's Method

Suppose that we have a first-order differential equation, which we write in the form

$$y' = f(x, y) \tag{1}$$

and a boundary condition that $x = x_p$ when $y = y_p$. We seek a solution $y = F(x)$ such that the graph of this function (Fig. 15-1) passes through $P(x_p, y_p)$ and has a slope at P given by (1), $m_p = f(x_p, y_p)$.

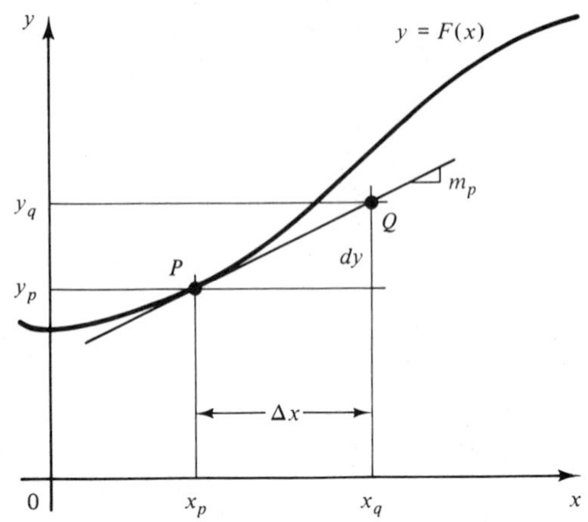

FIGURE 15-1. Euler's method.

Having the slope at P, we then step a distance Δx to the right. The rise dy of the tangent line is then

$$\text{rise} = (\text{slope})(\text{run}) = m_p\,\Delta x$$

This is no different than when, in Sec. 5-5, we used differentials to approximate a small change in a function.

Thus the point Q has the coordinates $(x_p + \Delta x,\ y_p + m_p\,\Delta x)$. *Point Q is not on the curve $y = F(x)$, but may be close enough to use as an approximation to the curve.*

From Q, we repeat the process enough times to reconstruct as much of the curve $y = F(x)$ as needed. Thus having the coordinates (x_p, y_p) at P and the slope m_p at P, we find the coordinates (x_q, y_q) of Q by the iteration formulas,

After a Swiss mathematician, Leonhard Euler (1707–1783).

Euler's Method	$x_q = x_p + \Delta x$ $y_q = y_p + m_p\,\Delta x$	**399**

EXAMPLE 1: Find an approximate solution to $y' = x^2/y$, with the boundary condition that $y = 2$ when $x = 3$. Calculate y for $x = 3$ to 10, in steps of 1.

Solution: The slope at (3, 2) is

$$m = y'(3, 2) = \frac{9}{2}$$

If $\Delta x = 1$, the rise is

$$dy = m\,\Delta x = 4.5(1) = 4.5$$

The ordinate of our next point is then 2 + 4.5 = 6.5, and our next point is (4, 6.5). Repeating,

$$m = y'(4, 6.5) = \frac{16}{6.5} = 2.462$$

$$dy = 2.452(1) = 2.462$$

So the next point is (5, 8.962). The remaining values are given in the following computer-generated table.

x	*Approximate y*	*Exact y*	*Error*
3	2.00000	2.00000	0.00000
4	6.50000	5.35413	1.14587
5	8.96154	8.32666	0.63487
6	11.75124	11.40176	0.34948
7	14.81475	14.65151	0.16324
8	18.12226	18.09236	0.02991
9	21.65383	21.72556	0.07173
10	25.39451	25.54734	0.15283

The solution to our differential equation, then, is in the form of a set of (x, y) pairs, the first two columns in the table above. We do not get an equation for our solution.

Reducing the step size with this method and those to follow will result in better accuracy (and more work).

We normally use numerical methods for a differential equation whose exact solution cannot be found. Here we have used an equation whose solution is known so that we may check our answer. The solution is $3y^2 = 2x^3 - 42$, which we use to compute the values shown in the third column of the table, with the difference between exact and approximate values in the third column. Note that the error at $x = 10$ is about 0.6%.

Euler's Method with Predictor-Corrector

The various *predictor–corrector* methods all have two main steps. First, the *predictor* step tries to predict the next point on the curve from preceding values, and then the *corrector* step tries to improve the predicted values.

Here Euler's method of the preceding section is the predictor, using the coordinates and slope at the present point P to estimate the next point Q. The corrector step then recomputes Q *using the average of the slopes at P and Q*. Our iteration formula is then

Modified Euler's Method	$x_q = x_p + \Delta x$ $y_q = y_p + \frac{m_p + m_q}{2} \Delta x$	**400**

EXAMPLE 2: Repeat Example 1 using the modified Euler's method.

Solution: Using data from before, the slope m_1 at (3, 2) was 4.5, the second point was predicted to be (4, 6.5), and the slope m_2 at (4, 6.5) was 2.462. The recomputed ordinate at the second point is then, by Eq. 400,

$$y_2 = 2 + \frac{4.5 + 2.462}{2}(1) = 5.481$$

so our corrected second point is (4, 5.481). The slope at this point is

$$m_2 = y'(4, 5.481) = 2.919$$

We use this to predict y_3.

$$y_3(\text{predicted}) = y_2 + m_2\,\Delta x = 5.481 + 2.919(1) = 8.400$$

The slope at (5, 8.400) is

$$m_3 = y'(5, 8.400) = 2.976$$

The corrected y_3 is then

$$y_3(\text{corrected}) = 5.481 + \frac{2.919 + 2.976}{2}(1) = 8.429$$

The remaining values are given in the following table.

x	*Approximate y*	*Exact y*	*Error*
3	2.00000	2.00000	0.00000
4	5.48077	5.35413	0.12664
5	8.42850	8.32666	0.10184
6	11.49126	11.40176	0.08950
7	14.73299	14.65151	0.08148
8	18.16790	18.09236	0.07555
9	21.79642	21.72556	0.07086
10	25.61434	25.54734	0.06700

Note that the error at $x = 10$ is about 0.26%, compared with 0.6% before.

Runge–Kutta Method

The Runge–Kutta method, like the modified Euler's, is a predictor–corrector method.

Prediction: Starting at *P*, Fig. 15-2, we use

The slope m_p to predict *R*,
the slope m_r to predict *S*,
the slope m_s to predict *Q*.

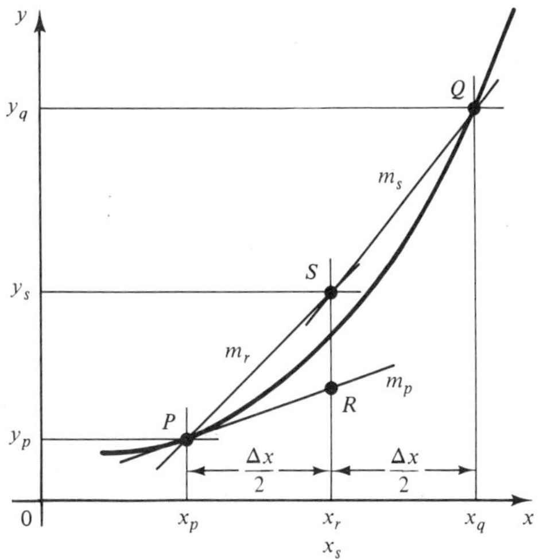

FIGURE 15-2. Runge Kutta method.

Correction: We then take a weighted average of the slopes at P, Q, R, and S, giving twice the weight to the slopes at the midpoints R and S than at the endpoints P and Q.

$$m_{avg} = \frac{1}{6}(m_p + 2m_r + 2m_s + m_q)$$

and use the average slope and the coordinates of P to find Q.

$$y_q = y_p + m_{avg}\,\Delta x$$

After two German mathematicians, Carl Runge (1856–1927) and Wilhelm Kutta (1867–1944).

Runge–Kutta Method	$x_q = x_p + \Delta x$ $y_q = y_p + m_{avg}\,\Delta x$ where $m_{avg} = \frac{1}{6}(m_p + 2m_r + 2m_s + m_q)$ and $m_p = f'(x_p, y_p)$ $m_r = f'\left(x_p + \frac{\Delta x}{2}, y_p + m_p\frac{\Delta x}{2}\right)$ $m_s = f'\left(x_p + \frac{\Delta x}{2}, y_p + m_r\frac{\Delta x}{2}\right)$ $m_q = f'(x_p + \Delta x, y_p + m_s\,\Delta x)$	401

EXAMPLE 3: Repeat Example 2 using the Runge–Kutta method and a step of 1.

Solution: Starting at $P(3, 2)$, with $m_p = 4.5$,

$$y_r = 2 + 4.5(0.5) = 4.25$$

$$m_r = f'(3.5, 4.25) = \frac{(3.5)^2}{4.25} = 2.882$$

$$y_s = 2 + 2.882(0.5) = 3.441$$

$$m_s = f'(3.5, 3.441) = \frac{(3.5)^2}{3.441} = 3.560$$

$$y_q = 2 + 3.560(1) = 5.560$$

$$m_q = f'(4, 5.560) = \frac{4^2}{5.560} = 2.878$$

$$m_{avg} = \frac{4.5 + 2(2.882) + 2(3.560) + 2.878}{6} = 3.377$$

$$y_q(\text{corrected}) = y_p + m_{avg}\,\Delta x = 2 + 3.377(1) = 5.377$$

So our second point is (3, 5.377). The rest of the calculation is given in the following table.

x	*Approximate y*	*Exact y*	*Error*
3	2.00000	2.000000	0.00000
4	5.37703	5.354126	0.02290
5	8.34141	8.326664	0.01475
6	11.41255	11.401760	0.01079
7	14.65992	14.651510	0.00841
8	18.09918	18.092360	0.00682
9	21.73125	21.725560	0.00569
10	25.55219	25.547340	0.00484

The error at $x = 10$ is about 0.02%, compared with 0.26% for the modified Euler method.

Comparison of the Three Methods

The following table shows the errors obtained when solving the same differential equation by our three different methods, each with a step size of 1.

x	*Euler*	*Modified Euler*	*Runge–Kutta*
3	0.00000	0.00000	0.00000
4	1.14587	0.12664	0.02290
5	0.63487	0.10184	0.01475
6	0.34948	0.08950	0.01079
7	0.16324	0.08148	0.00841
8	0.02991	0.07555	0.00682
9	0.07173	0.07086	0.00569
10	0.15283	0.06700	0.00484

Divergence

As with most numerical methods, the computation can diverge, with our answers getting further and further from the true values. The methods of this chapter can diverge if the step size Δx is too large, or the range of x over which we compute is too large. Numerical methods can also run into trouble when computing close to a discontinuity, a maximum or minimum point, or a point of inflection.

When using these methods on the computer (the only practical way) it is easy to change the step size. A good practice is to reduce the step size by a factor of 10 each time (1, 0.1, 0.01, . . .) until the answers do not change in the decimal place of interest. Further reduction of the step size will eventually result in numbers too small for the computer to handle, and strange results will occur.

EXERCISE 1

Solve each differential equation and find the value of y asked for. Use either Euler's method, the modified Euler method, or the Runge–Kutta method, with a step size of 0.1.

1. $y' + xy = 3$; $y(4) = 2$. Find $y(5)$.
2. $y' + 5y^2 = 3x$; $y(0) = 3$. Find $y(1)$.
3. $x^4y' + ye^x = xy^3$; $y(2) = 2$. Find $y(3)$.
4. $xy' + y^2 = 3 \ln x$; $y(0) = 3$. Find $y(1)$.
5. $1.94x^2y' + 8.23x \sin y - 2.99y = 0$; $y(0) = 3$. Find $y(1)$.
6. $y' \ln x + 3.99(x - y)^2 = 4.28y$; $y(1) = 1.76$. Find $y(2)$.
7. $(x + y)y' + 2.76x^2y^2 = 5.33y$; $y(0) = 1$. Find $y(1)$.
8. $y' + 297xye^x + 192x \cos y = 0$; $y(2) = 188$. Find $y(3)$.
9. $xy' + xy^2 = 8y \ln |xy|$; $y(1) = 3$. Find $y(2)$.
10. $x^2y' + 88.5x^2 \ln y + 32.7x = 0$; $y(1) = 23.9$. Find $y(2)$.

Computer

11. Write a program or use a spreadsheet to solve a first-order differential equation by Euler's method. Try it on any of the equations above.

12. Write a program or use a spreadsheet to solve a first-order differential equation by the modified Euler's method. Try it on any of the equations above.

13. Write a program or use a spreadsheet to solve a first-order differential equation by the Runge–Kutta method. Try it on any of the equations above.

14. If you have done one of the calculations above using a spreadsheet which has a graphing utility, use that utility to make a graph of x versus y over the given range.

15-2. SECOND-ORDER EQUATIONS

To solve a second-order differential equation numerically, we first transform the given equation into two first-order equations by substituting another variable for y'.

EXAMPLE 1: By making the substitution $y' = m$, we can transform the second-order equation

$$y'' + 3y' + 4xy = 0$$

into the two first-order equations

$$y' = m$$

and

$$\begin{aligned} m' &= -3m - 4xy \\ &= f(x, y, m) \end{aligned}$$

As before, we start with the given point $P(x_p, y_p)$, at which we must also know the slope m_p. We proceed from P to the next point $Q(x_q, y_q)$ as follows.

1. Compute the average first derivative, m_{avg}, over the interval Δx.
2. Compute the average second derivative m'_{avg} over that interval.
3. Find the coordinates x_q and y_q at Q by using the equations

$$x_q = x_p + \Delta x$$

and

$$y_q = y_p + m_{\text{avg}}\, \Delta x$$

4. Find the first derivative m_q at Q by using

$$m_q = m_p + m'_{avg}\Delta x$$

5. Go to step 1 and repeat for the next step.

We can find m_{avg} and m'_{avg} in steps 1 and 2 by the modified Euler's method, or by the Runge–Kutta method, which, as before, requires us to find values at the intermediate points R and S. The equations are similar to those we had for first-order equations

At P:	x_p, y_p, and m_p are given, $m'_p = f(x_p, y_p, m_p)$.	
At R:	$x_r = x_p + \frac{\Delta x}{2}$	$y_r = y_p + m_p\frac{\Delta x}{2}$
	$m_r = m_p + m'_p\frac{\Delta x}{2}$	$m'_r = f(x_r, y_r, m_r)$
At S:	$x_s = x_p + \frac{\Delta x}{2}$	$y_s = y_p + m_r\frac{\Delta x}{2}$
	$m_s = m_p + m'_r\frac{\Delta x}{2}$	$m'_s = f(x_s, y_s, m_s)$
At Q:	$x_q = x_p + \Delta x$	$y_q = y_p + m_s\Delta x$
	$m_q = m_p + m'_s\Delta x$	$m'_q = f(x_q, y_q, m_q)$

EXAMPLE 2: Find an approximate solution to the equation

$$y'' - 2y' + y = 2e^x$$

with boundary conditions $y' = 2e$ at (1, 0). Find points from 1 to 3 in steps of 0.1.

Solution: Replacing y' by m and solving for m' gives

$$m' = f(x, y, m)$$
$$= 2e^x - y + 2m$$

At P: The boundary values are $x_p = 1$, $y_p = 0$, and $m_p = 5.4366$. Then

$$m'_p = f(x_p, y_p, m_p) = 2e^1 - 0 + 2(5.4366) = 16.3097$$

At R:

$$x_r = x_p + \frac{\Delta x}{2} = 1 + 0.05 = 1.05$$

$$y_r = y_p + m_p\frac{\Delta x}{2} = 0 + 5.4366(0.05) = 0.2718$$

$$m_r = m_p + m'_p\frac{\Delta x}{2} = 5.4366 + 16.3097(0.05) = 6.2521$$

$$m'_r = 2e^{1.05} - 0.2718 + 2(6.2521) = 17.9477$$

At S: $x_s = 1.05$

$$y_s = y_p + m_r \frac{\Delta x}{2} = 0 + 6.2521(0.05) = 0.3126$$

$$m_s = m_p + m_r' \frac{\Delta x}{2} = 5.4366 + 17.9477(0.05) = 6.3360$$

$$m_s' = 2e^{1.05} - 0.3126 + 2(6.3360) = 18.0707$$

At Q: $x_q = x_p + \Delta x = 1.1$

$$y_q = y_p + m_s \Delta x = 0 + 6.3360(0.1) = 0.6336$$

$$m_q = m_p + m_s' \Delta x = 5.4366 + 18.0707(0.1) = 7.2437$$

$$m_q' = 2e^{1.1} - 0.6336 + 2(7.2437) = 19.8621$$

Then

$$m_{\text{avg}} = \frac{m_p + 2m_r + 2m_s + m_q}{6}$$

$$= \frac{5.4366 + 2(6.2521) + 2(6.3360) + 7.2437}{6} = 6.3094$$

and

$$m_{\text{avg}}' = \frac{(m_p' + 2m_r' + 2m_s' + m_q'}{6}$$

$$= \frac{16.3097 + 2(17.9477) + 2(18.0707) + 19.8621}{6} = 18.0348$$

So

$$y_q = y_p + m_{\text{avg}} \Delta x = 0 + 6.3094(0.1) = 0.63094$$

and

$$m_q = m_p + m_{\text{avg}}' \Delta x = 5.4366 + 18.0348(0.1) = 7.2401$$

The computation is then repeated. The remaining values are given in the following table, along with the exact y and exact slope, found by analytical solution of the given equation.

x	y	*Exact y*	*Slope*	*Exact slope*
1.0	0.000	0.000	5.437	5.437
1.1	0.631	0.631	7.240	7.240
1.2	1.461	1.461	9.429	9.429
1.3	2.532	2.532	12.072	12.072
1.4	3.893	3.893	15.248	15.248
1.5	5.602	5.602	19.047	19.047
1.6	7.727	7.727	23.576	23.576
1.7	10.346	10.346	28.957	28.957
1.8	13.551	13.551	35.330	35.330
1.9	17.450	17.450	42.856	42.857
2.0	22.167	22.167	51.723	51.723
2.1	27.846	27.847	62.144	62.145
2.2	34.656	34.656	74.366	74.366
2.3	42.789	42.789	88.670	88.670
2.4	52.470	52.470	105.381	105.382
2.5	63.958	63.958	124.870	124.870
2.6	77.551	77.551	147.562	147.562
2.7	93.593	93.593	173.943	173.944
2.8	112.480	112.481	204.570	204.571
2.9	134.669	134.670	240.079	240.080
3.0	160.683	160.684	281.196	281.197

EXERCISE 2

Solve each equation by the Runge–Kutta method and find the value of y at $x = 2$. Take a step size of 0.1 unit.

1. $y'' + y' + xy = 3,\qquad y'(1, 1) = 2$

2. $y'' - xy' + 3xy^2 = 2y,\qquad y'(1, 1) = 2$

3. $3y'' + y' + ye^x = xy^3,\qquad y'(1, 2) = 2$

4. $2y'' - xy' + y^2 = 5 \ln x,\qquad y'(1, 0) = 1$

5. $xy'' - 2.4x^2y' + 5.3x \sin y - 7.4y = 0,\qquad y'(1, 1) = 0.1$

6. $y'' + y' \ln x + 2.69(x - y)^2 = 6.26y,\qquad y'(1, 3.27) = 8.26$

7. $x^2y'' - (x + y)y' + 8.26x^2y^2 = 1.83y,\qquad y'(1, 1) = 2$

8. $y'' - y' + 183xye^x + 826x \cos y = 0,\qquad y'(1, 73.4) = 273$

9. $y'' - xy' + 59.2xy = 74.1y \ln xy,\qquad y'(1, 1) = 2$

10. $xy'' - x^2y' + 11.4x^2 \ln y + 62.2x = 0,\qquad y'(1, 8) = 52.1$

Computer

11. Write a program or use a spreadsheet to solve a second-order differential equation by the Runge–Kutta method. Try it on any of the equations above. If your spreadsheet has a graphing utility, use it to graph x versus y over the given range.

CHAPTER TEST

Solve each differential equation. Use any method with a step size of 0.1 unit.

1. $y' + xy = 4y;\ y(2) = 2.$ Find $y(3)$.
2. $xy' + 3y^2 = 5xy;\ y(1) = 5.$ Find $y(2)$.
3. $3y'' - y' + xy^2 = 2 \ln y;\ y'(1, 2) = 8.$ Find $y(2)$.
4. $xy'' - 7.2y^2y' + 2.8x \sin x - 2.2y = 0;\ y'(1, 5) = 8.2.$ Find $y(2)$.
5. $y'' + y' \ln y + 5.27(x - 2y)^2 = 2.84x;\ y'(1, 1) = 3.$ Find $y(2)$.
6. $x^2y'' - (y^2 + y)y' + 3.67x^2 = 5.28x;\ y'(1, 2) = 3.$ Find $y(2)$.
7. $y'' - xy' + 6e^x + 126 \cos y = 0;\ y'(1, 2) = 5.$ Find $y(2)$.
8. $y' + xy^2 = 8 \ln y;\ y(1) = 7.$ Find $y(2)$.
9. $7.35y' + 2.85x \sin y - 7.34x = 0;\ y(2) = 4.3.$ Find $y(3)$.
10. $xy'' - y' + 77.2y^2 = 28.4x \ln xy;\ y'(3, 5) = 3.$ Find $y(4)$.

Infinite Series

To find the sine of an angle or the logarithm of a number, you would probably use a calculator, computer, or a table. But where did the table come from? And how can the chips in a calculator or computer find sines or logs when all they can do are the four arithmetic operations of addition, subtraction, multiplication, and division?

We can, in fact, find the sine of an angle x (in radians) from a *polynomial,* such as

$$\sin x \cong x - \frac{x^3}{6} + \frac{x^5}{120}$$

and the natural logarithm from

$$\ln x \cong (x - 1) - \frac{(x - 1)^2}{2} + \frac{2(x - 1)^3}{6}$$

But where did these formulas come from? How accurate are they, and can the accuracy be improved? Are they good for all values of x, or for just a small range of x? Can similar formulas be derived for other functions, such as e^x? In this chapter we try to answer such questions.

The technique of representing a function by a polynomial is called *polynomial approximation,* and each of these formulas is really the first several terms of an infinitely long series. We'll learn how to write these series, and those for other functions, and how to estimate and improve the accuracy of computation.

Why bother with infinite series when we can get these functions so easily from calculator or computer? First, we may want to understand how a computer finds certain functions, even if we may never have to design the logic to do it. Perhaps more important, many numerical methods rely on polynomial approximation of functions. Finally, some limits, derivatives, or integrals are best found by this method.

We first consider series in general, then write polynomial approximations for some common functions and use them for computation, and finally look at other uses for series.

16-1. SEQUENCES AND SERIES

Sequence

A *sequence*

$$u_1, u_2, u_3, \ . \ . \ . \ u_n$$

is a set of quantities, called *terms,* arranged in the same order as the positive integers.

EXAMPLE 1

(a) The sequence $1, \frac{1}{2}, \frac{1}{3}, \frac{1}{4}, \frac{1}{5}, \ldots, \frac{1}{n}$ is called an *finite* sequence, because it has a finite number of terms (n).

(b) The sequence $1, \frac{1}{2}, \frac{1}{3}, \frac{1}{4}, \frac{1}{5}, \ldots, \frac{1}{n}, \ldots$ is called an *infinite* sequence. The three dots at the end indicate that the sequence continues indefinitely.

(c) The sequence 3, 7, 11, 15, . . . is called an *arithmetic* sequence, or *arithmetic progression,* because each term is equal to the sum of the preceding term and a constant. That constant (4, in this example), is called the *common difference.*

(d) The sequence 2, 6, 18, 54, . . . is called a *geometric* sequence, or *geometric progression,* because the ratio of a term and the one preceding it is a constant. That constant (3 in this example) is called the *common ratio.*

Series

A *series*

$u_1 + u_2 + u_3 + \cdots + u_n + \cdots$	**402**

is the indicated sum of a sequence.

EXAMPLE 2

(a) The series $1 + \frac{1}{2} + \frac{1}{3} + \frac{1}{4} + \frac{1}{5}$ is called a *finite* series. It is called a *positive* series because all its terms are positive.

(b) The series $x - x^2 + x^3 - \cdots x^n \cdots$ is an *infinite* series. It is also called an *alternating* series because the terms alternate in sign.

(c) The series $6 + 9 + 12 + 15 + \cdots$ is an infinite arithmetic series.

(d) $1 - 2 + 4 - 8 + 16 - \cdots$ is an infinite, alternating, geometric series.

Series of Constants and Series of Variables

The terms of a series may be either constants or variables. Thus Example 2(a), (c), and (d) are series of constant terms, while Example 2(b) is a series of variable terms. We will first cover series of constants, and then, in Sec. 16-2, study series of variables.

General Term

The *general term* u_n, in a series such as Eq. 402, is an expression involving n (where $n = 1, 2, 3, \ldots$) by which we can obtain any term of the series.

EXAMPLE 3: If the general term of a certain series is

$$u_n = \frac{n}{2n + 1}$$

find the first three terms of the series.

Solution: We have

$$u_1 = \frac{1}{2(1) + 1}, \qquad u_2 = \frac{2}{2(2) + 1}, \qquad u_3 = \frac{3}{2(3) + 1}$$

so our series is

$$\frac{1}{3} + \frac{2}{5} + \frac{3}{7} + \cdots + \frac{n}{2n + 1} + \cdots$$

Recursion Relations

We can find the terms of a series if we have an expression for the *n*th term. There are also series where each term is found from one or more immediately preceding terms. The relationship between a term and those preceding it is called a *recursion relation*.

EXAMPLE 4: Each term in the series

$$2 + 7 + 22 + 67 + \cdots$$

is found by multiplying the preceding term by 3 and adding 1. The recursion relation is then

$$u_{n+1} = 3u_n + 1$$

EXAMPLE 5: The recursion relation for the Fibonacci sequence

$$1, 1, 2, 3, 5, 8, 13, 21, 34, \ldots$$

is $u_{n+2} = u_n + u_{n+1}$.

Convergence and Divergence

In many applications, we first replace a given function by an infinite series, and then do any required operations on the series rather than the original function. But we cannot work with an infinite number of terms, nor do we want to work with many terms of the series. For practical applications, we want to represent our original function *by only the first few terms* of an infinite series. Thus we need a series in which the terms decrease in magnitude rapidly, and for which the sum of the first several terms is not too different from the sum of all the terms of the series. Such a series is said to *converge* upon some limit. A series that does not converge is said to *diverge*.

The Terms Must Approach Zero

For an infinite series to converge, the terms must get smaller as n increases, and approach zero as n becomes infinite. In symbols, if u_n is the general term of a series,

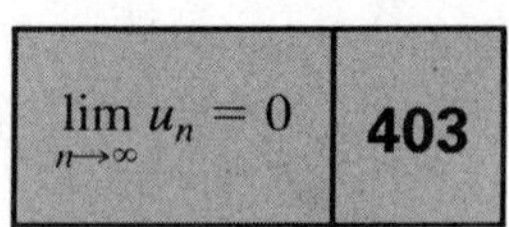

If the terms of an infinite series do not approach zero as n approaches infinity, the series diverges. However, if the terms do approach zero, we cannot say for sure that the series converges.

EXAMPLE 6: The infinite series

$$1 + \frac{2}{3} + \frac{3}{5} + \cdots + \frac{n}{2n-1} + \cdots$$

You may want to review limits (Sec. 3-1) before going further.

is divergent, because

$$\lim_{n\to\infty} \frac{n}{2n-1} = \lim_{n\to\infty} \frac{1}{2 - 1/n} = \frac{1}{2} \neq 0$$

EXAMPLE 7: The series

$$1 + \frac{1}{2} + \frac{1}{3} + \cdots + \frac{1}{n} + \cdots$$

has terms that approach zero as n approaches infinity. However, we cannot yet tell whether this series (called a *harmonic series*) converges. In fact, it does *not* converge, as we'll see in the following section.

Sum of an Infinite Series

Let S_n stand for the sum of the first n terms of an infinite series. Thus for the series

$$u_1 + u_2 + u_3 + \cdots + u_n + \cdots$$

we have

$$S_1 = u_1, \qquad S_2 = u_1 + u_2, \qquad S_3 = u_1 + u_2 + u_3, \qquad \text{etc.}$$

These are called *partial sums*. The infinite sequence of partial sums, $S_1, S_2, S_3, \ldots$ may or may not have a definite limit S as n becomes infinite. If a limit S exists, we call it the *sum of the series*.

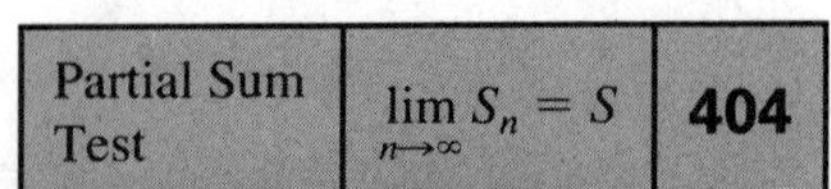

If an infinite series has a sum S, the series converges. Otherwise, the series diverges.

It is not hard to see that a *finite* series will have a finite sum. But is it possible that the sum of an infinite number of terms can be finite? Yes, as shown in the next example.

EXAMPLE 8: If the sum of the first n terms of a geometric series with common ratio r is given by $S_n = a(1 - r^n)/(1 - r)$, does the following geometric series converge?

$$1 + \frac{1}{2} + \frac{1}{2^2} + \frac{1}{2^3} + \cdots + \frac{1}{2^{n-1}} + \cdots$$

Solution: The sum of n terms of this series is

$$S_n = \frac{a(1 - r^n)}{1 - r} = \frac{1 - (\frac{1}{2})^n}{1 - \frac{1}{2}} = 2 - \frac{1}{2^{n-1}}$$

Since this limit of this sum is 2 as n approaches infinity, the given series is convergent.

Series by Computer

In Example 8 we were given an equation for the sum of the series to be tested. But how do we apply the partial sum test if we cannot find an expression for the sum? Let us generate the terms of a series on the computer and calculate the partial sum as we go. This will not *prove* convergence or divergence, since we cannot produce an infinite number of terms, but we can compute enough to give a good idea as to how a series is behaving.

EXAMPLE 9: Table 16-1 shows a computer printout of 80 terms of the series

$$\frac{1}{e} + \frac{2}{e^2} + \frac{3}{e^3} + \cdots + \frac{n}{e^n} + \cdots$$

Also shown for each term u_n is the partial sum S_n and the ratio of the term to the one preceding.

TABLE 16-1

n	*Term* u_n	*Sum* S_n	*Ratio*
1	0.367879	0.367879	
2	0.270671	0.638550	0.735759
3	0.149361	0.787911	0.551819
4	0.073263	0.861174	0.490506
5	0.033690	0.894864	0.459849
6	0.014873	0.909736	0.441455
7	0.006383	0.916119	0.429193
8	0.002684	0.918803	0.420434
9	0.001111	0.919914	0.413864
10	0.000454	0.920368	0.408755
.	.	.	.
.	.	.	.
.	.	.	.
.	.	.	.
75	0.000000	0.920674	0.372851
76	0.000000	0.920674	0.372785
77	0.000000	0.920674	0.372720
78	0.000000	0.920674	0.372657
79	0.000000	0.920674	0.372596
80	0.000000	0.920674	0.372536

We notice several things about this series. First, the terms u_n approach zero, but this alone does not tell us if the series converges. Second, the partial sums S_n approach a limit (0.920674), which shows that the series does in fact converge. Third, we note that the ratio of two successive terms approaches a limit that is less than 1. More about this later.

Our next computer example is for the harmonic series encountered earlier. Recall that the terms approached zero as n got larger, so that we could not tell then if the series converged.

EXAMPLE 10: Table 16-2 shows a computer run for the harmonic series $1 + \frac{1}{2} + \frac{1}{3} + \cdots + \frac{1}{n} + \cdots$.

TABLE 16-2

n	*Term* u_n	*Sum* S_n	*Ratio*
1	1.000000	1.000000	
2	0.500000	1.500000	0.500000
3	0.333333	1.833333	0.666667
4	0.250000	2.083334	0.750000
5	0.200000	2.283334	0.800000
6	0.166667	2.450000	0.833333
7	0.142857	2.592857	0.857143
8	0.125000	2.717857	0.875000
9	0.111111	2.828969	0.888889
10	0.100000	2.928969	0.900000
.	.	.	.
.	.	.	.
.	.	.	.
75	0.013333	4.901356	0.986667
76	0.013158	4.914514	0.986842
77	0.012987	4.927501	0.987013
78	0.012821	4.940322	0.987180
79	0.012658	4.952980	0.987342
80	0.012500	4.965480	0.987500
.	.	.	.
.	.	.	.
.	.	.	.
495	0.002020	6.782784	0.997980
496	0.002016	6.784800	0.997984
497	0.002012	6.786813	0.997988
498	0.002008	6.788821	0.997992
499	0.002004	6.790825	0.997996
500	0.002000	6.792825	0.998000

For the harmonic series we see that the terms u_n approach zero, as we had found earlier. However, the partial sum S_n does not approach a limit, as did our other series, but continues to grow even after 495 terms. Therefore, this series appears to diverge. We also note that the ratio of two successive terms is approaching a limit of 1.

Ratio Test

The series in our first computer example was seen to converge. For this series we saw that as n increased, the ratio of any term to the following term approached a limit less than 1. It can be proven that this is always so, and also that if the limit is greater than 1, the series diverges. However a limit of 1, as in Example 10, gives no information.

Also called the *Cauchy ratio test*. There are many tests for convergence which we cannot give here, but the ratio test will be the most useful for our later work.

Ratio Test	If $\lim\limits_{n\to\infty}\left\lvert\dfrac{u_{n+1}}{u_n}\right\rvert$ (a) is less than 1, the series converges. (b) is greater than 1, the series diverges. (c) is equal to 1, the test fails.	**405**

EXAMPLE 11: Use the ratio test on the series

$$\frac{1}{e}+\frac{2}{e^2}+\frac{3}{e^3}+\cdots+\frac{n}{e^n}+\cdots.$$

Solution: By the ratio test,

$$\frac{u_{n+1}}{u_n}=\frac{n+1}{e^{n+1}}\div\frac{n}{e^n}=\frac{n+1}{e^{n+1}}\cdot\frac{e^n}{n}=\frac{n+1}{ne}=\frac{1+1/n}{e}$$

so

$$\lim_{n\to\infty}\left\lvert\frac{u_{n-1}}{u_n}\right\rvert=\frac{1}{e}$$

Since $1/e$ is less than zero, the series is convergent.

EXERCISE 1

Write the first five terms of each series, given the general term.

1. $u_n = 3n$

2. $u_n = 2n + 3$

3. $u_n = \dfrac{n+1}{n^2}$

4. $u_n = \dfrac{2^n}{n}$

Deduce the general term of each series. Use it to predict the next two terms.

5. $2 + 4 + 6 + \cdots$

6. $1 + 8 + 27 + \cdots$

7. $\dfrac{2}{4}+\dfrac{4}{5}+\dfrac{8}{6}+\dfrac{16}{7}+\cdots$

8. $2 + 5 + 10 + \cdots$

Deduce a recursion relation for each series. Use it to predict the next two terms.

9. $1 + 5 + 9 + \cdots$

10. $\frac{1}{2} + \frac{3}{5} + \frac{5}{8} + \frac{7}{11} + \cdots$

11. $3 + 9 + 27 + 81 + \cdots$

12. $5 + 8 + 14 + 26 + \cdots$

Use the ratio test to determine if each series converges or diverges.

13. $1 + \frac{1}{2!} + \frac{2}{3!} + \frac{3}{4!} + \cdots$

14. $1 + \frac{3}{2!} + \frac{5}{3!} + \frac{7}{4!} + \cdots$

15. $1 + \frac{9}{8} + \frac{27}{16} + \frac{81}{128} + \cdots + \frac{3n}{n \cdot 2^n} + \cdots$

16. $3 - \frac{3^3}{3 \cdot 1!} + \frac{3^5}{5 \cdot 2!} - \frac{3^7}{7 \cdot 3!} + \cdots$

17. $1 + \frac{1}{2} + \frac{1}{3} + \frac{1}{4} + \cdots$

18. $1 + \frac{4}{7} + \frac{9}{49} + \frac{16}{343} + \cdots$

Computer

19. Write a program or use a spreadsheet to generate the terms of a series, given the general term or a recursion formula. Have the program compute and print each term, the partial sum, and the ratio of that term to the preceding one.

16-2. MACLAURIN'S SERIES

Power Series

An infinite series of the form

$$c_0 + c_1x + c_2x^2 + \cdots + c_nx^n + \cdots$$

in which the c's are constants and x is a variable, is called a *power series* in x. Note that this series contains *variables,* while all our preceding examples contained constants.

EXAMPLE 1: The series $x - \frac{x^2}{2} + \frac{x^3}{3} - \cdots + (-1)^{n-1}\frac{x^n}{n} + \cdots$ is a power series.

Another way of writing a power series is

$$c_0 + c_1(x - a) + c_2(x - a)^2 + \cdots + c_n(x - a)^n + \cdots$$

where a and the c's are constants.

Interval of Convergence

The range of x for which a particular power series converges is called the *interval of convergence*. We find this interval by means of the ratio test.

It is usual to test for convergence at the endpoints of the interval as well, but this requires tests that we have not covered.

EXAMPLE 2: Find the interval of convergence of the series given in Example 1.

Solution: Applying the ratio test,

$$\lim_{n\to\infty}\left|\frac{u_{n+1}}{u_n}\right| = \lim_{n\to\infty}\left|\frac{x^{n+1}}{n+1}\left(\frac{n}{x_n}\right)\right| = \lim_{n\to\infty}\left|\frac{n}{n+1}(x)\right| = |x|$$

Thus the series converges when x lies between 1 and -1 and diverges when x is greater than 1 or less than -1. Our interval of convergence is thus $-1 < x < 1$.

Representing a Function by a Series

We know that it is possible to represent certain functions by an infinite power series. Take the function

$$f(x) = \frac{1}{1+x}$$

By ordinary division (try it) we get the infinite power series

$$(1+x)^{-1} = 1 - x + x^2 - x^3 + \cdots$$

which can be shown to converge when x is less than 1. But what about other functions? Can we write a power series to represent any function? The answer is no, but many useful functions can be represented by a power series, as shown in the following section.

We require, of course, that the function being represented is continuous and differentiable over the interval of interest. In Chapter 17 we'll use Fourier series to represent discontinuous functions.

Maclaurin's Series

Let us assume that a certain function $f(x)$ can be represented by an infinite power series, so that

$$f(x) = c_o + c_1x + c_2x^2 + c_3x^3 + \cdots + c_{n-1}x^{n-1} + \cdots \qquad (1)$$

We now evaluate the constants in this equation. We get c_0 by setting x to zero. Thus

$$f(0) = c_0$$

Next, taking the derivative of (1) gives

$$f'(x) = c_1 + 2c_2x + 3c_3x^2 + 4c_4x^3 + \cdots$$

from which we find c_1.

$$f'(0) = c_1$$

We continue taking derivatives and evaluating each at $x = 0$. Thus

$$f''(x) = 2c_2 + 3(2)c_3x + 4(3)c_4c^2 + \cdots$$

$$f''(0) = 2c_2$$

$$f'''(x) = 3(2)c_3 + 4(3)\,(2)c_4x + \cdots$$

$$f'''(0) = 3(2)c_3$$

and so on. The constants in (1) are thus

$$c_0 = f(0), \qquad c_1 = f'(0), \qquad c_2 = \frac{f''(0)}{2}, \qquad c_3 = \frac{f'''(0)}{3(2)}$$

and, in general,

$$c_n = \frac{f^{(n)}(0)}{n!}$$

Substituting the c's back into (1) gives

Named for a Scottish mathematician and physicist, Colin Maclaurin (1698–1746).

Maclaurin's Series	$f(x) = f(0) + f'(0)x + \frac{f''(0)}{2!}x^2 + \cdots + \frac{f^{(n)}(0)}{n!}x^n + \cdots$	**406**

We now apply this to the two functions mentioned in the introduction to this chapter, the sine and the natural logarithm.

EXAMPLE 3: Write a Maclaurin's series for the function

$$f(x) = \sin x$$

Solution: We take successive derivatives and evaluate each at $x = 0$.

$$\begin{aligned} f(x) &= \sin x & f(0) &= 0 \\ f'(x) &= \cos x & f'(0) &= 1 \\ f''(x) &= -\sin x & f''(0) &= 0 \\ f'''(x) &= -\cos x & f'''(0) &= -1 \\ f^{(iv)}(x) &= \sin x & f^{(iv)}(0) &= 0 \\ &\text{etc.} & &\text{etc.} \end{aligned}$$

Substituting into Eq. 406,

$$\sin x = x - \frac{x^3}{3!} + \frac{x^5}{5!} - \frac{x^7}{7!} + \cdots \tag{210}$$

Now we turn to the natural logarithm.

EXAMPLE 4: Write a Maclaurin's series for $f(x) = \ln x$.

Solution: We proceed as before,

$$f(x) = \ln x; \qquad f(0) \text{ is undefined!}$$

This shows that in order to write a Maclaurin's series for a function, *the function and all its derivatives must exist at* $x = 0$. We'll write an expression for ln x in the next section, using a Taylor's series.

Computing with Maclaurin Series

We find the value of a function simply by substituting into the series representation of that function.

EXAMPLE 5: Find the sine of 0.5 rad using the first three terms of the Maclaurin's series. Work to seven decimal places.

Solution: Substituting into Eq. 210,

$$\sin 0.5 = 0.5 - \frac{(0.5)^3}{6} + \frac{(0.5)^5}{120}$$

$$= 0.5 - 0.0208333 + 0.0002604 = 0.4794271$$

By calculator (which we assume to be correct to seven decimal places) the sine of 0.5 rad is 0.4794255. Our answer is thus high by 0.0000016, or about 0.0003%.

The error caused by discarding terms of a series is called the *truncation error.*

Accuracy

In Example 5 we could estimate the truncation error by finding the exact value by calculator. But what if we were designing a computer to calculate sines, and must know *in advance* what the accuracy is?

Because the sine function is periodic, we need only calculate sines of angles up to 90°. Sines of larger angles can be found from these.

Table 16-3 shows the sines of the angles from 0 to 90°, computed using one, two, three, and four terms of the Maclaurin series (Eq. 406). In the last column is the sine, given by the computer, which we assume to be accurate to the number of places given. Notice that the sines obtained by series differ from the computer values. The percent error of each absolute value is given in Table 16-4 and plotted in Fig. 16-1.

TABLE 16-3 Sin x Calculated by Maclaurin Series

Angle	*1 Term*	*2 Terms*	*3 Terms*	*4 Terms*	*By computer*
0	0.000000	0.000000	0.000000	0.000000	0.000000
10	0.174533	0.173647	0.173648	0.173648	0.173648
20	0.349066	0.341977	0.342020	0.342020	0.342020
30	0.523599	0.499674	0.500002	0.500000	0.500000
40	0.698132	0.641422	0.642804	0.642788	0.642788
50	0.872665	0.761903	0.766120	0.766044	0.766045
60	1.047198	0.855801	0.866295	0.866021	0.866025
70	1.221731	0.917800	0.940482	0.939676	0.939693
80	1.396263	0.942582	0.986806	0.984753	0.984808
90	1.570796	0.924832	1.004525	0.999843	1.000000

Note that the error decreases rapidly as more terms of the series are used. Also note that the series gives the exact answer at $x = 0$ regardless of the number of terms, and that the error increases as we go farther from zero. With a Maclaurin's series, we say that the function is expanded about $x = 0$. In the next section we'll be able to expand a function about any point $x = a$ using a Taylor's series, and that the best accuracy, as here, will be near that point. In Sec. 16-4 we'll show how to estimate the error in the answer.

TABLE 16-4 Percent Error

Angle	*1 Term*	*2 Terms*	*3 Terms*	*4 Terms*
0	0.0000	0.0000	0.0000	0.0000
10	0.5095	0.0008	0.0000	0.0000
20	2.0600	0.0126	0.0000	0.0000
30	4.7198	0.0652	0.0004	0.0000
40	8.6100	0.2125	0.0025	0.0000
50	13.9183	0.5407	0.0099	0.0001
60	20.9200	1.1806	0.0312	0.0005
70	30.0138	2.3298	0.0840	0.0018
80	41.7803	4.2877	0.2029	0.0056
90	57.0796	7.5158	0.4525	0.0157

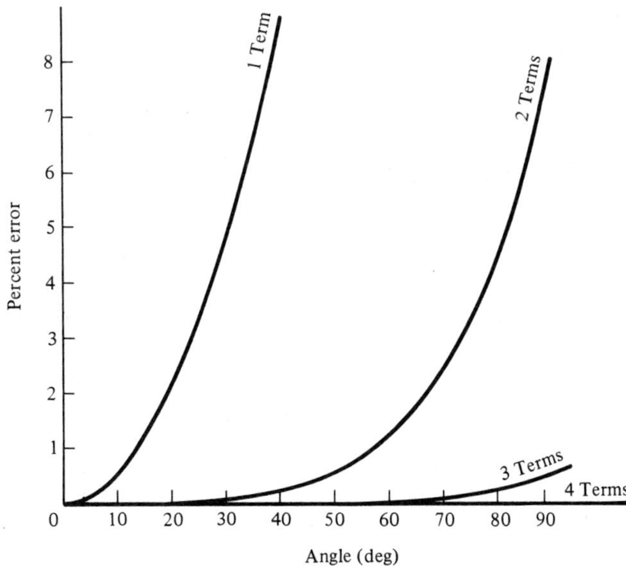

FIGURE 16-1. Percent error in the sine of an angle, using one, two, three, and four terms of a Maclaurin series.

EXERCISE 2

Use the ratio test to find the interval of convergence of each power series.

1. $1 + x + x^2 + \cdots + x^{n-1} + \cdots$
2. $1 + \dfrac{2x}{5} + \dfrac{3x^2}{5^2} + \cdots + \dfrac{nx^{n-1}}{5^{n-1}} + \cdots$
3. $x + 2!x^2 + 3!x^3 + \cdots + n!x^n + \cdots$
4. $2x + 4x^2 + 8x^3 + \cdots + 2^n x^n + \cdots$
5. $x - \dfrac{x^3}{3} + \dfrac{x^5}{5} - \dfrac{x^7}{7} + \cdots$
6. $\dfrac{x}{1!} + \dfrac{x^2}{2!} + \cdots + \dfrac{x^n}{n!} + \cdots$

Verify each Maclaurin expansion.

7. $\dfrac{1}{1+x} = 1 - x + x^2 - x^3 + \cdots$
8. $\sqrt{1+x} = 1 + \dfrac{x}{2} - \dfrac{x^2}{8} + \dfrac{x^3}{16} - \cdots$
9. $e^x = 1 + x + \dfrac{x^2}{2!} + \dfrac{x^3}{3!} + \cdots$ (Eq. 203)

10. $e^{-x} = 1 - x + \frac{x^2}{2!} - \frac{x^3}{3!} + \cdots$

11. $e^{-x^2} = 1 - x^2 + \frac{x^4}{2!} - \frac{x^6}{3!} + \cdots$

12. $xe^{2x} = x + 2x^2 + 2x^3 + \cdots$

13. $\ln(1 + x) = x - \frac{x^2}{2} + \frac{x^3}{3} - \frac{x^4}{4} + \cdots$

14. $\sin 2x = 2x - \frac{2^3x^3}{3!} + \frac{2^5x^5}{5!} - \frac{2^7x^7}{7!} + \cdots$

15. $\cos x = 1 - \frac{x^2}{2!} + \frac{x^4}{4!} - \frac{x^6}{6!} + \cdots$ (Eq. 211)

16. $\tan x = x + \frac{x^3}{3} + \frac{2x^5}{15} + \frac{17x^7}{315} + \cdots$

17. $\sec x = 1 + \frac{x^2}{2} + \frac{5x^4}{24} + \cdots$

18. $\cos x^2 = 1 - \frac{x^4}{2!} + \frac{x^8}{4!} - \frac{x^{12}}{6!} + \cdots$

19. $\sin^2 x = x^2 - \frac{x^4}{3} + \cdots$

20. $\ln \cos x = -\frac{x^2}{2} - \frac{x^4}{12} - \frac{x^6}{45} - \cdots$

Compute each number using three terms of the appropriate Maclaurin's series above.

21. $\sqrt{e}$ **22.** $\sin 1$ **23.** $\ln 1.2$

24. $\cos 8°$ **25.** $\sqrt{1.1}$ **26.** $\sec \frac{1}{4}$

16-3. TAYLOR'S SERIES

Writing a Taylor's Series

Named for Brook Taylor (1685–1731), an English analyst, geometer, and philosopher. Since the Maclaurin series is a special case of the Taylor's, they are both referred to as Taylor's series.

We have seen that certain functions, such as $f(x) = \ln x$, cannot be represented by a Maclaurin series. Further, a Maclaurin's series may converge too slowly to be useful. However, if we cannot represent a function by a power series in x (a Maclaurin series), we can often represent the function by a power series in $(x - a)$, called a *Taylor's series.*

Let us assume that a function $f(x)$ can be represented by a power series in $(x - a)$, where a is some constant.

$$f(x) = c_0 + c_1(x - a) + c_2(x - a)^2 + \cdots + c_n(x - a)^n + \cdots \quad (1)$$

As with the Maclaurin series, we take successive derivatives. We find the c's by evaluating each derivative at $x = a$.

$$f(x) = c_0 + c_1(x-a) + c_2(x-a)^2 + \cdots \qquad f(a) = c_0$$

$$f'(x) = c_1 + 2c_2(x-a) + \cdots \qquad f'(a) = c_1$$

$$f''(x) = 2c_2 + 3(2)c_3(x-a) + \cdots \qquad f''(a) = 2c_2$$

$$\vdots \qquad \qquad \vdots$$

$$f^{(n)}(x) = n!c_n + \cdots \qquad f^{(n)}(a) = n!c_n$$

The values of c obtained are substituted back into (1) and we get

Taylor's Series	$f(x) = f(a) + f'(a)(x-a) + \dfrac{f''(a)}{2!}(x-a)^2 + \cdots + \dfrac{f^{(n)}(a)}{n!}(x-a)^n + \cdots$	**407**

EXAMPLE 1: Write four terms of a Taylor's series for $f(x) = \ln x$, expanded about $a = 1$.

Solution: We take derivatives and evaluate each at $x = 1$.

$$f(x) = \ln x \qquad f(1) = 0$$

$$f'(x) = \frac{1}{x} \qquad f'(1) = 1$$

$$f''(x) = -x^{-2} \qquad f''(1) = -1$$

$$f'''(x) = 2x^{-3} \qquad f'''(1) = 2$$

$$f^{(iv)}(x) = -6x^{-4} \qquad f^{(iv)}(1) = -6$$

Substituting into Eq. 407 gives

$$\ln x = 0 + 1(x-1) + \frac{-1}{2}(x-1)^2 + \frac{2}{6}(x-1)^3 + \frac{-6}{24}(x-1)^4 + \cdots$$

Recall that we were not able to write a Maclaurin's series for ln x.

$$= (x-1) - \frac{(x-1)^2}{2} + \frac{(x-1)^3}{3} - \frac{(x-1)^4}{4} + \cdots$$

Computing with Taylor's Series

As with the Maclaurin's series, we simply substitute the required x into the proper Taylor's series. For best accuracy, we expand the Taylor's series about a point that is close to the value we are computing. Of course, a must be chosen such that $f(a)$ is known.

EXAMPLE 2: Evaluate ln 0.9 using four terms of a Taylor's series. Work to seven decimal places.

Solution: We choose $a = 1$, since 1 is close to 0.9 and ln 1 is known. Substituting into the series from Example 1, with $x = 0.9$,

$$x - 1 = 0.9 - 1 = -0.1$$

Then

$$\ln 0.9 = -0.1 - \frac{(-0.1)^2}{2} + \frac{(-0.1)^3}{3} - \frac{(-0.1)^4}{4} + \cdots$$

$$= -0.1 - \frac{0.01}{2} + \frac{-0.001}{3} - \frac{0.0001}{4} + \cdots$$

$$= -0.1 - 0.005 - 0.0003333 - 0.0000250 = -0.1053583$$

So $\ln 0.9 \cong -0.1053583$. By calculator, $\ln 0.9 = -0.1053605$, to seven decimal places, a difference of about 0.002% from our answer. In Sec. 16-4 we'll see how to estimate the error in our calculation without using the calculator value.

EXAMPLE 3: Calculate the sines of the angles from 0 to 90°, using two, three, and four terms of a Taylor's series expanded about 45°.

Solution: We take derivatives and evaluate each at $x = \pi/4$, remembering to work in radians,

$$f(x) = \sin x \qquad f\left(\frac{\pi}{4}\right) = \frac{\sqrt{2}}{2}$$

$$f'(x) = \cos x \qquad f'\left(\frac{\pi}{4}\right) = \frac{\sqrt{2}}{2}$$

$$f''(x) = -\sin x \qquad f''\left(\frac{\pi}{4}\right) = -\frac{\sqrt{2}}{2}$$

$$f'''(x) = -\cos x \qquad f'''\left(\frac{\pi}{4}\right. = -\frac{\sqrt{2}}{2}$$

Substituting into Eq. 407 gives

$$\sin x = \frac{\sqrt{2}}{2} + \frac{\sqrt{2}}{2}\left(x - \frac{\pi}{4}\right) - \frac{\sqrt{2}}{2(2!)}\left(x - \frac{\pi}{4}\right)^2 - \frac{\sqrt{2}}{2(3!)}\left(x - \frac{\pi}{4}\right)^3 \cdots$$

A computer calculation of sin x, using two, three, and four terms of this series, is given in table 16-5. The absolute value of the truncation error, as a percent, is shown in table 16-6 and plotted in Fig. 16-2.

TABLE 16-5 Sin *x* Calculated by Taylor's Series

Angle	*2 Terms*	*3 Terms*	*4 Terms*	*By computer*
0	0.151746	−0.066343	−0.009247	0.000000
10	0.275160	0.143229	0.170093	0.173648
20	0.398573	0.331262	0.341052	0.342020
30	0.521987	0.497755	0.499869	0.500000
40	0.645400	0.642708	0.642786	0.642788
50	0.768814	0.766121	0.766043	0.766045
60	0.892227	0.867995	0.865880	0.866025
70	1.015640	0.948329	0.938539	0.939693
80	1.139054	1.007123	0.980259	0.984808
90	1.262467	1.044378	0.987282	1.000000

TABLE 16-6. Percent Error

Angle	*2 Terms*	*3 Terms*	*4 Terms*
10	58.4582	17.5176	2.0473
20	16.5350	3.1456	0.2832
30	4.3973	0.4491	0.0262
40	0.4064	0.0124	0.0003
50	0.3615	0.0100	0.0002
60	3.0255	0.2274	0.0168
70	8.0822	0.9190	0.1228
80	15.6625	2.2659	0.4619
90	26.2467	4.4378	1.2718

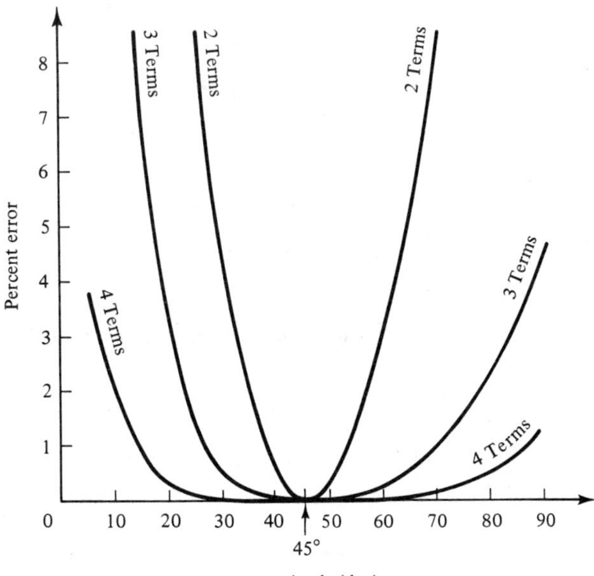

FIGURE 16-2. Error in sin x, computed with a Taylor's series expanded about 45°.

Note that we get the best accuracy near the point about which the series is expanded ($x = 45°$). Also note that at other values the Taylor series gives *less accurate* results than the Maclaurin's series with the same number of terms (see Fig. 16-1). That is because the Maclaurin's series converges much faster than the Taylor's.

EXERCISE 3

Verify each Taylor's series, expanded about the given value of a.

1. For $a = 1$, $\quad \dfrac{1}{x} = 1 - (x - 1) + (x - 1)^2 - (x - 1)^3 + \cdots$

2. For $a = \dfrac{\pi}{6}$, $\quad \sin x = \dfrac{1}{2} + \dfrac{\sqrt{3}}{2}\left(x - \dfrac{\pi}{6}\right) - \dfrac{1}{4}\left(x - \dfrac{\pi}{6}\right)^2 - \dfrac{\sqrt{3}}{12}\left(x - \dfrac{\pi}{6}\right)^3 + \cdots$

3. For $a = \dfrac{\pi}{4}$, $\quad \cos x = \dfrac{\sqrt{2}}{2}\left[1 - \left(x - \dfrac{\pi}{4}\right) - \dfrac{(x - \pi/4)^2}{2} + \cdots\right]$

4. For $a = \dfrac{\pi}{4}$, $\quad \tan x = 1 + 2\left(x - \dfrac{\pi}{4}\right) + 2\left(x - \dfrac{\pi}{4}\right)^2 + \cdots$

5. For $a = \dfrac{\pi}{3}$, $\quad \sec x = 2 + 2\sqrt{3}\left(x - \dfrac{\pi}{3}\right) + 7\left(x - \dfrac{\pi}{3}\right)^2 + \cdots$

6. For $a = 1$, $\quad \sqrt{x} = 1 + \dfrac{1}{2}(x - 1) - \dfrac{1}{8}(x - 1)^2 + \cdots$

7. For $a = -1$, $\quad \sqrt[3]{x} = -1 + \dfrac{x + 1}{3} + \dfrac{(x + 1)^2}{9} + \dfrac{5(x + 1)^3}{81} + \cdots$

8. For $a = 1$, $\quad e^x = e\left[1 + (x - 1) + \dfrac{(x - 1)^2}{2!} + \cdots\right]$

9. For $a = -1$, $\quad e^{-x} = e\left[1 - (x + 1) + \dfrac{(x + 1)^2}{2!} + \cdots\right]$

10. For $a = 1$, $\quad \ln x = (x - 1) - \dfrac{(x - 1)^2}{2!} + \dfrac{2(x - 1)^3}{3!} - \dfrac{6(x - 1)^4}{4!} + \cdots$

Compute each value using three terms of the appropriate Taylor's above. Compare your answer with that from a calculator.

11. $\dfrac{1}{1.1}$

12. $\sin 32°$

13. $\cos 46.5°$ (use $\cos 45° = 0.707107$)

14. $\tan 44.2°$

15. $\sec 61.4°$

16. $\sqrt{1.24}$

17. $\sqrt[3]{-0.8}$

18. $e^{1.1}$

19. $\ln(1.25)$

20. Calculate $\sin 65°$ using (a) three terms of a Maclaurin's series, (b) three terms of a Taylor's series expanded about $\pi/6$, and (c) a calculator. Compare the results of each.

16-4. ACCURACY OF COMPUTATION

We normally use only the first several terms of a series to represent a function. By discarding terms, we introduce an error into our computation. We now estimate such errors.

Alternating Series

A theorem by Leibniz, which we give without proof, says that in a convergent alternating series

$$u_1 - u_2 + u_3 - u_4 \cdots$$

the truncation error produced by discarding terms is smaller than the first term discarded.

EXAMPLE 1: What is the maximum error when computing sin 0.5 using only three terms of the Maclaurin's series (Eq. 210).

Solution: This series is alternating and convergent. The magnitude of the fourth term is $x^7/7!$ For $x = 0.5$ rad,

$$\text{maximum error} = \frac{x^7}{7!} = \frac{(0.5)^7}{5040} = 0.000002$$

When we computed sin 0.5 in Sec. 16-2 using three terms of the Maclaurin's series we got an error of 0.0000016, which agrees with this result. Further, since our first discarded term (the fourth) is negative, we would expect our computed answer to be too high, which it is.

Taylor's Series with Remainder

We can think of a Taylor's series as consisting of the first n terms plus a *remainder* R_n,

$$\text{exact } f(x) = \boxed{n \text{ terms}} + R_n$$

where the remainder R_n is given (without proof), by

There are several forms for the remainder. This one is called *Lagrange's form*, after Joseph Louis Lagrange (1736–1813).

Remainder after n Terms	$R_n = \dfrac{(x-a)^n}{n!} f^{(n)}(c)$ where c lies between a and x.	**408**

We cannot find R_n exactly because the number c is unknown, except that it lies between a and x. We can, however, choose c so that R_n is a maximum.

EXAMPLE 2: Estimate the maximum truncation error when using four terms of a Taylor's series at $a = 1$ to find ln (0.9).

Solution: We have already found that the sum of the first four terms is -0.1053583. Thus

$$\ln (0.9) = -0.1053583 + R_4$$

where

$$R_4 = \frac{(x-a)^4}{4!} f^{(\text{iv})}(c)$$

Here $x = 0.9$ and $a = 1$, so $x - a = 0.9 - 1 = -0.1$. Also, $4! = 24$, and the fourth derivative of $\ln x$ is $-6x^{-4}$, so $f^{(\text{iv})}(c) = -6/c^4$. Substituting,

$$R_4 = \frac{(-0.1)^4}{24}\left(\frac{-6}{c^4}\right) = -\frac{0.000025}{c^4}$$

Further, c must lie between a and x, or between 1 and 0.9.
The value of c within that interval that makes R_4 a maximum is 0.9, so

$$\text{maximum error} = -\frac{0.000025}{(0.9)^4} = -0.000038$$

We conclude that only five decimal places should be retained in our computed value of ln (0.9), giving $\ln (0.9) = -0.10536$. This is confirmed by calculator.

EXERCISE 4

Estimate the maximum truncation error for each computation.

1. Exercise 2, Problem 21

2. Exercise 2, Problem 22

3. Exercise 3, Problem 11

4. Exercise 3, Problem 12

5. Exercise 3, Problem 13

6. Exercise 3, Problem 16

7. Exercise 3, Problem 17

16-5. OPERATIONS WITH POWER SERIES

To expand a function in a power series, we can write a Maclaurin's or Taylor's series by the methods of Sec. 16-2 or 16-3. A faster way to get a new series is to modify or combine, when possible, series already derived.

Substitution

Given a series for $f(x)$, a new series can be obtained simply by substituting a different expression for x.

EXAMPLE 1: We get a series for $e^{\sin x}$ from the series

$$e^x = 1 + x + \frac{x^2}{2!} + \frac{x^3}{3!} + \cdots$$

by substituting $\sin x$ for x,

$$e^{\sin x} = 1 + \sin x + \frac{\sin^2 x}{2!} + \frac{\sin^3 x}{3!} + \cdots$$

Addition and Subtraction

Two power series may be added or subtracted, term by term, for those values of x for which both series converge.

EXAMPLE 2: Adding the series

$$\ln(1 + x) = x - \frac{x^2}{2} + \frac{x^3}{3} - \frac{x^4}{4} + \cdots$$

and

$$\ln(1 - x) = -x - \frac{x^2}{2} - \frac{x^3}{2} - \frac{x^4}{4} - \cdots$$

gives

$$\ln(1 - x^2) = -x^2 - \frac{x^4}{2} - \frac{x^6}{3} - \frac{x^8}{4} - \cdots$$

since $\ln(1 + x) + \ln(1 - x) = \ln[(1 + x)(1 - x)] = \ln(1 - x^2)$.

Multiplication and Division

A series can also be multiplied by a constant or by a polynomial. Thus a series for $x \ln x$ can be found by multiplying the series for $\ln x$ by x.

One power series may be multiplied or divided by another, term by term, for those values of x for which both converge, provided that division by zero does not occur.

EXAMPLE 3: Multiplying the series for e^x and $\sin x$ (Eqs. 203 and 210) gives

$$e^x \sin x = x - \frac{x^4}{3!} + \frac{x^7}{2!\,5!} - \frac{x^{10}}{3!\,7!} + \cdots$$

Differentiation and Integration

A power series may be differentiated or integrated term by term for values of x within its interval of convergence (but not, usually, at the endpoints).

EXAMPLE 4: The series $\frac{1}{1+x} = 1 - x + x^2 - x^3 + \cdots$ converges for $|x| < 1$. Integrating term by term between the limits 0 and x gives

$$\int_0^x \frac{dx}{1+x} = \left| x - \frac{x^2}{2} + \frac{x^3}{3} - \frac{x^4}{4} + \cdots \right|_0^x$$

But, from integral 7,

$$\int_0^x \frac{dx}{1+x} = \ln(1 + x)$$

So

$$\ln(1 + x) = x - \frac{x^2}{2} + \frac{x^3}{3} - \frac{x^4}{4} + \cdots$$

Evaluating Definite Integrals

If an expression cannot be integrated by our former methods, it may be possible to expand it in a power series and integrate term by term.

EXAMPLE 5: Find the approximate area under the curve $y = (\sin x)/x$ from $x = 0$ to $x = 1$.

Solution: The area is given by the definite integral

$$A = \int_0^1 \frac{\sin x}{x}\, dx$$

This expression is not in our table of integrals, nor have we any earlier method of dealing with it. Instead, let us take the series for sin x (Eq. 210) and divide it by x, getting

$$\frac{\sin x}{x} = 1 - \frac{x^2}{3!} + \frac{x^4}{5!} - \cdots$$

Integrating,

$$A = \int_0^1 \frac{\sin x}{x}\, dx = \left| x - \frac{x^3}{3(3!)} + \frac{x^5}{5(5!)} - \frac{x^7}{7(7!)} + \cdots \right|_0^1$$

$$\cong 1 - 0.05556 + 0.00167 - 0.00003 + \cdots = 0.94608$$

Euler's Formula

Here we show how Euler's formula (Eq. 228) may be derived by substituting into the power series for e^x. If we replace x by jx in Eq. 203, we get

$$\begin{aligned} e^{jx} &= 1 + jx + \frac{(jx)^2}{2!} + \frac{(jx)^3}{3!} + \frac{(jx)^4}{4!} + \cdots \\ &= 1 + jx + \frac{j^2x^2}{2!} + \frac{j^3x^3}{3!} + \frac{j^4x^4}{4!} + \cdots \end{aligned}$$

But by Eq. 213, $j^2 = -1$, $j^3 = -j$, $j^4 = j$, and so on. So

$$\begin{aligned} e^{jx} &= 1 + jx - \frac{x^2}{2!} - j\frac{x^3}{3!} + \frac{x^4}{4!} + \cdots \\ &= \left[1 - \frac{x^2}{2!} + \frac{x^4}{4!} - \cdots\right] + j\left[x - \frac{x^3}{3!} + \frac{x^5}{5!} - \cdots\right] \end{aligned}$$

But the series in the brackets are those for sin x and cos x, (Equations 210 and 211), so,

Euler's Formula	$e^{jx} = \cos x - j \sin x$	**228**

EXERCISE 5

Substitution

By substituting into the series for e^x (Eq. 203), verify the following.

1. $e^{-x} = 1 - x + \frac{x^2}{2!} - \frac{x^3}{3!} + \cdots$

2. $e^{2x} = 1 + 2x + 2x^2 + \frac{4x^3}{3} + \cdots$

3. $e^{-x^2} = 1 - x^2 + \frac{x^4}{2!} - \frac{x^6}{3!} + \cdots$

By substituting into the series for sin x (Eq. 210), verify the following.

4. $\sin 2x = 2x - \frac{2^3x^3}{3!} + \frac{2^5x^5}{5!} - \frac{2^7x^7}{7!} + \cdots$

5. $\sin x^2 = x^2 - \frac{x^6}{3!} + \frac{x^{10}}{5!} - \cdots$

Addition and Subtraction

Add or subtract two earlier series to verify the following.

6. $\sec x + \tan x = 1 + x + \frac{x^2}{2} + \frac{x^3}{x} + \frac{5x^4}{24} + \cdots$

7. $\sin x + \cos x = 1 + x - \frac{x^2}{2!} - \frac{x^3}{3!} + \frac{x^4}{4!} + \cdots$

8. $\sinh x = \frac{e^x - e^{-x}}{2} = x + \frac{x^3}{3!} + \frac{x^5}{5!} + \frac{x^7}{7!} + \cdots$

9. $\cosh x = \frac{e^x + x^{-x}}{2} = 1 + \frac{x^2}{2!} + \frac{x^4}{4!} + \frac{x^6}{6!} + \cdots$

10. Using the series for cos $2x$ and the identity $\sin^2 x = (1 - \cos 2x)/2$, show that $\sin^2 x = x^2 - \frac{x^4}{3} + \frac{2x^6}{45} - \frac{x^8}{315} + \cdots$.

11. Using the series for cos $2x$ and the identity $\cos^2 x = (1 + \cos 2x)/2$, show that $\cos^2 x = 1 - x^2 + \frac{x^4}{3} - \frac{2x^6}{45} \cdots$.

Multiplication and Division

Multiply or divide earlier series to verify the following.

12. $xe^{2x} = x + 2x^2 + 2x^3 + \cdots$

13. $xe^{-2x} = x - 2x^2 + 2x^3 - \cdots$

14. $e^x \cos x = 1 + x - \frac{x^3}{3} - \frac{x^4}{6} + \cdots$

15. $e^{-x} \cos 2x = 1 - x - \frac{3x^2}{2} + \frac{11x^3}{6} + \cdots$

16. $\frac{\sin x}{x} = 1 - \frac{x^2}{3!} + \frac{x^4}{5!} - \frac{x^6}{7!} + \cdots$

17. $\frac{e^x - 1}{x} = 1 + \frac{x}{2!} + \frac{x^2}{3!} + \frac{x^3}{4!} + \cdots$

18. $\tan x = \frac{\sin x}{\cos x} = x + \frac{x^3}{3} + \frac{2x^5}{15} + \frac{17x^7}{315} + \cdots$

19. Square the first four terms of the series for $\sin x$ to obtain the series for $\sin^2 x$ given in Problem 10.

20. Cube the first three terms of the series for $\cos x$ to obtain the series for $\cos^3 x$.

Differentiation and Integration

21. Find the series for $\cos x$ by differentiating the series for $\sin x$.

22. Differentiate the series for $\tan x$ to show that the series for $\sec^2 x$ is, $\sec^2 x = 1 + x^2 + 2x^4/3 + \cdots$.

23. Find the series for $1/(1 + x)$ by differentiating the series for $\ln (1 + x)$.

24. Find the series for $\ln \cos x$ by integrating the series for $\tan x$.

25. Find the series for $\ln (1 - x)$ by integrating the series for $1/(1 - x)$.

26. Find the series for $\ln (\sec x + \tan x)$ by integrating the series for $\sec x$.

Evaluate each integral by integrating the first three terms of each series.

27. $\int_1^2 \frac{e^x}{\sqrt{x}}\, dx$ **28.** $\int_0^1 \sin x^2\, dx$ **29.** $\int_0^{1/4} e^x \ln (x + 1)\, dx$

CHAPTER TEST

Compute each number using three terms of the appropriate Maclaurin's series.

1. e^3 **2.** $\sin 0.5$

Compute each value using three terms of the appropriate Taylor's series from Exercise 3.

3. $\frac{1}{2.1}$ **4.** $\sin 31.5°$

Estimate the maximum error for the computation in the following.

5. Problem 1

6. Problem 4

7. Exercise 3, problem 18

8. Exercise 3, problem 19

9. Use the ratio test to find the interval of convergence of the power series

$$1 + x + x^2 + x^3 + \cdots + x^n + \cdots$$

10. *Use the ratio test to determine if the following series converges or diverges.*

$$1 + \frac{3}{3!} + \frac{5}{5!} + \frac{7}{7!} + \cdots$$

Write the first five terms of each series, given the general term.

11. $u_n = 2n^2$

12. $u_n = 3n - 1$

Multiply or divide earlier series to verify the following.

13. $x^2e^{2x} = x^2 + 2x^3 + 2x^4 + \cdots$

14. $\dfrac{\sin x}{x^2} = \dfrac{1}{x} - \dfrac{x}{3!} + \dfrac{x^3}{5!} - \cdots$

Deduce the general term of each series. Use it to predict the next two terms.

15. $-1 + 2 + 7 + 14 + 23 + \cdots$

16. $4 + 7 + 10 + 13 + 16 + \cdots$

17. Evaluate using series, $\displaystyle\int_0^1 \frac{e^x - 1}{x}\,dx$

Deduce a recursion relation for each series. Use it to predict the next two terms.

18. $3 + 5 + 9 + 17 + \cdots$

19. $3 + 7 + 47 + \cdots$

20. By substituting into the series for $\cos x$ (Eq. 211) verify the series

$$\cos x^2 = 1 - \frac{x^4}{2!} + \frac{x^8}{4!} - \frac{x^{12}}{6!} + \cdots$$

21. Verify the Maclaurin expansion

$$(1 + x)^3 = 1 + 3x + 3x^2 + x^3$$

22. Verify the following Taylor's series, expanded about $a = 1$,

$$\sqrt{x^2 + 1} = \sqrt{2} + \frac{(x - 1)}{\sqrt{2}} + \frac{(x - 1)^2}{4\sqrt{2}} + \cdots$$

Fourier Series

In Chapter 16 we used power series to approximate continuous functions. But a power series cannot be written for a *discontinuous function,* such as a square wave, that is defined by a different expression in different parts of the interval. For these, we use a Fourier series.

Most periodic functions can be represented by an infinite series of sine and cosine terms if the amplitude and frequency of each term is properly chosen. In this chapter we learn how to write such a series. Then we show how to use any symmetry in the waveform to reduce the work of computation, and conclude with a method for writing a Fourier series for a waveform for which we have no equation, such as from an oscillogram trace.

17-1. WRITING A FOURIER SERIES

Definition

A *Fourier series* is an infinite series of sine and cosine terms of the form

Fourier Series	$f(x) = \frac{a_0}{2} + a_1 \cos x + a_2 \cos 2x + a_3 \cos 3x + \cdots + a_n \cos nx + \cdots$ $+ b_1 \sin x + b_2 \sin 2x + b_3 \sin 3x + \cdots + b_n \sin nx + \cdots$	**409**

Named for a French analyst and mathematical physicist, Jean Baptiste Joseph Fourier (1768–1830).

Here $a_0/2$ is the *constant term*, $a_1 \cos x$ and $b_1 \sin x$ are called the *fundamental* components, and the remaining terms are called the *harmonics:* second harmonics, third harmonics, and so on. (We have let the constant term be $a_0/2$ instead of just a_0 because it will lead to a simpler expression for a_0 later.)

The period of the fundamental terms is 2π. That is, the terms repeat every 2π radians. The harmonics, having a higher frequency than the fundamental, will repeat more often than 2π. Thus the fifth harmonic has a period of $2\pi/5$. Therefore, the entire series repeats every 2π radians. Thus the behavior of the entire series can be found by studying only one cycle, from 0 to 2π, or from $-\pi$ to π, or any other interval containing a full cycle.

This is an infinite series, but as with Taylor series, we use only a few terms for a given computation.

Finding the Coefficients

The period of the Fourier series is the same as that of the waveform for which it is written. It remains only to evaluate the coefficients: the a's and b's. We first find a_0.

Let us integrate both sides of the Fourier series over one cycle. We can choose any two limits of integration that are spaced 2π radians apart, such as $-\pi$ and π.

$$\begin{aligned}\int_{-\pi}^{\pi} f(x)dx = \frac{a_0}{2}\int_{-\pi}^{\pi} dx + a_1 \int_{-\pi}^{\pi} \cos x\, dx \\ + a_2 \int_{-\pi}^{\pi} \cos 2x\, dx + \cdots + a_n \int_{-\pi}^{\pi} \cos nx\, dx + \cdots \\ + b_1 \int_{-\pi}^{\pi} \sin x\, dx + b_2 \int_{-\pi}^{\pi} \sin 2x\, dx + \cdots \\ + b_n \int_{-\pi}^{\pi} \sin nx\, dx + \cdots \end{aligned} \tag{1}$$

But the integral of a sine wave or a cosine wave over one cycle is zero, so (1) reduces to

$$\int_{-\pi}^{\pi} f(x)\, dx = \frac{a_0}{2}\int_{-\pi}^{\pi} dx = \frac{a_0}{2} x \Big|_0^{2\pi} = a_0\pi$$

from which

$$a_0 = \frac{1}{\pi} \int_{-\pi}^{\pi} f(x)\, dx \qquad \textbf{410}$$

Next we find $a_1, a_2, \ldots, a_n$. Multiplying (1) by cos nx gives

$$\begin{aligned}\int_{-\pi}^{\pi} f(x) \cos nx\, dx = {} & \frac{a_0}{2} \int_{-\pi}^{\pi} \cos nx\, dx + a_1 \int_{-\pi}^{\pi} \cos x \cos nx\, dx \\ & + a_2 \int_{-\pi}^{\pi} \cos 2x \cos nx\, dx + \cdots + a_n \int_{-\pi}^{\pi} \cos nx \cos nx\, dx + \cdots \\ & + b_1 \int_{-\pi}^{\pi} \sin x \cos nx\, dx + b_2 \int_{-\pi}^{\pi} \sin 2x \cos nx\, dx + \cdots \\ & + b_n \int_{-\pi}^{\pi} \sin nx \cos nx\, dx + \cdots \qquad (2)\end{aligned}$$

As before, the integral of the cosine function over one cycle is zero. Further, the trigonometric identity (Eq. 181)

$$2 \cos A \cos B = \cos (A - B) + \cos (A + B)$$

shows that the product of two cosine functions can be written as the sum of two cosine functions. Thus the integral of the product of two cosines is equal to the sum of the integrals of two cosines, each of which is zero when integrated over one cycle. Thus (2) reduces to

$$\begin{aligned}\int_{-\pi}^{\pi} f(x) \cos nx\, dx = {} & a_n \int_{-\pi}^{\pi} \cos nx \cos nx\, dx + \cdots \\ & + b_1 \int_{-\pi}^{\pi} \sin x \cos nx\, dx \\ & + b_2 \int_{-\pi}^{\pi} \sin 2x \cos nx\, dx + \cdots \\ & + b_n \int_{-\pi}^{\pi} \sin nx \cos nx\, dx + \cdots \qquad (3)\end{aligned}$$

In a similar way, the identity (Eq. 182)

$$2 \sin A \cos B = \sin (A + B) + \sin (A - B)$$

shows that those terms containing the integral of the sine times the cosine will vanish. This leaves

$$\int_{-\pi}^{\pi} f(x) \cos nx \, dx = a_n \int_{-\pi}^{\pi} \cos^2 nx \, dx$$

$$= \frac{a_n}{n}\left[\frac{nx}{2} + \frac{\sin 2nx}{4}\right]_{-\pi}^{\pi} = \pi a_n$$

by integral 17, so

$$a_n = \frac{1}{\pi}\int_{-\pi}^{\pi} f(x) \cos nx \, dx \qquad \mathbf{411}$$

We now use a similar method to find the b coefficients. We multiply (1) by sin nx and get

$$\int_{-\pi}^{\pi} f(x) \sin nx \, dx = \frac{a_0}{2}\int_{-\pi}^{\pi} \sin nx \, dx + a_1 \int_{-\pi}^{\pi} \cos x \sin nx \, dx$$

$$+ a_2 \int_{-\pi}^{\pi} \cos 2x \sin nx \, dx + \cdots + a_n \int_{-\pi}^{\pi} \cos nx \sin nx \, dx + \cdots$$

$$+ b_1 \int_{-\pi}^{\pi} \sin x \sin nx \, dx + b_2 \int_{-\pi}^{\pi} \sin 2x \sin nx \, dx + \cdots$$

$$+ b_n \int_{-\pi}^{\pi} \sin nx \sin nx \, dx + \cdots \qquad (4)$$

The trigonometric identity (Eq. 180)

$$2 \sin A \sin B = \cos (A - B) - \cos (A + B)$$

shows that the product of two sines is equal to the difference of two cosines, each of which have a zero integral over a full cycle. Other integrals vanish for reasons given before, so (4) reduces to

$$\int_{-\pi}^{\pi} f(x) \sin nx \, dx = b_n \int_{-\pi}^{\pi} \sin^2 nx \, dx$$

$$= b_n \left[\frac{nx}{2} - \frac{\sin 2nx}{4}\right]_{-\pi}^{\pi} = \pi b_n$$

by integral 16, so

$$b_n = \frac{1}{\pi}\int_{-\pi}^{\pi} f(x) \sin nx \, dx \qquad \mathbf{412}$$

In summary,

<table>
<tr><td rowspan="4">Fourier Series</td><td colspan="2">A periodic function $f(x)$ can be replaced with the Fourier series
$$f(x) = \frac{a_0}{2} + a_1 \cos x + a_2 \cos 2x + a_3 \cos 3x + \cdots$$ $$+ b_1 \sin x + b_2 \sin 2x + b_3 \sin 3x + \cdots$$</td><td>409</td></tr>
<tr><td rowspan="3">where</td><td>$$a_0 = \frac{1}{\pi}\int_{-\pi}^{\pi} f(x)\, dx$$</td><td>410</td></tr>
<tr><td>$$a_n = \frac{1}{\pi}\int_{-\pi}^{\pi} f(x) \cos nx\, dx$$</td><td>411</td></tr>
<tr><td>$$b_n = \frac{1}{\pi}\int_{-\pi}^{\pi} f(x) \sin nx\, dx$$</td><td>412</td></tr>
</table>

Keep in mind that the limits can be any two values spaced 2π apart. We'll sometimes use limits of 0 and 2π. In Sec. 17-3 we'll rewrite the series for functions having any period.

EXAMPLE 1: Write a Fourier series to represent the square wave (Fig. 17-1a) for which

$$f(x) = \begin{cases} -1 & \text{for } -\pi \le x < 0 \\ 1 & \text{for } 0 \le x < \pi \end{cases}$$

Solution: Two equations describe $f(x)$ over a full cycle; thus we need to *integrate twice* to find each a or b, once from $-\pi$ to 0 and again from 0 to π. We first find a_0 from Eq. 410,

$$a_0 = \frac{1}{\pi}\int_{-\pi}^{\pi} f(x)\, \mathrm{dx} = \frac{1}{\pi}\int_{-\pi}^{0} (-1)dx + \frac{1}{\pi}\int_{0}^{\pi} (1)dx$$

$$= -\frac{x}{\pi}\bigg|_{-\pi}^{0} + \frac{x}{\pi}\bigg|_{0}^{\pi} = 0$$

We then get a_n from Eq. 411,

$$a_n = \frac{1}{\pi}\int_{-\pi}^{0} (-1) \cos nx\, dx + \frac{1}{\pi}\int_{0}^{\pi} (1) \cos nx\, dx$$

$$= -\frac{\sin nx}{\pi}\bigg|_{-\pi}^{0} + \frac{\sin nx}{\pi}\bigg|_{0}^{\pi} = 0$$

since the sine is zero for any angle that is a multiple of π. Thus all the

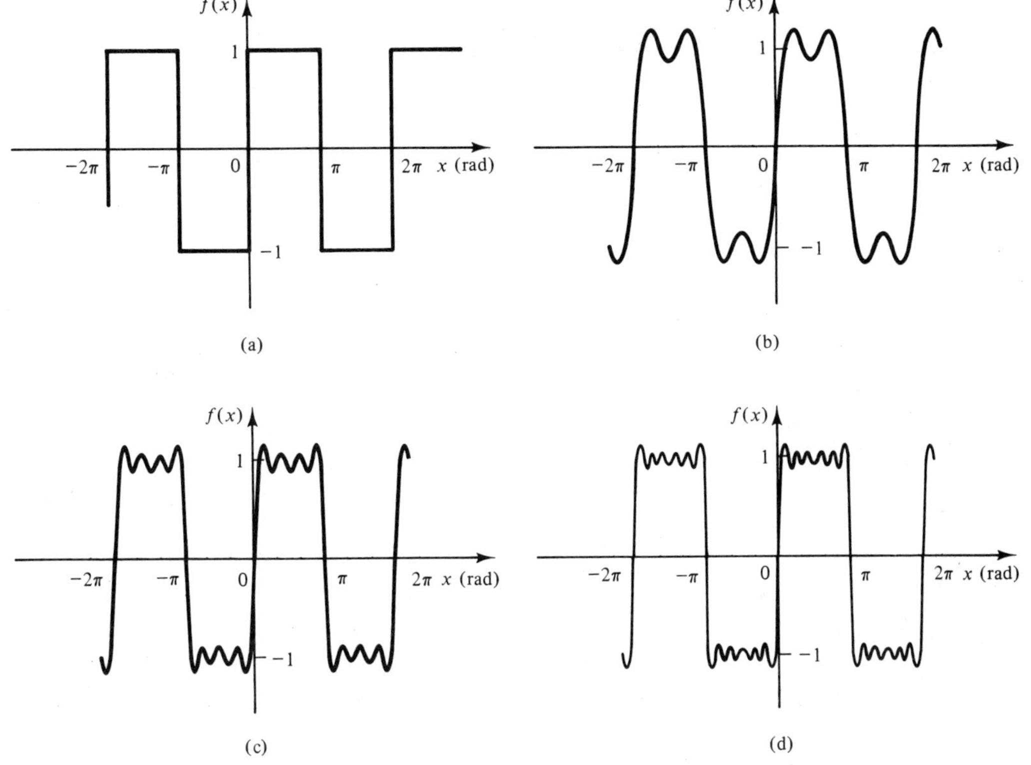

FIGURE 17-1. A square wave, synthesized using two, four, and six terms of a Fourier series.

In Sec. 17-2 we'll learn how to tell in advance if all the sine or cosine terms will vanish.

a coefficients are zero and our series contains no cosine terms. Then, by Eq. 412,

$$b_1 = \frac{1}{\pi}\int_{-\pi}^{0} (-1) \sin x \, dx + \frac{1}{\pi}\int_{0}^{\pi} (1) \sin x \, dx$$

$$= \frac{\cos x}{\pi}\bigg|_{-\pi}^{0} - \frac{\cos x}{\pi}\bigg|_{0}^{\pi} = \frac{4}{\pi}$$

$$b_2 = \frac{1}{\pi}\int_{-\pi}^{0} (-1) \sin 2x \, dx = \frac{1}{\pi}\int_{0}^{\pi} (1) \sin 2x \, dx$$

$$= \frac{\cos 2x}{2\pi}\bigg|_{-\pi}^{0} - \frac{\cos 2x}{2\pi}\bigg|_{0}^{\pi} = \frac{1}{2\pi}[(1 - 1) - (1 - 1)] = 0$$

Note that b would be zero for $n = 4$ or, in fact, all even values of n, since $\cos 2x = \cos 4x = \cos 6x$, and so on. Our series, then, contains no even harmonics. Continuing,

$$b_3 = \frac{1}{\pi}\int_{-\pi}^{0} (-1) \sin 3x \, dx + \frac{1}{\pi}\int_{0}^{\pi} (1) \sin 3x \, dx$$

$$= \frac{\cos 3x}{3\pi}\Bigg|_{-\pi}^{0} - \frac{\cos 3x}{3\pi}\Bigg|_{0}^{\pi} = \frac{4}{3\pi}$$

Continuing the same way gives values of b of $4/\pi$, $4/3\pi$, $4/5\pi$, and so on. Our final Fourier series is then

$$f(x) = \frac{4}{\pi} \sin x + \frac{4}{3\pi} \sin 3x + \frac{4}{5\pi} \sin 5x + \frac{4}{7\pi} \sin 7x + \cdots$$

Figure 17-1b shows the first two terms (the fundamental and the third harmonic) of our Fourier series. Figure 17-1c shows the first four terms (fundamental plus third, fifth, and seventh harmonics), and Fig. 17-1d shows the first six terms (fundamental plus third, fifth, seventh, ninth, and eleventh harmonics).

EXERCISE 1

Write seven terms of the Fourier series given the following coefficients.

1. $a_0 = 4$, $a_1 = 3$, $a_2 = 2$, $a_3 = 1$;
$b_1 = 4$, $b_2 = 3$, $b_3 = 2$

2. $a_0 = 1.6$, $a_1 = 5.2$, $a_2 = 3.1$, $a_3 = 1.4$;
$b_1 = 7.5$, $b_2 = 5.3$, $b_3 = 2.8$

Verify the Fourier series for each function in Table 17-1.

3. number 3 **4.** number 4 **5.** number 5

6. number 6 **7.** number 7 **8.** number 8

9. number 10

Computer

10. Write a program to synthesize a function by Fourier series. It should accept as input the first seven Fourier coefficients and print the value of the function at values of x from 0 to 360°.

17-2. WAVEFORM SYMMETRIES

Kinds of Symmetry

We have seen that for the square wave, all the a_n coefficients were zero, and thus the final series had no cosine terms. If we had a way of telling in advance that the cosine terms would vanish, we could save

TABLE 17-1 Fourier Series for Common Waveforms

Square wave

1. 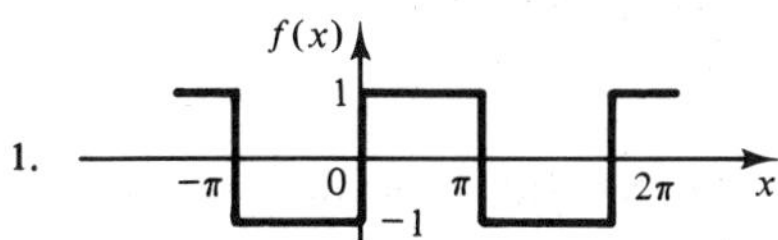

$$y = \frac{4}{\pi}\left(\sin x + \frac{1}{3}\sin 3x + \frac{1}{5}\sin 5x + \frac{1}{7}\sin 7x + \cdots\right)$$

2. $f(x)$, $-\pi$, 1, π, 0, -1, x

$$y = \frac{4}{\pi}\left(\cos x - \frac{1}{3}\cos 3x + \frac{1}{5}\cos 5x - \cdots\right)$$

3. $f(x)$, 2, $-\pi$, 0, π, 2π, x

$$y = \frac{4}{\pi}\left(\frac{\pi}{4} + \sin x + \frac{1}{3}\sin 3x + \frac{1}{5}\sin 5x + \cdots\right)$$

Triangular wave

4. 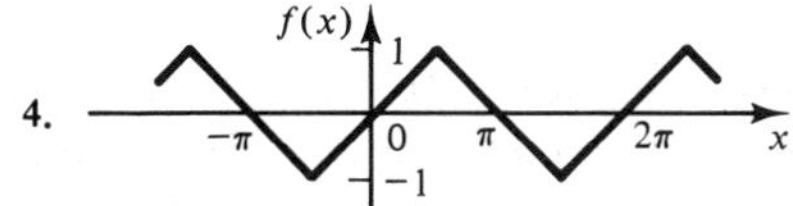

$$y = \frac{8}{x^2}\left(\sin x - \frac{1}{3^2}\sin 3x + \frac{1}{5^2}\sin 5x - \cdots\right)$$

Sawtooth wave

5. 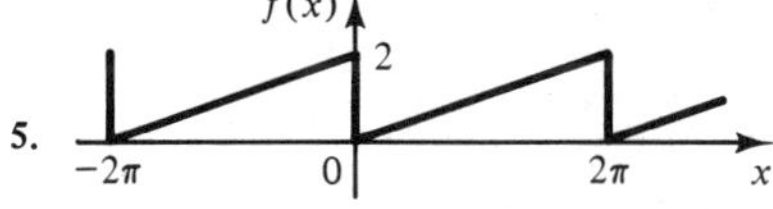

$$y = 1 - \frac{2}{\pi}\left(\sin x + \frac{1}{2}\sin 2x + \frac{1}{3}\sin 3x + \cdots\right)$$

6. 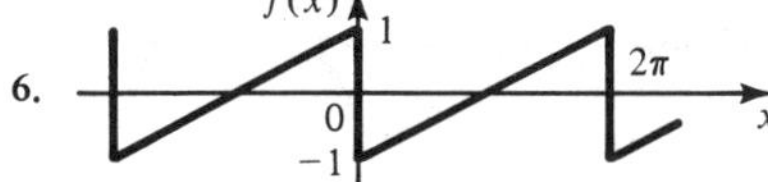

$$y = -\frac{2}{\pi}\left(\sin x + \frac{1}{2}\sin 2x + \frac{1}{3}\sin 3x + \cdots\right)$$

7. 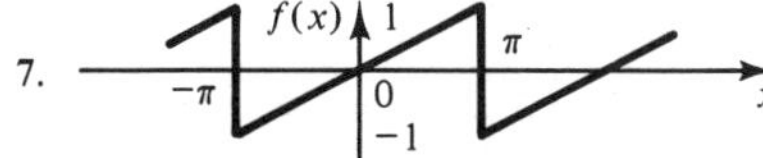

$$y = \frac{2}{\pi}\left(\sin x - \frac{1}{2}\sin 2x + \frac{1}{3}\sin 3x - \frac{1}{4}\sin 4x + \cdots\right)$$

8. 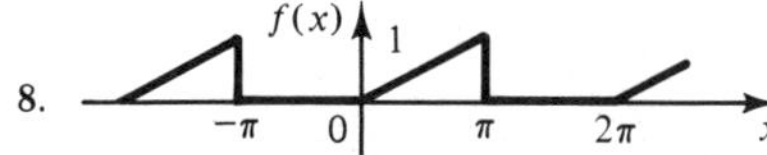

$$y = \frac{\pi}{4} - \frac{2}{\pi}\left(\cos x + \frac{1}{3^2}\cos 3x + \cdots\right) + \sin x - \frac{1}{2}\sin 2x + \cdots$$

9. 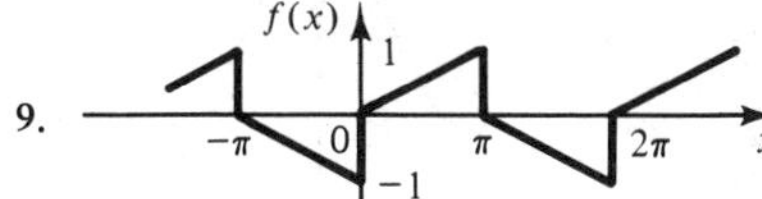

$$y = \frac{1}{2} - \frac{1}{4}\left(\cos x + \frac{1}{3^2}\cos 3x + \frac{1}{5^2}\cos 5x + \cdots\right)$$

Half-wave rectifier

10. 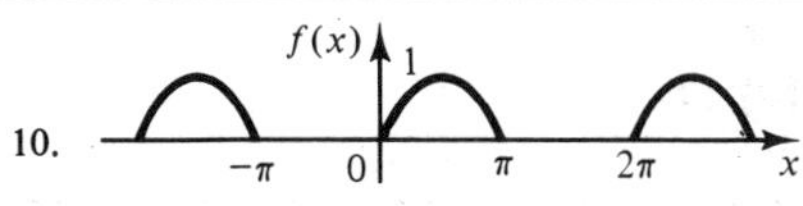

$$y = \frac{1}{\pi}\left(1 + \frac{\pi}{2}\sin x - \frac{2}{3}\cos 2x - \frac{2}{15}\cos 4x - \frac{2}{35}\cos 6x + \cdots\right)$$

11. 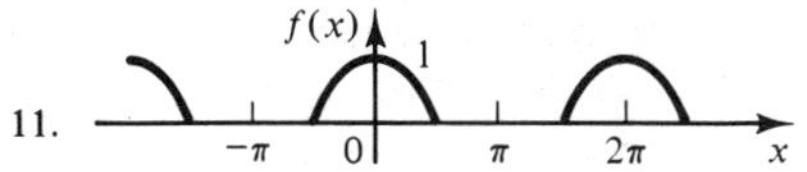

$$y = \frac{1}{\pi}\left(1 + \frac{\pi}{2}\cos x + \frac{2}{3}\cos 2x - \frac{2}{15}\cos 4x + \frac{2}{35}\cos 6x + \cdots\right)$$

Full-wave rectifier

12. 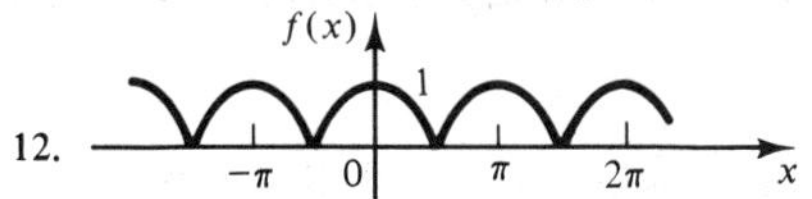

$$y = \frac{2}{\pi}\left(1 + \frac{2}{3}\cos x - \frac{2}{15}\cos 4x + \frac{2}{35}\cos 6x + \cdots\right)$$

work by not even bothering to look for them. We can tell in advance, by noticing the kind of symmetry the given function has. The three kinds of symmetry that are most useful here are:

1. *Symmetry about the origin.* Such a function has only sine terms and is called an *odd* function.
2. *Symmetry about the y axis.* Such a function has no sine terms and is called an *even* function.
3. *Half-wave symmetry.* Such a function has only odd harmonics.

We now cover each type of symmetry.

Odd Functions

When studying graphing in Chapter 4, we noted that a function $f(x)$ is symmetrical about the origin if it remained unchanged when we substitute both $-x$ for x and $-y$ for y. Another way of saying this is that $f(x) = -f(-x)$. They are called odd functions (Fig. 17-2a).

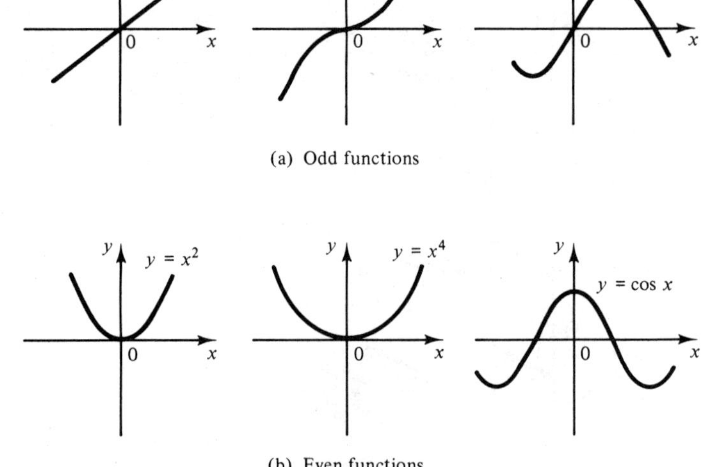

(a) Odd functions

(b) Even functions

FIGURE 17-2

EXAMPLE 1: The function $y = x^3$ is odd because $(-x)^3 = -x^3$. Similarly, x, x^5, x^7, . . . are odd, with the name "odd" coming from the exponent.

Even Functions

A function is symmetrical about the y axis (Fig. 17-2b) if it remains unchanged when we substitute $-x$ for x. In other words, $f(x) = f(-x)$. These are called even functions.

EXAMPLE 2: The function $y = x^2$ is even because $(-x)^2 = x^2$. Similarly, other even powers of x, such as x^4, x^6, . . . , are even.

Fourier Expansion of Odd and Even Functions

Next we note that every term in the power series expansion for the sine function

$$\sin x = x - \frac{x^3}{6} + \frac{x^5}{120} - \cdots \tag{210}$$

is an odd function, and that every term in the power series expansion for the cosine function

$$\cos x = 1 - \frac{x^2}{2} + \frac{x^4}{24} - \cdots \tag{211}$$

is an even function. Thus we conclude the following about the Fourier series for odd and even functions.

Odd and Even Functions	(a) Odd functions have only sine terms (and no constant term). (b) Even functions have only cosine terms (and may have a constant term).	417

EXAMPLE 3: In Table 17-1, functions 1, 4, 6, and 7 are odd. When deriving them from scratch, we would save work by not even looking for cosine terms. Functions 2, 11, and 12 are even, while the others are neither odd nor even.

Shift of Axes

When writing a Fourier series for a periodic waveform we are often free to choose the position of the y axis. Our choice of axis can change the function from even to odd, or vice versa.

EXAMPLE 4: In Table 17-1 notice that functions 1 and 2 are the same waveform but with the axis shifted by $\pi/2$ radians. Thus 1 is an even function, while 2 is odd. A similar shift is seen with the half-wave rectifier waveforms, 10 and 11.

A *vertical* shift of the x axis will affect the constant term in the series.

EXAMPLE 5: Waveform 3 in Table 17-1 is the same as waveform 1, except for a vertical shift of 1 unit. Thus the series for waveform 3 has a constant term of 1 unit. The same can be seen with waveforms 5 and 6.

Half-Wave Symmetry

When the negative half-cycle of a periodic wave has the same shape as the positive half-cycle (but is, of course, inverted), we say that the wave has *half-wave symmetry*. The sine wave, as well as waveforms 1, 2, 4, and 9 in Table 17-1, have half-wave symmetry.

A quick graphical way to test for half-wave symmetry is to measure the ordinates on the wave at two points spaced half a cycle apart. The two ordinates should be equal in magnitude but opposite in sign.

Figure 17-3 shows a sine wave and its second, third, and fourth harmonics. For each, one arrow shows the ordinate at an arbitrarily chosen point, and the other arrow shows the ordinate half a cycle away (π radians for the sine wave). For half-wave symmetry, the arrows should

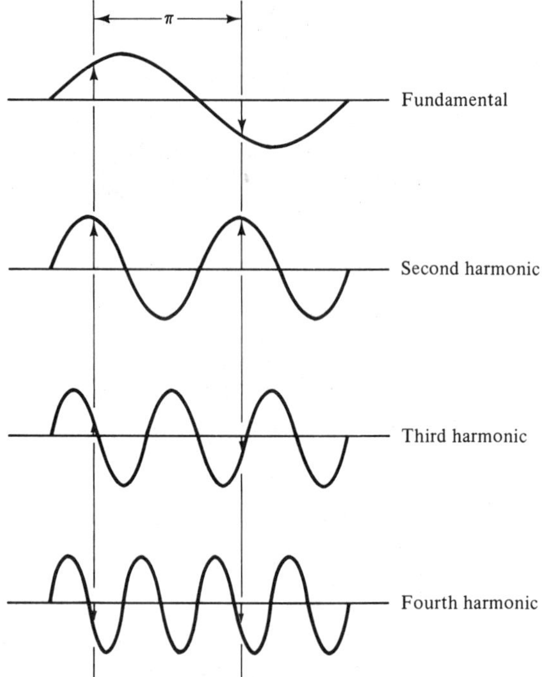

FIGURE 17-3. The fundamental and the third harmonic have half-wave symmetry.

be equal in length but opposite in direction. Note that the fundamental and the odd harmonics have half-wave symmetry, while the even harmonics do not. This is also true for harmonics higher than those shown. Thus for the Fourier series representing a waveform, we conclude:

Half-Wave Symmetry	A wave that has half-wave symmetry has only odd harmonics.	418

Common Error	Don't confuse odd functions with odd harmonics. Thus, sin $4x$ is an even harmonic but an odd function.

By recognizing that a waveform is odd or even or if it has half-wave symmetry, we can reduce the work of computation. Or, when possible, we try to choose our axes to cause symmetry, as in the following example.

EXAMPLE 6: Choose the y axis and write a Fourier series for the triangular wave in Fig. 17-4a.

Solution: The given waveform has no symmetry, so let us temporarily shift the x axis up 2 units, which will eliminate the constant term from our series. We will add 2 to our final series to account for this shift. We are free to locate the y axis, so let us place it as in Fig. 17-4b. Our function is now odd and has half-wave symmetry, so we expect a series of odd-harmonic sine terms, $y = b_1 \sin x + b_3 \sin 3x + b_5 \sin 5x + \cdots$. From Eq. 412,

$$b_n = \frac{1}{\pi}\int_{-\pi}^{\pi} f(x) \sin nx\, dx$$

From 0 to $\pi/2$ our waveform is a straight line through the origin with a slope of $2/\pi$, so

$$f(x) = \frac{2}{\pi}x \qquad \left(0 < x < \frac{\pi}{2}\right)$$

The equation of the waveform is different elsewhere in the interval $-\pi$ to π, but we'll see that we need only the portion from 0 to $\pi/2$.

If we graph $f(x)\sin nx$, for any odd value of n, say 3, we get a curve such as in Fig. 17-4c. The integral of $f(x)\sin nx$ corresponds to the shaded portion under the curve. Note that this area repeats, and that the area under the curve between 0 and $\pi/2$ is one-fourth the area from $-\pi$ to π. Thus we need only integrate from 0 to $\pi/2$ and multiply the result by 4. Thus

$$\begin{aligned} b_n &= \frac{1}{\pi}\int_{-\pi}^{\pi} f(x) \sin nx\, dx = \frac{4}{\pi}\int_{0}^{\pi/2} f(x) \sin nx\, dx \\ &= \frac{4}{\pi}\int_{0}^{\pi/2} \frac{2}{\pi}x \sin nx\, dx = \frac{8}{\pi^2}\int_{0}^{\pi/2} x \sin nx\, dx \\ &= \frac{8}{\pi^2}\left[\frac{1}{n^2}\sin nx - \frac{x}{n}\cos nx\right]_0^{\pi/2} = \frac{8}{\pi^2}\left[\frac{1}{n^2}\sin\frac{n\pi}{2} - \frac{\pi}{2n}\cos\frac{n\pi}{2}\right] \end{aligned}$$

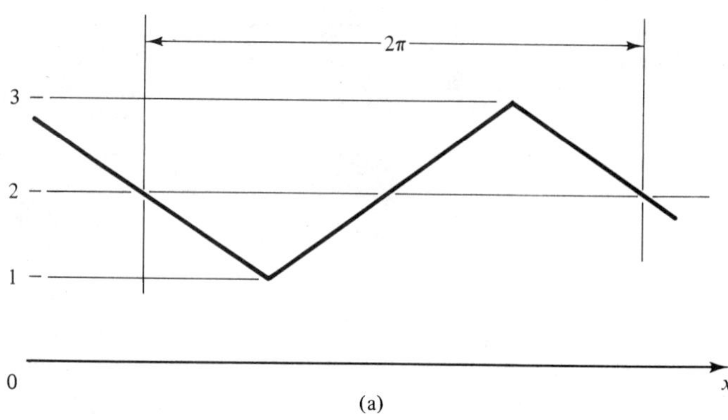

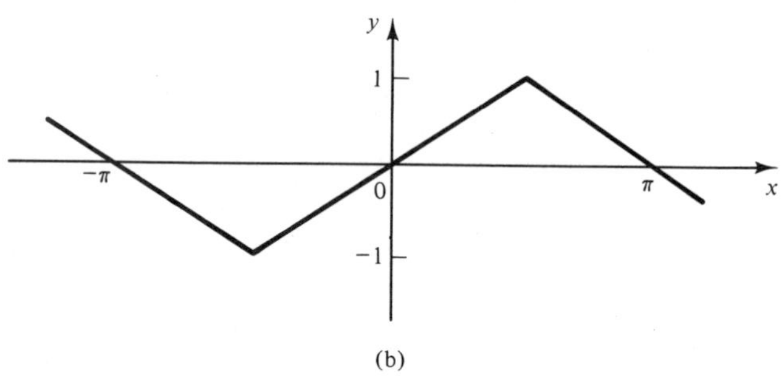

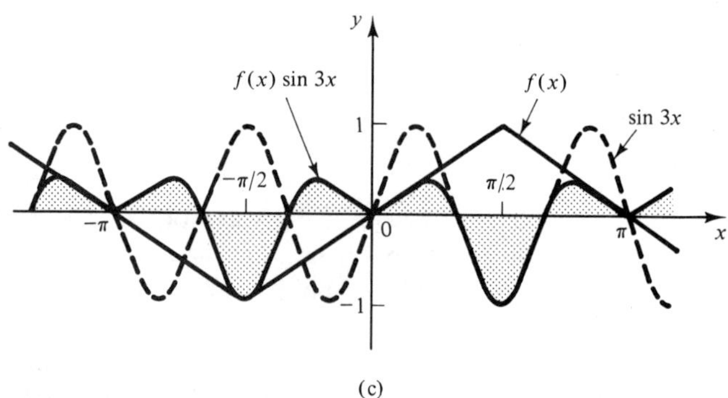

FIGURE 17-4. Synthesis of a triangular wave.

We have only odd harmonics, so $n = 1, 3, 5, 7, \cdots$. For these values of n, $\cos(n\pi/2) = 0$, so

$$b_n = \frac{8}{\pi^2}\left(\frac{1}{n^2}\sin\frac{n\pi}{2}\right)$$

from which

$$b_1 = \frac{8}{\pi^2}\left(\frac{1}{1^2}\right) = \frac{8}{\pi^2} \qquad b_3 = \frac{8}{\pi^2}\left(\frac{1}{3^2}\right)(-1) = -\frac{8}{9\pi^2}$$

$$b_5 = \frac{8}{\pi^2}\left(\frac{1}{5^2}\right)(1) = \frac{8}{25\pi^2} \qquad b_7 = \frac{8}{\pi^2}\left(\frac{1}{7^2}\right)(-1) = -\frac{8}{49\pi^2}$$

Our final series, remembering to add a constant term of 2 units for the shift of axis, is then

$$y = 2 + \frac{8}{\pi^2}\left(\sin x - \frac{1}{9}\sin 3x + \frac{1}{25}\sin 5x - \frac{1}{49}\sin 7x + \cdots\right)$$

This is the same as waveform 4 from Table 17-1, except for the 2 unit shift.

EXERCISE 2

Label each function as odd, even, or neither.

1. Fig. 17-5a **2.** Fig. 17-5b **3.** Fig. 17-5c

4. Fig. 17-5d **5.** Fig. 17-5e **6.** Fig. 17-5f

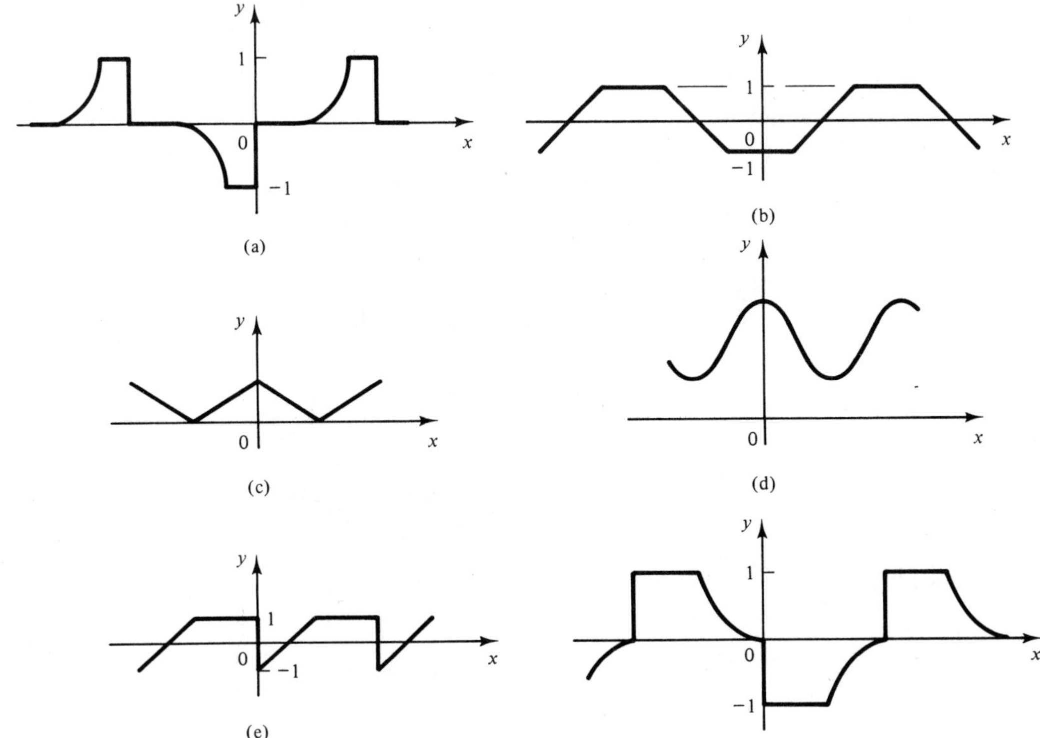

FIGURE 17-5

Which functions have half-wave symmetry?

7. Fig. 17-5a **8.** Fig. 17-5b **9.** Fig. 17-5c

10. Fig. 17-5d **11.** Fig. 17-5e **12.** Fig. 17-5f

Using symmetry to simplify your work, verify the Fourier series for the functions in Table 17-1.

13. number 2 **14.** number 9
15. number 11 **16.** number 12

17-3. WAVEFORMS WITH PERIOD OF 2L

So far we have considered only functions with a period of 2π. But a waveform could have some other period, say 8. Here we modify our formulas to suit any period.

Figure 17-6a shows a function $y = g(z)$ with a period of 2π. The Fourier coefficients for this function are

$$a_0 = \frac{1}{\pi}\int_{-\pi}^{\pi} g(z)\,dz \tag{1}$$

$$a_n = \frac{1}{\pi}\int_{-\pi}^{\pi} g(z)\cos nz\,dz \tag{2}$$

and

$$b_n = \frac{1}{\pi}\int_{-\pi}^{\pi} g(z)\sin nz\,dz \tag{3}$$

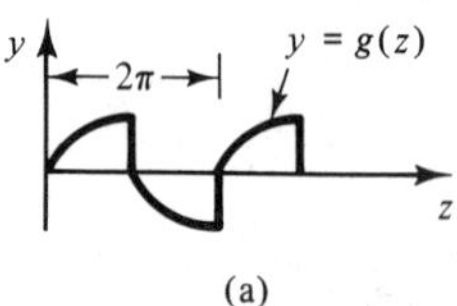

(a)

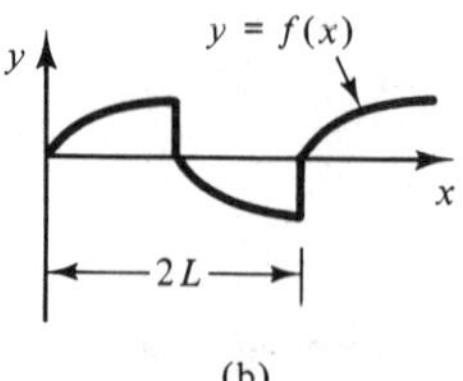

(b)

FIGURE 17-6

Figure 17-6b shows a function $y = f(x)$ with a period $2L$, where x and z are related by the proportion

$$\frac{x}{2L} = \frac{z}{2\pi}$$

or $z = \pi x/L$. Substituting into (1), with $g(z) = y = f(x)$ and $dz = d(\pi x/L) = (\pi/L)\,dx$,

$$a_0 = \frac{1}{\pi}\int_{-\pi}^{\pi} g(z)\,dz = \frac{1}{\pi}\int_{-L}^{L} f(x)\,\frac{\pi}{L}\,dx = \frac{1}{L}\int_{-L}^{L} f(x)\,dx$$

Next, from (2) we get

$$a_n = \frac{1}{\pi}\int_{-\pi}^{\pi} g(z)\cos nz\,dz = \frac{1}{\pi}\int_{-L}^{L} f(x)\left(\cos\frac{n\pi x}{L}\right)\frac{\pi}{L}\,dx$$
$$= \frac{1}{L}\int_{-L}^{L} f(x)\cos\frac{n\pi x}{L}\,dx$$

Similarly, from (3),

$$b_n = \frac{1}{\pi}\int_{-\pi}^{\pi} g(z)\sin nz\,dz = \frac{1}{\pi}\int_{-L}^{L} f(x)\left(\sin\frac{n\pi x}{L}\right)\frac{\pi}{L}\,dx$$
$$= \frac{1}{L}\int_{-L}^{L} f(x)\sin\frac{n\pi x}{L}\,dx$$

In summary,

Waveforms with Period of $2L$	A periodic function can be replaced by the Fourier series $f(x) = \frac{a_0}{2} + a_1\cos\frac{\pi x}{L} + a_2\cos\frac{2\pi x}{L} + a_3\cos\frac{3\pi x}{L} + \cdots + b_1\sin\frac{\pi x}{L} + b_2\sin\frac{2\pi x}{L} + b_3\sin\frac{3\pi x}{L} + \cdots$		**413**
	where	$a_0 = \frac{1}{L}\int_{-L}^{L} f(x)\,dx$	**414**
		$a_n = \frac{1}{L}\int_{-L}^{L} f(x)\cos\frac{n\pi x}{L}\,dx$	**415**
		$b_n = \frac{1}{L}\int_{-L}^{L} f(x)\sin\frac{n\pi x}{L}\,dx$	**416**

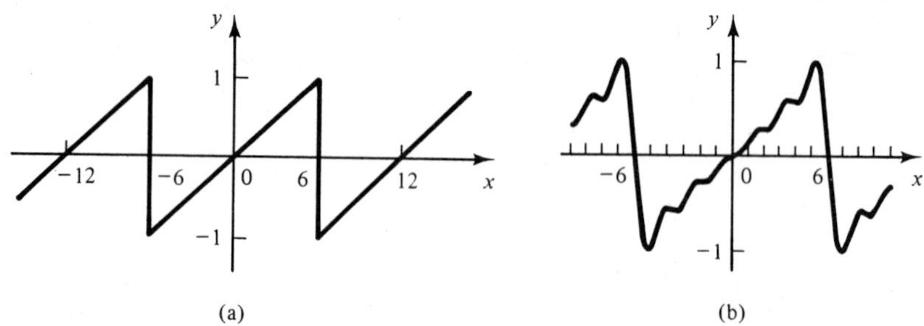

FIGURE 17-7. Synthesis of a sawtooth wave.

EXAMPLE 1: Compute the Fourier series for waveform in Fig. 17-7a.

Solution: Since the function is odd, we expect only sine terms, and no constant term. The equation of the waveform from −6 to 6 is $f(x) = x/6$. From Eq. 416, with $L = 6$,

$$b_n = \frac{1}{L}\int_{-L}^{L} f(x)\sin\frac{n\pi x}{L}\,dx = \frac{1}{6}\int_{-6}^{6}\frac{x}{6}\sin\frac{n\pi x}{6}\,dx$$

Integrating by rule 31 and substituting the limits gives

$$b_n = \frac{1}{(n\pi)^2}\left[\sin\frac{n\pi x}{6} - \frac{n\pi x}{6}\cos\frac{n\pi x}{6}\right]_{-6}^{6} = -\frac{2}{n\pi}\cos n\pi$$

from which

$$b_1 = \frac{2}{\pi},\quad b_2 = \frac{-1}{\pi},\quad b_3 = \frac{2}{3\pi},\quad b_4 = \frac{-1}{2\pi},\quad \text{etc.}$$

Our series is then

$$y = \frac{2}{\pi}\left(\sin\frac{\pi x}{6} - \frac{1}{2}\sin\frac{\pi x}{3} + \frac{1}{3}\sin\frac{\pi x}{2} - \frac{1}{4}\sin\frac{2\pi x}{3} + \cdots\right)$$

Figure 17-7b shows a graph of the first five terms of this series.

EXERCISE 3

Verify the Fourier series for each waveform.

1. Fig. 17-8a
2. Fig. 17-8b
3. Fig. 17-8c
4. Fig. 17-8d
5. Fig. 17-8e
6. Fig. 17-8f

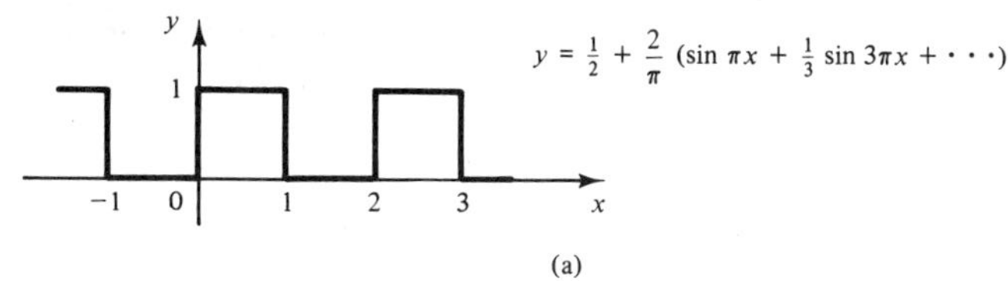

(a)

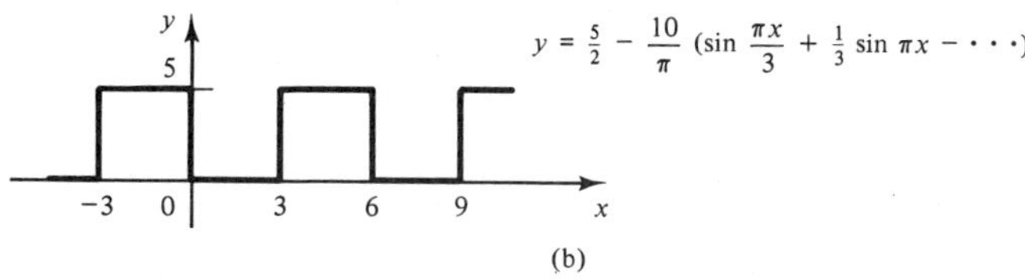

(b)

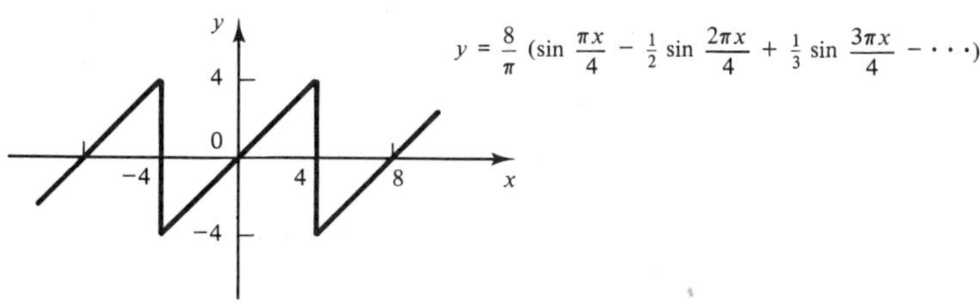

(c)

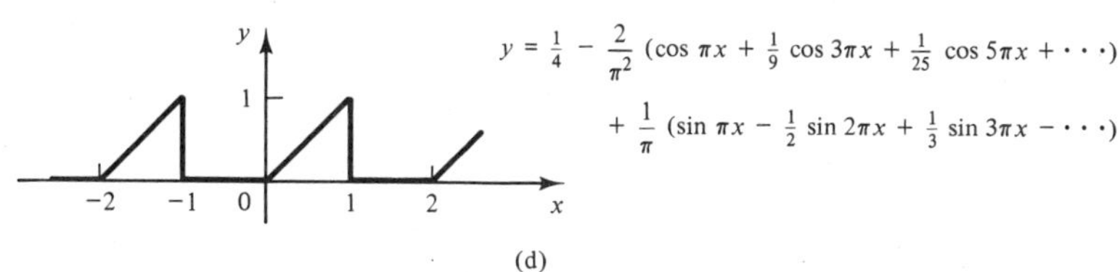

(d)

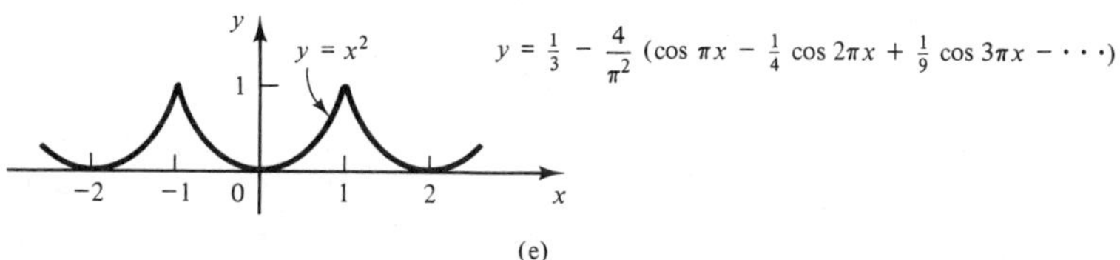

(e)

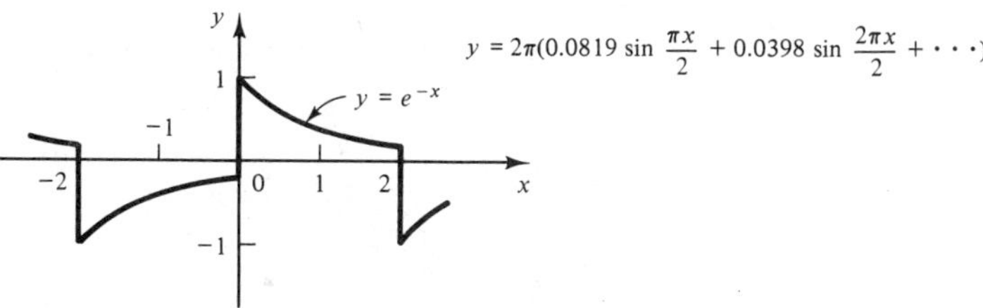

(f)

FIGURE 17-8

17-4. A NUMERICAL METHOD

So far we have written Fourier series for waveforms for which we have equations. But what about a wave whose shape we know from an oscilloscope trace or from a set of meter readings but for which we have no equation? Here we give a simple numerical method, easily programmed for the computer. It is based on approximate methods for finding the area under a curve, which we covered in Sec. 11-5.

EXAMPLE 1: Find the first eight terms of the Fourier series for the waveform in Fig. 17-9.

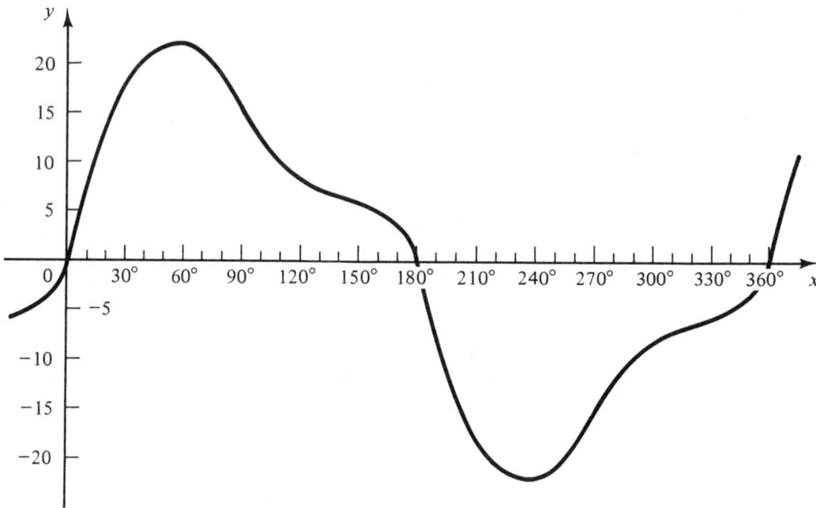

FIGURE 17-9

Solution: *Symmetry:* We first note that the wave does not appear to be odd or even, so we expect our series to contain both sine and cosine terms. There is half-wave symmetry, however, so we look only for odd harmonics. The first eight terms, then, will be

$$y = a_1 \cos x + a_3 \cos 3x + a_5 \cos 5x + a_7 \cos 7x$$
$$+ b_1 \sin x + b_3 \sin 3x + b_5 \sin 5x + b_7 \sin 7x$$

Finding the coefficients: From Eq. 411 the first coefficient is

$$a_1 = \frac{1}{\pi} \int_0^{2\pi} y \cos x \, dx$$

But the value of the integral $\int y \cos x \, dx$ is equal to the area under the $y \cos x$ curve, which we find by approximate integration. We choose the average ordinate method here. Taking advantage of half-wave symmetry, we integrate over only a half-cycle, and double the result.

If y_{avg} is the average ordinate of the $y \cos x$ curve over a half-cycle (0 to π), then by Eq. 378,

$$\int_0^{2\pi} y \cos x \, dx = 2\pi y_{avg}$$

So

$$a_1 = \frac{1}{\pi}\int_0^{2\pi} y \cos x \, dx = 2y_{avg}$$

We divide the half-cycle into a number of intervals, say at every 10°. At each x we measure the ordinate y and calculate $y \cos x$. For example, at $x = 20°$, y is measured at 14.8. So

$$y \cos x = 14.8 \cos 20° = 14.8(0.9397) = 13.907$$

The computed values of $y \cos x$ for the remaining angles are shown in Table 17-2.

We then add all the values of $y \cos x$ over the half-cycle and get an average value by dividing that sum (56.883) by the total number of data points, 19. Thus

$$y_{avg} = \frac{56.883}{19} = 2.994$$

Our first coefficient is then

$$a_1 = 2y_{avg} = 2(2.994) = 5.988$$

The remainder of the computation is similar, and the results are shown in Table 17-2. Our final Fourier series is then

$$y = 5.988 \cos x - 3.756 \cos 3x - 0.176 \cos 5x - 0.512 \cos 7x$$
$$+ 16.050 \sin x + 2.940 \sin 3x + 0.866 \sin 5x + 0.750 \sin 7x$$

Check: To test our result we compute y by series and get the values listed in Table 17-3 and graphed in Fig. 17-10. For better accuracy we could compute more terms of the series or use finer spacing when finding y_{avg}. This would not be much more work when done by computer.

TABLE 17-2

x	y	$\sin x$	$y \sin x$	$\cos x$	$y \cos x$	$\sin 3x$	$y \sin 3x$	$\cos 3x$	$y \cos 3x$
0	0.0	0.0000	0.000	1.0000	0.000	0.0000	0.000	1.0000	0.000
10	10.0	0.1736	1.736	0.9848	9.848	0.5000	5.000	0.8660	8.660
20	14.8	0.3420	5.062	0.9397	13.907	0.8660	12.817	0.5000	7.400
30	16.1	0.5000	8.050	0.8660	13.943	1.0000	16.100	0.0000	0.000
40	20.0	0.6428	12.856	0.7660	15.321	0.8660	17.320	−0.5000	−10.000
50	21.0	0.7660	16.087	0.6428	13.499	0.5000	10.500	−0.8660	−18.187
60	21.2	0.8660	18.360	0.5000	10.600	0.0000	−0.000	−1.0000	−21.200
70	20.3	0.9397	19.076	0.3420	6.943	−0.5000	−10.150	−0.8660	−17.580
80	17.7	0.9848	17.431	0.1736	3.074	−0.8660	−15.329	−0.5000	−8.850
90	14.1	1.0000	14.100	0.0000	0.000	−1.0000	−14.100	0.0000	0.000
100	11.2	0.9848	11.030	−0.1737	−1.945	−0.8660	−9.699	0.5000	5.600
110	8.8	0.9397	8.269	−0.3420	−3.010	−0.5000	−4.400	0.8660	7.621
120	7.5	0.8660	6.495	−0.5000	−3.750	0.0000	0.000	1.0000	7.500
130	6.6	0.7660	5.056	−0.6428	−4.242	0.5000	3.300	0.8660	5.716
140	6.1	0.6428	3.921	−0.7660	−4.673	0.8660	5.283	0.5000	3.050
150	5.5	0.5000	2.750	−0.8660	−4.763	1.000	5.500	−0.0000	−0.000
160	4.6	0.3420	1.573	−0.9397	−4.323	0.8660	3.984	−0.5000	−2.300
170	3.6	0.1736	0.625	−0.9848	−3.545	0.5000	1.800	−0.8660	−3.118
180	0.0	0.0000	0.000	−1.0000	0.000	−0.0000	0.000	−1.0000	0.000
		Sum:	152.477		56.883		27.926		−35.688
		y_{avg}:	8.025		2.994		1.470		−1.878
			$b_1 = 16.050$		$a_1 = 5.988$		$b_3 = 2.940$		$a_3 = -3.756$

TABLE 17-3

x	Original y	y from series
0	0.0	1.5
10	10.0	8.0
20	14.8	13.5
30	16.1	16.8
40	20.0	18.4
50	21.0	19.5
60	21.2	20.2
70	20.3	19.5
80	17.7	16.8
90	14.1	13.2
100	11.2	10.3
110	8.8	8.6
120	7.5	7.4
130	6.6	6.2
140	6.1	5.3
150	5.5	5.2
160	4.6	5.2
170	3.6	3.3
180	0.0	−1.5

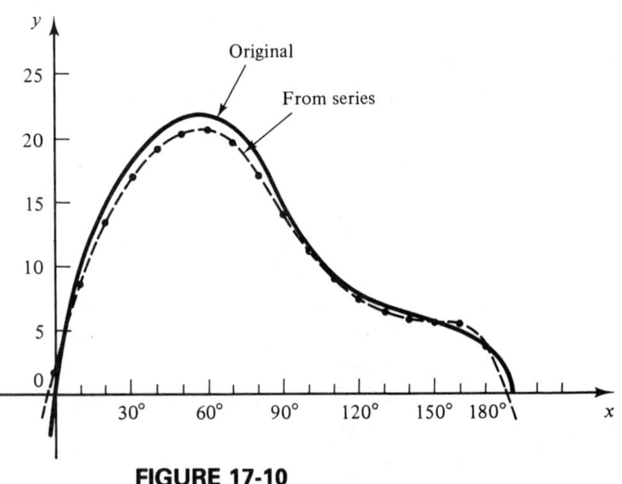

FIGURE 17-10

TABLE 17-2 *(cont.)*

sin 5x	*y sin 5x*	*cos 5x*	*y cos 5x*	*sin 7x*	*y sin 7x*	*cos 7x*	*y cos 7x*
0.0000	0.000	1.0000	0.000	0.0000	0.000	1.0000	0.000
0.7660	7.660	0.6428	6.428	0.9397	9.397	0.3420	3.420
0.9848	14.575	−0.1737	−2.570	0.6428	9.513	−0.7660	−11.338
0.5000	8.050	−0.8660	−13.943	−0.5000	−8.050	−0.8660	−13.943
−0.3420	−6.841	−0.9397	−18.794	−0.9848	−19.696	0.1737	3.473
−0.9397	−19.734	−0.3420	−7.182	−0.1736	−3.646	0.9848	20.681
−0.8660	−18.360	0.5000	10.600	0.8660	18.360	0.5000	10.600
−0.1736	−3.525	0.9848	19.992	0.7660	15.550	−0.6428	−13.049
0.6428	11.378	0.7660	13.559	−0.3420	−6.054	−0.9397	−16.632
1.0000	14.100	−0.0000	−0.000	−1.0000	−14.100	0.0000	0.000
0.6428	7.199	−0.7661	−8.580	−0.3420	−3.830	0.9397	10.525
−0.1737	−1.528	−0.9848	−8.666	0.7661	6.741	0.6428	5.656
−0.8660	−6.495	−0.5000	−3.750	0.8660	6.495	−0.5000	−3.750
−0.9397	−6.202	0.3420	2.257	−0.1737	−1.146	−0.9848	−6.500
−0.3420	−2.086	0.9397	5.732	−0.9848	−6.007	−0.1736	−1.059
0.5000	2.750	0.8660	4.763	−0.5000	−2.750	0.8660	4.763
0.9848	4.530	0.1736	0.799	0.6428	2.957	0.7660	3.524
0.7660	2.758	−0.6428	−2.314	0.9397	3.383	−0.3421	−1.231
−0.0000	0.000	−1.0000	0.000	−0.0001	0.000	−1.0000	0.000
	8.230		−1.670		7.116		−4.860
	0.433		−0.088		0.375		−0.256
	$b_5 = 0.866$		$a_5 = -0.176$		$b_7 = 0.750$		$a_7 = -0.512$

EXERCISE 4

Find the first six terms of the Fourier series for each waveform. Assume half-wave symmetry.

1.

x	0°	20°	40°	60°	80°	100°	120°	140°	160°	180°
y	0	2.1	4.2	4.5	7.0	10.1	14.3	14.8	13.9	0

2.

x	0°	20°	40°	60°	80°	100°	120°	140°	160°	180°
y	0	31	52	55	80	111	153	108	49	0

3.

x	0°	20°	40°	60°	80°	100°	120°	140°	160°	180°
y	0	12.1	24.2	34.5	47.0	30.1	24.3	11.8	9.9	0

4.

x	0°	20°	40°	60°	80°	100°	120°	140°	160°	180°
y	0	4.6	9.4	15.4	12.3	10.8	14.1	16.8	10.9	0

Computer

5. Write a program or use a spreadsheet to compute the coefficients of the first eight terms of a Fourier series for a given set of experimental data. Test your program on the problems above.

CHAPTER TEST

Label each function as odd, even, or neither.

1. Fig. 17-11a
2. Fig. 17-11b
3. Fig. 17-11c
4. Fig. 17-11d

Which functions have half-wave symmetry?

5. Fig. 17-11a
6. Fig. 17-11b
7. Fig. 17-11c
8. Fig. 17-11d

Write a Fourier series for each waveform.

9. Fig. 17-11a
10. Fig. 17-11b
11. Fig. 17-11c
12. Fig. 17-11d

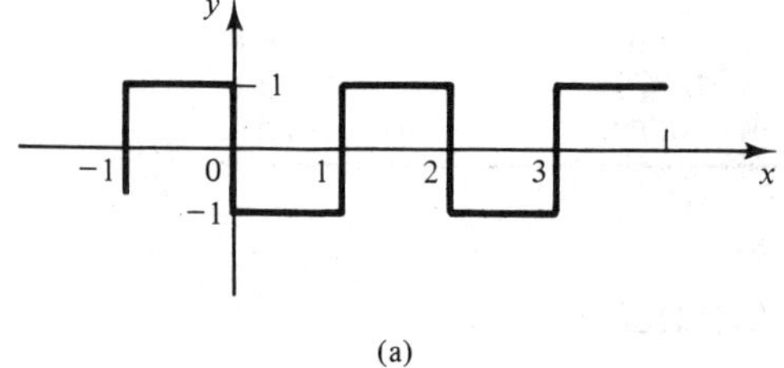

(a)

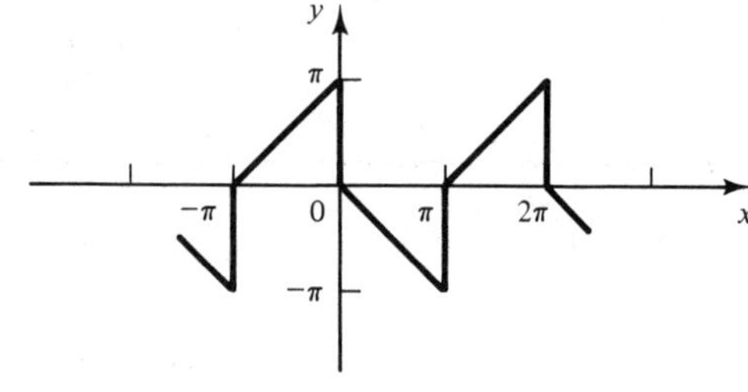

(b)

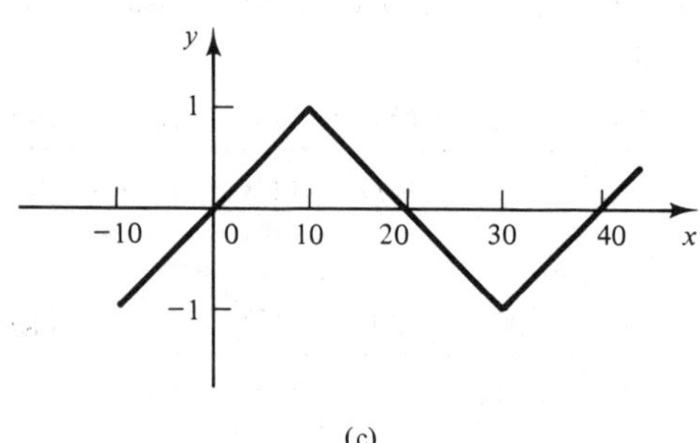

(c)

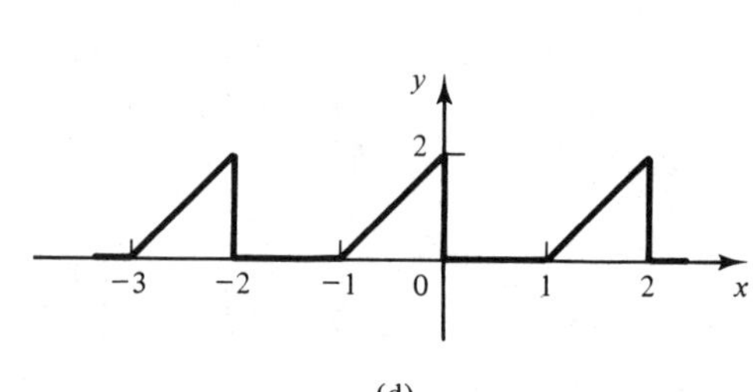

(d)

FIGURE 17-11

13. Find the first six terms of the Fourier series for the given waveform. Assume half-wave symmetry.

x	0°	20°	40°	60°	80°	100°	120°	140°	160°	180°
y	0	3.2	5.5	6.2	8.3	11.7	15.1	14.3	12.6	0

Summary of Facts & Formulas

Many of the formulas in this Summary are included for reference, even though they may not appear elsewhere in the text.

	No.				
ALGEBRAIC LAWS	1	Commutative Law	Addition	$a + b = b + a$	
	2		Multiplication	$ab = ba$	
	3	Associative Law	Addition	$a + (b + c) = (a + b) + c = (a + c) + b$	
	4		Multiplication	$a(bc) = (ab)c = (ac)b = abc$	
	5	Distributive Law		$a(b + c) = ab + ac$	
RULES OF SIGNS	6	Addition and Subtraction	$a + (-b) = a - (+b) = a - b$		
	7		$a + (+b) = a - (-b) = a + b$		
	8	Multiplication	$(+a)(+b) = (-a)(-b) = +ab$		
	9		$(+a)(-b) = (-a)(+b) = -(+a)(+b) = -ab$		
	10	Division	$\frac{+a}{+b} = \frac{-a}{-b} = -\frac{-a}{+b} = -\frac{+a}{-b} = \frac{a}{b}$		
	11		$\frac{+a}{-b} = \frac{-a}{+b} = -\frac{-a}{-b} = -\frac{a}{b}$		
PERCENTAGE	12	Amount = base × rate, $A = BP$			
	13	Percent change = $\frac{\text{new value} - \text{original value}}{\text{original value}} \times 100\%$			
	14	Percent error = $\frac{\text{relative error}}{\text{true value}} \times 100\%$			
	15	Percent concentration of ingredient A = $\frac{\text{amount of } A}{\text{amount of mixture}} \times 100\%$			
	16	Percent efficiency = $\frac{\text{output}}{\text{input}} \times 100\%$			

	No.				
BINARY NUMBERS	17	Largest n-Bit Binary Number		$2^n - 1$	
	18	Addition		$0 + 0 = 0$ $0 + 1 = 1$ $1 + 0 = 1$ $1 + 1 = 0$ with carry to next column	
	19	Subtraction		$0 - 0 = 0$ $0 - 1 = 1$ with borrow from next column $1 - 0 = 1$ $1 - 1 = 0$ $10 - 1 = 1$	
	20			Subtracting B from A is equivalent to adding the two's complement of B to A	
	21	Multiplication		$0 \times 0 = 0$ $0 \times 1 = 0$ $1 \times 0 = 0$ $1 \times 1 = 1$	
	22	Division		$0 \div 0$ is not defined $1 \div 0$ is not defined $0 \div 1 = 0$ $1 \div 1 = 1$	
	23	Complements	n-Bit One's Complement of x	$(2^n - 1) - x$	
	24			The one's complement of a number can be written by changing the 1's to 0's and the 0's to 1's	
	25		n-Bit Two's Complement of x	$2^n - x$	
	26			To find the two's complement of a number first write the one's complement, and add 1	
	27	Negative Binary Numbers		If M is a positive binary number, then $-M$ is the two's complement of M	

	No.				
EXPONENTS	28	Definition		$a^n = \underbrace{a \cdot a \cdot a \cdot \ldots \cdot a}_{n \text{ factors}}$	
EXPONENTS	29	Laws of Exponents	Products	$x^a \cdot x^b = x^{a+b}$	
EXPONENTS	30	Laws of Exponents	Quotients	$\frac{x^a}{x^b} = x^{a-b}$	
EXPONENTS	31	Laws of Exponents	Powers	$(x^a)^b = x^{ab} = (x^b)^a$	
EXPONENTS	32	Laws of Exponents	Product Raised to a Power	$(xy)^n = x^n \cdot y^n$	
EXPONENTS	33	Laws of Exponents	Quotient Raised to a Power	$\left(\frac{x}{y}\right)^n = \frac{x^n}{y^n}$	
EXPONENTS	34	Laws of Exponents	Zero Exponent	$x^0 = 1$	
EXPONENTS	35	Laws of Exponents	Negative Exponent	$x^{-a} = \frac{1}{x^a}$	
EXPONENTS	36	Fractional Exponents		$a^{1/n} = \sqrt[n]{a}$	
EXPONENTS	37	Fractional Exponents		$a^{m/n} = \sqrt[n]{a^m} = (\sqrt[n]{a})^m$	
RADICALS	38	Rules of Radicals	Root of a Product	$\sqrt[n]{ab} = \sqrt[n]{a}\,\sqrt[n]{b}$	
RADICALS	39	Rules of Radicals	Root of a Quotient	$\sqrt[n]{\frac{a}{b}} = \frac{\sqrt[n]{a}}{\sqrt[n]{b}}$	
RADICALS	40	Rules of Radicals	Root of a Power	$\sqrt[n]{a^m} = (\sqrt[n]{a})^m$	
SPECIAL PRODUCTS AND FACTORING	41	Binomials	Difference of Two Squares	$a^2 - b^2 = (a - b)(a + b)$	
SPECIAL PRODUCTS AND FACTORING	42	Binomials	Sum of Two Cubes	$a^3 + b^3 = (a + b)(a^2 - ab + b^2)$	
SPECIAL PRODUCTS AND FACTORING	43	Binomials	Difference of Two Cubes	$a^3 - b^3 = (a - b)(a^2 + ab + b^2)$	
SPECIAL PRODUCTS AND FACTORING	44	Trinomials	Test for Factorability	$ax^2 + bx + c$ is factorable if $b^2 - 4ac$ is a perfect square	
SPECIAL PRODUCTS AND FACTORING	45	Trinomials	Leading Coefficient = 1	$x^2 + (a + b)x + ab = (x + a)(x + b)$	
SPECIAL PRODUCTS AND FACTORING	46	Trinomials	General Quadratic Trinomial	$acx^2 + (ad + bc)x + bd = (ax + b)(cx + d)$	
SPECIAL PRODUCTS AND FACTORING	47	Trinomials	Perfect Square Trinomials	$a^2 + 2ab + b^2 = (a + b)^2$	
SPECIAL PRODUCTS AND FACTORING	48	Trinomials	Perfect Square Trinomials	$a^2 - 2ab + b^2 = (a - b)^2$	
SPECIAL PRODUCTS AND FACTORING	49	Factoring by Grouping		$ac + ad + bc + bd = (a + b)(c + d)$	

	No.				
FRACTIONS	50	Simplifying	$\frac{ad}{bd} = \frac{a}{b}$		
	51	Multiplication	$\frac{a}{b} \cdot \frac{c}{d} = \frac{ac}{bd}$		
	52	Division	$\frac{a}{b} \div \frac{c}{d} = \frac{a}{b} \cdot \frac{d}{c} = \frac{ad}{bc}$		
	53	Addition and Subtraction	Same Denominators	$\frac{a}{b} \pm \frac{c}{b} = \frac{a \pm c}{b}$	
	54		Different Denominators	$\frac{a}{b} \pm \frac{c}{d} = \frac{ad}{bd} \pm \frac{bc}{bd} = \frac{ad \pm bc}{bd}$	
PROPORTION	55	In the Proportion $a:b = c:d$	The product of the means equals the product of the extremes	$ad = bc$	
	56		The extremes may be interchanged	$d:b = c:a$	
	57		The means may be interchanged	$a:c = b:d$	
	58		The means may be interchanged with the extremes	$b:a = d:c$	
	59	Mean Proportional	In the Proportion	$a:b = b:c$	
			Geometric Mean	$b = \pm\sqrt{ac}$	
VARIATION	60	k = Constant of Proportionality	Direct	$y \propto x$ or $y = kx$	
	61		Inverse	$y \propto \frac{1}{x}$ or $y = \frac{k}{x}$	
	62		Joint	$y \propto xw$ or $y = kxw$	

	No.				
SYSTEMS OF LINEAR EQUATIONS	63	Algebraic solution	$a_1x + b_1y = c_1$ $a_2x + b_2y = c_2$	$x = \dfrac{b_2c_1 - b_1c_2}{a_1b_2 - a_2b_1}$, and $y = \dfrac{a_1c_2 - a_2c_1}{a_1b_2 - a_2b_1}$	where $a_1b_2 - a_2b_1 \neq 0$
	64	Algebraic solution	$a_1x + b_1y + c_1z = k_1$ $a_2x + b_2y + c_2z = k_2$ $a_3x + b_3y + c_3z = k_3$	$x = \dfrac{b_2c_3k_1 + b_1c_2k_3 + b_3c_1k_2 - b_2c_1k_3 - b_3c_2k_1 - b_1c_3k_2}{a_1b_2c_3 + a_3b_1c_2 + a_2b_3c_1 - a_3b_2c_1 - a_1b_3c_2 - a_2b_1c_3}$ $y = \dfrac{a_1c_3k_2 + a_3c_2k_1 + a_2c_1k_3 - a_3c_1k_2 - a_1c_2k_3 - a_2c_3k_1}{a_1b_2c_3 + a_3b_1c_2 + a_2b_3c_1 - a_3b_2c_1 - a_1b_3c_2 - a_2b_1c_3}$ $z = \dfrac{a_1b_2k_3 + a_3b_1k_2 + a_2b_3k_1 - a_3b_2k_1 - a_1b_3k_2 - a_2b_1k_3}{a_1b_2c_3 + a_3b_1c_2 + a_2b_3c_1 - a_3b_2c_1 - a_1b_3c_2 - a_2b_1c_3}$	
	65	Determinants — Value of a Determinant	Second Order	$\begin{vmatrix} a_1 & b_1 \\ a_2 & b_2 \end{vmatrix} = a_1b_2 - a_2b_1$	
	66	Determinants — Value of a Determinant	Third Order	$\begin{vmatrix} a_1 & b_1 & c_1 \\ a_2 & b_2 & c_2 \\ a_3 & b_3 & c_3 \end{vmatrix} = a_1b_2c_3 + a_3b_1c_2 + a_2b_3c_1 - (a_3b_2c_1 + a_1b_3c_2 + a_2b_1c_3)$	
	67	Determinants — Value of a Determinant	Minors	The signed minor of element b in the determinant $\begin{vmatrix} a & b & c \\ d & e & f \\ g & h & i \end{vmatrix}$ is $-\begin{vmatrix} d & f \\ g & i \end{vmatrix}$	
	68	Determinants — Value of a Determinant	Minors	To find the value of a determinant: 1. Choose any row or any column to develop by minors. 2. Write the product of every element in that row or column and its signed minor. 3. Add these products to get the value of the determinant.	
	69	Determinants — Cramer's Rule		The solution for any variable is a fraction whose denominator is the determinant of the coefficients, and whose numerator is also the determinant of the coefficients, except that the column of coefficients of the variable being solved for is replaced by the column of constants	
	70	Determinants — Cramer's Rule	Two Equations	$x = \dfrac{\begin{vmatrix} c_1 & b_1 \\ c_2 & b_2 \end{vmatrix}}{\begin{vmatrix} a_1 & b_1 \\ a_2 & b_2 \end{vmatrix}}$ and $y = \dfrac{\begin{vmatrix} a_1 & c_1 \\ a_2 & c_2 \end{vmatrix}}{\begin{vmatrix} a_1 & b_1 \\ a_2 & b_2 \end{vmatrix}}$	
	71	Determinants — Cramer's Rule	Three Equations	$x = \dfrac{\begin{vmatrix} k_1 & b_1 & c_1 \\ k_2 & b_2 & c_2 \\ k_3 & b_3 & c_3 \end{vmatrix}}{\Delta}$, $y = \dfrac{\begin{vmatrix} a_1 & k_1 & c_1 \\ a_2 & k_2 & c_2 \\ a_3 & k_3 & c_3 \end{vmatrix}}{\Delta}$, $z = \dfrac{\begin{vmatrix} a_1 & b_1 & k_1 \\ a_2 & b_2 & k_2 \\ a_3 & b_3 & k_3 \end{vmatrix}}{\Delta}$ Where $\Delta = \begin{vmatrix} a_1 & b_1 & c_1 \\ a_2 & b_2 & c_2 \\ a_3 & b_3 & c_3 \end{vmatrix} \neq 0$	

Section	No.	Group	Topic	Formula / Statement	
SYSTEMS OF LINEAR EQUATIONS (Continued)	72	Properties of Determinants	Zero Row or Column	If all elements in a row (or column) are zero, the value of the determinant is zero	
	73		Identical Rows or Columns	The value of a determinant is zero if two rows (or columns) are identical	
	74		Zeros below the Principal Diagonal	If all elements below the principal diagonal are zeros, then the value of the determinant is the product of the elements along the principal diagonal	
	75		Interchanging Rows with Columns	The value of a determinant is unchanged if we change the rows to columns and the columns to rows	
	76		Interchange of Rows or Columns	A determinant will change sign when we interchange two rows (or columns)	
	77		Multiplying by a Constant	If each element in a row (or column) is multiplied by some constant, the value of the determinant is multiplied by that constant	
	78		Multiples of One Row or Column Added to Another	The value of a determinant is unchanged when the elements of a row (or column) are multiplied by some factor, and then added to the corresponding elements of another row or column	
MATRICES	79	Addition	Commutative Law	$\mathbf{A} + \mathbf{B} = \mathbf{B} + \mathbf{A}$	
	80		Associative Law	$\mathbf{A} + (\mathbf{B} + \mathbf{C}) = (\mathbf{A} + \mathbf{B}) + \mathbf{C} = (\mathbf{A} + \mathbf{C}) + \mathbf{B}$	
	81		Addition and Subtraction	$\begin{pmatrix} a & b \\ c & d \end{pmatrix} + \begin{pmatrix} w & x \\ y & z \end{pmatrix} = \begin{pmatrix} a+w & b+x \\ c+y & d+z \end{pmatrix}$	
	82	Multiplication	Conformable Matrices	The product **AB** of two matrices **A** and **B** is defined only when the number of columns in **A** equals the number of rows in **B**	
	83		Commutative Law	$\mathbf{AB} \neq \mathbf{BA}$ Matrix multiplication is <u>not</u> commutative	
	84		Associative Law	$\mathbf{A}(\mathbf{BC}) = (\mathbf{AB})\mathbf{C} = \mathbf{ABC}$	
	85		Distributive Law	$\mathbf{A}(\mathbf{B} + \mathbf{C}) = \mathbf{AB} + \mathbf{BC}$	
	86		Dimensions of the Product	$(m \times p)(p \times n) = (m \times n)$	
	87		Product of a Scalar and a Matrix	$k \begin{pmatrix} a & b \\ c & d \end{pmatrix} = \begin{pmatrix} ka & kb \\ kc & kd \end{pmatrix}$	
	88		Scalar Product of a Row Vector and a Column Vector	$(1 \times 2)\ (2 \times 1) \quad (1 \times 1)$ $(a \quad b) \begin{pmatrix} x \\ y \end{pmatrix} = (ax + by)$	
	89		Tensor Product of a Column Vector and a Row Vector	$(2 \times 1)\ (1 \times 2) \quad (2 \times 2)$ $\begin{pmatrix} a \\ b \end{pmatrix} (x \quad y) = \begin{pmatrix} ax & ay \\ bx & by \end{pmatrix}$	
	90		Product of a Row Vector and a Matrix	$(1 \times 2) \quad (2 \times 3) \quad (1 \times 3)$ $(a \quad b) \begin{pmatrix} u & v & w \\ x & y & z \end{pmatrix} = (au + bx \quad av + by \quad aw + bz)$	

Section	No.	Subsection	Topic	Formula
MATRICES (Continued)	91	Matrix Multiplication (Continued)	Product of a Matrix and a Column Vector	(2 X 2) (2 X 1) (2 X 1) $\begin{pmatrix} a & b \\ c & d \end{pmatrix} \begin{pmatrix} x \\ y \end{pmatrix} = \begin{pmatrix} ax + by \\ cx + dy \end{pmatrix}$
	92		Product of Two Matrices	(2 X 3) (3 X 2) (2 X 2) $\begin{pmatrix} a & b & c \\ d & e & f \end{pmatrix} \begin{pmatrix} u & x \\ v & y \\ w & z \end{pmatrix} = \begin{pmatrix} au + bv + cw & ax + by + cz \\ du + ev + fw & dx + ey + fz \end{pmatrix}$
	93		Product of a Matrix and Its Inverse	$\mathbf{AA}^{-1} = \mathbf{A}^{-1}\mathbf{A} = \mathbf{I}$
	94		Multiplying by the Unit Matrix	$\mathbf{AI} = \mathbf{IA} = \mathbf{A}$
	95	Solving Systems of Equations	Matrix Form for a System of Equation	$\mathbf{AX} = \mathbf{B}$
	96		Elementary Transformations of a Matrix	1. Interchange any rows 2. Multiply a row by a nonzero constant 3. Add a constant multiple of one row to another row
	97		Gauss Elimination	When we transform the coefficient matrix **A** into the unit matrix **I**, the column vector **B** gets transformed into the solution set
	98		Solving a Set of Equations Using the Inverse	$\mathbf{X} = \mathbf{A}^{-1}\mathbf{B}$
QUADRATICS	99	General Form		$ax^2 + bx + c = 0$
	100	Quadratic Formula		$x = \dfrac{-b \pm \sqrt{b^2 - 4ac}}{2a}$
	101	Nature of the Roots	If a, b, and c are real, and	if $b^2 - 4ac > 0$ the roots are real and unequal if $b^2 - 4ac = 0$ the roots are real and equal if $b^2 - 4ac < 0$ the roots are not real
	102	Polynomial of Degree n		$a_0x^n + a_1x^{n-1} + \cdots + a_{n-1}x + a_n$
	103	Factor Theorem		If a polynomial equation $f(x) = 0$ has a root r, then $(x - r)$ is a factor of the polynomial $f(x)$; conversely, if $(x - r)$ is a factor of a polynomial $f(x)$, then r is a root of $f(x) = 0$

	No.				Page
INTERSECTING LINES	104	Opposite angles of two intersecting straight lines are equal			
	105	If two parallel straight lines are cut by a transversal, corresponding angles are equal and alternate interior angles are equal			
	106	If two lines are cut by a number of parallels, the corresponding segments are proportional			
QUADRILATERALS	107	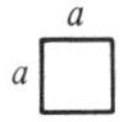 a, a	Square	Area = a^2	
	108	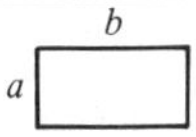 b, a	Rectangle	Area = ab	
	109	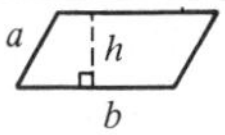 a, h, b	Parallelogram: Diagonals bisect each other	Area = bh	
	110	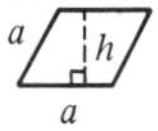 a, h, a	Rhombus: Diagonals intersect at right angles	Area = ah	
	111	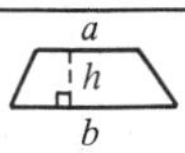 a, h, b	Trapezoid	Area = $\dfrac{(a+b)h}{2}$	
POLYGON	112	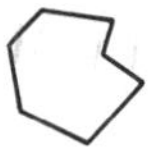 n sides	Sum of Angles = $(n-2)\ 180°$		
CIRCLES	113	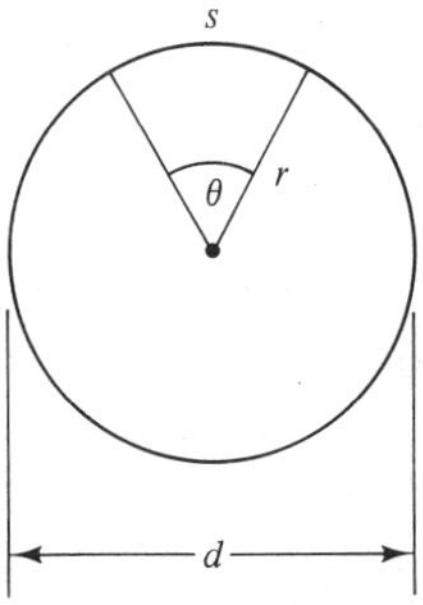 s, θ, r, d	Circumference = $2\pi r = \pi d$		
	114		Area = $\pi r^2 = \dfrac{\pi d^2}{4}$		280
	115		Central angle θ (radians) = $\dfrac{s}{r}$		
	116		Area of sector = $\dfrac{rs}{2} = \dfrac{r^2\theta}{2}$		
	117		1 revolution = 2π radians = 360°		
	118		Any angle inscribed in a semicircle is a right angle		
	119	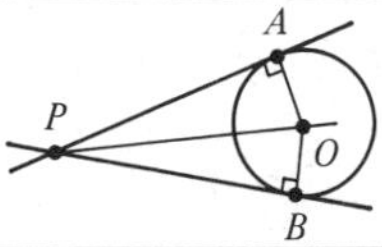 A, P, O, B	Tangents to a Circle	Tangent AP is perpendicular to radius OA	
	120			Tangent AP = tangent BP OP bisects angle APB	
	121	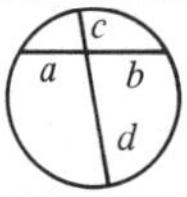 c, a, b, d	Intersecting Chords	$ab = cd$	

	No.				
SOLIDS	122	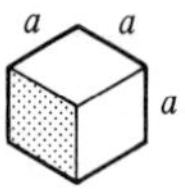 a a a	Cube	Volume = a^3	
	123			Surface area = $6a^2$	
	124	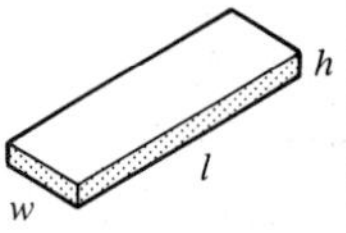 h l w	Rectangular Parallel-epiped	Volume = lwh	
	125			Surface area = $2(lw + hw + lh)$	
	126	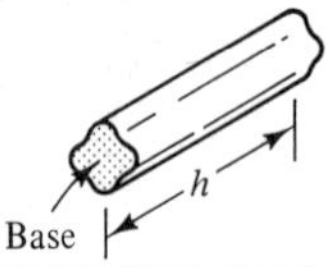 Base h	Any Cylinder or Prism	Volume = (area of base)(altitude)	
	127		Right Cylinder or Prism	Lateral area (not incl. bases) = (perimeter of base)(altitude)	
	128	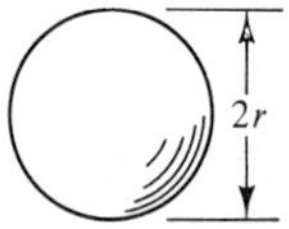 $2r$	Sphere	Volume = $\frac{4}{3}\pi r^3$	
	129			Surface area = $4\pi r^2$	
	130		Any Cone or Pyramid	Volume = $\frac{1}{3}$(area of base)(altitude)	
	131		Right Circular Cone or Regular Pyramid	Lateral area = $\frac{1}{2}$(perimeter of base) × (slant height)	
	132	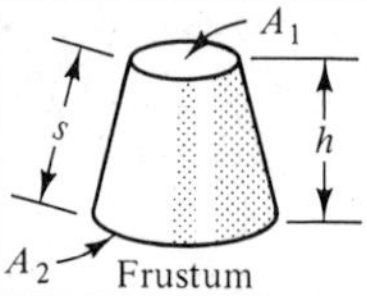 A_1 s h A_2 Frustum	Any Cone or Pyramid	Volume = $\frac{h}{3}(A_1 + A_2 + \sqrt{A_1A_2})$	
	133		Right Circular Cone or Regular Pyramid	Lateral area = $\frac{s}{2}$(sum of base perimeters) = $\frac{s}{2}(P_1 + P_2)$	
SIMILAR FIGURES	134	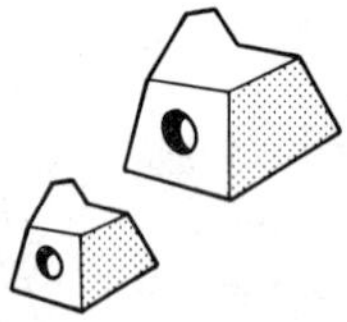	Corresponding dimensions of plane or solid similar figures are in proportion		
	135		Areas of similar plane or solid figures are proportional to the squares of any two corresponding dimensions		
	136		Volumes of similar solid figures are proportional to the cubes of any two corresponding dimensions		

	No.				Page
ANY TRIANGLE	137	Areas 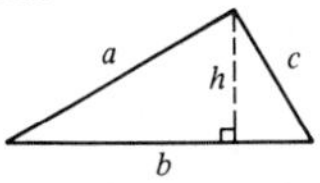		Area = $\frac{1}{2}bh$	280
	138		Hero's Formula:	Area = $\sqrt{s(s-a)(s-b)(s-c)}$ where $s = \frac{1}{2}(a+b+c)$	
	139	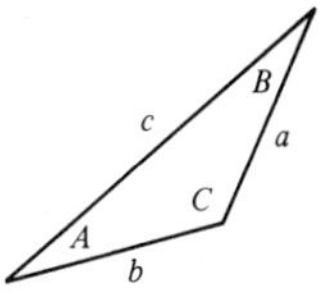	Sum of the Angles	$A + B + C = 180°$	
	140		Law of Sines	$\frac{a}{\sin A} = \frac{b}{\sin B} = \frac{c}{\sin C}$	
	141		Law of Cosines	$a^2 = b^2 + c^2 - 2bc \cos A$ $b^2 = a^2 + c^2 - 2ac \cos B$ $c^2 = a^2 + b^2 - 2ab \cos C$	
	142	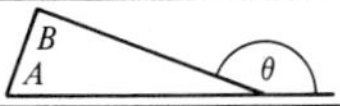	Exterior Angle	$\theta = A + B$	
SIMILAR TRIANGLES	143	If two angles of a triangle equal two angles of another triangle, the triangles are similar			
	144	Corresponding sides of similar triangles are in proportion			
RIGHT TRIANGLES	145	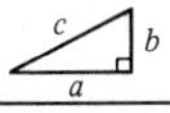	Pythagorean Theorem	$a^2 + b^2 = c^2$	
	146	Trigonometric Ratios 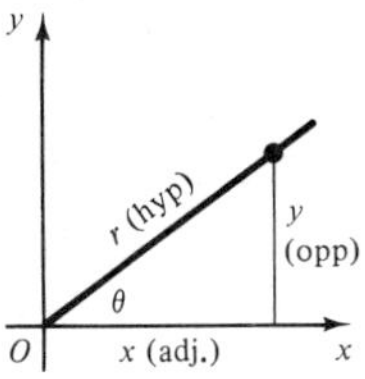	Sine	$\sin\theta = \frac{y}{r} = \frac{\text{opposite side}}{\text{hypotenuse}}$	
	147		Cosine	$\cos\theta = \frac{x}{r} = \frac{\text{adjacent side}}{\text{hypotenuse}}$	
	148		Tangent	$\tan\theta = \frac{y}{x} = \frac{\text{opposite side}}{\text{adjacent side}}$	
	149		Cotangent	$\cot\theta = \frac{x}{y} = \frac{\text{adjacent side}}{\text{opposite side}}$	
	150		Secant	$\sec\theta = \frac{r}{x} = \frac{\text{hypotenuse}}{\text{adjacent side}}$	
	151		Cosecant	$\csc\theta = \frac{r}{y} = \frac{\text{hypotenuse}}{\text{opposite side}}$	
	152	Reciprocal Relationships	(a) $\sin\theta = \frac{1}{\csc\theta}$	(b) $\cos\theta = \frac{1}{\sec\theta}$ (c) $\tan\theta = \frac{1}{\cot\theta}$	
	153	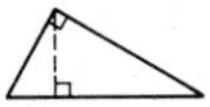	In a right triangle, the altitude to the hypotenuse forms two right triangles which are similar to each other and to the original triangle		
	154	A and B Are Complementary Angles 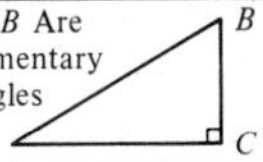	Cofunctions	(a) $\sin A = \cos B$ (b) $\cos A = \sin B$ (c) $\tan A = \cot B$ (d) $\cot A = \tan B$ (e) $\sec A = \csc B$ (f) $\csc A = \sec B$	
CONGRUENT TRIANGLES	155	Two Triangles Are Congruent If	Two angles and a side of one are equal to two angles and the corresponding side of the other (ASA), (AAS)		
	156		Two sides and the included angle of one are equal, respectively to two sides and the included angle of the other (SAS)		
	157		Three sides of one are equal to the three sides of the other (SSS)		
COORDINATE SYSTEMS	158	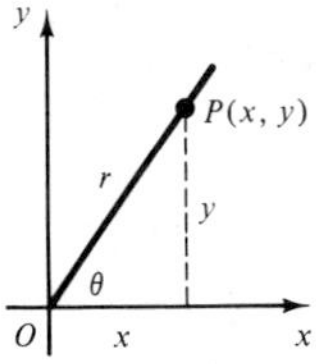	Rectangular	$x = r\cos\theta$	
	159			$y = r\sin\theta$	
	160		Polar	$r = \sqrt{x^2 + y^2}$	
	161			$\theta = \arctan\frac{y}{x}$	

TRIGONOMETRIC IDENTITIES

No.		Identity	
162	Quotient Relations	$\tan \theta = \dfrac{\sin \alpha}{\cos \alpha}$	
163		$\cot \alpha = \dfrac{\cos \alpha}{\sin \alpha}$	
164	Pythagorean Relations	$\sin^2 \alpha + \cos^2 \alpha = 1$	
165		$1 + \tan^2 \alpha = \sec^2 \alpha$	
166		$1 + \cot^2 \alpha = \csc^2 \alpha$	
167	Sum or Difference of Two Angles	$\sin(\alpha \pm \beta) = \sin \alpha \cos \beta \pm \cos \alpha \sin \beta$	
168		$\cos(\alpha \pm \beta) = \cos \alpha \cos \beta \mp \sin \alpha \sin \beta$	
169		$\tan(\alpha \pm \beta) = \dfrac{\tan \alpha \pm \tan \beta}{1 \mp \tan \alpha \tan \beta}$	
170	Double-Angle Relations	$\sin 2\alpha = 2 \sin \alpha \cos \alpha$	
171		(a) $\cos 2\alpha = \cos^2 \alpha - \sin^2 \alpha$ (b) $\cos 2\alpha = 1 - 2 \sin^2 \alpha$ (c) $\cos 2\alpha = 2 \cos^2 \alpha - 1$	
172		$\tan 2\alpha = \dfrac{2 \tan \alpha}{1 - \tan^2 \alpha}$	
173	Half-Angle Relations	$\sin \dfrac{\alpha}{2} = \pm \sqrt{\dfrac{1 - \cos \alpha}{2}}$	
174		$\cos \dfrac{\alpha}{2} = \mp \sqrt{\dfrac{1 + \cos \alpha}{2}}$	
175		(a) $\tan \dfrac{\alpha}{2} = \dfrac{1 - \cos \alpha}{\sin \alpha}$ (b) $\tan \dfrac{\alpha}{2} = \dfrac{\sin \alpha}{1 + \cos \alpha}$ (c) $\tan \dfrac{\alpha}{2} = \pm \sqrt{\dfrac{1 - \cos \alpha}{1 + \cos \alpha}}$	
176	Sum or Difference of Two Functions	$\sin \alpha + \sin \beta = 2 \sin \frac{1}{2}(\alpha + \beta) \cos \frac{1}{2}(\alpha - \beta)$	
177		$\sin \alpha - \sin \beta = 2 \cos \frac{1}{2}(\alpha + \beta) \sin \frac{1}{2}(\alpha - \beta)$	
178		$\cos \alpha + \cos \beta = 2 \cos \frac{1}{2}(\alpha + \beta) \cos \frac{1}{2}(\alpha - \beta)$	
179		$\cos \alpha - \cos \beta = 2 \sin \frac{1}{2}(\alpha + \beta) \sin \frac{1}{2}(\alpha - \beta)$	
180	Product of Two Functions	$\sin \alpha \sin \beta = \frac{1}{2} \cos(\alpha - \beta) - \frac{1}{2} \cos(\alpha + \beta)$	
181		$\cos \alpha \cos \beta = \frac{1}{2} \cos(\alpha - \beta) + \frac{1}{2} \cos(\alpha + \beta)$	
182		$\sin \alpha \cos \beta = \frac{1}{2} \sin(\alpha + \beta) + \frac{1}{2} \sin(\alpha - \beta)$	
183	Inverse Trigonometric Functions	$\text{arcsine}\, \theta = \arctan \dfrac{\theta}{\sqrt{1 - \theta^2}}$	
184		$\arccos \theta = \arctan \dfrac{\sqrt{1 - \theta^2}}{\theta}$	

	No.				
LOGARITHMS	185	Exponential to Logarithmic Form		If $b^x = y$ then $x = \log_b y$ $(y > 0,\ b > 0)$	
	186	Laws of Logarithms	Products	$\log_b MN = \log_b M + \log_b N$	
	187		Quotients	$\log_b \frac{M}{N} = \log_b M - \log_b N$	
	188		Powers	$\log_b M^p = p \log_b M$	
	189		Roots	$\log_b \sqrt[q]{M} = \frac{1}{q} \log_b M$	
	190		Log of 1	$\log_b 1 = 0$	
	191		Log of the Base	$\log_b b = 1$	
	192	Log of the Base Raised to a Power		$\log_b b^n = n$	
	193	Change of Base		$\log N = \frac{\ln N}{\ln 10} = \frac{\ln N}{2.3026}$	

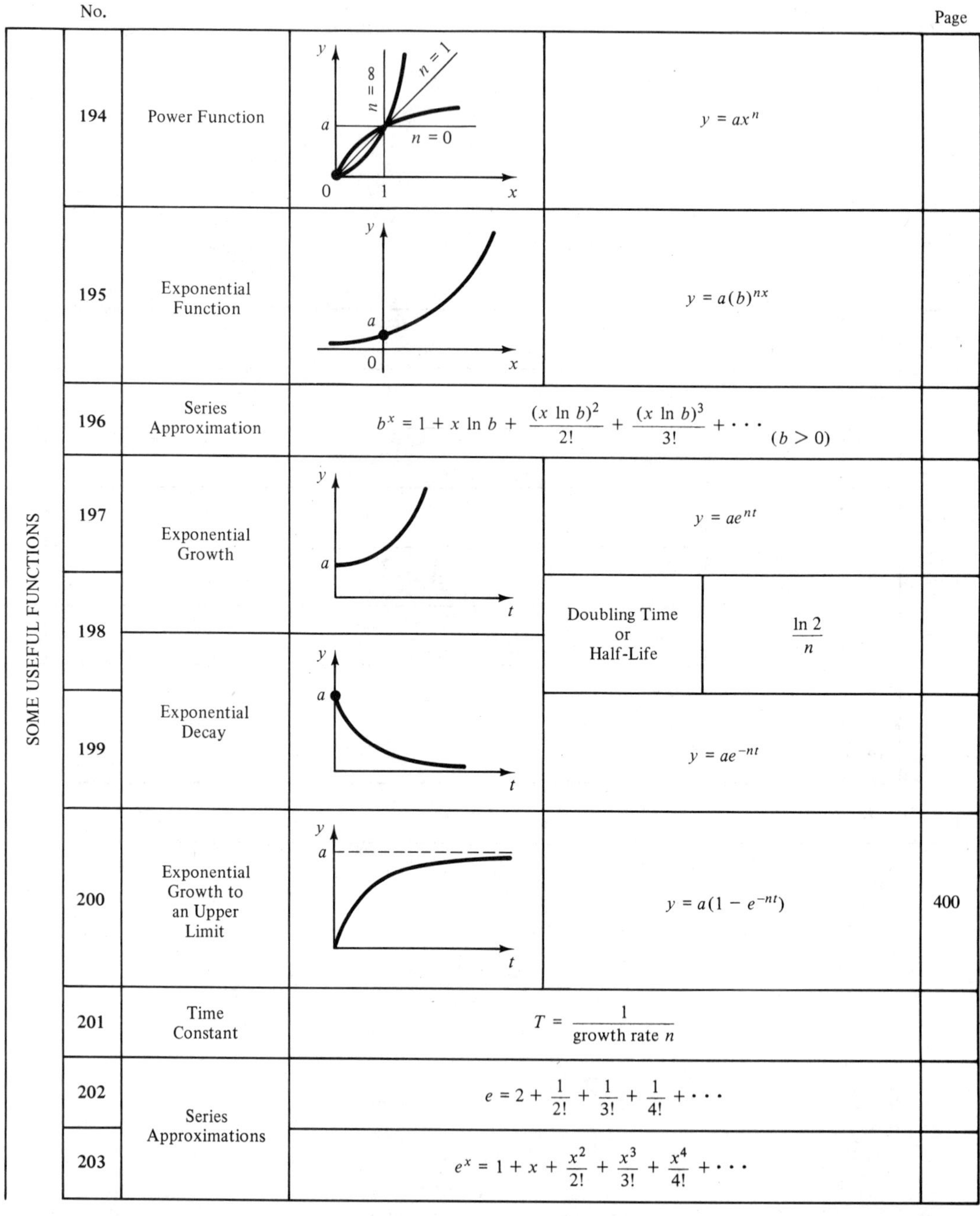

	No.				Page
SOME USEFUL FUNCTIONS	194	Power Function		$y = ax^n$	
	195	Exponential Function		$y = a(b)^{nx}$	
	196	Series Approximation	$b^x = 1 + x \ln b + \frac{(x \ln b)^2}{2!} + \frac{(x \ln b)^3}{3!} + \cdots \quad (b > 0)$		
	197	Exponential Growth		$y = ae^{nt}$	
	198			Doubling Time or Half-Life: $\frac{\ln 2}{n}$	
	199	Exponential Decay		$y = ae^{-nt}$	
	200	Exponential Growth to an Upper Limit		$y = a(1 - e^{-nt})$	400
	201	Time Constant	$T = \frac{1}{\text{growth rate } n}$		
	202	Series Approximations	$e = 2 + \frac{1}{2!} + \frac{1}{3!} + \frac{1}{4!} + \cdots$		
	203		$e^x = 1 + x + \frac{x^2}{2!} + \frac{x^3}{3!} + \frac{x^4}{4!} + \cdots$		

SOME USEFUL FUNCTIONS (Continued)

No.				
204	Logarithmic Function	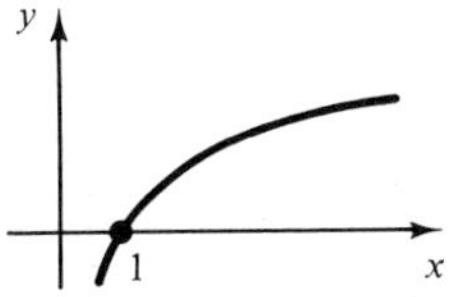	$y = \log_b x$ $x > 0,\ b > 0$	
205	Series Approximation	$\ln x = 2a + \dfrac{2a^3}{3} + \dfrac{2a^5}{5} + \dfrac{2a^7}{7} + \cdots$ where $a = \dfrac{x-1}{x+1}$		
206	Sine Wave of Amplitude a	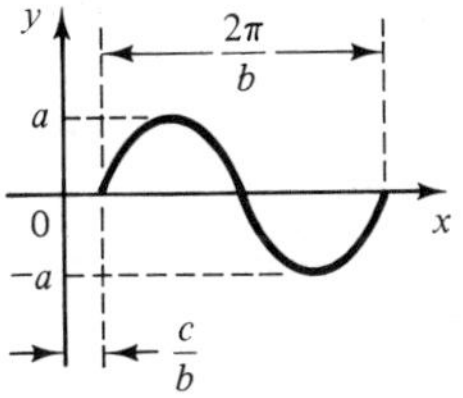	$y = a \sin(bx + c)$	
207			Period $= \dfrac{360}{b}$ deg/cycle $= \dfrac{2\pi}{b}$ rad/cycle	
208			Frequency $= \dfrac{b}{360}$ cycle/deg $= \dfrac{b}{2\pi}$ cycle/rad	
209			Phase Displacement $= \dfrac{c}{b}$	
210	Series Approximations	$\sin x = x - \dfrac{x^3}{3!} + \dfrac{x^5}{5!} - \dfrac{x^7}{7!} + \cdots$		
211		$\cos x = 1 - \dfrac{x^2}{2!} + \dfrac{x^4}{4!} - \dfrac{x^6}{6!} + \cdots$		
212	Binomial Formula	$(a+b)^n = a^n + na^{n-1}b + \dfrac{n(n-1)}{2}a^{n-2}b^2 + \cdots$ $+ \dfrac{n(n-1)\cdots(n-r+2)}{(r-1)!}a^{n-r+1}b^{r-1} + \cdots + nab^{n-1} + b^n$		

	No.					Page
COMPLEX NUMBERS	213	Powers of j			$j = \sqrt{-1}, \quad j^2 = -1, \quad j^3 = -j, \quad j^4 = 1, \quad j^5 = i$, etc.	
	214		Rectangular Form		$a + jb$	
	215			Sums	$(a + jb) + (c + jd) = (a + c) + j(b + d)$	
	216			Differences	$(a + jb) - (c + jd) = (a - c) + j(b - d)$	
	217			Products	$(a + jb)(c + jd) = (ac - bd) + j(ad + bc)$	
	218			Quotients	$\frac{a + jb}{c + jd} = \frac{ac + bd}{c^2 + d^2} + j\,\frac{bc - ad}{c^2 + d^2}$	
	219		Trigonometric Form		$a + jb = r(\cos\theta + j\sin\theta)$	
	220			where	$a = r\cos\theta$	
	221	Imaginary			$b = r\sin\theta$	
	222	$(a + jb)$			$r = \sqrt{a^2 + b^2}$	
	223	r b			$\theta = \arctan\frac{b}{a}$	
	224	θ a Real	Polar Form		$r\underline{/\theta} = a + jb$	
	225			Products	$r\underline{/\theta} \cdot r'\underline{/\theta'} = rr'\underline{/\theta + \theta'}$	
	226			Quotients	$r\underline{/\theta} \div r'\underline{/\theta'} = \frac{r}{r'}\underline{/\theta - \theta'}$	
	227			Roots and Powers	DeMoivre's Theorem: $(r\underline{/\theta})^n = r^n\underline{/n\theta}$	
	228		Exponential Form	Euler's Formula	$re^{j\theta} = r(\cos\theta + j\sin\theta)$	417, 511
	229			Products	$r_1e^{j\theta_1} \cdot r_2e^{j\theta_2} = r_1r_2e^{j(\theta_1 + \theta_2)}$	
	230			Quotients	$\frac{r_1e^{j\theta_1}}{r_2e^{j\theta_2}} = \frac{r_1}{r_2}e^{j(\theta_1 - \theta_2)}$	
	231			Powers and Roots	$(re^{j\theta})^n = r^ne^{jn\theta}$	

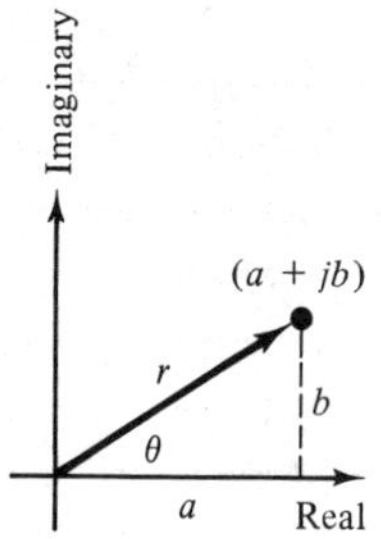

	No.			Truth Table	Venn Diagram	Switch Diagram	Logic Gate
BOOLEAN ALGEBRA AND SETS	232	Boolean Operators	AND	A B $A \cdot B$ 0 0 0 0 1 0 1 0 0 1 1 1	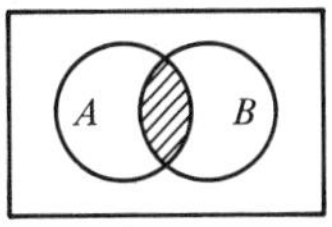 A B Intersection $A \cap B = \{x : x \in A, x \in B\}$	 A B In Series	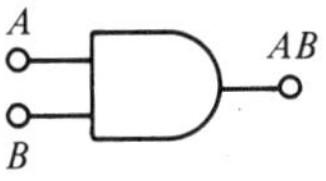 A B AB AND Gate
	233		OR	A B $A + B$ 0 0 0 0 1 1 1 0 1 1 1 1	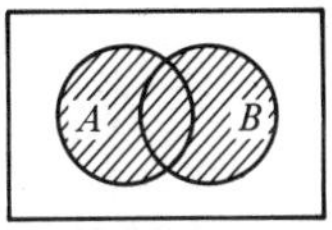 A B Union $A \cup B = \{x : x \in A \text{ or } x \in B\}$	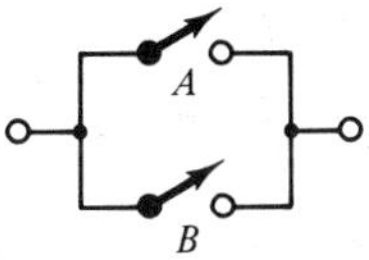 A B In Parallel	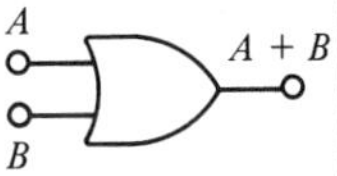 A B $A + B$ OR Gate
	234		NOT	A $\overline{A}$ 0 1 1 0	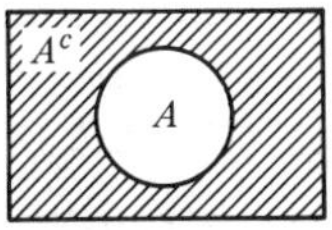 A^c A Complement $A^c = \{x : x \in U, x \in A\}$	A $\overline{A}$	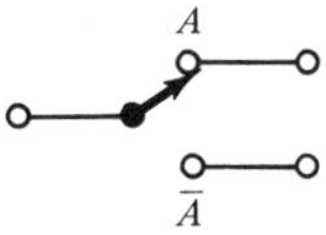 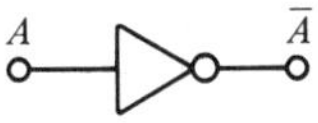 A $\overline{A}$ Inverter
	235		Exclusive OR	A B $A \oplus B$ 0 0 0 0 1 1 1 0 1 1 1 0	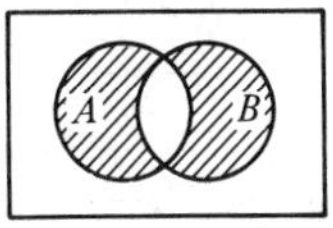 A B $A \oplus B = \{x : x \in A \text{ or } x \in B, x \in A \cap B\}$	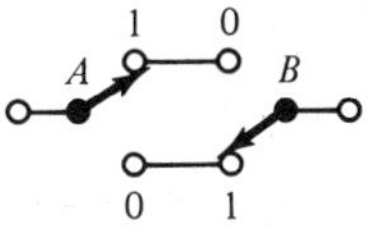 1 0 A B 0 1	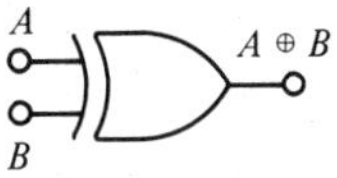 A B $A \oplus B$ Exclusive OR Gate

BOOLEAN ALGEBRA (Continued)

Laws of Boolean Algebra

No.		AND	OR
236	Cummutative Laws	(a) $AB \equiv BA$	(b) $A + B \equiv B + A$
237	Boundedness Laws	(a) $A \cdot 0 \equiv 0$	(b) $A + 1 \equiv 1$
238	Identity Laws	(a) $A \cdot 1 \equiv A$	(b) $A + 0 \equiv A$
239	Idempotent Laws	(a) $AA \equiv A$	(b) $A + A \equiv A$
240	Complement Laws	(a) $A \cdot \overline{A} \equiv 0$	(b) $A + \overline{A} \equiv 1$
241	Associative Laws	(a) $A(BC) \equiv (AB)C$	(b) $A + (B + C) \equiv (A + B) + C$
242	Distributive Laws	(a) $A(B + C) \equiv AB + AC$	(b) $A + BC \equiv (A + B)(A + C)$
243		(a) $A(\overline{A} + B) \equiv AB$	(b) $A + \overline{A} \cdot B \equiv A + B$
244	Absorption Laws	(a) $A(A + B) \equiv A$	(b) $A + (AB) \equiv A$
245	DeMorgan's Laws	(a) $\overline{A \cdot B} \equiv \overline{A} + \overline{B}$	(b) $\overline{A + B} \equiv \overline{A} \cdot \overline{B}$
246	Involution Law	$\overline{\overline{A}} \equiv A$	

	No.				Page
THE STRAIGHT LINE	247		Distance Formula	$d = \sqrt{(\Delta x)^2 + (\Delta y)^2} = \sqrt{(x_2 - x_1)^2 + (y_2 - y_1)^2}$	5
	248		Slope m	$m = \dfrac{\text{rise}}{\text{run}} = \dfrac{\Delta y}{\Delta x} = \dfrac{y_2 - y_1}{x_2 - x_1}$	7
	249		Slope m	$m = \tan$ (angle of inclination) $= \tan \theta$ $0 \leq \theta < 180°$	9
	250	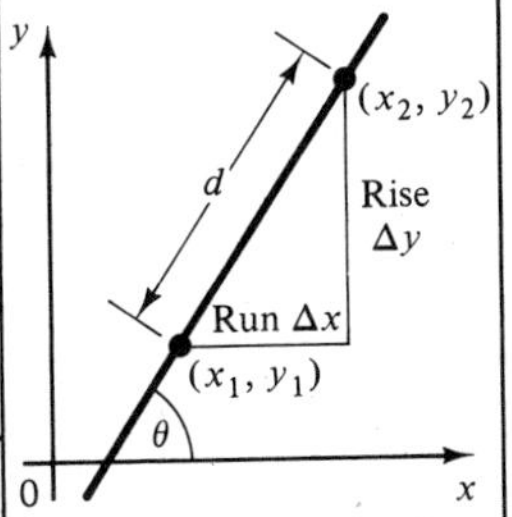 Equation of Straight Line	General Form	$Ax + By + C = 0$	19
	251	Equation of Straight Line	Parallel to x axis	$y = b$	23
	252	Equation of Straight Line	Parallel to y axis	$x = a$	24
	253	Equation of Straight Line	Slope-Intercept Form	$y = mx + b$	18
	254	Equation of Straight Line	Two-Point Form	$\dfrac{y - y_1}{x - x_1} = \dfrac{y_2 - y_1}{x_2 - x_1}$	22
	255	Equation of Straight Line	Point-Slope Form	$m = \dfrac{y - y_1}{x - x_1}$	20
	256	Equation of Straight Line	Intercept Form	$\dfrac{x}{a} + \dfrac{y}{b} = 1$	
	257	Equation of Straight Line	Polar Form	$r \cos(\theta - \beta) = p$	
	258	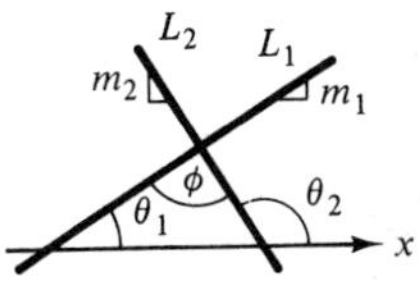	If L_1 and L_2 are parallel, then	$m_1 = m_2$	11
	259		If L_1 and L_2 are perpendicular, then	$m_1 = -\dfrac{1}{m_2}$	12
	260		Angle of Intersection	$\tan \phi = \dfrac{m_2 - m_1}{1 + m_1 m_2}$	13
CONIC SECTIONS	261	Any Conic	General Second-Degree Equation	$Ax^2 + Bxy + Cy^2 + Dx + Ey + F = 0$	36
	262	Any Conic	Translation of Axes	To translate or shift the axes of a curve to the left by a distance h and downward by a distance k, replace x by $(x - h)$ and y by $(y - k)$ in the equation of the curve.	35
	263	Any Conic	Eccentricity 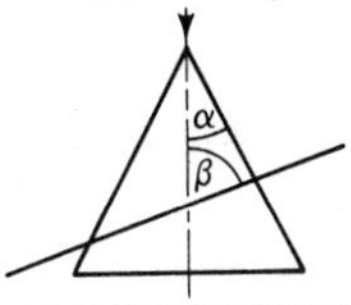	$e = \dfrac{\cos \beta}{\cos \alpha}$	
	264	Any Conic	Eccentricity	$e = 0$ for the Circle $0 < e < 1$ for the Ellipse $e = 1$ for the Parabola $e > 1$ for the Hyperbola	
	265	Any Conic	Definition of a Conic	$PF = e \cdot PD$	
	266	Any Conic	Polar Equation for the Conics	$r = \dfrac{ke}{1 - e \cos \theta}$	

CONIC SECTIONS (Continued)

No.					Page
267	Circle of Radius r	The set of points in a plane equidistant from a fixed point.			34
268	Circle of Radius r	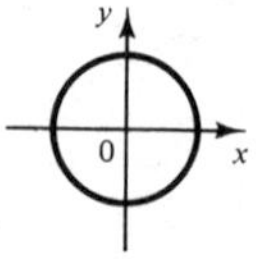	Standard Form	$x^2 + y^2 = r^2$	34
269	Circle of Radius r	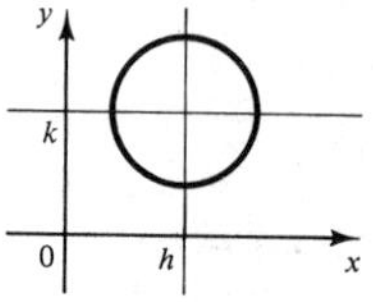	Standard Form	$(x - h)^2 + (y - k)^2 = r^2$	34
270	Circle of Radius r	General Form		$x^2 + y^2 + Dx + Ey + F = 0$	36
271	Parabola	The set of points in a plane such that the distance PF from each point to a fixed point (the focus) is equal to the distance PD to a fixed line (the directrix). $PF = PD$			41
272	Parabola	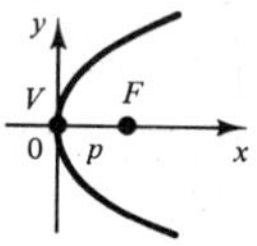	Standard Form	$y^2 = 4px$	43
273	Parabola	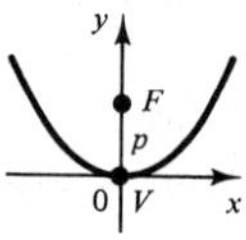	Standard Form	$x^2 = 4py$	44
274	Parabola	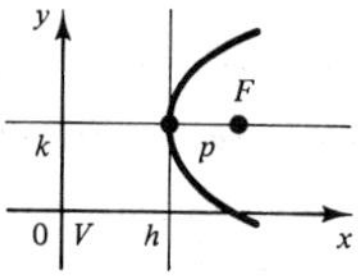	Standard Form	$(y - k)^2 = 4p(x - h)$	45
275	Parabola	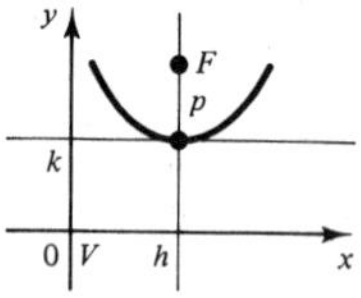	Standard Form	$(x - h)^2 = 4p(y - k)$	45
276	Parabola	General Form		$Cy^2 + Dx + Ey + F = 0$ or $Ax^2 + Dx + Ey + F = 0$	47
277	Parabola	Length of Latus Rectum		$L = \|4p\|$	45
278	Parabola	Area	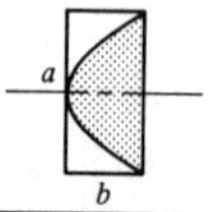	Area $= \frac{2}{3}ab$	280

CONIC SECTIONS (Continued)

No.					Page
279	Ellipse	The set of points in a plane such that the sum of the distances PF and PF' from each point to two fixed points (the foci) is constant, and equal to the length of the major axis. $PF + PF' = 2a$			54
280	Ellipse	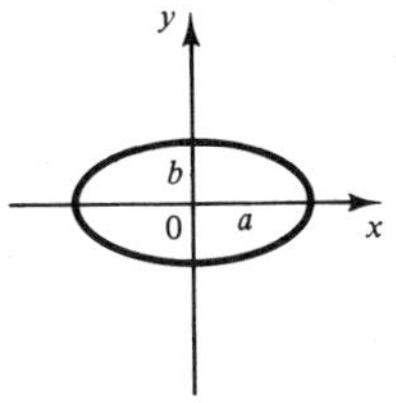	Standard Form	$\frac{x^2}{a^2} + \frac{y^2}{b^2} = 1$ $a > b$	58
281	Ellipse	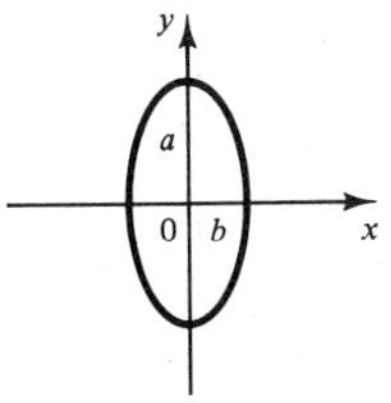	Standard Form	$\frac{y^2}{a^2} + \frac{x^2}{b^2} = 1$ $a > b$	59
282	Ellipse	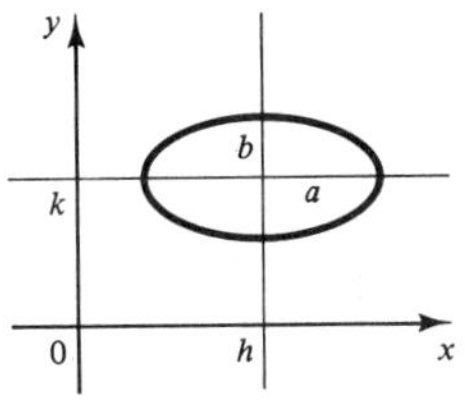	Standard Form	$\frac{(x-h)^2}{a^2} + \frac{(y-k)^2}{b^2} = 1$ $a > b$	60
283	Ellipse	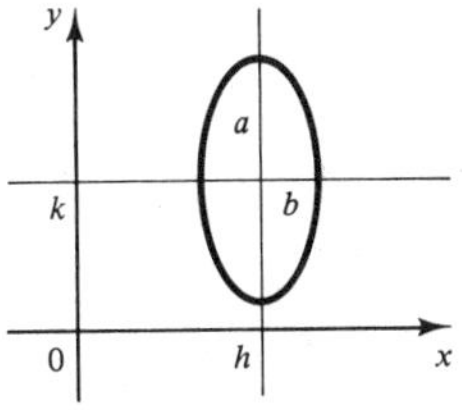	Standard Form	$\frac{(y-k)^2}{a^2} + \frac{(x-h)^2}{b^2} = 1$ $a > b$	61
284	Ellipse	General Form		$Ax^2 + Cy^2 + Dx + Ey + F = 0$ $A \neq C$, but have same signs	62
285	Ellipse	Distance from Center to Focus		$c = \sqrt{a^2 - b^2}$	56
286	Ellipse	Focal Width		$L = \frac{2b^2}{a}$	64
287	Ellipse	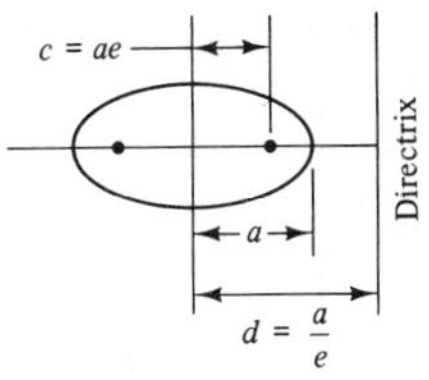		Eccentricity $e = \frac{a}{d} = \frac{c}{a}$	
288	Ellipse	Area = πab			280

CONIC SECTIONS (Continued)

No.				Page
289	Hyperbola	The set of points in a plane such that the difference of the distances PF and PF' from each point to two fixed points (the foci) is constant, and equal to the distance between the vertices. $PF' - PF = 2a$		67
290	Hyperbola	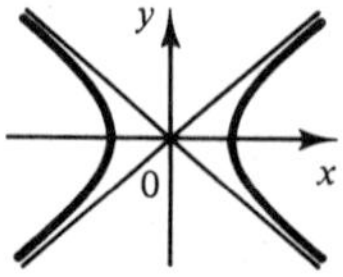 Standard Form	$\frac{x^2}{a^2} - \frac{y^2}{b^2} = 1$	70
291	Hyperbola	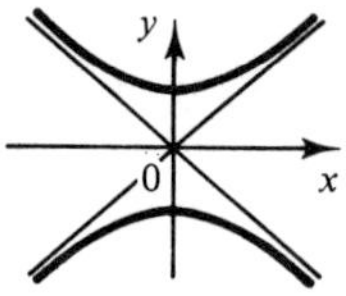 Standard Form	$\frac{y^2}{a^2} - \frac{x^2}{b^2} = 1$	70
292	Hyperbola	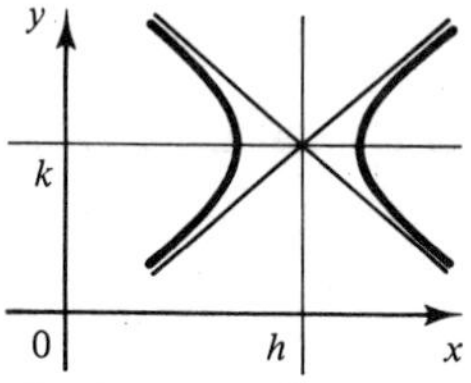 Standard Form	$\frac{(x-h)^2}{a^2} - \frac{(y-k)^2}{b^2} = 1$	73
293	Hyperbola	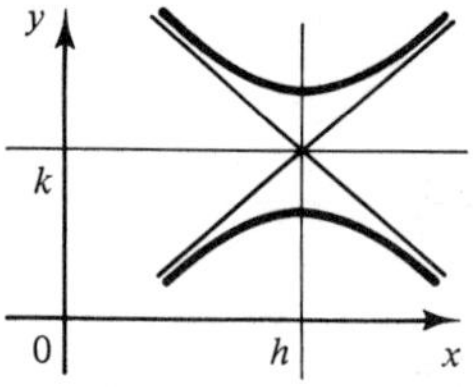 Standard Form	$\frac{(y-k)^2}{a^2} - \frac{(x-h)^2}{b^2} = 1$	73
294	Hyperbola	General Form	$Ax^2 + Cy^2 + Dx + Ey + F = 0$ $A \neq C$, and have opposite signs	74
295	Hyperbola	Distance to Focus	$c = \sqrt{a^2 + b^2}$	69
296	Hyperbola	Slope of Asymptotes — Transverse Axis Horizontal	Slope $= \pm \frac{b}{a}$	70
297	Hyperbola	Slope of Asymptotes — Transverse Axis Vertical	Slope $= \pm \frac{a}{b}$	70
298	Hyperbola	Length of Latus Rectum	$L = \frac{2b^2}{a}$	71
299	Hyperbola	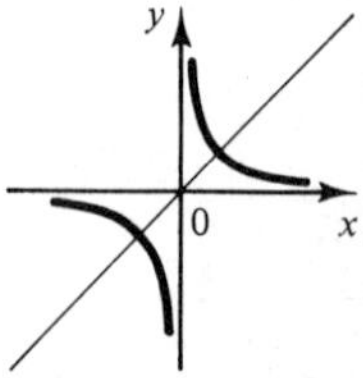	Axes Rotated 45°: $xy = k$	76

	No.				Page
DIFFERENTIAL CALCULUS	300	Limit Notation		$\lim_{x \to a} f(x) = L$	83
	301	Increments		$\Delta x = x_2 - x_1, \qquad \Delta y = y_2 - y_1$	4
	302	Definition of the Derivative		$\frac{dy}{dx} = \lim_{\Delta x \to 0} \frac{\Delta y}{\Delta x} = \lim_{\Delta x \to 0} \frac{f(x + \Delta x) - f(x)}{\Delta x}$	94
	303	The Chain Rule		$\frac{dy}{dx} = \frac{dy}{du} \cdot \frac{du}{dx}$	109
	304	Rules for Derivatives	Of a Constant	$\frac{d(c)}{dx} = 0$	101
	305		Of a Power Function	$\frac{d}{dx} x^n = nx^{n-1}$	102
	306		Of a Constant Times a Function	$\frac{d(cu)}{dx} = c \frac{du}{dx}$	116
	307		Of a Constant Times a Power of x	$\frac{d}{dx} cx^n = cnx^{n-1}$	102
	308		Of a Sum	$\frac{d}{dx}(u + v + w) = \frac{du}{dx} + \frac{dv}{dx} + \frac{dw}{dx}$	105
	309		Of a Function Raised to a Power	$\frac{d(u^n)}{dx} = nu^{n-1} \frac{du}{dx}$	109
	310		Of a Product	$\frac{d(uv)}{dx} = u \frac{dv}{dx} + v \frac{du}{dx}$	114
	311		Of a Product of Three Factors	$\frac{d(uvw)}{dx} = uv \frac{dw}{dx} + uw \frac{dv}{dx} + vw \frac{du}{dx}$	115
	312		Of a Product of n Factors	The derivative is an expression of n terms, each term being the product of $n - 1$ of the factors and the derivative of the other factor	115
	313		Of a Quotient	$\frac{d}{dx}\left(\frac{u}{v}\right) = \frac{v \frac{du}{dx} - u \frac{dv}{dx}}{v^2}$	117

No.					Page
314	DIFFERENTIAL CALCULUS (Continued)	Rules for Derivatives (cont.)	Of the Trigonometric Functions	$\frac{d(\sin u)}{dx} = \cos u \frac{du}{dx}$	197, 202
315				$\frac{d(\cos u)}{dx} = -\sin u \frac{du}{dx}$	198, 202
316				$\frac{d(\tan u)}{dx} = \sec^2 u \frac{du}{dx}$	202
317				$\frac{d(\cot u)}{dx} = -\csc^2 u \frac{du}{dx}$	202
318				$\frac{d(\sec u)}{dx} = \sec u \tan u \frac{du}{dx}$	202
319				$\frac{d(\csc u)}{dx} = -\csc u \cot u \frac{du}{dx}$	202
320			Of the Inverse Trigonometric Functions	$\frac{d(\sin^{-1} u)}{dx} = \frac{1}{\sqrt{1-u^2}} \frac{du}{dx} \quad -1 < u < 1$	207
321				$\frac{d(\cos^{-1} u)}{dx} = \frac{-1}{\sqrt{1-u^2}} \frac{du}{dx} \quad -1 < u < 1$	207
322				$\frac{d(\tan^{-1} u)}{dx} = \frac{1}{1+u^2} \frac{du}{dx}$	207
323				$\frac{d(\cot^{-1} u)}{dx} = \frac{-1}{1+u^2} \frac{du}{dx}$	207
324				$\frac{d(\sec^{-1} u)}{dx} = \frac{1}{u\sqrt{u^2-1}} \frac{du}{dx} \quad \lvert u \rvert > 1$	207
325				$\frac{d(\csc^{-1} u)}{dx} = \frac{-1}{u\sqrt{u^2-1}} \frac{du}{dx} \quad \lvert u \rvert > 1$	207
326			Of Logarithmic and Exponential Functions	(a) $\frac{d}{dx}(\log_b u) = \frac{1}{u} \log_b e \frac{du}{dx}$ (b) $\frac{d}{dx}(\log_b u) = \frac{1}{u \ln b} \frac{du}{dx}$	210, 211
327				$\frac{d}{dx}(\ln u) = \frac{1}{u} \frac{du}{dx}$	211
328				$\frac{d}{dx} b^u = b^u \frac{du}{dx} \ln b$	217
329				$\frac{d}{dx} e^u = e^u \frac{du}{dx}$	217
330		Graphical Applications	Maximum and Minimum Points	To find maximum and minimum points (and other stationary points) set the first derivative equal to zero and solve for x.	133
331			First-Derivative Test	The first derivative is negative to the left of, and positive to the right of, a minimum point. The reverse is true for a maximum point.	133
332			Second Derivative Test	If the first derivative at some point is zero, then, if the second derivative is 1. Positive, the point is a minimum 2. Negative, the point is a maximum 3. Zero, the test fails	135
333			Inflection Points	To find points of inflection, set the second derivative to zero and solve for x. Test by seeing if the second derivative changes sign a small distance to either side of the point.	138
334			Newton's Method	$x_{n+1} = x_n - \frac{f(x_n)}{f'(x_n)}$	141
335		Differentials	Differential of y	$dy = f'(x)\, dx$	186
336			Approximations Using Differentials	$\Delta y \cong \frac{dy}{dx} \Delta x$	188

	No.					Page
INTEGRAL CALCULUS	337	The Indefinite Integral			$\int F'(x)\,dx = F(x) + C$	225
	338	The Indefinite Integral			$\int f(x)\,dx = F(x) + C$	225
	339	The Definite Integral	The Fundamental Theorem		$A = \int_a^b f(x)\,dx = F(b) - F(a)$	259, 267
	340		Defined by Riemann Sums		$A = \lim\limits_{\Delta x \to 0} \sum\limits_{i=1}^{n} f(x_i^*)\,\Delta x = \int_a^b f(x)\,dx$	265
	341		Properties of the Definite Integral		$\int_a^b c\,f(x)\,dx = c\int_a^b f(x)\,dx$	
	342				$\int_a^b [f(x) + g(x)]\,dx = \int_a^b f(x)\,dx + \int_a^b g(x)\,dx$	
	343				$\int_a^b f(x)\,dx = -\int_b^a f(x)\,dx$	
	344				$\int_a^b f(x)\,dx = \int_a^c f(x)\,dx + \int_c^b f(x)\,dx$	
	345	Approximate Integration		Midpoint Method	$A \cong \sum\limits_{i=1}^{n} f(x_i^*)\,\Delta x$ where $f(x_i^*)$ is the height of the ith panel at its midpoint	264
	346			Average Ordinate Method	$A \cong y_{avg}\,(b - a)$	366
	347			Trapezoid Rule	With unequal spacing $A \cong \frac{1}{2}[(x_1 - x_0)(y_1 + y_0) + (x_2 - x_1)(y_2 - y_1) + \cdots + (x_n - x_{n-1})(y_n + y_{n-1})]$	367
	348				With equal spacing, $h = x_1 - x_0$: $A \cong h[\frac{1}{2}(y_0 + y_n) + y_1 + y_2 + \cdots + y_{n-1}]$	367
	349			Prismoidal Formula	$A = \frac{h}{3}(y_0 + 4y_1 + y_2)$	369
	350			Simpson's Rule	$A \cong \frac{h}{3}(y_0 + 4y_1 + 2y_2 + 4y_3 + \cdots + 4y_{n-1} + y_n)$	370

	No.					Page
APPLICATIONS OF THE DEFINITE INTEGRAL	351	Volumes of Solids of Revolution	Disk Method	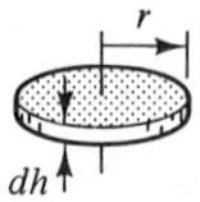	Volume $= dV = \pi r^2\, dh$	288
	352			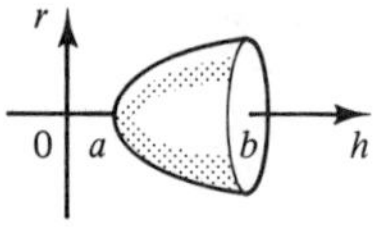	$V = \pi \int_a^b r^2\, dh$	289
	353		Ring Method	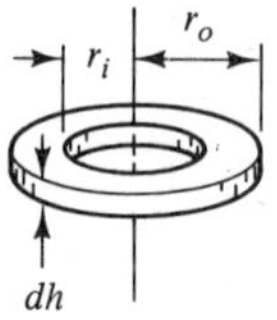	$dV = \pi(r_o^2 - r_i^2)\, dh$	288
	354			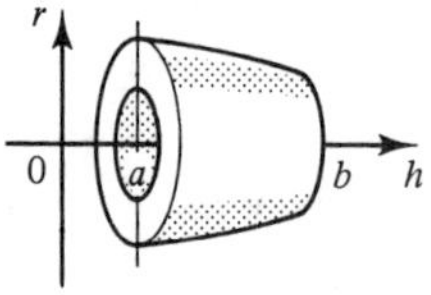	$V = \pi \int_a^b (r_o^2 - r_i)^2\, dh$	292
	355		Shell Method	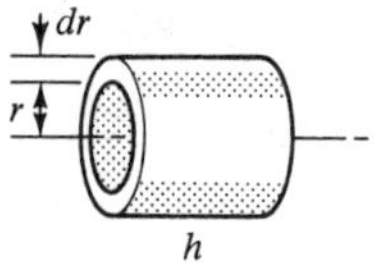	$dV = 2\pi rh\, dr$	288
	356			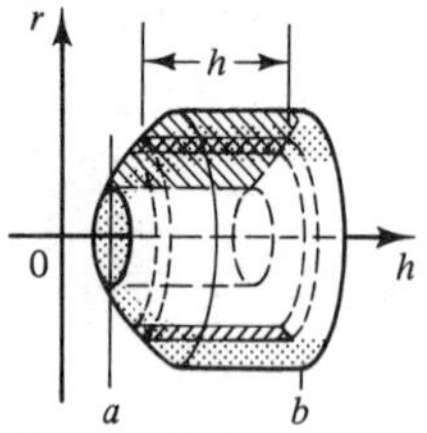	$V = 2\pi \int_a^b rh\, dr$	291

	No.				Page
APPLICATIONS OF THE DEFINITE INTEGRAL	357	Length of Arc	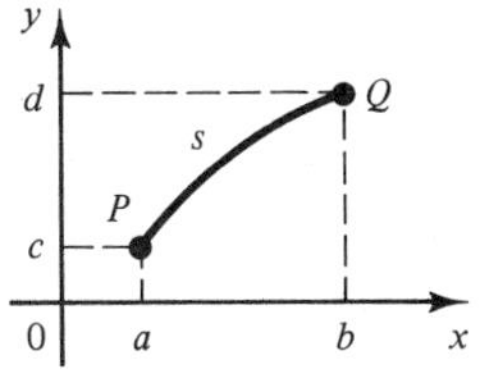	$s = \int_a^b \sqrt{1 + \left(\frac{dy}{dx}\right)^2}\, dx$	298
	358			$s = \int_c^d \sqrt{1 + \left(\frac{dx}{dy}\right)^2}\, dy$	299
	359	Surface Area	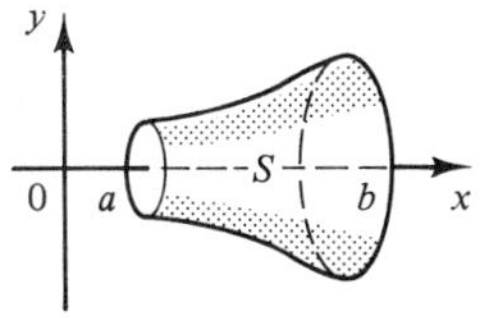	About x Axis: $S = 2\pi \int_a^b y \sqrt{1 + \left(\frac{dy}{dx}\right)^2}\, dx$	302
	360		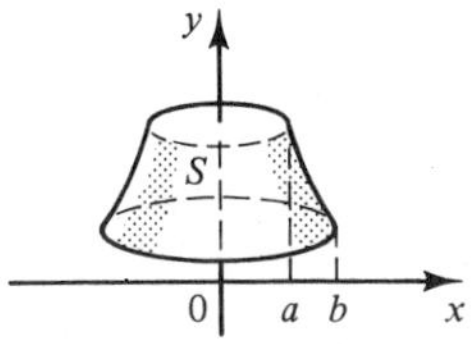	About y Axis: $S = 2\pi \int_a^b x \sqrt{1 + \left(\frac{dy}{dx}\right)^2}\, dx$	303
	361	Centroids	Of Plane Area: 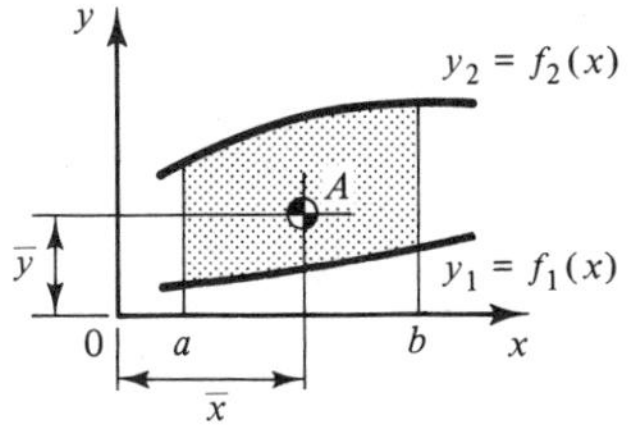	$\bar{x} = \frac{1}{A} \int_a^b x(y_2 - y_1)\, dx$	314
	362			$\bar{y} = \frac{1}{2A} \int_a^b (y_1 + y_2)(y_2 - y_1)\, dx$	314
	363		Of Solid of Revolution of Volume V: 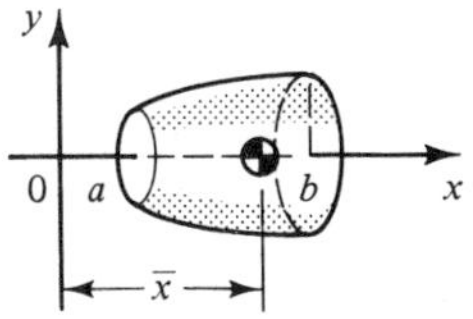	About x Axis: $\bar{x} = \frac{\pi}{V} \int_a^b xy^2\, dx$	318
	364		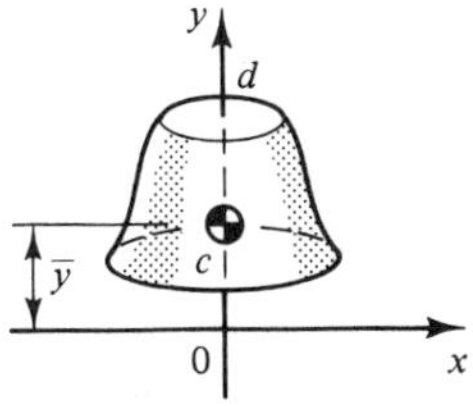	About y Axis: $\bar{y} = \frac{\pi}{V} \int_c^d yx^2\, dy$	319

APPLICATIONS OF THE DEFINITE INTEGRAL (Continued)

No.						Page
365	Moment of Inertia	Of Areas	Thin Strip: 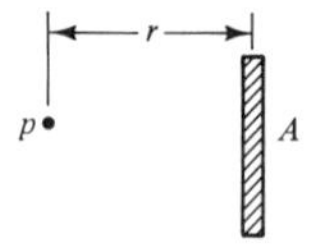 r, p, A		$I_p = Ar^2$	331
366	Moment of Inertia	Of Areas	Extended Area: 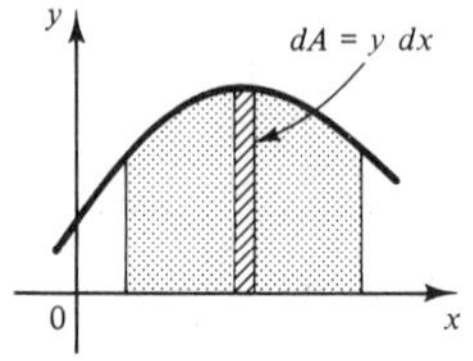 $dA = y\,dx$; y, 0, x		$I_x = \frac{1}{3}\int y^3\,dx$	334
367	Moment of Inertia	Of Areas			$I_y = \int x^2 y\,dx$	333
368	Moment of Inertia	Of Areas			Polar $I_0 = I_x + I_y$	
369	Moment of Inertia	Of Areas			Radius of Gyration: $r = \sqrt{\frac{I}{A}}$	333
370	Moment of Inertia	Of Areas	s, A, B, A		Parallel Axis Theorem: $I_B = I_A + As^2$	
371	Moment of Inertia	Of Masses	r, p, M		$I_p = Mr^2$	
372	Moment of Inertia	Of Masses	Disk: 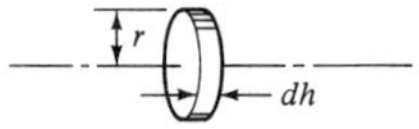 r, dh	About Axis of Revolution (Polar Moment of Inertia)	$dI = \frac{m\pi}{2} r^4\,dh$	288
373	Moment of Inertia	Of Masses	Ring: 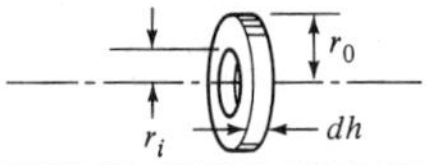 r_0, r_i, dh	About Axis of Revolution (Polar Moment of Inertia)	$dI = \frac{m\pi}{2}(r_0^4 - r_i^4)\,dh$	288
374	Moment of Inertia	Of Masses	Shell: 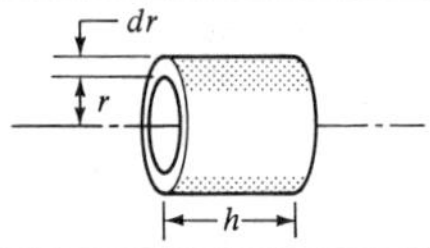 dr, r, h	About Axis of Revolution (Polar Moment of Inertia)	$dI = 2\pi m r^3 h\,dr$	288, 335
375	Moment of Inertia	Of Masses	Solid of Revolution: 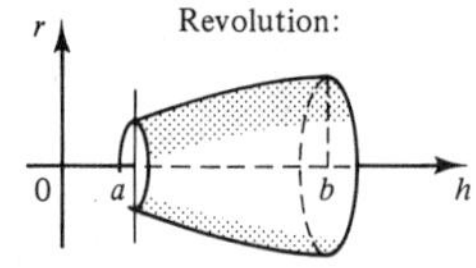 r, 0, a, b, h	About Axis of Revolution (Polar Moment of Inertia)	Disk Method: $I = \frac{m\pi}{2}\int_a^b r^4\,dh$	337
376	Moment of Inertia	Of Masses		About Axis of Revolution (Polar Moment of Inertia)	Shell Method: $I = 2\pi m \int r^3 h\,dr$	335
377	Moment of Inertia	Of Masses	s, B, A		Parallel Axis Theorem: $I_B = I_A + Ms^2$	
378	Average and rms Values		y, $y = f(x)$, 0, a, b, x		Average Ordinate: $y_{\text{avg}} = \frac{1}{b-a}\int_a^b f(x)\,dx$	305
379	Average and rms Values				Root-Mean-Square Value: $\text{rms} = \sqrt{\frac{1}{b-a}\int_a^b [f(x)]^2\,dx}$	306

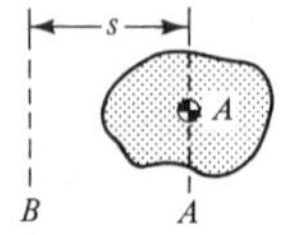

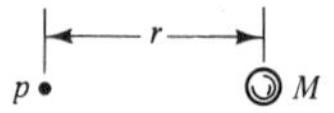

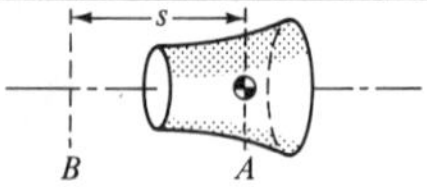

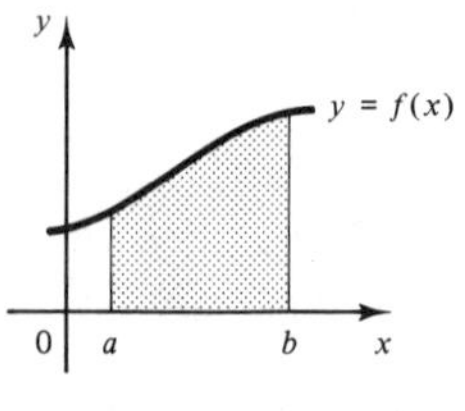

	No.						Page
DIFFERENTIAL EQUATIONS	380	First-Order	Variables Separable		$f(y)\,dy = g(x)\,dx$		379
	381		Integrable Combinations		$x\,dy + y\,dx = d(xy)$		384
	382				$\frac{x\,dy - y\,dx}{x^2} = d\left(\frac{y}{x}\right)$		384
	383				$\frac{y\,dx - x\,dy}{y^2} = d\left(\frac{x}{y}\right)$		384
	384				$\frac{x\,dy - y\,dx}{x^2 + y^2} = d\left(\tan^{-1}\frac{y}{x}\right)$		384
	385		Homogeneous		$M\,dx + N\,dy = 0$ (Substitute $y = vx$)		386
	386		First-Order Linear	Form	$y' + Py = Q$		389
	387			Integrating Factor	$R = e^{\int P dx}$		390
	388			Solution	$ye^{\int P dx} = \int Qe^{\int P dx}\,dx$		391
	389	Second-Order	Right Size Zero	Form	$ay'' + by' + cy = 0$		413
	390			Auxiliary Equation	$am^2 + bm + c = 0$		414
	☒			Form of Solution	Roots of Auxiliary Equation	Solution	
	391				Real and Unequal	$y = c_1 e^{m_1 x} + c_2 e^{m_2 x}$	414, 418
	392				Real and Equal	$y = c_1 e^{mx} + c_2 x e^{mx}$	416, 418
	393				Non-real	(a) $y = e^{ax}(C_1 \cos bx + C_2 \sin bx)$ or (b) $y = Ce^{ax} \sin(bx + \phi)$	417 / 418
	394		Right Side Not Zero	Form	$ay'' + by' + cy = f(x)$		411
	395			Complete Solution	$y = y_c + y_p$ (y_c: complementary function; y_p: particular integral)		422
	396	Bernoulli's Equation			$\frac{dy}{dx} + Py = Qy^n$ (Substitute $z = y^{1-n}$)		392

DIFFERENTIAL EQUATIONS (Continued)

No.				Formula	Page
397	Laplace Transform		Definition	$\mathcal{L}[f(t)] = \int_0^\infty f(t)e^{-st}\,dt$	452
398	Laplace Transform		Inverse Transform	$\mathcal{L}^{-1}[F(s)] = f(t)$	458
399	Numerical Solution	Euler's Method		$x_q = x_p + \Delta x$ $y_q = y_p + m_p \Delta x$	476
400	Numerical Solution	Modified Euler's Method		$x_j = x_i + \Delta x$ $y_j = y_i + \left(\frac{m_i + m_j}{2}\right)\Delta x$	477
401	Numerical Solution	Runge-Kutta Method		$x_q = x_p + \Delta x$ $y_q = y_p + m_{\text{avg}}\,\Delta x$ where $m_{\text{avg}} = \left(\frac{1}{6}\right)(m_p + 2m_r + 2m_s + m_q)$ and $m_p = f'(x_p, y_p)$ $m_r = f'\left(\frac{x_p + \Delta x}{2}, \frac{y_p + m_p\,\Delta x}{2}\right)$ $m_s = f'\left(\frac{x_p + \Delta x}{2}, \frac{y_p + m_r\,\Delta x}{2}\right)$ $m_q = f'(x_p + \Delta x, y_p + m_s\,\Delta x)$	479

	No.					Page
INFINITE SERIES	402	Power Series	Notation		$u_1 + u_2 + u_3 + \cdots + u_n + \cdots$	490
	403		Tests for Convergence	Terms Approach Zero	$\lim_{n \to \infty} u_n = 0$	492
	404			Partial Sum Test	$\lim_{n \to \infty} S_n = S$	492
	405			Ratio Test	If $\lim_{n \to \infty} \left\| \dfrac{u_{n-1}}{u_n} \right\|$ (a) is less than 1, the series converges. (b) is greater than 1, the series diverges. (c) is equal to 1, the test fails.	495
	406		Maclaurin's Series		$f(x) = f(0) + f'(0)x + \dfrac{f''(0)}{2!} x^2 + \cdots + \dfrac{f^{(n)}(0)}{n!} x^n + \cdots$	498
	407		Taylor's Series		$f(x) = f(a) + f'(a)(x - a) + \dfrac{f''(a)}{2!} (x - a)^2 + \cdots + \dfrac{f^{(n)}(a)}{n!} (x - a)^n + \cdots$	503
	408			Remainder after n Terms	$R_n = \dfrac{(x - a)^n}{n!} f^{(n)}(c)$ where c lies between a and x.	507
	409	Fourier Series	Period of 2π		$f(x) = a_0/2 + a_1 \cos x + a_2 \cos 2x + a_3 \cos 3x + \cdots + a_n \cos nx + \cdots$ $+ b_1 \sin x + b_2 \sin 2x + b_3 \sin 3x + \cdots + b_n \sin nx + \cdots$	517, 520
	410			where	$a_0 = \dfrac{1}{\pi} \int_{-\pi}^{\pi} f(t)\, dt$	518, 520
	411				$a_n = \dfrac{1}{\pi} \int_{-\pi}^{\pi} f(t) \cos nx\, dt$	519, 520
	412				$b_n = \dfrac{1}{\pi} \int_{-\pi}^{\pi} f(t) \sin nx\, dt$	519, 520
	413		Period of $2L$		$f(x) = \dfrac{a_0}{2} + a_1 \cos \dfrac{\pi x}{L} + a_2 \cos \dfrac{2\pi x}{L} + a_3 \cos \dfrac{3\pi x}{L} + \cdots$ $+ b_1 \sin \dfrac{\pi x}{L} + b_2 \sin \dfrac{2\pi x}{L} + b_3 \sin \dfrac{3\pi x}{L} + \cdots$	531
	414			where	$a_0 = \dfrac{1}{L} \int_{-L}^{L} f(x)\, dx$	531
	415				$a_n = \dfrac{1}{L} \int_{-L}^{L} f(x) \cos \dfrac{n\pi x}{L}\, dx$	531
	416				$b_n = \dfrac{1}{L} \int_{-L}^{L} f(x) \sin \dfrac{n\pi x}{L}\, dx$	531
	417		Waveform Symmetries	Odd and Even Functions	(a) Odd functions have only sine terms (and no constant term) (b) Even functions have only cosine terms (and may have a constant term).	525
	418			Half-Wave Symmetry	A wave that has half-wave symmetry has only odd harmonics.	527

APPLICATIONS

<table>
<tr><th></th><th>No.</th><th></th><th colspan="2"></th><th>Page</th></tr>
<tr><td rowspan="4">MIXTURES</td><td>A1</td><td rowspan="2">Mixture Containing Ingredients $A, B, C, \ldots$</td><td colspan="2">Total amount of mixture = amount of A + amount of B + · · ·</td><td></td></tr>
<tr><td>A2</td><td colspan="2">Final amount of each ingredient =
initial amount + amount added − amount removed</td><td></td></tr>
<tr><td>A3</td><td>Combination of Two Mixtures</td><td colspan="2">Final amount of A = amount of A from mixture 1 +
amount of A from mixture 2</td><td></td></tr>
<tr><td>A4</td><td>Fluid Flow</td><td colspan="2">Amount of flow = flow rate × elapsed time
$A = QT$</td><td></td></tr>
<tr><td rowspan="3">WORK</td><td>A5</td><td colspan="3">Amount done = rate of work × time worked</td><td></td></tr>
<tr><td>A6</td><td rowspan="2">d, F</td><td>Constant Force</td><td>Work = force × distance = Fd</td><td>326</td></tr>
<tr><td>A7</td><td>Variable Force</td><td>$\text{Work} = \int_a^b F(x)\,dx$</td><td>326</td></tr>
<tr><td rowspan="4">FINANCIAL</td><td>A8</td><td>Unit Cost</td><td colspan="2">$\text{Unit cost} = \dfrac{\text{total cost}}{\text{number of units}}$</td><td></td></tr>
<tr><td>A9</td><td rowspan="3">Interest:
Principal a Invested at Rate n for t years Accumulates to Amount y</td><td>Simple</td><td>$y = a(1 + nt)$</td><td></td></tr>
<tr><td>A10</td><td>Compounded Annually</td><td>$y = a(1 + n)^t$</td><td></td></tr>
<tr><td>A11</td><td>Compounded m times/yr</td><td>$y = a\left(1 + \dfrac{n}{m}\right)^{mt}$</td><td></td></tr>
<tr><td rowspan="5">STATICS</td><td>A12</td><td>a, F, d</td><td>Moment about Point a</td><td>$M_a = Fd$</td><td></td></tr>
<tr><td>A13</td><td rowspan="3">Equations of Equilibrium (Newton's First Law)</td><td colspan="2">The sum of all horizontal forces = 0</td><td></td></tr>
<tr><td>A14</td><td colspan="2">The sum of all vertical forces = 0</td><td></td></tr>
<tr><td>A15</td><td colspan="2">The sum of all moments about any point = 0</td><td></td></tr>
<tr><td>A16</td><td>f, N</td><td>Coefficient of Friction</td><td>$\mu = \dfrac{f}{N}$</td><td></td></tr>
</table>

	No.					Page
MOTION	A17		Uniform Motion (Constant Speed)		Distance = rate × time $D = Rt$	
	A18		Uniformly Accelerated (Constant Acceleration a, Initial Velocity v_0) For free fall, $a = g = 9.807\ \text{m/s}^2 = 32.2\ \text{ft/s}^2$	Displacement at Time t	$s = v_0 t + \frac{at^2}{2}$	249
	A19			Velocity at Time t	$v = v_0 + at$	249
	A20			Newton's Second Law	$F = ma$	
	A21	Linear Motion	Nonuniform Motion	Average Speed	$\text{Average speed} = \frac{\text{total distance traveled}}{\text{total time elapsed}}$	
	A22			Displacement	$s = \int v\, dt$	247
	A23			Instantaneous Velocity	$v = \frac{ds}{dt}$	159
	A24				$v = \int a\, dt$	247
	A25			Instantaneous Acceleration	$a = \frac{dv}{dt} = \frac{d^2 s}{dt^2}$	159
	A26	Rotation	Uniform Motion	Angular Displacement	$\theta = \omega t$	
	A27			Linear Speed of Point at Radius r	$v = \omega r$	
	A28		Nonuniform Motion	Angular Displacement	$\theta = \int \omega\, dt$	251
	A29			Angular Velocity	$\omega = \frac{d\theta}{dt}$	163
	A30				$\omega = \int \alpha\, dt$	251
	A31			Angular Acceleration	$\alpha = \frac{d\omega}{dt} = \frac{d^2\theta}{dt^2}$	164
	A32	Curvilinear Motion:	x and y Components	Displacement	(a) $x = \int v_x\, dt$ (b) $y = \int v_y\, dt$	250
	A33			Velocity	(a) $v_x = \frac{dx}{dt}$ (b) $v_y = \frac{dy}{dt}$	161
	A34				(a) $v_x = \int a_x\, dt$ (b) $v_y = \int a_y\, dt$	250
	A35			Acceleration	(a) $a_x = \frac{dv_x}{dt} = \frac{d^2 x}{dt^2}$ (b) $a_y = \frac{dv_y}{dt} = \frac{d^2 y}{dt^2}$	161

<table>
<tr><th></th><th>No.</th><th></th><th></th><th></th><th colspan="2"></th><th>Page</th></tr>
<tr><td rowspan="7">MECHANICAL VIBRATIONS</td><td>A36</td><td rowspan="7">k
W
x
$P \sin \omega t$
Coefficient of friction = c</td><td rowspan="6">Free Vibrations ($P = 0$)</td><td rowspan="3">Simple Harmonic Motion
(No Damping)</td><td colspan="2">$x = x_0 \cos \omega_n t$</td><td>430</td></tr>
<tr><td>A37</td><td>Undamped Angular Velocity</td><td>$\omega_n = \sqrt{\dfrac{kg}{W}}$</td><td>431</td></tr>
<tr><td>A38</td><td>Natural Frequency</td><td>$f_n = \dfrac{\omega_n}{2\pi}$</td><td>431</td></tr>
<tr><td>A39</td><td rowspan="2">Under-damped</td><td colspan="2">$x = x_0 e^{-at} \cos \omega_d t$</td><td>429</td></tr>
<tr><td>A40</td><td>Damped Angular Velocity</td><td>$\omega_d = \sqrt{\omega_n^2 - \dfrac{c^2 g^2}{\omega^2}}$</td><td>429</td></tr>
<tr><td>A41</td><td>Overdamped</td><td colspan="2">$x = C_1 e^{m_1 t} + C_2 e^{m_2 t}$</td><td>432</td></tr>
<tr><td>A42</td><td>Forced Vibrations</td><td>Maximum Deflection</td><td colspan="2">$x_0 = \dfrac{Pg}{W \sqrt{4a^2 \omega^2 + (\omega_n^2 - \omega^2)^2}}$</td><td>435</td></tr>
</table>

	No.				Page
MATERIAL PROPERTIES	A43	Density		Density = $\frac{\text{weight}}{\text{volume}}$ or $\frac{\text{mass}}{\text{volume}}$	
	A44	Mass		Mass = $\frac{\text{weight}}{\text{acceleration due to gravity}}$	
	A45	Specific Gravity		SG = $\frac{\text{density of substance}}{\text{density of water}}$	
	A46	 Pressure	Total Force on a Surface	Force = pressure × area	322
	A47	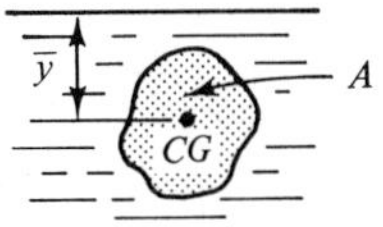	Force on a Submerged Surface	$F = \delta \int y \, dA$	323
	A48			$F = \delta \bar{y} A$	323
	A49	pH		pH = −10 log concentration	
TEMPERATURE	A50	Conversions between Degrees Celsuis (C) and Degrees Fahrenheit (F)		$C = \frac{5}{9}(F - 32)$	
	A51			$F = \frac{9}{5}C + 32$	
STRENGTH OF MATERIALS	A52	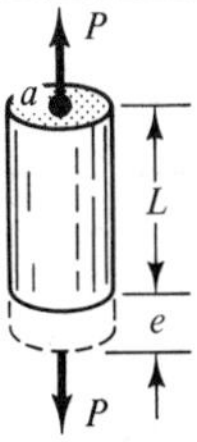 Tension or Compression	Normal Stress	$a = \frac{P}{a}$	
	A53		Strain	$\epsilon = \frac{e}{L}$	
	A54		Modulus of Elasticity and Hooke's Law	$E = \frac{PL}{ae}$	
	A55			$E = \frac{\sigma}{\epsilon}$	
	A56	Thermal Expansion 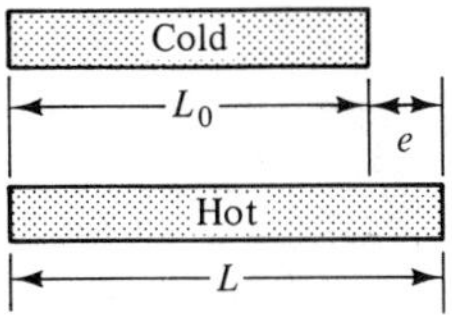 Temperature change = Δt Coefficient of thermal expansion = α	Elongation	$e = \alpha L \ \Delta t$	
	A57		New Length	$L = L_0 (1 + \alpha \Delta t)$	
	A58		Strain	$\epsilon = \frac{e}{L} = \alpha \Delta t$	
	A59		Stress, if Restrained	$\sigma = E\epsilon = E\alpha \ \Delta t$	
	A60		Force, if Restrained	$P = a\sigma = aE\alpha \ \Delta t$	
	A61	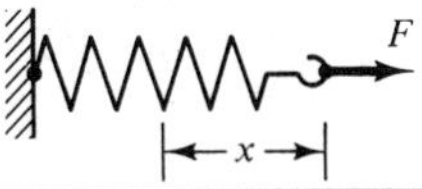	Force needed to Deform a Spring	F = spring constant × distance = kx	

ELECTRICAL TECHNOLOGY

No.				Page
A62	Ohm's Law	Current $= \dfrac{\text{voltage}}{\text{resistance}}$	$I = \dfrac{V}{R}$	
A63	Combinations of Resistors	In Series	$R = R_1 + R_2 + R_3 + \cdots$	
A64		In Parallel	$\dfrac{1}{R} = \dfrac{1}{R_1} + \dfrac{1}{R_2} + \dfrac{1}{R_3} + \cdots$	
A65	Power Dissipated in a Resistor 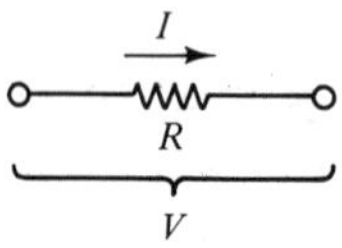		Power $= P = VI$	
A66			$P = \dfrac{V^2}{R}$	
A67			$P = I^2 R$	
A68	Kirchhoff's Laws	Loops	The sum of the voltage rises and drops around any closed loop is zero	
A69		Nodes	The sum of the currents entering and leaving any node is zero	
A70	Resistance Change with Temperature		$R = R_1[1 + \alpha(t - t_1)]$	
A71	Resistance of a Wire 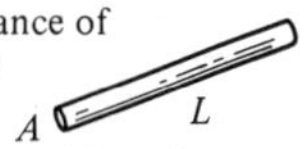		$R = \dfrac{\rho L}{A}$	
A72	Combinations of Capacitors	In Series	$\dfrac{1}{C} = \dfrac{1}{C_1} + \dfrac{1}{C_2} + \dfrac{1}{C_3} + \cdots$	
A73		In Parallel	$C = C_1 + C_2 + C_3 + \cdots$	
A74	Charge on a Capacitor at Voltage V		$Q = CV$	

ELECTRICAL TECHNOLOGY (Continued)

No.			Sinusoidal Form	Complex Form	Page
A75	Alternating Voltage		$v = V_m \sin(\omega t + \theta_1)$	$\mathbf{V} = V_m \angle\theta_1$	
A76	Alternating Current		$i = I_m \sin(\omega t + \theta_2)$	$\mathbf{I} = I_m \angle\theta_2$	
A77	Period		$P = \frac{2\pi}{\omega}$ seconds		
A78	Frequency		$f = \frac{1}{P} = \frac{\omega}{2\pi}$ hertz		
A79	Current		$i = \frac{dq}{dt}$		
A80	Charge		$q = \int i\,dt$		252
A81	Capacitor	Instantaneous Current	$i = C\frac{dv}{dt}$		156
A82	Capacitor	Instantaneous Voltage	$v = \frac{1}{C}\int i\,dt$ volts		253
A83	Capacitor	Current when Charging or Discharging	Series RC Circuit 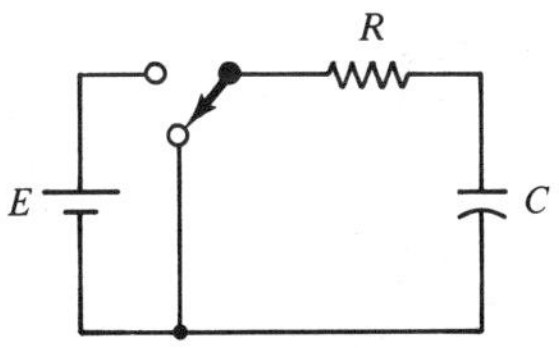 (E, R, C)	$i = \frac{E}{R} e^{-t/RC}$	405
A84	Capacitor	Voltage when Discharging		$v = Ee^{-t/RC}$	405
A85	Inductor	Instantaneous Current	$i = \frac{1}{L}\int v\,dt$		254
A86	Inductor	Instantaneous Voltage	$v = L\frac{di}{dt}$		156
A87	Inductor	Current when Charging	Series RL Circuit 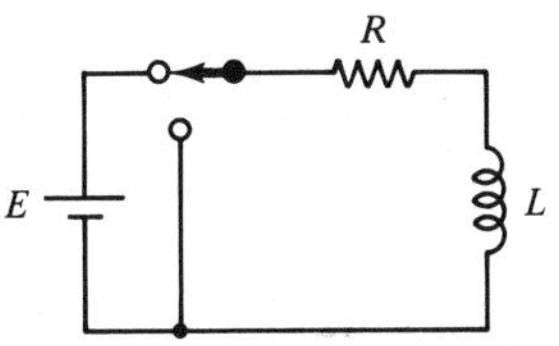 (E, R, L)	$i = \frac{E}{R}(1 - e^{-Rt/L})$	404
A88	Inductor	Voltage when Charging or Discharging		$v = Ee^{-Rt/L}$	404

ELECTRICAL TECHNOLOGY (Continued)	No.	Topic	Source	Item	Formula	Page
	A89	Series RLC Circuit 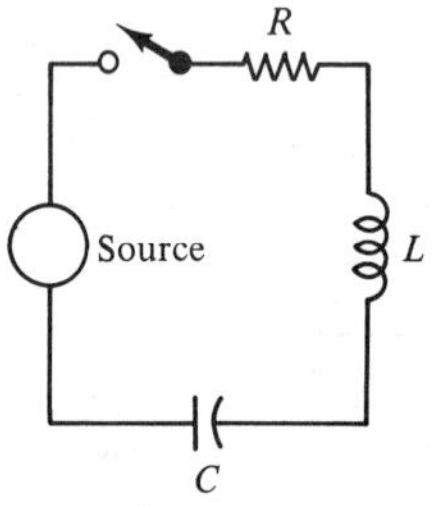 (Source, switch, R, L, C)	DC Source	Resonant Frequency	$\omega_n = \sqrt{\frac{1}{LC}}$	447
	A90		DC Source	No Resistance: The Series $L\,C$ Circuit:	$i = \frac{E}{\omega_n L} \sin \omega_n t$	441
	A91		DC Source	Underdamped	$i = \frac{E}{\omega_d L} e^{-at} \sin \omega_d t$	440
	A92		DC Source	Underdamped	where $\omega_d = \sqrt{\omega_n^2 - \frac{R^2}{4L^2}}$	440
	A93		DC Source	Overdamped	$i = \frac{E}{2j\omega_d L} \left[e^{(-a + j\omega_d)t} - e^{(-a - j\omega_d)t}\right]$	442
	A94		AC Source	Inductive Reactance	$X_L = \omega L$	
	A95		AC Source	Capacitive Reactance	$X_C = \frac{1}{\omega C}$	
	A96		AC Source	Total Reactance	$X = X_L - X_C$	
	A97		AC Source	Magnitude of Impedance	$\mid Z \mid = \sqrt{R^2 + X^2} = \sqrt{R^2 + \left(\omega L - \frac{1}{\omega C}\right)^2}$	
	A98		AC Source	Phase Angle	$\phi = \arctan \frac{X}{R}$	
	A99		AC Source	Complex Impedance	$Z = R + jx = Z\angle\phi = Ze^{j\phi}$	
	A100	Ohm's Law for AC			$\mathbf{V} = \mathbf{ZI}$	
	A101	Decibels Gained or Lost			$G = 10 \log_{10} \frac{P_2}{P_1}$ dB	

Conversion Factors

Unit	Equals	
	Length	
1 angstrom	1×10^{-10}	meter
	1×10^{-4}	micrometer (micron)
1 centimeter	10^{-2}	meter
	0.3937	inch
1 foot	12	inches
	0.3048	meter
1 inch	25.4	millimeters
	2.54	centimeters
1 kilometer	3281	feet
	0.5400	nautical mile
	0.6214	statute mile
	1094	yards
1 light-year	9.461×10^{12}	kilometers
	5.879×10^{12}	statute miles
1 meter	10^{10}	angstroms
	3.281	feet
	39.37	inches
	1.094	yards
1 micron	10^{4}	angstroms
	10^{-4}	centimeter
	10^{-6}	meter
1 nautical mile (International)	8.439	cables
	6076	feet
	1852	meters
	1.151	statute miles
1 statute mile	5280	feet
	8	furlongs
	1.609	kilometers
	0.8690	nautical mile
1 yard	3	feet
	0.9144	meter
	Angles	
1 degree	60	minutes
	0.01745	radian
	3600	seconds
	2.778×10^{-3}	revolution
1 minute of arc	0.01667	degree
	2.909×10^{-4}	radian
	60	seconds
1 radian	0.1592	revolution
	57.296	degrees
	3438	minutes
1 second of arc	2.778×10^{-4}	degree
	0.01667	minute

Source: Adapted from P. Calter, *Schaum's Outline of Technical Mathematics*, McGraw-Hill Book Company, New York, 1979.

Unit	Equals	
Area		
1 acre	4047	square meters
	43 560	square feet
1 are	0.024 71	acre
	1	square dekameter
	100	square meters
1 hectare	2.471	acres
	100	ares
	10 000	square meters
1 square foot	144	square inches
	0.092 90	square meter
1 square inch	6.452	square centimeters
1 square kilometer	247.1	acres
1 square meter	10.76	square feet
1 square mile	640	acres
	2.788×10^7	square feet
	2.590	square kilometers
Volume		
1 board-foot	144	cubic inches
1 bushel (U.S.)	1.244	cubic feet
	35.24	liters
1 cord	128	cubic feet
	3.625	cubic meters
1 cubic foot	7.481	gallons (U.S. liquid)
	28.32	liters
1 cubic inch	0.01639	liter
	16.39	milliliters
1 cubic meter	35.31	cubic feet
	10^6	cubic centimeter
1 cubic millimeter	6.102×10^{-5}	cubic inch
1 cubic yard	27	cubic feet
	0.7646	cubic meter
1 gallon (imperial)	277.4	cubic inches
	4.546	liters
1 gallon (U.S. liquid)	231	cubic inches
	3.785	liters
1 kiloliter	35.31	cubic feet
	1.000	cubic meter
	1.308	cubic yards
	220	imperial gallons
1 liter	10^3	cubic centimeters
	10^6	cubic millimeters
	10^{-3}	cubic meter
	61.02	cubic inches

Unit	Equals	
	Area	
1 gram	10^{-3}	kilogram
	6.854×10^{-5}	slug
1 kilogram	1000	grams
	0.06854	slug
1 slug	14.59	kilograms
	14,590	grams
1 metric ton	1000	kilograms
	Force	
1 dyne	10^{-5}	newton
1 newton	10^{5}	dynes
	0.2248	pound
	3.597	ounces
1 pound	4.448	newtons
	16	ounces
1 ton	2000	pounds
	Velocity	
1 foot/minute	0.3048	meter/minute
	0.011 364	mile/hour
1 foot/second	1.097	kilometers/hour
	18.29	meters/minute
	0.6818	mile/hour
1 kilometer/hour	3281	feet/hour
	54.68	feet/minute
	0.6214	mile/hour
1 kilometer/minute	3281	feet/minute
	37.28	miles/hour
1 knot	6076	feet/hour
	101.3	feet/minute
	1.852	kilometers/hour
	30.87	meters/minute
	1.151	miles/hour
1 meter/hour	3.281	feet/hour
1 mile/hour	1.467	feet/second
	1.609	kilometers/hour
	Power	
1 British thermal unit/hour	0.2929	watt
1 Btu/pound	2.324	joules/gram
1 Btu-second	1.414	horsepower
	1.054	kilowatts
	1054	watts

Unit	Equals	
	Power (continued)	
1 horsepower	42.44	Btu/minute
	550	footpounds/second
	746	watts
1 kilowatt	3414	Btu/hour
	737.6	footpounds/second
	1.341	horsepower
	10^3	joules/second
	999.8	international watt
1 watt	44.25	footpounds/minute
	1	joule/second
	Pressure	
1 atmosphere	1.013	bars
	14.70	pounds/square inch
	760	torrs
	101	kilopascals
1 bar	10^6	baryes
	14.50	pounds/square inch
1 barye	10^{-6}	bar
1 inch of mercury	0.033 86	bar
	70.73	pounds/square foot
1 pascal	1	newton/square meter
1 pound/square inch	0.068 03	atmosphere
	Energy	
1 British thermal unit	1054	joules
	1054	wattseconds
1 foot-pound	1.356	joules
	1.356	newtonmeters
1 joule	0.7376	foot-pound
	1	wattsecond
1 kilowatthour	3410	British thermal units
	1.341	horsepowerhours
1 newtonmeter	0.7376	footpounds
1 watthour	3.414	British thermal units
	2655	footpounds
	3600	joules

Table of Integrals

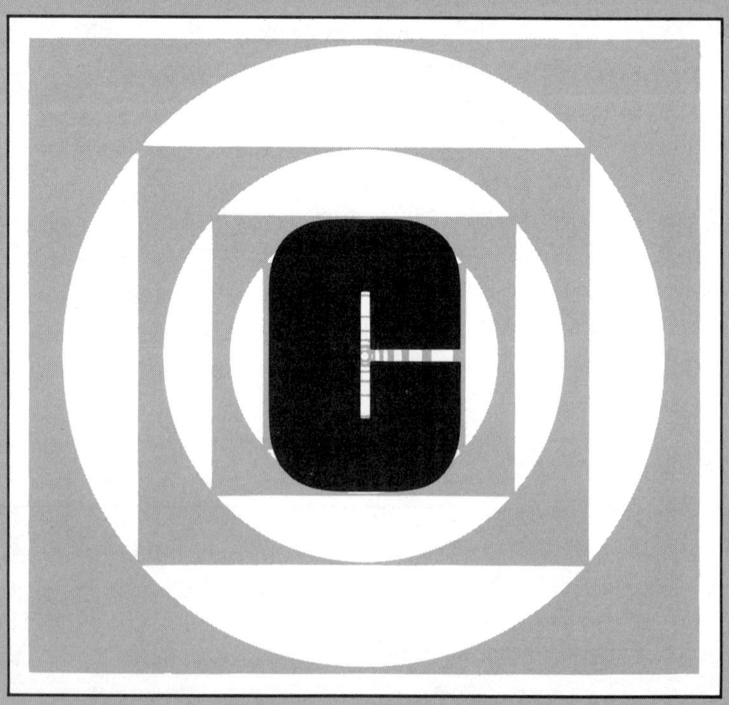

Note: Many integrals have alternate forms that are not shown here. Don't be surprised if another table of integrals gives an expression that looks very different than one listed here.

Basic forms	1. $\int du = u + C$
	2. $\int af(x)\,dx = a\int f(x)\,dx = aF(x) + C$
	3. $\int [f(x) + g(x) + h(x) + \cdots]\,dx = \int f(x)\,dx + \int g(x)\,dx + \int h(x)\,dx + \cdots + C$
	4. $\int x^n\,dx = \dfrac{x^{n+1}}{n+1} + C \qquad (n \neq -1)$
	5. $\int u^n\,du = \dfrac{u^{n+1}}{n+1} + C \qquad (n \neq -1)$
	6. $\int u\,dv = uv - \int v\,du$
	7. $\int \dfrac{du}{u} = \ln\lvert u\rvert + C$
	8. $\int e^u\,du = e^u + C$
	9. $\int b^u\,du = \dfrac{b^u}{\ln b} + C \qquad (b > 0,\ b \neq 1)$
Trigonometric functions	10. $\int \sin u\,du = -\cos u + C$
	11. $\int \cos u\,du = \sin u + C$
	12. $\int \tan u\,du = -\ln\lvert\cos u\rvert + C$
	13. $\int \cot u\,du = \ln\lvert\sin u\rvert + C$
	14. $\int \sec u\,du = \ln\lvert\sec u + \tan u\rvert + C$
	15. $\int \csc u\,du = \ln\lvert\csc u - \cot u\rvert + C$
Squares of the trigonometric functions	16. $\int \sin^2 u\,du = \dfrac{u}{2} - \dfrac{\sin 2u}{4} + C$
	17. $\int \cos^2 u\,du = \dfrac{u}{2} + \dfrac{\sin 2u}{4} + C$
	18. $\int \tan^2 u\,du = \tan u - u + C$
	19. $\int \cot^2 u\,du = -\cot u - u + C$
	20. $\int \sec^2 u\,du = \tan u + C$
	21. $\int \csc^2 u\,du = -\cot u + C$

Cubes of the trigonometric functions

22. $\int \sin^3 u\, du = \frac{\cos^3 u}{3} - \cos u + C$

23. $\int \cos^3 u\, du = \sin u - \frac{\sin^3 u}{3} + C$

24. $\int \tan^3 u\, dx = \frac{1}{2} \tan^2 u + \ln|\cos u| + C$

25. $\int \cot^3 u\, dx = -\frac{1}{2} \cot^2 u - \ln|\sin u| + C$

26. $\int \sec^3 u\, du = \frac{1}{2} \sec u \tan u + \frac{1}{2} \mathrm{l}\cdot \quad \mathrm{c}\, u + \tan u| + C$

27. $\int \csc^3 u\, du = -\frac{1}{2} \csc u \cot u + \frac{1}{2} \ln|\csc u - \cot| u + C$

Miscellaneous trigonometric forms

28. $\int \sec u \tan u\, du = \sec u + C$

29. $\int \csc u \cot u\, du = -\csc u + C$

30. $\int \sin^2 u \cos^2 u\, du = \frac{u}{8} - \frac{1}{32} \sin 4u + C$

31. $\int u \sin u\, du = \sin u - u \cos u + C$

32. $\int u \cos u\, du = \cos u + u \sin u + C$

33. $\int u^2 \sin u\, du = 2u \sin u - (u^2 - 2) \cos u + C$

34. $\int u^2 \cos u\, du = 2u \cos u + (u^2 - 2) \sin u + C$

35. $\int \mathrm{Sin}^{-1} u\, du = u\, \mathrm{Sin}^{-1} u + \sqrt{1 - u^2} + C$

36. $\int \mathrm{Tan}^{-1} u\, du = u\, \mathrm{Tan}^{-1} u - \ln \sqrt{1 + u^2} + C$

Exponential and logarithmic forms

37. $\int u e^{au}\, du = \frac{e^{au}}{a^2} (au - 1) + C$

38. $\int u^2 e^{au}\, du = \frac{e^{au}}{a^3} (a^2 u^2 - 2au + 2) + C$

39. $\int u^n e^u\, du = u^n e^u - n \int u^{n-1} e^u\, du$

40. $\int \frac{e^u\, du}{u^n} = \frac{-e^u}{(n-1)u^{n-1}} + \frac{1}{n-1} \int \frac{e^u\, du}{u^{n-1}} \qquad (n \neq 1)$

41. $\int e^{au} \sin bu\, du = \frac{e^{au}}{a^2 + b^2} (a \sin bu - b \cos bu) + C$

42. $\int e^{au} \cos bu\, du = \frac{e^{au}}{a^2 + b^2} (a \cos bu + b \sin bu) + C$

43. $\int \ln u\, du = u(\ln u - 1) + C$

44. $\int u^n \ln|u|\, du = u^{n+1} \left[\frac{\ln|u|}{n+1} - \frac{1}{(n+1)^2}\right] + C \qquad (n \neq -1)$

Forms involving $a + bu$

45. $\int \frac{u\, du}{a + bu} = \frac{1}{b^2} [a + bu - a \ln|a + bu|] + C$

46. $\int \frac{u^2\, du}{a + bu} = \frac{1}{b^3} \left[\frac{1}{2} (a + bu)^2 - 2a(a + bu) + a^2 \ln|a + bu|\right] + C$

47. $\int \frac{u\, du}{(a + bu)^2} = \frac{1}{b^2} \left[\frac{a}{a + bu} + \ln|a + bu|\right] + C$

Forms involving $a + bu$

48. $\displaystyle\int \frac{u^2\,du}{(a+bu)^2} = \frac{1}{b^3}\left[a + bu - \frac{a^2}{a+bu} - 2a\ln|a+bu|\right] + C$

49. $\displaystyle\int \frac{du}{u(a+bu)} = \frac{-1}{a}\ln\left|\frac{a+bu}{u}\right| + C$

50. $\displaystyle\int \frac{du}{u^2(a+bu)} = \frac{-1}{au} + \frac{b}{a^2}\ln\left|\frac{a+bu}{u}\right| + C$

51. $\displaystyle\int \frac{du}{u(a+bu)^2} = \frac{1}{a(a+bu)} - \frac{1}{a^2}\ln\left|\frac{a+bu}{u}\right| + C$

52. $\displaystyle\int u\sqrt{a+bu}\,du = \frac{2(3bu-2a)}{15b^2}(a+bu)^{3/2} + C$

53. $\displaystyle\int u^2\sqrt{a+bu}\,du = \frac{2(15b^2u^2 - 12abu + 8a^2)}{105b^3}(a+bu)^{3/2} + C$

54. $\displaystyle\int \frac{u\,du}{\sqrt{a+bu}} = \frac{2(bu-2a)}{3b^2}\sqrt{a+bu} + C$

55. $\displaystyle\int \frac{u^2\,du}{\sqrt{a+bu}} = \frac{2(3b^2u^2 - 4abu + 8a^2)}{15b^3}\sqrt{a+bu} + C$

Forms involving $u^2 \pm a^2$ $(a > 0)$

56. $\displaystyle\int \frac{du}{a^2+b^2u^2} = \frac{1}{ab}\tan^{-1}\frac{bu}{a} + C$

57. $\displaystyle\int \frac{du}{u^2-a^2} = \frac{1}{2a}\ln\left|\frac{u-a}{u+a}\right| + C$

58. $\displaystyle\int \frac{u^2\,du}{u^2-a^2} = u + \frac{a}{2}\ln\left|\frac{u-a}{u+a}\right| + C$

59. $\displaystyle\int \frac{u^2\,du}{u^2+a^2} = u - a\,\mathrm{Tan}^{-1}\frac{u}{a} + C$

60. $\displaystyle\int \frac{du}{u(u^2\pm a^2)} = \frac{\pm 1}{2a^2}\ln\left|\frac{u^2}{u^2\pm a^2}\right| + C$

Forms involving $\sqrt{a^2-u^2}$ $(a > 0)$

61. $\displaystyle\int \frac{du}{\sqrt{a^2-u^2}} = \mathrm{Sin}^{-1}\frac{u}{a} + C$

62. $\displaystyle\int \frac{du}{\sqrt{u^2\pm a^2}} = \ln|u + \sqrt{u^2\pm a^2}| + C$

63. $\displaystyle\int \frac{u^2\,du}{\sqrt{u^2\pm a^2}} = \frac{u}{2}\sqrt{u^2\pm a^2} \mp \frac{a^2}{2}\ln|u + \sqrt{u^2\pm a^2}| + C$

64. $\displaystyle\int \frac{du}{u\sqrt{u^2+a^2}} = \frac{1}{a}\ln\left|\frac{u}{a+\sqrt{u^2+a^2}}\right| + C$

65. $\displaystyle\int \frac{du}{u\sqrt{u^2-a^2}} = \frac{1}{a}\mathrm{Sec}^{-1}\frac{u}{a} + C$

66. $\displaystyle\int \sqrt{u^2\pm a^2}\,du = \frac{u}{2}\sqrt{u^2\pm a^2} \pm \frac{a^2}{2}\ln|u + \sqrt{u^2\pm a^2}| + C$

67. $\displaystyle\int \frac{\sqrt{u^2+a^2}\,du}{u} = \sqrt{u^2+a^2} - a\ln\left|\frac{a+\sqrt{u^2+a^2}}{u}\right| + C$

68. $\displaystyle\int \frac{\sqrt{u^2-a^2}\,du}{u} = \sqrt{u^2-a^2} - a\,\mathrm{Sec}^{-1}\frac{u}{a} + C$

69. $\displaystyle\int \sqrt{a^2-u^2}\,du = \frac{u}{2}\sqrt{a^2-u^2} + \frac{a^2}{2}\mathrm{Sin}^{-1}\frac{u}{a} + C$

70. $\displaystyle\int u^2\sqrt{a^2-u^2}\,du = \frac{-u}{4}(a^2-u^2)^{3/2} + \frac{a^2u}{8}\sqrt{a^2-u^2} + \frac{a^4}{8}\mathrm{Sin}^{-1}\frac{u}{a} + C$

71. $\displaystyle\int \frac{\sqrt{a^2-u^2}\,du}{u} = \sqrt{a^2-u^2} - a\ln\left|\frac{a+\sqrt{a^2-u^2}}{u}\right| + C$

72. $\displaystyle\int \frac{\sqrt{a^2-u^2}\,du}{u^2} = \frac{-\sqrt{a^2-u^2}}{u} - \mathrm{Sin}^{-1}\frac{u}{a} + C$

73. $\displaystyle\int \frac{u^2\,du}{\sqrt{a^2-u^2}} = \frac{-u}{2}\sqrt{a^2-u^2} + \frac{a^2}{2}\mathrm{Sin}^{-1}\frac{u}{a} + C$

Summary of BASIC

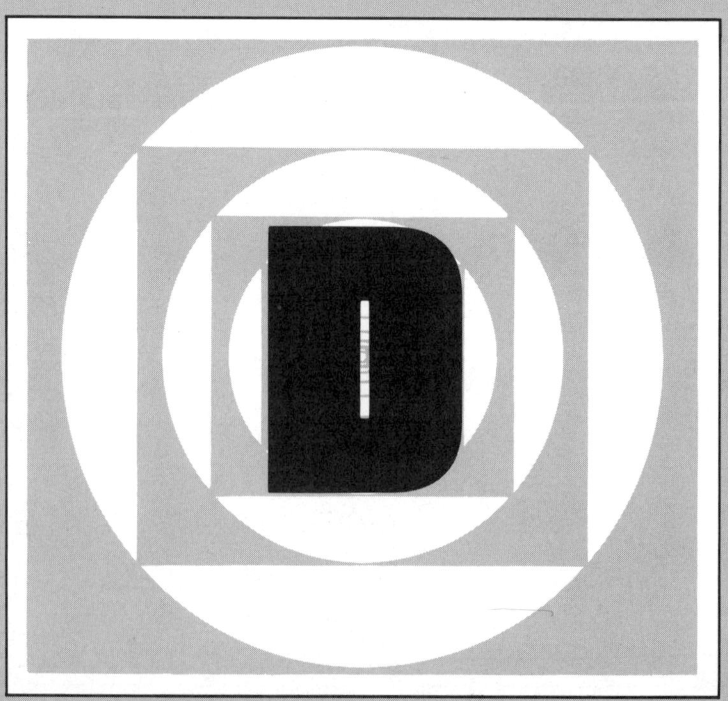

This brief listing contains commands, statements, and functions. *Commands,* such as RUN, or LIST are not part of a program, but tell the computer what to do with a program. Commands are typed without line numbers and are executed immediately. Program *statements,* such as PRINT or GO TO, tell the computer what to do during a run of the program. *Functions,* such as COS( ) or EXP( ), tell the computer what operation to perform on the quantity enclosed in parentheses (called the *argument*).

ABS( )

A function which returns the absolute value of the expression in parentheses.
Example: The lines

```
10 LET X = -5
20 PRINT ABS(X)
```

will cause the value 5 to be printed.

ATN( )

A function which returns the angle (in radians) whose tangent is specified.
Example: The statement PRINT ATN(2) will cause the arctangent of 2 (1.1071 radians) to be printed.

CONT

A command which causes a program to continue running after a STOP or CTRL-C (interrupt) has been executed.

COS( )

A function which returns the cosine of the angle (in radians).
Example: The statement PRINT COS(1.2) will cause the number 0.362358 (the cosine of 1.2 radians) to be printed.

DATA

A statement used to store numbers and strings, for later access by the READ statement.
Example: The statement DATA 5, 2, 9, 3 stores the numbers 5, 2, 9, and 3 for later access, and the statement

DATA JOHN, MARY, BILL

stores the given three names for later access. See READ.

DEF FN — A statement which defines a new function written by the user.
Examples: DEF FNA(X) = 3*X + 2
DEF FNG(X,Y,Z) = (5*X − 2*Y^2)*Z

DIM — A statement which reserves space for lists or tables.
Example: DIM A(50), DIM P$(20), B(25,30)

END — A statement which stops the run and closes all files. It is the last line in a program.

EXP( ) — A function which raises e (the base of natural logarithms) to the power specified.
Example: The statement PRINT EXP(2.5) will compute and print the value of $e^{2.5}$, (12.182).

FOR . . . NEXT — A pair of statements used to set up a loop.
Examples: The lines

```
10 FOR N = 1 TO 10
20 PRINT "HELLO"
30 NEXT N
```

will cause the word "HELLO" to be printed ten times.
The lines

```
10 FOR X = 20 TO 30 STEP .5
20 PRINT X,
30 NEXT X
```

will cause the numbers 20, 20.5, 21, . . . 30 to be printed.

GOSUB . . . RETURN — A pair of statements used to branch to and return from a subroutine.
Example: The line GOSUB 500 causes a branching to the subroutine starting on line 500. The RETURN statement placed at the end of the subroutine returns us to the line immediately following the GOSUB statement.

GOTO — Causes a branch to the specified line number.
Example: GOTO 150

IF . . . THEN — A statement which causes a branching to another line if the specified condition is met.
Example: The line

```
40 IF A > B THEN 200
```

will cause branching to line 200 if A is greater than B. If not, we go to the next line.

INPUT — A statement which allows input of data from the keyboard during a run.
Example: The statement INPUT X will cause the run to stop, and a question mark will be printed. The operator must then type a number, which will be accepted as X by the program.

Example: The statement
20 INPUT "WHAT ARE THE TWO NAMES"; A$, B$
will print the *prompt string* WHAT ARE THE TWO NAMES, then a question mark, and then stop the run until you enter two strings, separated by a comma.

INT( ) — A function which returns the integer part of a number.
Example: The statement PRINT INT(5.995) will cause the integer 5 to be printed.

KILL — A command which deletes a file from the disk.
Example: KILL "CIRCLE.BAS"

LEFT$( ) — A function which returns the leftmost characters of the specified string.
Example: The line 50 PRINT LEFT$(A$,5) will cause the five leftmost characters of A$ to be printed.

LEN( ) — A function which returns the number of characters, incuding blanks, in the specified string.
Example: 10 PRINT LEN(B$)

LET — A statement used to assign a value to a variable.
Examples:
```
30 LET X = 5
40 LET Y = 3*X-2
50 LET SUM = X + Y
```
(The LET statement is optional in many versions of BASIC.)

LIST — A command to list all or part of a program.
Examples: LIST lists the entire program.
LIST 100–150 lists lines 100 to 150 only.
LIST–200 lists up to line 200.
LIST 300– lists from line 300 to the end.

LLIST — A command to list the program on the printer.

LOAD — A command to load a program or file from the disk.
Example: LOAD "B:QUADRATIC" will load the program QUADRATIC from drive B.

LOG( ) — A function which returns the natural logarithm of the argument.

LPRINT — See PRINT.

MID$( ) — A function which returns a string of characters from the middle of the specified string.
Example: In the program
```
10 A$ = "COMPUTER"
20 PRINT MID$(A$,4,3)
```

line 20 will extract and print a string of length 3 from A$, starting with the 4th character. Thus, "PUT" will be printed.

NEW — A command used to clear the memory before entering a new program.

ON . . . GO TO — A statement which allows branching to one of several line numbers.
Example: Given the line
10 ON X GO TO 150, 300, 450
we branch to line 150 if X = 1, to line 300 if X = 2, and to line 450 if X = 3.

PRINT — A statement which causes information to be printed.
Examples:
20 PRINT X causes the value of X to be printed.
30 PRINT A, B, C causes the values of A, B, and C to be printed, each in a separate column.
40 PRINT A; B; C causes the values of A, B, and C to be printed with just a single space between them.
50 PRINT "HELLO" causes the word "HELLO" to be printed.
The statement LPRINT will cause the printing to occur at the printer, rather than at the video terminal.

PRINT USING — A statement which specifies a format for the printing of strings or numbers.
Example: 20 PRINT USING "###.##"; X
will cause the number X to be printed with three digits before the decimal point and two digits following the decimal point. There are many other ways to format numbers and strings. Consult your users' manual.

RANDOMIZE — A statement which allows you to obtain a different list of random numbers each time you use RND( ). Put the line 10 RANDOMIZE at the start of each program that uses random numbers. Then when you RUN, you will be asked to enter a number. Each such number will cause a different set of random numbers to be generated.

READ — A statement which reads values from a DATA statement and assigns them to variables.
Example: 10 DATA JOHN, 24
20 READ N$, A
Line 20 will read the string "JOHN" and assign it to N$, and read the number 24 and assign it to N.

REM — This statement allows us to place remarks in a program. REM's are ignored during a RUN.

Example: 10 REM THIS PROGRAM COMPUTES AVERAGES

RENUM — A command to renumber a program.
Example: RENUM renumbers a program with the line numbers, 10, 20, 30, . . .
RENUM 100, 300, 20 will start renumbering your program at the old line 300, which it changes to 100, and continues in increments of 20.

RESTORE — A statement which allows data to be reread.

RETURN — see GOSUB

RIGHT$( ) — A statement which returns the rightmost characters from the specified string. Similar to LEFT$( ).

RND — A function which returns a random number between 0 and 1.
Example: The program

```
10 FOR N = 1 TO 10
20 PRINT RND
30 NEXT N
```

will cause 10 random numbers to be printed.

RUN — A command to run the program in memory.

SAVE — A command to save a program on disk.
Example: SAVE "B:ROOTS"
will save the program "ROOTS" on the disk in drive B.

SGN( ) — A function which returns a value of +1, 0, or −1, depending on whether the argument is positive, zero, or negative, respectively.

SIN( ) — A function which returns the sine of the angle (in radians).

SQR( ) — A function which returns the square root of the argument.

STOP — A statement which stops a run, but does not close files. To resume the run, type CONT.

SYSTEM — A command to exit BASIC and return to the operating system.

TAB( ) — A function which tabs to the print position specified.
Example: The line 30 PRINT TAB(25); "HELLO" will cause the word HELLO to be printed starting at a position 25 characters from the left edge of the screen or paper.

TAN( ) — A function which returns the tangent of the angle (in radians).

Answers to Selected Problems

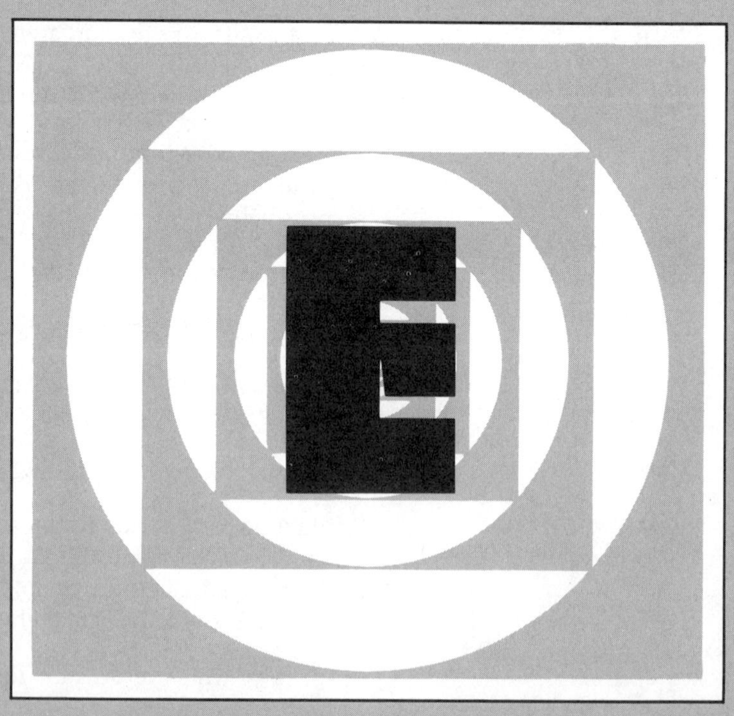

CHAPTER ONE: The Straight Line

Exercise 1, Page 5

1. 3	**3.** 9	**5.** 6.6	**7.** 4
9. 1	**11.** 3.34	**13.** 7.2	**15.** 12.5
17. 131	**19.** 2	**21.** 3	**23.** 1
25. 26.4 ft	**27.** 5.10, 8.06, 9.22, area = 20.6		

Exercise 2, Page 16

1. $-\frac{1}{10}$	**3.** $\dfrac{3-a}{a+2}$	**5.** 4.04	**7.** 0.787
9. −3.77	**11.** −0.439	**13.** 71.6°	**15.** 61.5°
17. 0°	**19.** $-\frac{1}{5}$	**21.** −0.21	**23.** 0.351
25. −0.620	**27.** 1.062	**29.** 41.3°	**31.** 45°
33. 45°	**35.** 64°	**37.**	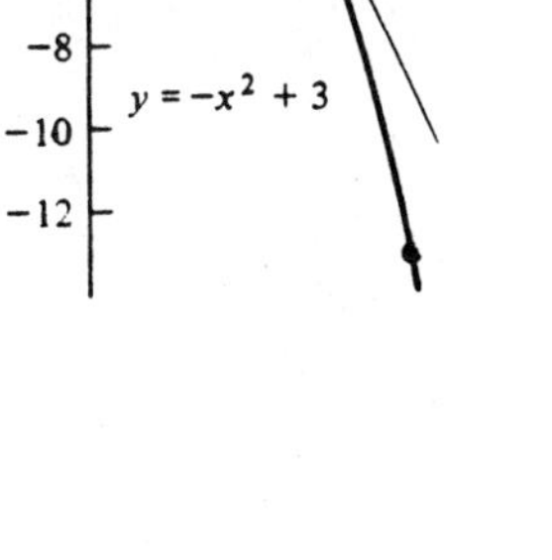
39. 2008 ft	**41.** 1259 ft		
43. 0.911°	**45.** 33.7°		

Exercise 3, Page 24

1. $m = 3, b = -5$
3. $m = -\frac{1}{2}, b = -\frac{1}{4}$
5. $m = 0, b = 6$
7. $m = -\frac{1}{2}, b = \frac{3}{2}$
9. $m = \frac{2}{3}, b = \frac{11}{3}$
11. $4x - y - 3 = 0$
13. $2.25x - y - 1.48 = 0$
15. $3x - 4y + 11 = 0$
17. $4x + y + 3 = 0$
19. $3x - 5y - 15 = 0$
21. $5x - y + 3 = 0$
23. $x + 3y + 15 = 0$
25. $5x - y + 15 = 0$
27. $x + 5y - 6 = 0$
29. $1.1x - y + 4 = 0$
31. $x - 2.38y - 6.38 = 0$
33. $y = 2$
35. $x - 2y = 0$
37. $x + 2 = 0$
39. $R = R_1[1 + \alpha(t - t_1)]$
41. 57.4°
43. 58.0 lb
47. $P = 20.6 + 0.432x$, 21.8 ft
49. $t = 25 - 0.789x$, 31.7 cm, −0.789
51.

```
10 '   FALSEPOS
20 '
30 '   FINDS ROOTS OF EQUATIONS BY
40 '   THE FALSE POSITION METHOD
50 '
60 '   WRITE THE EQUATION ON LINE 80
70 '
80  DEF FNA(X) = X^2 - 5
90  PRINT "ENTER TWO X VALUES THAT"
100 PRINT "BRACKET THE ROOT"
110 INPUT X1, X2
120 Y1 = FNA(X1)
130 Y2 = FNA(X2)
140'
150'---COMPUTE NEW X---
160 X3 = X2 - Y2*(X2 - X1)/(Y2 - Y1)
170 Y3 = FNA(X3)
180 IF ABS(X3 - X0)<.0001 THEN 260
190'
200'---WHICH X TO REPLACE?---
210 IF SGN(Y3) = SGN(Y1) THEN X1 = X3
220 GOTO 240
230 X2 = X3
240 X0 = X3
250 GOTO 120
260 PRINT "THE ROOT IS ";X3
```

Chapter Test, Page 29

1. 4
3. −1.44
5. 147.2°
7. $\dfrac{-2b}{a}$
9. $m = \frac{3}{2}, b = -\frac{7}{2}$
11. $7x - 3y + 32 = 0$
13. $5x - y + 27 = 0$
15. $7x - 3y + 21 = 0$
17. $y = 5$
19. −3
21. $x + 3 = 0$

CHAPTER TWO: The Conic Sections

Exercise 1, Page 38

1. $x^2 + y^2 = 49$
3. $(x - 2)^2 + (y - 3)^2 = 25$
5. $(x - 5)^2 + (y + 3)^2 = 16$
7. $C(0, 0), r = 7$
9. $C(2, -4), r = 4$
11. $C(3, -5), r = 6$
13. $C(4, 0), r = 4$
15. $C(5, -6), r = 6$
17. $x^2 + y^2 - 9x - 5y + 14 = 0$
19. $x^2 + y^2 - 4x - 4y = 17$
21. $4x + 3y = 25$
23. $4x - 3y + 10 = 0$
25. (3, 0), (2, 0), (0, 1), (0, 6)
27. (2, 3) and $(-\frac{3}{2}, -\frac{1}{2})$
31. 8.0 ft
33. 7.08 ft

Exercise 2, Page 48

1. $F(2, 0), L = 8$
3. $F(0, -\frac{3}{7}), L = \frac{12}{7}$

5. $x^2 + 8y = 0$

7. $3y^2 = 4x$

9. $V(3, 5)$, $F(5, 6)$, $L = 12$, $y = 5$

11. $V(2, -1)$, $F(\frac{13}{8}, -1)$, $L = \frac{3}{2}$, $y = -1$

13. $V(\frac{3}{2}, \frac{5}{4})$, $F(\frac{3}{2}, 1)$, $L = 1$, $2x = 3$

15. $(y - 2)^2 = 8(x - 1)$

17. $y^2 + 4x - 2y + 9 = 0$

19. $2y^2 + 4y - x - 4 = 0$

21. 74.2 ft

23. 2760 m

25. $x^2 = 1250y$

27. 2.08 ft

29. $x^2 = 488y$

31. $(x - 16)^2 = -3072(y - \frac{1}{12})$

Exercise 3, Page 65

1. $a = 5, b = 4, f = 3$

3. $2, \sqrt{3}, 1$

5. $(0, \pm 4), (0, \pm 2)$

7. $\dfrac{x^2}{25} + \dfrac{y^2}{9} = 1$

9. $\dfrac{x^2}{36} + \dfrac{y^2}{4} = 1$

11. $25x^2 + 169y^2 = 4225$

13. $3x^2 + 7y^2 = 115$

15. $C(2, -2)$, $V(6, -2)$, $(-2, -2)$, $F(4.65, -2)$, $(-0.65, -2)$

17. $C(-2, 3)$, $V(1, 3)$, $(-5, 3)$, $F(0, 3)$, $(-4, 3)$

19. $C(1, -1)$, $V(5, -1)$, $(-3, -1)$, $F(4, -1)$, $(-2, -1)$

21. $C(-3, 2)$, $V(-3, 7)$, $(-3, -3)$, $F(-3, 6)$, $(-3, -2)$

23. $\dfrac{x^2}{9} + \dfrac{(y - 3)^2}{36} = 1$

25. $\dfrac{(x + 2)^2}{12} + \dfrac{(y + 3)^2}{16} = 1$

27. $(1, 2), (1, -2)$

29. 60.8 cm

31. 20 cm

33. 6.93 ft

Exercise 4, Page 77

1. $V(\pm 4, 0)$, $F(\pm 6.40, 0)$, slope $= \pm\frac{5}{4}$

3. $V(\pm 3, 0)$, $F(\pm 5, 0)$, slope $= \pm\frac{4}{3}$

5. $V(\pm 4, 0)$, $F(\pm 4.47, 0)$, slope $= \pm\frac{1}{2}$

7. $\dfrac{x^2}{25} - \dfrac{y^2}{144} = 1$

9. $\dfrac{x^2}{9} - \dfrac{y^2}{7} = 1$

11. $\dfrac{y^2}{16} - \dfrac{x^2}{16} = 1$

13. $\dfrac{y^2}{25} - \dfrac{3x^2}{64} = 1$

15. $C(2, -1)$, $a = 5$, $b = 4$, $F(7.83, -1)$, $F'(-3.83, 1)$, $V(7, -1)$, $V'(-3, -1)$, slope $= \pm\frac{5}{4}$

17. $C(2, -3)$, $a = 3$, $b = 4$, $F(7, -3)$, $F'(-3, -3)$, $V(5, -3)$, $V'(-1, -3)$, slope $= \pm\frac{4}{3}$

19. $C(-1, -1)$, $a = 2$, $b = \sqrt{5}$, $F(-1, 2)$, $F'(-1, -4)$, $V(-1, 1)$, $V'(-1, -3)$, slope $= \pm 2/\sqrt{5}$

21. $\dfrac{(x - 3)^2}{2} - \dfrac{(y + 1)^2}{2} = 1$

23. $\dfrac{(x - 1)^2}{4} - \dfrac{(y + 2)^2}{12} = 1$

25. $xy = 36$

27. $\dfrac{x^2}{324} - \dfrac{y^2}{352} = 1$

29. $pv = 25{,}000$

Chapter Test, Page 79

1. hyperbola, $C(1, 2)$, $a = 2$, $b = 1$, $F(3.24, 2)$, $(-1.24, 2)$, $V(3, 2)$, $(-1, 2)$

3. circle, $C(0, 4)$, $r = 4$

5. hyperbola, $C(0, 0)$, $F(\pm 5, 0)$, $a = 3$, $b = 4$, $V(\pm 3, 0)$

7. $x^2 + 2y^2 = 100$

9. $y^2 = -17x$

11. $(x + 5)^2 + y^2 = 25$

13. $\dfrac{(x + 2)^2}{25} + \dfrac{(y - 1)^2}{9} = 1$

15. x int $= 4$, y int $= 8$ and -2

17. 3.0 m

19. 4.64 m

CHAPTER THREE: Derivatives of Algebraic Functions

Exercise 1, Page 90

1. 1 **3.** $\frac{1}{5}$ **5.** 10 **7.** 4

9. -8 **11.** $-\frac{4}{5}$ **13.** $-\frac{1}{4}$ **15.** ∞

17. ∞ **19.** $+\infty$ **21.** 2 **23.** $\frac{1}{5}$

25. 0 **27.** x^2 **29.** 0 **31.** 3

33. $3x - \dfrac{1}{x^2 - 4}$ **35.** $3x^2$ **37.** $2x - 2$

39.

```
10  ' LIMIT
20  '
30  ' This program finds the limit
40  ' of sin X/X as X approaches
50  ' zero.
60  '
70  PRINT "X", "sin X", "sin X/X"
80  LET X = 4
90  FOR N = 1 TO 10
100 LET X = X/10
110 PRINT X, SIN(X), SIN(X)/X
120 NEXT N
130 END
```

X	sin X	sin X/X
.4	.3894184	.9735459
.04	3.998933E-02	.9997334
.004	3.99999E-03	.9999974
.0004	.0004	1
.00004	.00004	1
.000004	.000004	1
.0000004	.0000004	1

Exercise 2, Page 100

1. 2 **3.** $2x$ **5.** $3x^2$ **7.** $-\dfrac{3}{x^2}$

9. $-\dfrac{1}{2\sqrt{3 - x}}$ **11.** -2 **13.** $\frac{3}{4}$ **15.** 2

17. 12 **19.** $4x$ **21.** -4 **23.** -16

25. 3 **27.** -5 **29.** 3

31.

```
10  ' TANGENT
20  '
30  ' This program finds the slope of the secant line between two points
40  ' on a curve.
50  '
60  D = 1
70  DEF FNA(X) = X^2 - 2*X
80  X = 3 : Y = 3
90  PRINT "Delta X", "X + Delta X", "Y + Delta Y", "Delta Y", "Delta Y/Delta X"
100 PRINT "_______________________________________________________________"
110 M1 = M
120 D = D / 2
130 X1 = X + D
140 Y1 = FNA(X1)
150 M = (Y1 - Y) / D
160 IF ABS(M1 - M) < .01 THEN 190
170 PRINT D, X1, Y1, Y1 - Y, M
180 GOTO 110
190 END
```

Exercise 3, Page 106

1. 0
3. $7x^6$
5. $6x$
7. $-\dfrac{5}{x^6}$
9. $-\dfrac{1}{x^2}$
11. $-\dfrac{9}{x^4}$
13. $2.5x^{-2/3}$
15. $\dfrac{2}{\sqrt{x}}$
17. $-\dfrac{51\sqrt{x}}{2}$
19. -2
21. $3 - 3x^2$
23. $9x^2 + 14x - 2$
25. a
27. $x - x^6$
29. $\dfrac{3}{2x^{1/4}} - \dfrac{1}{x^{5/4}}$
31. $\dfrac{8x^{1/3}}{3} - \dfrac{2}{x^{1/3}}$
33. $-\dfrac{4}{x^2}$
35. $6x^2$
37. 54
39. -8
41. $15x^4 + 2$
43. $\dfrac{10.4}{x^3}$
45. $6x + 2$
47. 4
49. 3
51. 13.5
53. $10t - 3$
55. $175t^2 - 63.8$
57. $\dfrac{3\sqrt{5}\,w^{1/2}}{2}$
59. $-\dfrac{341}{t^5}$

Exercise 4, Page 112

1. $10(2x + 1)^4$
3. $24x(3x^2 + 2)^3 - 2$
5. $-\dfrac{3}{(2 - 5x)^{2/5}}$
7. $-\dfrac{4.30x}{(x^2 + a^2)^2}$
9. $-\dfrac{6x}{(x^2 + 2)^2}$
11. $\dfrac{2b}{x^2}\left(a - \dfrac{b}{x}\right)$
13. $-\dfrac{3x}{\sqrt{1 - 3x^2}}$
15. $-\dfrac{1}{\sqrt{1 - 2x}}$
17. $-\dfrac{3}{(4 - 9x)^{2/3}}$
19. $-\dfrac{1}{2\sqrt{(x + 1)^3}}$
21. $2(3x^5 + 2x)(15x^4 + 2)$
23. $\dfrac{28.8(4.8 - 7.2x^{-2})}{x^3}$
25. $(5t^2 - 3t + 4)(20t - 6)$
27. $\dfrac{7.6 - 49.8t^2}{(8.3t^3 - 3.8t)^3}$
29. $118{,}000$
31. 540

Exercise 5, Page 118

1. $3x^2 - 3$
3. $42x + 44$
5. $5x^4 - 12x^2 + 4$
7. $(x^2 - 5x)^2(8x - 7)(64x^2 - 242x + 105)$
9. $\dfrac{x}{\sqrt{1 + 2x}} + \sqrt{1 + 2x}$
11. $\dfrac{-2x^2}{\sqrt{3 - 4x}} + 2x\sqrt{3 - 4x}$
13. $\dfrac{bx}{2\sqrt{a + bx}} + \sqrt{a + bx}$
15. $\dfrac{2(3x + 1)^3}{\sqrt{4x - 2}} + 9(3x + 1)^2\sqrt{4x - 2}$
17. $(5t^2 - 3t)(2t + 8) + (t + 4)^2(10t - 3)$
19. $(81.3t^3 - 73.8t)(2t - 94.4) + (t - 47.2)^2(244t^2 - 73.8)$
21. $(2x^5 + 5x)^2 + (2x - 6)(2x^5 + 5x)(10x^4 + 5)$
23. $3x(x + 1)^2(x - 2)^2 + (2x^2 + 2x)(x - 2)^3 + (x + 1)^2(x - 2)^3$
25. $\sqrt{x - 1}\,(x + 1)^{3/2} + \dfrac{3}{2}(x + 2)\sqrt{x^2 - 1} + \dfrac{(x + 1)^{3/2}(x + 2)}{2\sqrt{x - 1}}$

27. $\dfrac{2}{(x+2)^2}$

29. $\dfrac{8x}{(4-x^2)^2}$

31. $\dfrac{-5}{(x-3)^2}$

33. $\dfrac{2-4x}{(x-1)^3}$

35. $\dfrac{1}{2\sqrt{x}\,(\sqrt{x}+1)^2}$

37. $-\dfrac{a^2}{(z^2-a^2)^{3/2}}$

39. $-\frac{41}{90}$

41. 0

Exercise 6, Page 122

1. $6u^2\dfrac{du}{dw}$

3. $2y\dfrac{dy}{du}+3u^2$

5. $2x^3y\dfrac{dy}{dx}+3x^2y^2$

7. $\dfrac{3z\,dz/dt}{\sqrt{3z^2+5}}$

9. $\frac{5}{2}$

11. $\dfrac{-y}{x}$

13. $2a/y$

15. $\dfrac{ay-x^2}{y^2-ax}$

17. $\dfrac{3x^2+1}{3y^2+1}$

19. $\dfrac{8xy}{3y-8x^2+4y^2}$

21. 0.436

23. $\frac{14}{15}$

Exercise 7, Page 124

1. $36x^2-6x$

3. $\dfrac{8}{(x+2)^3}$

5. $-\dfrac{16x^2}{(5-4x^2)^{3/2}}-\dfrac{4}{(5-4x^2)^{1/2}}$

7. $4(x-3)^2+8(x-6)(x-3)$

9. 1.888

Chapter Test, Page 124

1. $\dfrac{2x}{(x^2+1)\sqrt{x^4-1}}$

3. $-\sqrt[3]{\dfrac{y}{x}}$

5. $\frac{8}{5}$

7. $-\dfrac{b^2x}{a^2y}$

9. -0.03516

11. $5-6x$

13. $\frac{3}{64}$

15. 178

17. 3

19. $\dfrac{4-4x^2}{(x+3)^5}+\dfrac{2x}{(x+3)^4}$

21. $(2w-8)\,dw/dx$

23. $-24x$

25. $175t^2-63.8$

CHAPTER FOUR: Graphical Applications of the Derivative

Exercise 1, Page 130

1. $2x-y+1=0$
   $x+2y-7=0$

3. $12x-y-13=0$
   $x+12y-134=0$

5. $3x+4y=25$
   $4x-3y=0$

7. (3, 0)

9. $2x+y=1$ and $2x+y+7=0$

11. $2x+5y=1.55$

13. -10

15. 63.4°, 19.4°, 15.3°

Exercise 2, Page 137

1. rising for all x
3. increasing
5. decreasing
7. upward
9. downward
11. min (0, 0)
13. max (3, 13)
15. max (0, 36), min (4.67, −14.8)
17. max (1, 1) and (−1, 1), min (0, 0)
19. min (1, −3)
21. max (0, 0), min (−1, −5) and (2, −32)
23. max at (−3, 32), min at (1, 0)
25. max when $x = -\frac{5}{7}$, min at $x = 1$
27. max (0, 2), min (0, −2)
29. max $(\frac{1}{2}, -4.23)$, min $(\frac{1}{2}, 4.23)$

Exercise 3, Page 139

1. (0, 0)
3. (0, 0)
5. (3, −126)
7. (−1, −5) and (1, −5)
9. (0, −17) and (−1, −28)

Exercise 4, Page 143

1. 3.42
3. 3.06
5. 1.67
7. −1.72
9. 1.89 in.
11. 3.38 m

13.

```
10  '  NEWTON
20  '
30  '  This program uses Newton's
40  '  method to find the root of
50  '  an equation.
60  '
70  '  Enter the degree of accuracy
80  '  on line 150, the equation on
90  '  line 160, the derivative of
100 '  the equation on line 170 and
110 '  the appropriate value of the
120 '  root on line 180.
130 '
140 PRINT "Correction", "X"
150 A = .001
160 DEF FNA(X) = X^4 + 8 * X - 12
170 DEF FND(X) = 4 * X^3 + 8
180 X = 1
190 H = FNA(X) / FND(X)
200 X = X - H
210 PRINT H, X
220 IF ABS(H) > A THEN GOTO 190
230 END
```

Exercise 5, Page 150

1.

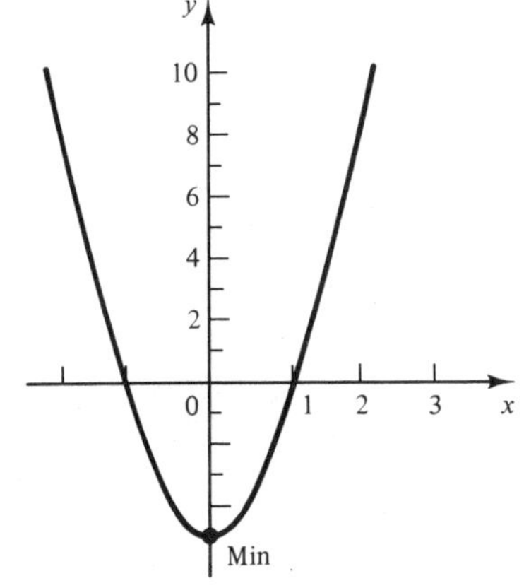

3.

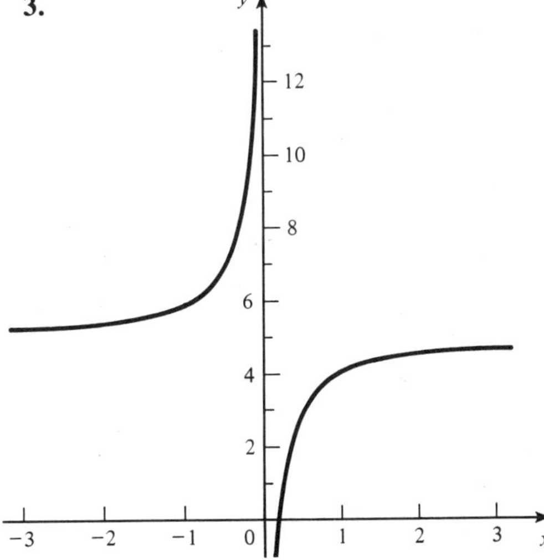

5.

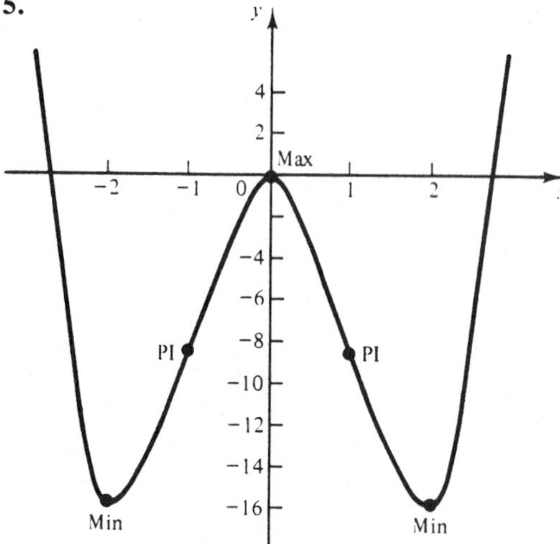

y
4
2
Max
−2
−1
0
1
2
x
−4
−6
−8
PI
PI
−10
−12
−14
−16
Min
Min

7.

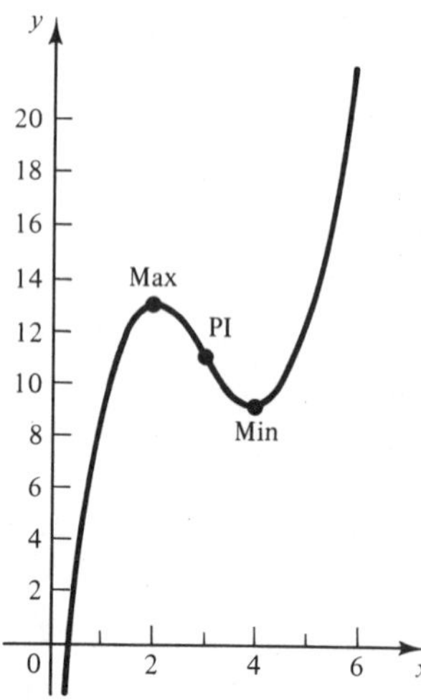

y
20
18
16
14
Max
PI
12
10
Min
8
6
4
2
0
2
4
6
x

9.

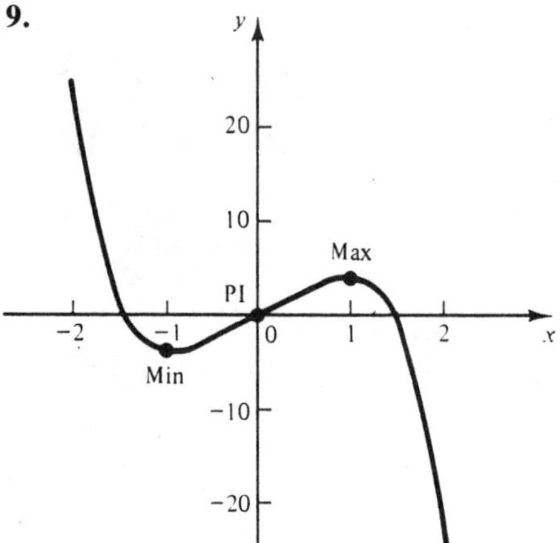

y
20
10
Max
PI
−2
−1
0
1
2
x
Min
−10
−20

11.

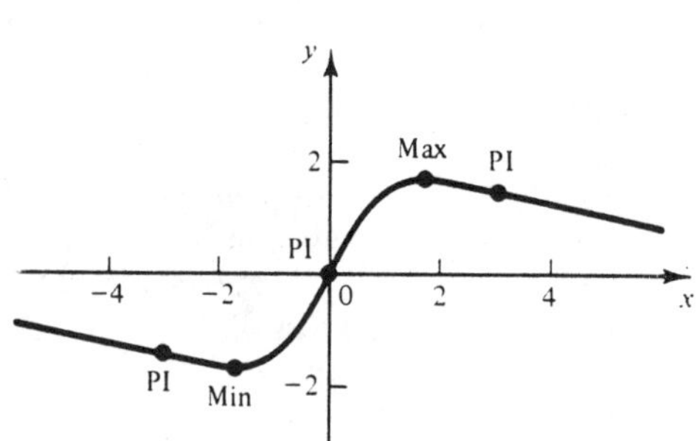

y
2
Max
PI
PI
−4
−2
0
2
4
x
PI
Min
−2

13.

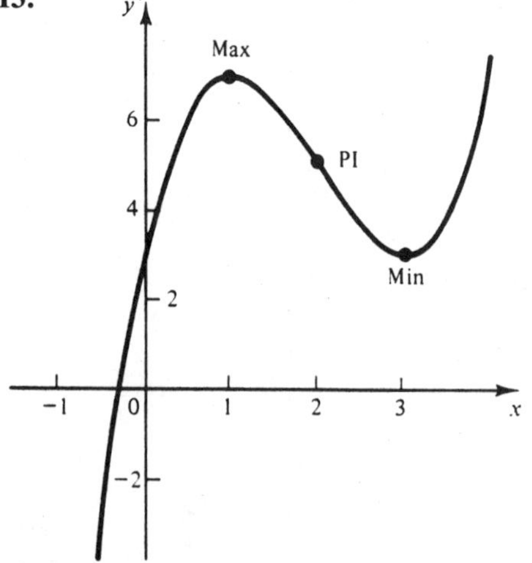

y
Max
6
PI
4
Min
2
−1
0
1
2
3
x
−2

15.

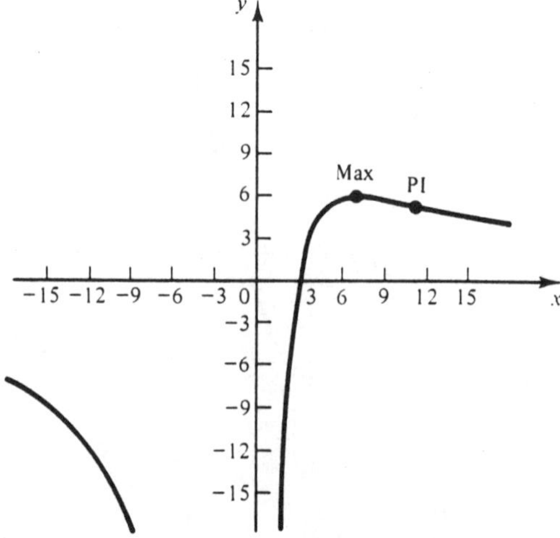

y
15
12
9
Max
PI
6
3
−15
−12
−9
−6
−3
0
3
6
9
12
15
x
−3
−6
−9
−12
−15

17.

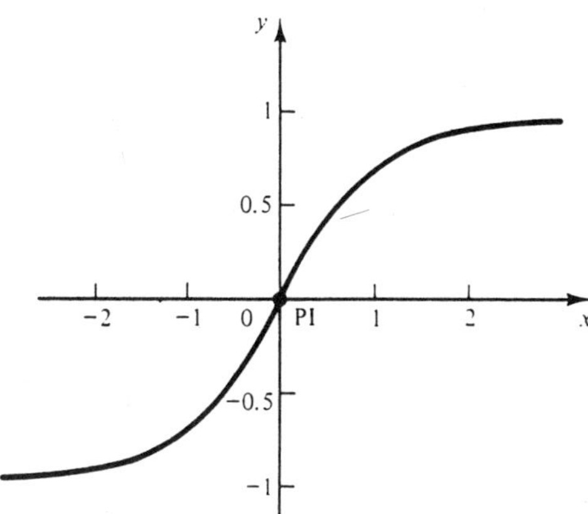
y
1
0.5
−2
−1
0
PI
1
2
x
−0.5
−1

19.

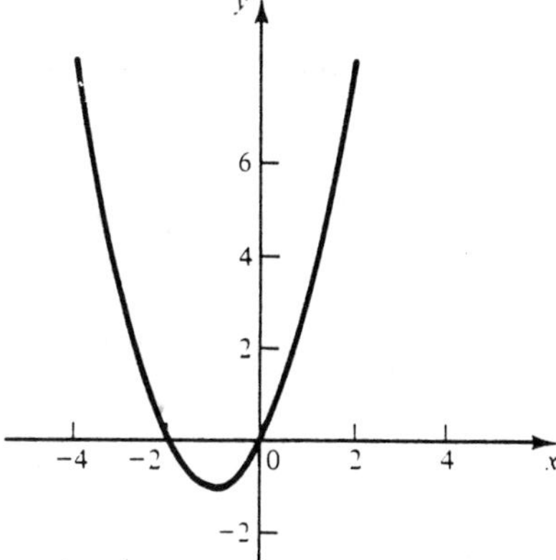
y
6
4
2
−4
−2
0
2
4
x
−2

21.

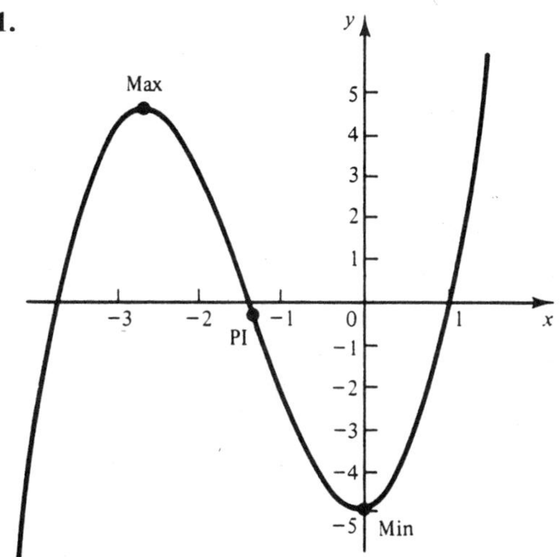
y
Max
5
4
3
2
1
−3
−2
PI
−1
0
1
x
−1
−2
−3
−4
−5
Min

23.

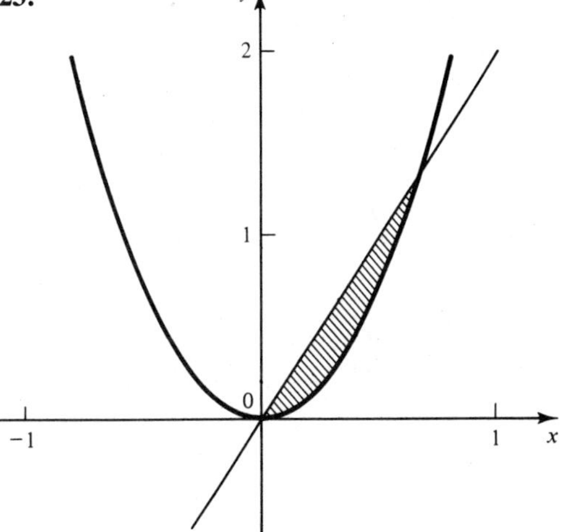
y
2
1
0
−1
1
x

25.

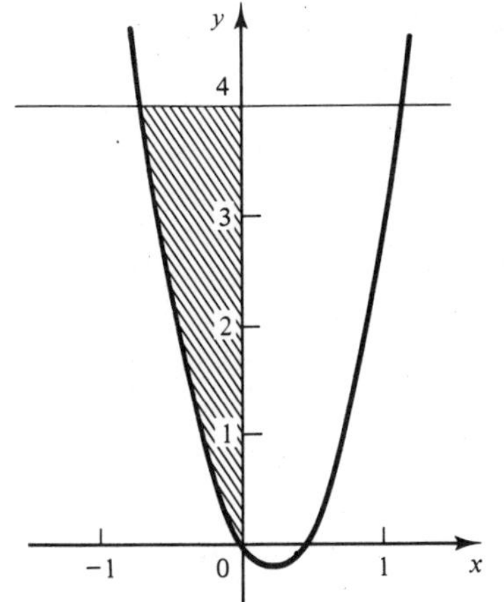
y
4
3
2
1
−1
0
1
x

Chapter Test, Page 150

1. −2.22, 0.54, 1.68 **3.** max (1, 2) **5.** $7x - y - 9 = 0$
$x + 7y - 37 = 0$

7. $-\frac{7}{3}$ **9.** rising for $x > 0$ **11.** down

13. $34x - y = 44$ and $x + 34y = 818$ **15.** 71.6°

CHAPTER FIVE: More Applications of the Derivative

Exercise 1, Page 157

1. −0.175 lb/in.²/in.³ **3.** 12.6 m³/m

5. 0.054 s/in. **7.** $\dfrac{dy}{dx} = \dfrac{wx}{6EI}(x^2 + 3L^2 - 3Lx)$

9. $i = 6.96t - 1.64$ A **11.** 50.9 V

13. 804 A **15.** 2.48 mA

17. 95.6 kV

Exercise 2, Page 164

1. $v = 0,\ a = -16$ **3.** $v = -3,\ a = 18$

5. $v = -8,\ a = -32$ **7.** 32 ft/s²

9. 60 ft/s **11.** 2000 ft/s, 62,500 ft, 1680 ft/s

13. $s = 0,\ a = 32$ units/s² **15.** $v_x = 6y + 5,\ v_y = -6t,\ a_x = 6,\ a_y = -6$

17. 8.00 ft/s at −1.79°

19. $v_x = 16$ units/s, $v_y = -\frac{1}{2}$ units/s, $a_x = 8$ units/s², $a_y = \frac{3}{4}$ units/s², $a = 8.04$ units/s²

21. 12 units/s **23.** (2, 4)

25. 6000 rad/s²

Exercise 3, Page 171

1. 60 m/s **3.** 3.58 mi/h **5.** 2.4 m/s

7. 4.47 ft/s **9.** 0.83 ft/s **11.** 170 km/h

13. 8.33 ft/s **15.** 4 ft/s **17.** 0.101 in./s

19. 0.133 in./min **21.** −4 in./min **23.** 1.19 m/min

25. 0.239 m/min **27.** 7.0 lb/in.²/s **29.** 2.99 in./s

Exercise 4, Page 180

1. 1 **3.** $6\frac{2}{3}$ and $13\frac{1}{3}$ **5.** 18 m × 24 m

7. 9 × 18 yards **9.** $h = 3.26$ cm, $r = 6.52$ cm

11. 25 in.² **13.** 6 **15.** 3.46 in.

17. (2, 2) **19.** 5 × 4.33 in. **21.** 9.9 × 14.1

23. $r = 4.08,\ h = 5.77$ **25.** 12 ft

27. $r = 5.24$ m, total height = 10.48 m **29.** 10.4 × 14.7 in.

31. 59.0 in. **33.** 7.5 A **35.** 5.0 A

37. 0.65

Exercise 5, Page 190

1. $3x^2\,dx$
3. $\dfrac{2\,dx}{(x+1)^2}$
5. $2(x+1)(2x+3)^2(5x+6)\,dx$
7. $\dfrac{2x-13}{2\sqrt{x-4}(3-2x)^2}\,dx$
9. $-\dfrac{x^2}{2y^2}\,dx$
11. $-\dfrac{2\sqrt{y}}{3\sqrt{x}}\,dx$
13. 0.016
15. 0.60
17. 0.00153 s
19. $V = 2\pi rht$
21. $3\pi r^2 t$
23. 0.2
25. 0.75 ft^2
27. $\frac{1}{2}\%$

Chapter Test, Page 192

1. 188 ft/s
3. 4.0 km/h
5. 3.33 cm^2/min
7. $v = \dfrac{s}{2}$
9. 6.93
11. −2.26 lb/in.^{2} per second
13. $t = 4$, $s = 108$ and $t = -2$, $s = 0$
15. 4
17. 4 and 6
19. 10 in. × 10 in. × 5 in.
23. 36.4 A
25. **(a)** 142 rad/s, **(b)** 37.0 rad/s^2

CHAPTER SIX: Derivatives of Trigonometric, Logarithmic, and Exponential Functions

Exercise 1, Page 200

1. $\cos x$
3. $-3 \sin x \cos^2 x$
5. $3 \cos 3x$
7. $\cos^2 x - \sin^2 x$
9. $3.75 (\cos x - x \sin x)$
11. $-2 \sin (\pi - x) \cos (\pi - x)$
13. $2 \cos 2x \cos x - \sin 2x \sin x$
15. $2 \sin x \cos^2 x - \sin^3 x$
17. $1.23 (2 \cos 3x \sin x \cos x - 3 \sin^2 x \sin 3x)$
19. $\dfrac{-\sin 2t}{\sqrt{\cos 2t}}$
21. $-k^2 \sin kx$
23. $-2 \sin x - x \cos x$
25. $-\dfrac{\pi^2}{4}$
27. $\dfrac{y \cos x + \cos y - y}{x \sin y - \sin x + x}$
29. $\sec (x + y) - 1$
31. −0.4161
33. 1.381
35. max $\left(\dfrac{\pi}{2}, 1\right)$, min $\left(\dfrac{3\pi}{2}, -1\right)$, $PI(\pi, 0)$
37. max (2.50, 5), min (5.64, −5), inflection points, (0.927, 0), (4.07, 0)
39. 0.515
41. 1.404

Exercise 2, Page 203

1. $2 \sec^2 2x$
3. $-15 \csc 3x \cot 3x$
5. $6.60x \sec^2 x^2$
7. $-21x^2 \csc x^3 \cot x^3$
9. $x \sec^2 x + 1$
11. $5 \csc 6x - 30x \csc 6x \cot 6x$
13. $2 \sin \theta \sec^2 2\theta + \tan 2\theta \cos \theta$
15. $5 \csc 3t \sec^2 t - 15 \tan t \csc 3t \cot 3t$
17. −294
19. 853
21. $6 \sec^2 x \tan x$
23. 9.92

25. $-y \sec x \csc x$

27. -1

29. $1.85x - y = 0.293$

31. max $(\pi/4, 0.571)$, min $(3\pi/4, 5.71)$, *PI* $(0, 0)$, *PI* $(\pi, 2\pi)$

33. $v = 23.5$ units/s, $a = -19.1$ cm/s^2

35. -3.58 deg/min

37. 15.6 ft

39. 31°

41. 25 ft

Exercise 3, Page 208

1. $\text{Sin}^{-1} x + \dfrac{x}{\sqrt{1 - x^2}}$

3. $\dfrac{-1}{\sqrt{a^2 - x^2}}$

5. $\dfrac{\cos x + \sin x}{\sqrt{1 + \sin 2x}}$

7. $-\dfrac{t^2}{\sqrt{1 - t^2}} + 2t \cos^{-1} t$

9. $\dfrac{1}{1 + 2x + 2x^2}$

11. $\dfrac{-a}{a^2 + x^2}$

13. $\dfrac{-1}{x\sqrt{4x^2 - 1}}$

15. $2t \text{ Arcsin } \dfrac{t}{2} + \dfrac{t^2}{\sqrt{4 - t^2}}$

17. $\dfrac{1}{\sqrt{a^2 - x^2}}$

19. -0.285

21. -0.054

23. 1.3°

Exercise 4, Page 214

1. $\dfrac{2}{x} \log e$

3. $\dfrac{3}{x \ln b}$

5. $\dfrac{(5 + 9x) \log e}{5x + 6x^2}$

7. $\left(\dfrac{x^2}{2}\right) \log e + \log \left(\dfrac{2}{x}\right)$

9. $2.75 + \dfrac{3}{x}$

11. $\dfrac{2x - 3}{x^2 - 3x}$

13. $8.25 + 2.75 \ln 1.02x^3$

15. $\dfrac{1}{x^2(x + 5)} - \dfrac{2 \ln (x + 5)}{x^3}$

17. $\dfrac{1}{2t - 10}$

19. $\cot x$

21. $\cos x(1 + \ln \sin x)$

23. $\dfrac{\sin x}{1 + \ln y}$

25. $\dfrac{x + y - 1}{x + y + 1}$

27. $-y$

29. $\dfrac{a^2}{\sqrt{x^2 - a^2}}$

31. $x^{\sin x}[(\sin x)/x + \cos x \ln x]$

33. $(\text{Arccos } x)^x \left(\ln \text{Arccos } x - \dfrac{x}{\sqrt{1 - x^2} \text{ Arccos } x}\right)$

35. 1

37. 0.3474

39. 128°

41. min $\left(\dfrac{1}{e}, -\dfrac{1}{e}\right)$

43. min (e, e), *PI* $(e^2, \frac{1}{2}e^2)$

45. 1.37

47. 0.607

49. -0.022

51. -0.0054 dB/day

Exercise 5, Page 218

1. $2(3^{2x}) \ln 3$

3. $10^{2x+3}(1 + 2x \ln 10)$

5. $2x(2^{x^2}) \ln 2$

7. $2e^{2x}$

9. e^{x+e^x}

11. $\dfrac{-xe^{\sqrt{1-x^2}}}{\sqrt{1 - x^2}}$

13. $-\dfrac{2}{e^x}$

15. $xe^{3x}(3x + 2)$

17. $\dfrac{e^x(x - 1)}{x^2}$

19. $\dfrac{3e^{-x} - e^x}{x^3}$

21. $2(x + xe^x + e^x + e^{2x})$

23. $\dfrac{(1 - e^x)(2xe^x - e^x - 1)}{x^2}$

25. $-e^{x-y}$

27. $\dfrac{\cos(x + y)}{e^y - \cos(x + y)}$

29. $3e^x \sin^2 e^x \cos e^x$

31. $e^x(b \cos bx - b \sin bx + \cos x + \sin bx)$

33. 0

35. $e^x\left(\ln x + \dfrac{1}{x}\right)$

37. $e^x\left(\dfrac{1}{x} + \ln x\right)$

39. $-2e^t \sin t$

41. $2e^x \cos x$

43. 0.015

45. 4.965

47. (0.402, 4.474)

49. max $\left(\dfrac{\pi}{4}, 3.22\right)$, min $\left(\dfrac{5\pi}{4}, -0.139\right)$

51. \$915.30/yr

53. 74.2

55. -563 rev/min^2

57. 2.2×10^7

59. -245 in. Hg/h

61. 0.044 lb/ft^3

Chapter Test, Page 221

1. $\frac{1}{2}(e^{x/a} + e^{-x/a})$

3. $\dfrac{4}{\sqrt{x}} \sec^2 \sqrt{x}$

5. $\dfrac{4x}{16x^2 + 1} + \tan^{-1} 4x$

7. $\dfrac{\cos 2x}{y}$

9. $x \cos x + \sin x$

11. $3x^2 \cos x - x^3 \sin x$

13. $\dfrac{x \cos x - \sin x}{x^2}$

15. $\dfrac{1 + 3x^2}{x^3 + x} \log e$

17. $\dfrac{1}{\sqrt{x^2 + a^2}}$

19. $-3 \csc 3x \cot 3x$

21. $\sec^2 x$

23. $\dfrac{6x^2 + 1}{2x^3 + x}$

25. $2x \cos^{-1} x - \dfrac{x^2}{\sqrt{1 - x^2}}$

27. (8.96, 0.291)

29. max at $x = \dfrac{\pi}{4}$

31. 0.053

33. $y - \dfrac{1}{2} = \dfrac{\sqrt{3}}{2}\left(x - \dfrac{\pi}{6}\right)$

35. $\dfrac{\pi}{4}$

37. 0.517

39. 1049°F at 5.49 h

CHAPTER SEVEN: Integration

Exercise 1, Page 231

1. $x + C$

3. $\frac{1}{2}x^2 + C$

5. $-\dfrac{1}{x} + C$

7. $\frac{3}{2}x^2 + C$

9. $\frac{3}{4}x^4 + C$

11. $\dfrac{1}{n + 1} x^{n+1} + C$

13. $2\sqrt{x} + C$

15. $x^4 + C$

17. $x^{7/2} + C$

19. $x^2 + 2x + C$

21. $x^{4/3} + C$

23. $u^2 + C$

25. $(\frac{8}{3})s^{3/2} + C$

27. $\frac{6x^{5/2}}{5} - \frac{4x^{3/2}}{3} + C$

29. $2x^2 - 4\sqrt{x} + C$

31. $-\frac{(1-s)^4}{4} + C$

33. $y = \frac{4x^3}{3} + C$

35. $y = -\frac{1}{2x^2} + C$

37. $y = \frac{3t^{1/3}}{2} + C$

Exercise 2, Page 235

1. $\frac{1}{4}(x^4 + 1)^4 + C$

3. $(x^2 + 2x)^2 + C$

5. $\frac{1}{1-x} + C$

7. $(x^3 + 1)^3 + C$

9. $-\frac{2}{3}(1 - y^3)^{1/2} + C$

11. $\frac{2}{a^2 - x^2} + C$

13. $\frac{1}{3}\tan^3 x + C$

15. $y = -\frac{1}{8}(1 - x^2)^4 + C$

17. $y = \frac{1}{8}(2x^3 + x^2)^4 + C$

19. $v = -(\frac{5}{3})(7 - t^2)^{3/2} + C$

Exercise 3, Page 238

1. $y = 3x + C$

3. $y = x^3 + C$

5. $x^3 - 3y + 2 = 0$

7. 17

9. $6y = x^3 - 6x - 9$

11. $xy + 6x = 6$

Exercise 4, Page 245

1. $3 \ln x + C$

3. $x + \ln |x| + C$

5. $\frac{5}{3} \ln (z^3 - 3) + C$

7. $\frac{a^{5x}}{5 \ln a} + C$

9. $\frac{5^{7x}}{7 \ln x} + C$

11. $\frac{a^{3x}}{3 \ln a} + C$

13. $4e^x + C$

15. $\frac{e^{x^2}}{2} + C$

17. $\frac{1}{6}e^{3x^2} + C$

19. $2e^{\sqrt{x}} + C$

21. $\frac{1}{2}e^{2x} - 2e^x + x + C$

23. $2e^{\sqrt{x+2}} + C$

25. $ae^{x/a} - ae^{-x/a} + C$

27. $-\frac{1}{3}\cos 3x + C$

29. $-\frac{1}{5}\ln |\cos 5\theta| + C$

31. $\frac{1}{4}\ln |\sec 4x + \tan 4x| + C$

33. $-\frac{1}{3}\ln |\cos 9\theta| + C$

35. $-\frac{1}{2}\cos x^2 + C$

37. $-\frac{1}{3}\ln |\cos \theta^3| + C$

39. $-\cos (x + 1) + C$

41. $\frac{1}{5}\ln |\cos (4 - 5\theta| + C$

43. $\frac{1}{8}\ln |\sec (4x^2 - 3) + \tan (4x^2 - 3)| + C$

45. $\frac{1}{2}\ln (y^2 + 4y) + C$

47. $\frac{x}{2}\sqrt{25 - 9x^2} + \frac{25}{6}\text{Arcsin}\left(\frac{3x}{5}\right) + C$

49. $\frac{1}{3}\tan^{-1}\frac{x}{3} + C$

51. $\frac{1}{12}\text{Arctan}\left(\frac{4x}{3}\right) + C$

53. $\frac{1}{3}\ln (2 + x^3) + C$

55. $\frac{1}{4}\ln \left|\frac{x-2}{x+2}\right| + C$

57. $\frac{x}{4}\sqrt{x^2 - 4} - \ln |x + \sqrt{x^2 - 4}| + C$

59. $\frac{5}{2}\text{Arcsin}\, x^2 + C$

61. $\frac{1}{12}\ln \left|\frac{2 + 3t}{2 - 3t}\right| + C$

Exercise 5, Page 251

1. $v = 4t - \frac{1}{3}t^3 - 1$

3. **(a)** $x = \frac{1}{3}t^3 - 4t$, $y = 2t^2$ **(b)** 9.6

5. $s = 20t - 16t^2$

7. 1667 cm/s and 100 cm/s

9. 344 cm/s and 2250 cm/s

11. 18,000 rev

13. 15.7 rad/s

Exercise 6, Page 254

1. 12.1 coul
3. 15.2 coul
5. 2.26 V
7. 4.25 A
9. 20.2 A

Chapter Test, Page 255

1. $-\frac{1}{3}\ln|\cos 3\theta| + C$

3. $\dfrac{a^{2x}}{2\ln a} + C$

5. $\frac{1}{5}\left(e^{5x} + \dfrac{a^{5x}}{\ln a}\right) + C$

7. $\frac{1}{2}(e^{2x+1} + x^2) + C$

9. $\dfrac{x^2}{2} + x - \dfrac{1}{2x^2} + C$

11. $-\dfrac{1}{e^x} + C$

13. $-\frac{1}{3}\cot 3x + C$

15. $-\dfrac{2}{t} + C$

17. $\frac{1}{2}\ln|\sec 2\theta + \tan 2\theta| + C$

19. $\frac{1}{9}(x^3 - 4)^3 + C$

21. $\frac{1}{2}\ln|x^2 + 3| + C$

23. $y = \dfrac{x^3}{6} + \dfrac{9x}{2} - 18$

25. $v = \dfrac{3t^2}{2}, s = \dfrac{t^3}{2} + s_0$

27. 19,360 rad/s, 15,380 rev

29. $s_x = \dfrac{t^3}{6} + 4t + 1, s_y = \dfrac{5t^3}{6} + 15t + 1$

CHAPTER EIGHT: The Definite Integral

Exercise 1, Page 261

1. $\frac{3}{2}$
3. 60.7
5. $\frac{112}{3}$
7. $\frac{218}{3}$
9. 2.30
11. 0.112
13. 0.732
15. 0.859
17. 2
19. 2

Exercise 2, Page 268

1. 15
3. 84
5. 40
7. 176
9. 1.575
11. 100
13. $113\frac{1}{3}$
15. 64
17. 4.24
19. 9.83
21. 15.75

23.

```
10 '          PANELS
20 '
30 ' COMPUTES AREA UNDER A CURVE
40 ' BY MIDPOINT METHOD
50 '
60 DEF FNA(X) = X^2 + 1
70 INPUT "LOWER LIMIT"; A
80 INPUT "UPPER LIMIT"; B
90 INPUT "NUMBER OF PANELS";N
100 W = (B-A)/N
110 FOR X=A+W/2 TO B-W/2 STEP W
120 Y = FNA(X)
130 AREA = AREA + Y*W
140 NEXT X
150 PRINT "PANEL WIDTH IS "; W
160 PRINT "AREA IS "; AREA
```

Exercise 3, page 278

1. 2.797 **3.** 16 **5.** $42\frac{2}{3}$ **7.** $1\frac{1}{6}$
9. $10\frac{2}{3}$ **11.** 72.4 **13.** 32.83 **15.** $2\frac{2}{3}$
17. 2 **19.** $8/\pi$ **21.** 3 **23.** 9
25. 25.6 **27.** $5\frac{1}{3}$ **37.** 90.1 ft^2 **39.** 24.9 ft^3
41. 500 ft^2 **43.** 35 ft^2

Chapter Test, Page 283

1. 0.693 **3.** 4 **5.** $81\frac{2}{3}$ **7.** $-\frac{1}{2}$
9. 170 **11.** 280 **13.** 12π **15.** 19.8
17. 25.6 **19.** 2.13

CHAPTER NINE: Applications of the Definite Integral

Exercise 1, Page 295

1. $\dfrac{128\pi}{7}$ **3.** 16.8 **5.** 0.666 **7.** 0.479
9. 4.42 **11.** 50.3 **13.** 32π **15.** 19.7
17. 91.9 **19.** 3.35 **21.** 1309 **23.** 1.65 m^3
25. 0.683 m^3

Exercise 2, Page 300

1. 0.704 **3.** 1.32 **5.** $\frac{3}{2}$ **7.** 24.8
9. $\frac{19}{5}$ **11.** 9.07 **13.** 1096 ft **15.** 223 ft
17. 30.3 ft

Exercise 3, Page 304

1. 32.1 **3.** 154 **5.** 36.2 **7.** 58.6
9. 72.6 **11.** 1570 **13.** 36.3 **15.** $4\pi r^2$
17. 36.2 ft^2 **19.** $16,010

Exercise 4, Page 307

1. 12 **3.** 4.08 **5.** $\frac{1}{2}$ **7.** 7.81
9. 19.1 **11.** 1.17

Chapter Test, Page 307

1. 31.4 **3.** 51.5 **5.** 151 **7.** 197
9. 310 **11.** 407 **13.** $\frac{1}{2}$

CHAPTER TEN: Centroids and Moments

Exercise 1, Page 319

1. 5/4, −3/2 **3.** 3.22 **5.** (2.4, 0) **7.** 1.07
9. 0.93 **11.** 1.067 **13.** $(\frac{9}{20}, \frac{9}{20})$ **15.** 0.299

17. 1
19. 5
21. 5
23. $\frac{2p}{3}$
25. $\frac{h}{4}$
27. 4.13 ft
29. 36.1 mm

Exercise 2, Page 325

1. 88,200 lb
3. 41,200 lb
5. 62.4 lb
7. 2160 lb

Exercise 3, Page 329

1. 25.0 ft lb
3. 32.0 in. lb
5. 50,200 ft lb
7. 5.71×10^5 ft lb
9. 723,000 ft lb
11. $k/100$ ft lb
13. 13,800 ft lb

Exercise 4, Page 338

1. $\frac{1}{12}$
3. 2.62
5. 19.5
7. $\frac{2}{9}$
9. 0.471
11. $10.1m$
13. $89.4m$
15. $\frac{3Mr^2}{10}$
17. $\frac{4Mp^2}{3}$

Chapter Test, Page 339

1. 48,300 ft lb
3. 1123 lb
5. 1.8 ft lb
7. 2560 lb
9. 4.24
11. (1.70, 1.27)

CHAPTER ELEVEN: Methods of Integration

Exercise 1, Page 346

1. $\sin x - x \cos x + C$
3. $x \tan x + \ln|\cos x| + C$
5. $x \sin x + \cos x + C$
7. $\frac{x^2}{4} - \frac{x}{12} \sin 6x - \frac{1}{72} \cos 6x + C$
9. 1.24
11. $-\frac{1}{x+1}[\ln|x+1| + 1] + C$
13. 5217
15. $-\frac{x^2}{3}(1-x^2)^{3/2} - \frac{2}{15}(1-x^2)^{5/2} + C$ or $-\frac{1}{15}(1-x^2)^{3/2}(3x^2+2) + C$
17. $\frac{x}{\sqrt{1-x^2}} - \sin^{-1} x + C$
19. $\sin x\,(\ln|\sin x| - 1) + C$
21. $-x^2e^{-x} - 2xw^{-x} - 2e^{-x} + C$
23. $\frac{e^x(\sin x - \cos x)}{2} + C$
25. -0.081
27. $\frac{e^{-x}}{\pi^2+1}(\pi \sin \pi x - \cos x)$

Exercise 2, Page 353

1. $\ln\left|\frac{x(x-2)}{(x+1)^2}\right| + C$
3. -0.1054
5. $\ln\left|\frac{(x-1)^{1/2}}{x^{1/3}(x+3)^{1/6}}\right| + C$

7. -0.739

9. $\dfrac{9-4x}{2(x-2)^2}+C$

11. 0.0637

13. $-\dfrac{1}{x}-\text{Arctan}\,x+C$

15. $\ln 7$

17. $\dfrac{1}{2}\text{Arctan}\,(1+x^2)+\dfrac{1}{4}\ln\left|\dfrac{x+1}{x-1}\right|+C$

19. 0.667

21. $\dfrac{x^2}{2}-4\ln(x^2+4)-\dfrac{8}{x+4}+C$

Exercise 3, Page 357

1. $2\sqrt{x}-2\ln|1+\sqrt{x}\,|+C$

3. $\dfrac{3\ln|x^{2/3}-1|}{2}+C$

5. $\frac{3}{5}\sqrt[3]{(1+x)^5}-\frac{3}{2}\sqrt[3]{(1+x)^2}+C$ or $\frac{3}{10}(1+x)^{2/3}(2x-3)+C$

7. $-\frac{2}{147}(4+7x)\sqrt{2-7x}+C$

9. $\dfrac{6x^2+6x+1}{12(4x+1)^{3/2}}+C$

11. $2\sqrt{x+2}+\sqrt{2}\tan^{-1}\sqrt{\dfrac{x+2}{2}}+C$

13. $\ln\left|\dfrac{1-\sqrt{1-x^2}}{x}\right|+C$

15. $\dfrac{3(1+x)^{4/3}(4x-3)}{28}+C$

17. 0.2375

19. 0.1042

Exercise 4, Page 361

1. $\dfrac{x}{\sqrt{5-x^2}}+C$

3. $\dfrac{1}{2}\ln\left|\dfrac{x}{2+\sqrt{x^2+4}}\right|+C$ or $\dfrac{1}{2}\ln\left|\dfrac{\sqrt{x^2+4}-2}{x}\right|+C$

5. $-\dfrac{\sqrt{4-x^2}}{4x}+C$

7. $-\dfrac{x}{2}\sqrt{4-x^2}+2\sin^{-1}\dfrac{x}{2}+C$

9. $-\dfrac{\sqrt{16-x^2}}{x}-\sin^{-1}\dfrac{x}{4}+C$

11. 16.49

13. $\dfrac{\sqrt{x^2-7}}{7x}+C$

Exercise 5, Page 365

1. $\frac{1}{2}$

3. e

5. π

7. 2

9. diverges

11. π

Exercise 6, Page 370

1. 1.79

3. 66

5. 2

7. 0.33

9. 5960

11. 260

13. 13,000 ft^2

15. 90,000 ft^2

17. 86.6 kW

19.
```
10  '  AREA
20  '
30  '  This program uses the trap-
40  '  ezoid rule to find the area
50  '  under a number of given
60  '  points.
70  '
80  '  Enter the points(X, Y)in the
90  '  data statements.
100 '
110 INPUT "How many points are there"; S
120 READ X, Y
130 FOR N = 2 TO S
140 READ X1, Y1
150 A = A + (.5 * (X1 - X) * (Y1 + Y))
160 LET X = X1: Y = Y1
170 NEXT N
180 DATA 0, 4, 1, 3, 2, 4, 3, 7, 4, 12, 5, 19
190 DATA 6, 28, 7, 39, 8, 52, 9, 67, 10, 84
200 PRINT "The area is"; A
210 END
```

Chapter Test, Page 373

1. $-\frac{1}{3}\cos^3 x - \frac{1}{5}\cot^5 x + C$

3. $\frac{1}{12}\ln\left|\frac{3x-2}{3x+2}\right| + C$

5. $\frac{3}{2}\ln|x^2+9| - \frac{1}{3}\tan^{-1}\frac{x}{3} + C$

7. $\ln\left|\frac{\sqrt{1-x^2}-1}{x}\right| + C$

9. $\tan^{-1}(2x-1) + C$

11. $x\tan x + \ln|\cos x| + C$

13. $\frac{1}{2}\ln|x^2+x+1| + \sqrt{3}\tan^{-1}\left(\frac{2x+1}{\sqrt{3}}\right) + C$

15. -0.4139

17. 1.493

19. $\frac{1}{6}\ln\left|\frac{3x-1}{3x+1}\right| + C$

21. $\frac{1}{4(1-x)^4} - \frac{1}{3(1-x)^3} + C$ or $\frac{4x-1}{12(1-x)^4} + C$

23. $-\csc x - \sin x + C$

25. 0.2877

27. 2.171

29. $3\sqrt[3]{2}$

31. diverges

CHAPTER TWELVE: First-Order Differential Equations

Exercise 1, Page 378

1. first order, first degree, ordinary

3. third order, first degree, ordinary

5. second order, fourth degree, ordinary

7. $y = \frac{7x^2}{2} + C$

9. $y = \frac{2x^2}{3} - \frac{5x}{3} + C$

11. $y = \frac{x^3}{3} + C$

Exercise 2, Page 382

1. $y = \pm\sqrt{x^2+C}$

3. $\ln|y^3| = x^3 + C$

5. $4x^3 - 3y^3 = C$

7. $y = \frac{\sqrt{x^2+1}}{C}$

9. $\arctan x + \arctan y = C$

11. $\sqrt{1+x^2} + \ln y = C$

13. $\arctan x - \arctan y = C$

15. $2x + xy - 2y = C$

17. $x^2 + Cy^2 = C - 1$

19. $y = C\sqrt{1+e^{2x}}$

21. $e^{2x} + e^{2y} = C$

23. $(1+x)\sin y = C$

25. $\cos x\cos y = C$

27. $2\sin^2 x - \sin y = C$

29. $y^3 - x^3 = 1$

31. $\sin x + \cos y = 2$

Exercise 3, Page 385

1. $xy - 7x = C$
3. $xy - 3x = C$
5. $4xy - x^2 = C$
7. $y/x = 3x + C$
9. $x^3 + 2xy = C$
11. $x^2 - 4xy + y^2 = C$
13. $x/y = 2x + C$
15. $2y^3 = x + Cy$
17. $4xy^2 - 3x^2 = C$
19. $2x^2 - xy = 15$
21. $3y^3 - y = 2x$
23. $3x^2 + 3y^2 - 4xy = 8$

Exercise 4, Page 388

1. $x(x - 3y)^2 = C$
3. $x \ln y - y = Cx$
5. $y^3 = x^3(3 \ln x + C)$
7. $x(x - 3y)^2 = 4$
9. $(y - 2x)^2(x + y) = 27$

Exercise 5, Page 393

1. $y = 2x + \dfrac{C}{x}$
3. $y = x^3 + \dfrac{C}{x}$
5. $y = 1 + Ce^{-x^3/3}$
7. $y = \dfrac{C}{\sqrt{x}} - \dfrac{3}{x}$
9. $y = Cx^2 + x$
11. $y(1 + x^2)^2 = x^2 + 2 \ln x + C$
13. $y = \dfrac{C}{xe^{x^2/2}}$
15. $y = \dfrac{e^x}{2} + \dfrac{C}{e^x}$
17. $y = (4x + C)e^{2x}$
19. $x^2y = 2 \ln^2 x + C$
21. $y = 1/2 \sin x - 1/2 \cos x + C/e^x$
23. $y = \sin x + \cos x + Ce^{-x}$
25. $xy(C - 3x^2/2) = 1$
27. $y = \dfrac{1}{x^2 + Ce^{-x}}$
29. $xy - 2x^2 = 3$
31. $y = \dfrac{5x}{2} - \dfrac{1}{2x}$
33. $y = \tan x + \sqrt{2} \sin x$

Exercise 6, page 398

1. $y = x^2 + 3x - 1$
3. $y^2 = x^2 + 7$
5. $y = 0.812e^x - x - 1$
7. $xy = 4$
9. $y = \dfrac{e^x + e^{-x}}{2}$
11. $y = Cx$
13. $3x^2y + x^3 = k$

Exercise 7, Page 401

3. 158×10^6 bbl/day
5. 9.71°F
7. 1240°F
13. 15.0 ft/s
15. 2.03 s
17. 2.36 s

Exercise 8, Page 407

7. 231 mA, 0.882 V
9. 46.4 mA
11. 237 mA
13. 139 mA

Chapter Test, Page 408

1. $2xy = e^x + Ce^{-x}$
3. $y = \sin x + \cos x + Ce^{-x}$
5. $x^3 + 2xy = C$
7. $x \ln y - y = Cx$

9. $Cy + y \ln|1 - x| = 1$

11. $x^2 + \dfrac{x}{y} + \ln y = C$

13. $\sin x + \cos x = 1.707$

15. $x + 2y = 2xy^2$

17. 14.6 rev/s

19. $-2 \ln|x| + C$

21. $y = 2x \ln x + 2x$

23. 4740 g

CHAPTER THIRTEEN: Second-Order Differential Equations

Exercise 1, Page 413

1. $y = \dfrac{5x^2}{2} + C_1x + C_2$

3. $y = 3e^x + C_1x + C_2$

5. $y = \dfrac{x^4}{12} + x$

Exercise 2, Page 420

1. $y = C_1e^x + C_2e^{5x}$

3. $y = C_1e^x + C_2e^{2x}$

5. $y = C_1e^{3x} + C_2e^{-2x}$

7. $y = C_1 + C_2e^{0.4x}$

9. $y = C_1e^{2x/3} + C_2e^{-3x/2}$

11. $y = (C_1 + C_2x)e^{2x}$

13. $y = C_1e^x + C_2xe^x$

15. $y = (C_1 + C_2x)e^{-2x}$

17. $y = (C_1 + C_2x)e^{-x}$

19. $y = e^{-2x}(C_1 \cos 3x + C_2 \sin 3x)$

21. $y = e^{3x}(C_1 \cos 4x + C_2 \sin 4x)$

23. $y = C_1 \cos 2x + C_2 \sin 2x$

25. $y = e^{2x}(C_1 \cos x + C_2 \sin x)$

27. $y = 3xe^{-3x}$

29. $y = (5 - 14x)e^x$

31. $y = 1 + e^{2x}$

33. $y = \dfrac{e^{2x} + 3e^{-2x}}{4}$

35. $y = e^{-x} \sin x$

37. $y = C_1e^x + C_2e^{-x} + C_3e^{2x}$

39. $y = C_1e^x + C_2e^{2x} + C_3e^{3x}$

41. $y = C_1e^x + C_2e^{-x} + C_3e^{3x}$

43. $y = (C_1 + C_2)e^{x/2} + C_3e^{-x}$

Exercise 3, Page 426

1. $y = C_1e^{2x} + C_2e^{-2x} - 3$

3. $y = C_1e^{2x} + C_2e^{-x} - 2x + 1$

5. $y = C_1e^{2x} + C_2e^{-2x} - x^3/4 - 5x/8$

7. $y = C_1e^{3x} + C_2e^{-x} - 3e^x$

9. $y = C_1e^{2x} + C_2e^{-2x} + e^x - x$

11. $y = e^{-2x}(C_1 + C_2x) + (2e^{2x} + x - 1)/4$

13. $y = C_1 \cos 2x + C_2 \sin 2x - \dfrac{x \cos 2x}{4}$

15. $y = (C_1 + C_2x)e^{-x} + \dfrac{\sin x}{2}$

17. $y = C_1 \cos x + C_2 \sin x + \cos 2x + x \sin x$

19. $y = C_1 \cos x + C_2 \sin x + \dfrac{e^x(\sin x - 2 \cos x)}{5}$

21. $y = e^{2x}\left[C_1 \sin x + C_2 \cos x - \dfrac{x \cos x}{2}\right]$

23. $y = \dfrac{e^{4x}}{2} - 2x - \frac{1}{2}$

25. $y = \dfrac{1 - \cos 2x}{2}$

27. $y = e^x(x^2 - 1)$

29. $y = x \cos x$

Exercise 4, Page 436

1. **(a)** 34.5 ms, **(b)** 3.75 in.

3. $x = \sin 15t$

5. $x = \cos 6.95t$

7. $x = 1.50 \cos 2t$
11. -0.714 in.
13. $x = 7.10e^{-t} + 5.60e^{-4t}$
15. (b) 17.9 lb
17. 7.29 Hz
19. 2.41 in.
21. 2.46 in.

9.
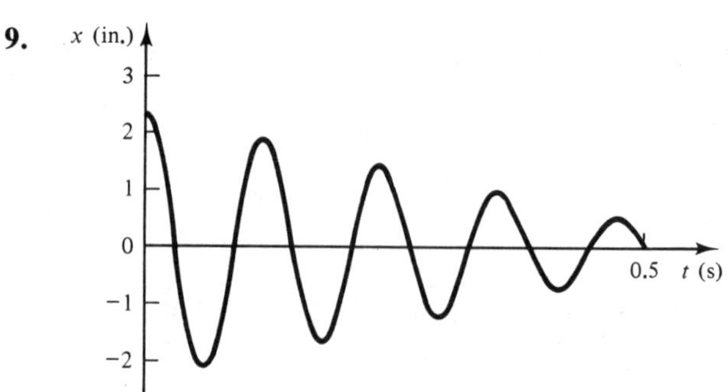

Exercise 5, Page 447

3. $q = 255 \cos 1000t$ μC
11. $i = 161 \sin 377t$ mA
13. 9.84 Hz or 61.7 rad/s
15. 0.496 μF
17. 10.0 A

Chapter Test, Page 448

1. $y = -(\frac{1}{9}) \sin 3x + C_1x + C_2$
3. $y = C_1 \cos \sqrt{5}\,x + C_2 \sin \sqrt{5}\,x$
5. $y = C_1e^{2x} + C_2e^{-x} - \frac{5}{2}x$
7. $y = C_1e^{5x} + C_2e^{-x} - \frac{1}{5}e^{4x}$
9. $5y = e^{5x} - 6x + 4$
11. $y = C_1e^{9x} + C_2e^{-2x}$
13. $y = C_1e^{2x} + C_2e^{3x} + C_3e^{x}$
15. $C_1 \cos \sqrt{2}\,x + C_2 \sin \sqrt{2}\,x - \frac{3}{2} \sin 2x$
17. $y = 4e^{-x} - 3e^{-2x}$
19. $y = C_1e^{x} + C_2e^{-4x}$
21. $y = C_1e^{-0.519} + C_2e^{0.192}$
23. $5y = 2(e^{-4x} - e^{x})$
25. $y = C_1e^{-3x} + C_2xe^{-3x}$
27. $y = C_1e^{x/3} + C_2xe^{x/3}$
29. $x = 1.5 \cos 7.43t$
31. $98.1 \sin 510t$ mA
33. $i = 4.66e^{-18.75t} \sin 258t$ A

CHAPTER FOURTEEN: The Laplace Transform

Exercise 1, Page 457

1. $\dfrac{6}{s}$
3. $\dfrac{2}{s^3}$
5. $\dfrac{s}{s^2 + 25}$
7. $\dfrac{2}{s^3} + \dfrac{4}{s}$
9. $\dfrac{3}{s - 1} + \dfrac{2}{s + 1}$
11. $\dfrac{2}{s^2 + 4} + \dfrac{s}{s^2 + 9}$
13. $5\dfrac{(s - 3)}{(s - 3)^2 + 25}$
15. $\dfrac{5}{s - 1} - \dfrac{6s}{(s^2 + 9)^2}$
17. $s\mathcal{L}[y] - 1 + 2\mathcal{L}[y]$
19. $s\mathcal{L}[y] - 4\mathcal{L}[y]$
21. $s^2\mathcal{L}[y] + 3s\mathcal{L}[y] - s - 6 - \mathcal{L}[y]$
23. $(2s^2 + 3s)\mathcal{L}[y] - 4s - 12 + \mathcal{L}[y]$

Exercise 2, Page 460

1. 1
3. t^2
5. $2 \sin 2t$

7. $3 \cos \sqrt{2}\, t)$

9. $(13t + 1)e^{9t}$

11. $\dfrac{5(e^{-2t} \sin 3t)}{3}$

13. $\cos t + 1$

15. $e^{t} + e^{-2t} + t - 2$

17. $\dfrac{2(e^{t} - e^{-2t})}{3}$

19. $6e^{-3t} - 4e^{-2t}$

Exercise 3, Page 463

1. $y = e^{3t}$

3. $y = e^{t/2} - \dfrac{t}{2} - 1$

5. $y = \dfrac{21e^{2t/3}}{4} - \dfrac{t^2}{2} - \dfrac{3t}{2} - \dfrac{9}{4}$

7. $y = \dfrac{3[e^{-2t} - (\sin 2t)/2]}{2}$

9. $y = \dfrac{e^{t} - e^{-3t}}{2}$

11. $y = t - \frac{1}{3}e^{-3t} + \frac{1}{3}$

13. $y = 4t - 5.66 \sin 0.707t + 3 \cos 0.707t$

15. $y = t/3 - \frac{4}{9} + \frac{9}{2}e^{-t} - \frac{37}{8}e^{-3t}$

17. $y = 12 - 9.9e^{-t/3} - 0.1 \cos t - 0.3 \sin t$

19. $y = \frac{1}{2}t^2e^{2t}$

21. $y = 0.111e^{-t}\cos \sqrt{2}\, t + 0.0393e^{-t}\sin \sqrt{2}\, t - 0.111e^{t} + 0.167te^{t}$

23. 5.68 m/s

25. 5.7×10^{29}

27. $x = 2.28e^{-6.25t} - 2.28 \cos 2.25t + 6.36 \sin 2.25t$

Exercise 4, Page 471

1. $i = 41.8(1 - e^{-16t})$ mA

3. $i = 16.4(1 - e^{-16t})$ mA

5. $q = -26.3 \cos 35t + 343 \sin 35t + 26.3e^{-900t} \cos 118t + 98e^{-900t} \sin 118t\ \mu\text{C}$

7. $i = 6.03(1 - e^{-33.5t})$

Chapter Test, Page 472

1. $\dfrac{2}{s^2}$

3. $\dfrac{s}{s^2 + 9}$

5. $\dfrac{3}{(s - 2)^2}$

7. $\dfrac{3}{s + 1} - \dfrac{4}{s^3}$

9. $(s + 3)\mathscr{L}[y] - 1$

11. $(3s + 2)\mathscr{L}[y]$

13. $(s^2 + 3s + 4)\mathscr{L}\,[y] - s - 6$

15. $e^{3t}(1 + 3t)$

17. $\dfrac{t \sin \sqrt{5}\, t}{\sqrt{5}}$

19. $5e^{4t} - 3e^{2t}$

21. $3e^{-t}(1 - t)$

23. $4e^{-2t}(1 - 2t)$

25. $y = -\dfrac{2}{27} - \dfrac{2t}{9} - \dfrac{t^3}{3} + \dfrac{2e^{3t}}{27}$

27. $y = 2 - 2e^{-t}(1 + t)$

29. $y = 2t + 3 \cos \sqrt{2}\, t - 2 \sin \sqrt{2}\, t$

31. $v = 7.82(1 - e^{-4.12t})$

33. $i = 0.671(1 - e^{-18.1t})$

35. $i = 40.7(e^{-200t} - e^{-24{,}800t})$ mA

CHAPTER FIFTEEN: Numerical Solution of Differential Equations

Exercise 1, Page 481

1. 0.644

3. 1.899

5. 5.824

7. 3.123

9. 13.751

11.
```
10 '          EULER
20 '
30 'THIS PROGRAM SOLVES A DIFFERENTIAL
40 'EQUATION USING EULER'S METHOD
50 '
60 DEF FNM(X, Y) = (2.99*Y-8.23*X*SIN(Y))/1.94*X^2
70 W=.1 :X0=0: Y0=3
80 Y1=Y0
90 PRINT "  X           APPROX Y"
100 PRINT USING " #.#         ##.###";X0,Y0
110 FOR X1 = X0 TO 1 STEP W
120 Y2 = Y1 + W*FNM(X1,Y1)
130 PRINT USING "##.#         ##.###";X1+W,Y2
140 Y1=Y2
150 NEXT X1
```

13.
```
10 '                    RUNGE
20 '
30 'THIS PROGRAM SOLVES A DIFFERENTIAL EQUATION
40 'USING THE RUNGE-KUTTA METHOD
50 '
60 DEF FNM(X, Y) = (2.99*Y-8.23*X*SIN(Y))/1.94*X^2
70 W=.1:X0=0:  Y0=3
80 Y1=Y0
90 PRINT "X", "APPROX Y"
100 PRINT X0, Y0
110 FOR X1 = X0 TO 1 STEP W
120 X2 = X1+W
130 K1=FNM(X1,Y1)
140 K2=FNM(X1+W/2, Y1+K1*W/2)
150 K3=FNM(X1+W/2,Y1+K2*W/2)
160 K4=FNM(X1+W,Y1+K3*W)
170 Y2 = Y1 + W*(K1+2*K2+2*K3+K4)/6
180 PRINT X2   ,Y2
190 Y1=Y2
200 NEXT X1
```

Exercise 2, Page 485

1. 2.502 **3.** 12.090 **5.** 9.033

7. −1.267 **9.** 3.267

11.
```
10 '        ORDER_2
20 '
30 'SOLVES A SECOND ORDER D.E.
40 'BY THE RUNGE-KUTTA METHOD
50 '
60 H = .1
70 X=1: Y=0: M=2
80 DEF FNA(X,Y,M)=3-X*Y-M
90 A$="#.##      ###.####"
```

```
100 '
110 PRINT "X         APPROX Y"
120 PRINT "-----------------------------------------------"
130 PRINT USING A$; X,Y
140 '
150 'COMPUTE RUNGE NUMBERS
160 M1=M
170 MM1=FNA(X,Y,M)
180 M2=M+MM1*H/2
190 MM2=FNA(X+H/2,Y+H* M1/2,M2)
200 M3=M+MM2*H/2
210 MM3=FNA(X+H/2,Y+H*M2/2,M3)
220 M4=M+MM3*H
230 MM4=FNA(X+H,Y+H*M3,M4)
240 '
250 'COMPUTE AVERAGE SLOPES
260 MAVG=(M1+2*M2+2*M3+M4)/6
270 MMAVG=(MM1+2*MM2+2*MM3+MM4)/6
280 '
290 'COMPUTE NEXT VALUES
300 XNEXT=X+H
310 YNEXT=Y+H*MAVG
320 MNEXT=M+H*MMAVG
330 '
340 IF XNEXT>3 THEN STOP
350 PRINT USING A$; XNEXT, YNEXT
360 'SWAP VALUES
370 X=XNEXT
380 Y=YNEXT
390 M=MNEXT
400 GOTO 150
```

Chapter Test, Page 486

1. 8.913 **3.** 6.167 **5.** 0.844 **7.** 5.184 **9.** 7.065

CHAPTER SIXTEEN: Infinite Series

Exercise 1, Page 495

1. $3 + 6 + 9 + 12 + 15 + \cdots$

3. $2 + \frac{3}{4} + \frac{4}{9} + \frac{5}{16} + \frac{6}{25} + \cdots$

5. $2n$

7. $\dfrac{2^n}{n + 3}$

9. $u_{n-1} + 4$

11. 3^n

13. converges

15. diverges

17. test fails

19.

```
10 '        SERIES
20 '
30 'THIS PROGRAM GENERATES AN INFINITE SERIES
```

```
40 'AND COMPUTES PARTIAL SUMS AND RATIOS
50 '
60 E=EXP(1)
70 LAST=1
80 A$="###     #.######     #.######     #.######"
90 PRINT " N","TERM","SUM","RATIO"
100 PRINT "-------------------------------------------------------------------------------------------"
110 FOR N=1 TO 85
120 T=N/E^N
130 S=S + T
140 R=T/LAST
150 IF N<11 THEN 170
160 IF N/10<>INT(N/10) THEN 180
170 PRINT USING A$;N,T, S, R
180 LAST=T
190 NEXT N
```

Exercise 2, Page 501

1. $-1 < x < 1$ **3.** $x = 0$ **5.** $-1 < x < 1$
21. 1.625 **23.** 0.182 **25.** 1.0488

Exercise 4, Page 508

1. 0.04 **3.** 0.001 **5.** 0.000003 **7.** 0.0005

Exercise 5, Page 512

27. 2.98 **29.** 0.0308

Chapter Test, Page 513

1. 8.5 **3.** 0.4756 **5.** 74.2
7. 0.0006 **9.** $x < 1$
11. $2 + 8 + 18 + 32 + 50 + \cdots$ **15.** $n^2 - 2$
17. 0.72 **19.** $u^2 - 2$

CHAPTER SEVENTEEN: Fourier Series

Exercise 1, Page 522

1. $2 + 3 \cos x + 2 \cos 2x + \cos 3x + 4 \sin x + 3 \sin 2x + 2 \sin 3x$

Exercise 2, Page 529

1. neither **3.** even **5.** neither **7.** yes **9.** no **11.** no

Exercise 4, Page 537

1. $y = -4.92 \cos x + 2.16 \cos 3x + 0.95 \cos 5x + 10.17 \sin x + 3.09 \sin 3x + 0.78 \sin 5x$
3. $B_n = 31.5, -3.31, 1.59, 0.65$
$A_n = 3.94, -4.76, 1.21, -2.06$

5.
```
10 '    FS_2
20 '
30 ' THIS PROGRAM WILL COMPUTE THE COEFFICIENTS OF THE FIRST 8 TERMS
40 ' OF A FOURIER SERIES FOR A GIVEN SET OF EXPERIMENTAL DATA.
50 ' THE PROGRAM ASSUMES HALF-WAVE SYMMETRY, A PERIOD OF TWO PIE,
60 '
70   DIM X(20),Y(20),XD(20)
80   INPUT "ENTER THE STEP SIZE BETWEEN X VALUES";D
90   FOR Z = 0 TO 180/D
100            READ Y(Z+1)
110            DATA 0,10,14.8,16.1,20,21,21.2,20.3,17.7,14.1,11.2,8.8,7.5,6.6
120            DATA 6.1,5.5,4.6,3.6,0
130            K=K+1
140  NEXT Z
150  X(1) = 0
160  XD(1) = 0
170  D = D*3.14/180
180  FOR Z = 1 TO K
190             XD(Z+1) = XD(Z) + D
200             X(Z+1) = X(Z)+D1
210  NEXT Z
220  PRINT " X   Y   Y SIN X   Y COS X   Y SIN 3X   Y COS 3X   Y SIN 5X   Y COS 5X";
230  PRINT " YSIN7X   YCOS7X"
240  PRINT " -------------------------------------------------------------------------------------"
250    FOR Z = 1 TO K
260            S = Y(Z)*SIN(X(Z))
270                          STOTAL = STOTAL + S
280            C = Y(Z)*COS(X(Z))
290                          CTOTAL = CTOTAL + C
300            S3 = Y(Z)*SIN(3*X(Z))
310                          S3TOTAL = S3TOTAL + S3
320            C3 = Y(Z)*COS(3*X(Z))
330                          C3TOTAL = C3TOTAL + C3
340            S5 = Y(Z)*SIN(5*X(Z))
350                          S5TOTAL = S5TOTAL + S5
360            C5 = Y(Z)*COS(5*X(Z))
370                          C5TOTAL = C5TOTAL + C5
380            S7 = Y(Z)*SIN(7*X(Z))
390                          S7TOTAL = S7TOTAL + S7
400            C7 = Y(Z)*COS(7*X(Z))
410                          C7TOTAL = C7TOTAL + C7
420          PRINT USING "###  ##.#  ###.##  ###.##";XD(Z), Y(Z), S, C;
430          PRINT USING " ###.##  ###.##  ###.##  ###.##";S3,C3,S5,C5;
440          PRINT USING " ###.##  ###.##";S7,C7
450    NEXT Z
460  PRINT "---------------------------------------------------------------------------------------"
470    PRINT " SUMS";
480    PRINT USING " ###.##  ###.##";STOTAL, CTOTAL;
490    PRINT USING " ###.##  ###.##";S3TOTAL, C3TOTAL;
500    PRINT USING " ###.##  ###.##";S5TOTAL, C5TOTAL;
```

```
510   PRINT USING "###.##  ###.##";S7TOTAL, C7TOTAL
520   PRINT
530   PRINT "  Y-AVG";
540   PRINT USING "      ###.##  ###.##";STOTAL/K, CTOTAL/K;
550   PRINT USING " ###.##  ###.##";S3TOTAL/K, C3TOTAL/K;
560   PRINT USING " ###.##  ###.##";S5TOTAL/K, C5TOTAL/K;
570   PRINT USING " ###.##  ###.##";S7TOTAL/K, C7TOTAL/K
580   PRINT: PRINT
590               PRINT "            A(N) AND B(N) COEFFICIENTS"
600   PRINT " =====================================================";
610   PRINT "            B1      A1      A3      B3      A5      B5";
620   PRINT "   A7    B7"
630   PRINT USING "      ###.##  ###.##";2*STOTAL/K, 2*CTOTAL/K;
640   PRINT USING " ###.##  ###.##";2*S3TOTAL/K, 2*C3TOTAL/K;
650   PRINT USING " ###.##  ###.##";2*S5TOTAL/K, 2*C5TOTAL/K;
660   PRINT USING " ###.##  ###.##";2*S7TOTAL/K, 2*C7TOTAL/K
```

Chapter Test, Page 538

1. odd **3.** odd **5.** yes **7.** yes

9. $-\frac{4}{\pi}\sin \pi x - \frac{4}{3\pi}\sin 3\pi x - \frac{4}{5\pi}\sin 5\pi x - \cdots$

11. $\frac{8}{\pi^2}\left(\sin\frac{\pi x}{20} - \frac{1}{9}\sin\frac{3\pi x}{20} + \frac{1}{25}\sin\frac{5\pi x}{20} - \cdots\right)$

13. $y = -4.11 \cos x + 2.07 \cos 3x + 0.57 \cos 5x + 11.26 \sin x + 2.69 \sin 3x + 0.63 \sin 5x$

Index to Applications

ASTRONOMY, SPACE

Astronomical unit, 67
Orbit of comet, 50, 67
Rocket nose cone, 296, 304, 321

COMPUTER

Approximate integration by computer, 373
Derivatives by computer, 100
False position method for roots, 29
Finding Fourier coefficients, 532, 538
Generating series by computer, 496
Identifying conics by computer, 79
Limits by computer, 89, 91
Midpoint method for areas, 269
Newton's method, 144
Numerical solution of differential equations, 482, 485
Sides of a triangle, 7
Summary of BASIC, 586

DYNAMICS

Curvilinear motion, 160
Freely falling body, 249
Integrating for displacement and velocity, 247
Maximum deflection of vibrating body, 435
Mechanical vibrations, 427, 436, 464
Moment of inertia, 331
Motion formulas, 571
Motion of a point, 158
Pendulum, 157, 191, 220
Projectiles, 50, 51, 205
Resonance, 435
Rotation, 163, 166, 250, 252
Simple harmonic motion, 204, 430, 436
Speed decay in rotating wheels, 220, 402
Uniform motion, 91
Uniformly accelerated motion, 27
Velocity and acceleration found by derivative, 159, 164
Work done by a force, 326

ELECTRICAL

Alternating current, 405, 408, 443
Charge, 252
Current and voltage in a capacitor, 155, 158, 193, 220, 253, 404
Current and voltage in an inductor, 156, 158, 193, 254, 404
Decibels gained or lost, 216
Effective values of voltage and current, 306
Electrical formulas, 574
Fourier analysis of waveforms, 515
Impedance triangle, 406, 445
Instantaneous current, 155
Maximum power to a load, 185, 193
Parabolic antenna, 52
Power dissipated in a resistor, 190
Rectifier waves, 523
Resistance change with temperature, 26
Resonance, 447
Root mean square value, 306
Series circuits, LC, 441, 447
RC, 404, 464
RL, 403, 466–67
RLC, 437, 447, 470
Square, triangular, and sawtooth waves, 523

Transient and steady-state currents, 445
Underwater cable, 215

ENERGY

Oil consumption, 401
Parabolic solar collector, 53
Work formulas, 570

FINANCIAL

Financial formulas, 29, 570
Interest, compound, 220
Maximizing profit and minimizing cost, 183, 192

FLUIDS

Density of seawater at various depths, 221
Flow in pipes, culverts, and gutters, 67, 182, 280
Flow to or from tanks, 168, 173
Gas laws, 157, 163, 192, 372
pH, 216
Pressure on a submerged surface, 27, 322
Turbines, 27, 192
Volume of tanks, 372
Work done in compressing a gas, 330, 372
Work needed to empty a tank, 327, 329, 339

LIGHT

Cassegrain telescope, 78
Cylindrical mirror, 280, 300
Elliptical mirror, 66
Hyperbolic mirror, 78, 297, 304
Illumination on a surface, 157, 184, 205
Inverse square law, 157
Lenses, 282
Light output of lamp, 372
Parabolic reflector, 52, 280, 321
Shadows, 172
Window area, 283, 300

MACHINE SHOP

Finding missing dimensions, 6, 40

MATERIALS

Hardening of concrete, 402
Materials formulas, 573

MECHANISMS

Belts and pulleys, 204
Instantaneous velocity and acceleration, 160
Lever, 185
Screw efficiency, 185
Spring deflection, 26
Work done in deflecting a spring, 327, 329, 339

NAVIGATION

Position found from shore stations, 79
Rate of separation of moving vehicles, 170

STATICS

Center of gravity, 310
First moment, 311
Force needed to hold a rope passing over a beam, 221
Friction, 205
Gravitational attraction, 330
Moment of inertia, 331
Polar moment of inertia, 335
Statics formulas, 570

STRENGTH OF MATERIALS

Beams, 54, 157–58, 178, 184, 282
Columns, 280
Elastic curve, 157–58
Hooke's law, 28
Strength of materials formulas, 573

STRUCTURES

Bridge cable, 52, 300, 307
Catenary, 220
Circular arch, 40
Elliptical bridge arch, 66, 80
Elliptical column, 281
Girder length, 6
Gothic arch, 41
Heights of antennas, poles, towers, 192
Parabolic arch, 51, 300
Parabolic beam, 282
Trusses and frameworks, 171

SURVEYING

Elevations found by barometer readings, 216, 221
Elevations on sloping terrain, highways, tunnels, etc., 17
Grade, 17
Street layout, 41
Vertical highway curves, 52, 300

THERMAL

Cooling of ingot, 402
Heat loss from a pipe, 216
Temperature gradient, 28, 157
Temperature change in ovens and furnaces, 154, 220, 222, 402
Thermal expansion, 27, 173

General Index

Absolute maximum and minimum, 131
Acceleration, 158
Addition and subtraction of series, 509
Algebraic substitution, integration by, 354
Alternating current calculations, 405, 408, 443
Alternating series (*see* Series)
Analytic geometry, 2
Angle(s):
  of inclination, 9
  of intersection between curves, 129
  of intersection between lines, 13
Angular acceleration, 163
Angular displacement, 163, 251
  damped, 429
Antiderivative, 225
Approximate methods (*see* Numerical methods)
Approximations using differentials, 188
Arc length, by integration, 297
Arch, circular, 40
  elliptical, 66, 80
  parabolic, 51, 300
Area(s):
  by integration, 262, 270
  of surface of revolution, 301
Arithmetic progression, 489
Astronomy (*see* Index to Applications)
Asymptote, 70, 145
Auxiliary equation, 414
Average ordinate, 305
Average ordinate method, 366
Average value of a function, 304
Axis (Axes):
  major and minor, 55
  semimajor and semiminor, 55
  of symmetry, 42
  translation or shift of, 35
  transverse and conjugate, 68

BASIC, summary of, 586
Bernoulli, James, 392
Bernoulli's equation, 392
Binomial formula, 102
Boundary conditions, 236, 378

Catenary, 220
Cauchy ratio test, 495
Center, of a circle, 34
  of an ellipse, 55
  of a hyperbola, 68
Center of gravity, 310
Centroid, 310
  by integration, 313
  of solids of revolution, 317
Chain rule, 108
Charge, 252
Circle, 33
  area by integration, 280
Circular motion (*see* Rotation)
Common difference, 489
Common ratio, 489
Complementary function, 421
Completing the square, 37, 47, 62, 459
Composite functions, 108
Computer (*see* Index to Applications)
Computer methods (*see* Numerical methods)
Conic sections, 32
Conjugate axis, 68
Constant of integration, 225, 235
Construction, of an ellipse, 66
  of a parabola, 49

Continuity and discontinuity, 251, 276, 364
Convergence and divergence, 481
  interval of, 497
  of a series, 491
Conversion factors, table of, 577
Curve sketching, 144
Curves, families of (*see* Families of curves)
Curvilinear motion, 160

Damping, 428
Decay, exponential (*see* Exponential growth and decay)
Definite integral (*see* Integral)
Degree of a differential equation, 376
Delta method, 95
Derivative(s), differentiation:
  of composite functions, 108
  of a constant, 101
  definition of, 94
  by the delta method, 95
  of exponential functions, 216
  of higher order, 123
  of implicit relations, 119
  of inverse trigonometric functions, 206
  logarithmic, 213
  of logarithmic functions, 209
  as an operator, 98
  of a power function, 109
  of a power of x, 102
  of a power series, 510
  of a product, 113
  of a quotient, 116
  of a sum, 104
  symbols for, 94, 98
  of trigonometric functions, 195
Differential equation, 226, 375
  Bernoulli's, 392
  first-order, 375
    exact, 383
    geometric applications of, 394
    homogeneous, 386
    linear, 388
    numerical solution of, 475
    variables separable, 379
  second-order, 411
    homogeneous, 412
    nonhomogeneous, 412
    numerical solution of, 485
    right side not zero, 421
    right side zero, 413
    variables separable, 411
  solution by Laplace transform, 461
  third-order, 419
Differential form, 187, 376
Differential, 186
  integral of, 226
Directed distance, 3
Directrix, 41
Discontinuity, 99
Displacement found by integration, 288
Disk method for volumes, 288
Distance, directed, 3
Distance formula, 3
Divergence and convergence (*see* Convergence)
Division of series, 510
Dynamics (*see* Index to Applications)

e (base of natural logarithms), 210
Effective (rms) value, 306
Electrical applications, of the derivative, 155
  of differential equations, 403, 437
  of the integral, 253
  of the Laplace transform, 464
  (*see also* Index to Applications)
Ellipse, 33, 54
  area of, 280
  construction of, 66
Energy (*see* Index to Applications)
Equation(s):
  approximate solution of (*see* Numerical methods)
  auxiliary, 414
  of a circle, 34
  differential (*see* Differential equations)
  of an ellipse, 58, 62
  of a hyperbola, 68, 74
  of a parabola, 42
  parametric, 161
  second degree, 36
  of a straight line, 17
  summary of (table), 539
Error estimation by differentials, 189
Euler, Leonhard, 210
Euler's formula, 416, 511
Euler's method, 475
Even function, 145, 524
Exact differential equation, 383
Exponential function, derivative of, 216
  integral of, 241
Exponential growth and decay, 399, 464

Factor(s), conversion (table), 577
  integrating, 389
False position method, 29
Families of curves, 235, 397
Financial problems (*see* Index to Applications)
Finite series (*see* Series)
First moment, 311
First-order linear differential equation, 388
  integrating factor for, 389
Fluid pressure, 322
Fluids Problems (*see* Index to Applications)
Focal radius, 56
Focal width (*see* Latus rectum)
Focus, focal point:
  of ellipse, 55
  of hyperbola, 67
  of parabola, 41
Formulas, summary of (table), 539
Fourier, Jean, 517
Fourier series, 515
  numerical method for, 523
Four-step rule, 95
Fraction(s):
  improper, 347
  partial (*see* Partial fractions)
  proper, 346
Freely falling body, 249
Frequency, natural, 431
  resonant, 447
Function(s):
  average value of, 304
  complementary, 421
  increasing and decreasing, 131
  odd and even, 145, 524
  root mean square (rms) value of, 306
  trial, 423
Fundamental component, 517
Fundamental theorem of calculus, 266

General second-degree equation, 36
General solution of a differential equation, 378
General term of a series, 490
Geometric progression, 489
Geometry, analytic, 2
Graphs, graphing:
  of the hyperbola, 71
  methods for, 144
  of regions, 148
Graphical applications of the derivative, 125
  of differential equations, 394

Gravity, center of, 310
Growth, exponential (*see* Exponential growth and decay)
Gyration, radius of, 332

Half-wave symmetry, 526
Harmonic, 517
Harmonic series (*see* Series)
Hemisphere, centroid of, 318
Homogeneous differential equation, 386
Hyperbola, 33, 67
  graphing of, 71

Implicit relations, derivatives of, 119
Improper integrals, 362
Inclination, angle of, 9
Increasing and decreasing functions, 131, 145
Increment, 3
Indefinite integral, 224
Inertia, moment of, 331
Infinite series (*see* Series)
Inflection points, 138, 145
Initial condition, 236, 378, 454
Integral(s), integration:
  by algebraic substitution, 354
  approximate methods, 365
  areas by, 262, 270
  constant of, 225, 235
  definite, 258
    by series, 510
  of a differential, 226
  of exponential functions, 241
  improper, 362, 452
  indefinite, 224
  methods of, 340
  particular, 421
  by parts, 341
  of a power function, 232
  of a power series, 510
  of rational fractions, 346
  by rationalization, 354
  rules for, 227
  sign, 225
  successive, 236
  table of, 582
  by tables, 238
  of trigonometric functions, 242
  by trigonometric substitution, 357
  variable of, 225
Integrand, 225
  discontinuous, 364
Integrating factor, 389
Intercept, 18, 145
Intersecting circles, 40
Intersection, angle of, 13
Interval of convergence, 497
Inverse Laplace transform, 458
Inverse trigonometric functions, derivatives of, 206

Kutta, Wilhelm, 479

Lagrange, Louis, 507
Laplace, Pierre, 452
Laplace transform(s):
  applications of, 464
  inverse of, 458
  solving differential equations by, 461
  table of, 455
Latus rectum of an ellipse, 64
  of a hyperbola, 71
  of a parabola, 45
Length of arc, 297
Light and optics problems (*see* Index to Applications)
Limit, 82
Line, straight, 2
Logarithmic differentiation, 213
Logarithmic function, derivative of, 209

Machine shop problems (*see* Index to Applications)
Maclaurin, Colin, 498
Maclaurin's series (*see* Series)
Major axis, 55
Mass, center of, 310
Materials problems (*see* Index to Applications)
Maximum and minimum points, 131
  as graphing aid, 145
  tests for, 133, 135–36
Maximum-minimum problems, 175
Mean value theorem, 267
Mechanical vibrations, 427
  free, 427
  overdamped free, 432
  simple harmonic motion, 430
  underdamped forced, 433
  underdamped free, 428
Mechanisms (*see* Index to Applications)
Midpoint method, 264
Minor axis, 55
Modified Euler's method, 477
Moment, first, 311
Moment, second, (moment of inertia), 331
  by integration, 333
  polar, 335
Motion, circular, 250
  curvilinear, 160, 249
  nonuniform, 158
  in a resisting fluid, 400
  uniform, 91
Multiplication and division of series, 510

Natural frequency, 431
Navigation (*see* Index to Applications)
Newton, Isaac, 140
Newton's method for finding roots, 140
Normal to a curve, 14, 30, 127, 396
Numerical methods:
  for Fourier analysis, 534
  of integration:
    average ordinate method, 366
    midpoint method, 264
    prismoidal formula, 368
    Simpson's rule, 370
    trapezoid rule, 366
  for roots of equations:
    false position method, 29
    Newton's method, 140
  for solving differential equations:
    Euler's method, 475
      modified, 477
    Runge-Kutta method, 478, 483

Odd function, 145, 524
Operator notation, 98, 412
Ordinary differential equation, 376
Ordinate, average, 306
Orthogonal trajectory, 397

Panels, 263
Parabola, 33, 41
  area of, 280
  construction of, 49
Partial differential equation, 376
Partial fractions, 349, 459
Partial sum test, 492
Particular integral, 421
Particular solution to a differential equation, 378, 381
Parts, integration by, 341
Perpendicular lines, slopes of, 11
Point(s):
  of inflection, 138
  maximum and minimum, 131
  stationary, 132
Point-slope form, 20
Polar moment of inertia, 335

Positive series (*see* Series)
Power function, derivative of, 109
integral of, 232
Power series (*see* Series)
Predictor-corrector methods, 477, 482
Pressure of fluid, 322
Prismoidal formula, 368
Products, derivative of, 113
integral of, 232
Progressions, 489

Quotient, derivative of, 116

Radius, 34
Radius, focal, 56
Radius of gyration, 332
Rate(s), average, 92
of change, 91, 154
instantaneous, 93
related, 166
Ratio test, 495
Rational fractions, integration of, 346
Rectangle:
area by integration, 280
moment of inertia, 331
Rectifier waves, 523
Rectifying a curve, 299
Recursion relation, 491
Region, graphing of, 148
Related rate problems, 166
Relation, recursion, 491
Resonance, electrical, 447
mechanical, 435
Resonant frequency, 447
Riemann, Georg, 264
Riemann sums, 264
Ring method for volumes, 292
Rise, 7
Root-mean-square (rms) value, 306
Rotation, 163, 250
Run, 7
Runge, Carl, 479
Runge-Kutta method, 478

Second-degree equations, 36
Second derivative, 123
Second-derivative test, 135
Second-order differential equation, 411
Sections, conic, 32
Segment, line, 2
Semimajor and semiminor axes, 55
Sequence, 489
Series:
accuracy of, 499, 505
alternating, 490, 507
by computer, 493
convergence of, 491
finite, 490
Fourier (*see* Fourier series)
general term of, 490
harmonic, 492, 494
infinite, 487
Maclaurin's, 496
operations with, 509
positive, 490
power, 496
sum of, 492
Taylor's, 502
with remainder, 507
Series circuits:
LC, 441
RC, 404
RL, 403
RLC, 437
Shell method, for moment of inertia, 335
for volumes, 291
Shift of axes, 35, 525
Simple harmonic motion, 204, 430
Simpson, Thomas, 370
Simpson's rule, 370
Slope(s), 7
of a curve, 15
of horizontal and vertical lines, 8
of parallel and perpendicular lines, 11
Slope-intercept form, 18
Solid of revolution, 286
centroid of, 317
volume of, 288
Speed, 159
Square, area by integration, 280
Square wave, 521, 523
Statics (*see* Index to Applications)
Stationary point, 132
Straight line, 2
Strength of materials (*see* Index to Applications)
Structures (*see* Index to Applications)
Substitution;
integration by, 354, 357
into a power series, 509
Successive integration, 236
Sum, derivative of, 104
Riemann, 264
of a series, 492
Surface area by integration, 301
Surface of revolution, 301
Surveying (*see* Index to Applications)
Symmetry, 145
axis of, 42
half-wave, 526
of waveforms, 522

Tables:
of conversion factors, 577
of facts and formulas, 539
of integrals, 582
Tangent to a curve, 14, 30, 84, 93, 126, 396
Taylor, Brook, 502
Taylor's series (*see* Series)
Thermal problems (*see* Index to Applications)
Third-order differential equation, 419
Trajectories, orthogonal, 397
Transform, Laplace (*see* Laplace transform)
Translation of axes, 35, 525
Transverse axis, 68
Trapezoid rule, 366
Trial function, 423
Triangle area by integration, 280
Triangular wave, 523, 528
Trigonometric functions, derivative of, 195
integral of, 242
Trigonometric substitution, integration by, 357
Truncation error, 499, 507
Two-point form, 22

Undetermined coefficients, method of, 351, 423
Uniform motion, 91

Variable of integration, 225
Variables separable, 379
Velocity, 158
by integration, 247
Vertex:
of ellipse, 55
of hyperbola, 68
of parabola, 42
Vibrations, mechanical (*see* Mechanical vibrations)
Volume(s):
approximate, by differentials, 189
by integration, 286

Washer method for volumes, 292
Waveform symmetry, 522
Waveforms, Fourier series for, 523
Work, 326

y intercept, 18